Springer Proceedings in Mathematics & Statistics

Volume 304

Springer Proceedings in Mathematics & Statistics

This book series features volumes composed of selected contributions from workshops and conferences in all areas of current research in mathematics and statistics, including operation research and optimization. In addition to an overall evaluation of the interest, scientific quality, and timeliness of each proposal at the hands of the publisher, individual contributions are all refereed to the high quality standards of leading journals in the field. Thus, this series provides the research community with well-edited, authoritative reports on developments in the most exciting areas of mathematical and statistical research today.

More information about this series at http://www.springer.com/series/10533

Martin T. Barlow · Gordon Slade
Editors

Random Graphs, Phase Transitions, and the Gaussian Free Field

PIMS-CRM Summer School in Probability,
Vancouver, Canada, June 5–30, 2017

Pacific Institute for the Mathematical Sciences

Springer

Editors
Martin T. Barlow
Department of Mathematics
The University of British Columbia
Vancouver, BC, Canada

Gordon Slade
Department of Mathematics
The University of British Columbia
Vancouver, BC, Canada

ISSN 2194-1009 ISSN 2194-1017 (electronic)
Springer Proceedings in Mathematics & Statistics
ISBN 978-3-030-32010-2 ISBN 978-3-030-32011-9 (eBook)
https://doi.org/10.1007/978-3-030-32011-9

Mathematics Subject Classification (2010): 05C80, 60G15, 60K35, 82B20, 82B43

This Springer imprint is published by the registered company Springer Nature Switzerland AG
The registered company address is: Gewerbestrasse 11, 6330 Cham, Switzerland

Organization

The 2017 PIMS-CRM Summer School in Probability was organized by the Pacific Institute for the Mathematical Sciences at the University of British Columbia, in cooperation with the Centre de Recherches Mathématiques in Montréal.

Organizing Commitee

Omer Angel, University of British Columbia, Canada
Mathav Murugan, University of British Columbia, Canada
Edwin Perkins, University of British Columbia, Canada
Gordon Slade, University of British Columbia, Canada

Scientific Commitee

Louigi Addario-Berry, McGill University, Canada
Siva Athreya, Indian Statistical Institute Bangalore, India
María Emilia Caballero, Universidad Nacional Autónoma de México, Mexico
Dayue Chen, Peking University, China
Zhen-Qing Chen, University of Washington, USA
Takashi Kumagai, Research Institute for Mathematical Sciences, Japan
Jean-François Le Gall, Université Paris-Sud Orsay, France
Jeremy Quastel, University of Toronto, Canada
Maria Eulália Vares, Universidade Federal do Rio de Janeiro, Brazil

PIMS Workshop Coordinator

Ruth Situma, University of British Columbia, Canada

Participant Lecture Coordinator

Alma Saraí Hernández Torres, University of British Columbia, Canada

Sponsors

Pacific Institute for the Mathematical Sciences, Vancouver, BC, Canada
Centre de Recherches Mathématiques, Montréal, QC, Canada
Department of Mathematics, University of British Columbia, Vancouver, BC, Canada
National Science Foundation, USA
International Association of Mathematical Physics

Coffee break between morning classes. Photo: Miguel Eichelberger (PIMS)

Hike to Stawamus Chief

Foreword

Probability theory is flourishing. Randomness is prominent in most scientific disciplines, including physics, biology, and computer science. Pure mathematicians have understood that their subjects become richer when given a probabilistic slant—this is true in analysis, in combinatorics, in differential equations, and in number theory. The combination of inspiration from applications and connections with other branches of mathematics has greatly enriched probability theory and will continue to do so.

The best known theorems in probability are the law of large numbers and the central limit theorem. The former asserts that the running average of a sequence X_n of independent identically distributed random variables (such as coin flips with value 0 or 1) converges to the mean as n tends to infinity. The latter asserts that the properly scaled fluctuations around the mean converge to a Gaussian distribution. These two theorems illustrate three of the main themes of current research in probability:

1. The search for *limit laws* as system size tends to infinity,
2. The appearance of *continuous distributions* in the description of the limit, even when the distribution of X_n is discrete, and
3. The *universality* of the limit: the Gaussian distribution describes the fluctuations about the mean regardless of the distribution of the X_n (as long as it has finite variance).

Much of modern probability theory, and each of the three contributions in this volume, explore these themes in the context of random geometry. Random geometry is the study of random variables with spatial significance and of the interpretation of the geometric meaning of the collective values of these random variables. This leads to a fourth theme:

4. A *phase transition* can occur, in which the nature of the geometry undergoes abrupt qualitative change as a parameter is varied.

Two closely related models of phase transition were introduced independently in the late 1950s: the random graph and percolation. For the first 30 years, the development of these two subjects was largely separate in distinct communities with different priorities. Over the last 30 years, there has been a drawing together of perspective and approach.

The random graph model is defined by fixing n vertices and independently linking pairs of vertices with probability p. This random linking gives rise to connected clusters. The interest is in the behaviour in the limit as $n \to \infty$. For $p = c/n$, with high probability the largest connected cluster has size of order $\log n$ if $c < 1$, of order n if $c > 1$, and of order $n^{2/3}$ when p has its critical value $p_c = 1/n$. The behaviour near the critical point is magnified by choosing p to lie in the critical window parametrized by $p = \frac{1}{n}(1 + \frac{\lambda}{n^{1/3}})$.

The lecture notes of Christina Goldschmidt concern models of random graphs and random trees, and their limits as $n \to \infty$, in the context of random metric spaces. A connected component of the random graph on n vertices is a finite metric space whose points are vertices and the distance between two points is the graph distance (least number of edges to connect the two vertices). Remarkably, the set of all isometry classes of compact metric spaces can itself be made into a metric space, the Gromov–Hausdorff space. The Gromov–Hausdorff metric provides a topology on this space, so it is possible to speak of convergence of a sequence of metric spaces. The centrepiece in Goldschmidt's notes is a theorem which asserts that, in the critical window of the random graph, the sequence of metric spaces defined by the connected components, ordered by size and appropriately rescaled, converges to a limiting sequence of compact metric spaces that themselves emerge from the continuum random tree. The continuum random tree is the limit of all critical branching processes with finite offspring variance, and background on this topic is provided by Goldschmidt's notes. All four of the above themes are illustrated in this theorem: near the point of phase transition in the random graph there is a continuous limit law defined in terms of a universal distribution. Of course the limit law is vastly more sophisticated than the Gaussian distribution occurring in the central limit, although it is based on Brownian motion and hence has a Gaussian in the background. Goldschmidt further illustrates the universality via the study of the critical configuration model, another popular model of phase transition in random graphs.

The random graph is a model of random subgraphs of the complete graph. As such, there is initially no geometric structure: each pair of points in the complete graph is separated by unit distance. The percolation model differs in being defined from the start on an infinite graph, with the infinite graph having geometric structure. The most commonly studied example is bond percolation on the integer lattice $\mathbb{Z}^d$ in dimensions $d \geq 2$. Now the edges (bonds) are the pairs of sites in $\mathbb{Z}^d$ that differ by ± 1 in exactly one coordinate. Bonds are independently declared to be *occupied* with probability p, and the focus is on the connected clusters of occupied bonds. Percolation undergoes a phase transition: there is a critical value $p_c \in (0, 1)$ such that for $p < p_c$ all clusters are almost surely finite, whereas for $p > p_c$ there is almost surely exactly one infinite cluster. For $d = 2$ or $d \geq 11$, or for $d > 6$ for a closely related model, it has been proven that there is no infinite cluster when $p = p_c$. Although about 60 years have gone by since it was established that percolation undergoes a phase transition, it has not yet been proved that at p_c there is no infinite cluster in intermediate dimensions, including for $d = 3$. This is a striking example of an open problem in probability: it is longstanding, occupies a central position in the subject, is universally believed to be true, yet is resistant to attack.

The percolation model is the $q = 1$ case of a one-parameter family of related models, called the random cluster model with parameter $q \in (0, \infty)$. The random cluster model is a variant of percolation in which bond occupations are no longer independent when $q \neq 1$. It draws significance from the fact that it provides a geometric representation of an important class of spin systems called q-state Potts models ($q = 2, 3, 4, \ldots$). As q varies continuously over the interval $(1, \infty)$, the random cluster model interpolates between the Potts models at integer values of q.

The case $q = 2$ is the famous Ising model of ferromagnetism, which involves spin variables with two possible values. A Hamiltonian, or energy function, favours spin configurations with greater spin alignment, via a Boltzmann weight in a Gibbs measure. The energetic encouragement for spin alignment is modulated by a parameter that corresponds physically to temperature. At high temperature, the encouragement to align is less than at low temperatures. A phase transition occurs as the temperature is lowered past a critical temperature T_c. Typical configurations are disordered for temperatures above T_c, whereas long-range alignment of spins occurs for temperatures below T_c. For $q = 3, 4, 5, \ldots$, the q-state Potts model is a spin model where each spin can assume one of q distinct values, rather than two values as in the Ising model. Potts models also experience a phase transition. The lecture notes of Hugo Duminil-Copin provide a panorama of some of the most recent results concerning the phase transition in Potts models, including percolation and the Ising model. The geometric interpretation provided by the random cluster model is a fundamental tool. The random current representation is a second important geometric representation for the special case of the Ising model.

The results discussed in Duminil-Copin's lecture notes include: the sharpness of the phase transition in the random cluster model for all $q \geq 1$, the computation of the critical temperature for the random cluster model on the square lattice $\mathbb{Z}^2$ for all $q \geq 1$, the continuity of the magnetization at the critical temperature for the Ising model in all dimensions $d \geq 2$ (for the physically most interesting and mathematically most difficult case of dimension $d = 3$ the first proof of this was only published in 2015), and the continuity of the phase transition of the planar random cluster model for $q \in [1, 4]$ and its discontinuity for $q > 4$. The role of SLE (Schramm–Loewner Evolution) in describing continuum limits of the two-dimensional Ising model at the critical point is also presented, via a connection with discrete complex analysis and the introduction of a parafermionic observable. As a warm-up for the parafermionic observable, the connective constant for self-avoiding walk on the hexagonal lattice is shown to equal $\sqrt{2 + \sqrt{2}}$. To prove all these results, an extensive toolbox of methods is put to use, both new and traditional ones.

The Ising and Potts models are examples of spin systems with discrete spins, in which a spin assumes one of q possible values. Spin systems with continuous spins are also widely studied, including the ϕ^4 and $O(n)$ spin systems. The simplest such system is the discrete Gaussian free field (DGFF). When formulated on a two-dimensional lattice, the DGFF is a model of a random surface. Marek Biskup's lectures are devoted to the study of these random surfaces over a finite subset of $\mathbb{Z}^2$, as the subset increases indefinitely in size.

The DGFF is a multivariate Gaussian field on $\mathbb{Z}^2$ whose covariance is given by the Green function for the discrete Laplacian with zero boundary values outside a finite domain. It can equivalently be defined as a Gibbs measure for real-valued spins with Hamiltonian given by the Dirichlet energy. The Hamiltonian favours spin configurations with smaller differences between neighbouring spins, which is a smoothing effect. Nevertheless, typical configurations of the DGFF are irregular and spiky. Biskup's lectures are concerned particularly with the spikes: the extreme values of the DGFF.

More generally, he focuses on the intermediate level sets, which are the level sets of the DGFF at heights that scale proportionally to the absolute maximum of the DGFF as the domain size increases. Naturally, this requires an understanding of the asymptotic behaviour of the maximum value itself. The main result is a description of the scaling limit of the intermediate level sets as the lattice domain with increasingly finer lattice spacing converges to a domain in the plane $\mathbb{C}$. The plane is considered as the complex plane here rather than merely as $\mathbb{R}^2$ because there is a conformal invariance property of the limit. In fact, the limit is intimately related to Liouville Quantum Gravity. Also presented is an analysis of the extremal level sets, consisting of those points in the domain whose spin values differ from the absolute maximum by a bounded amount. A thorough study of the probability distribution of the extremal level sets is laid out, including the locations of local maxima and the shape of the field in their vicinity. Finally, a random walk in a random environment determined by a configuration of the DGFF is analysed in detail.

Several concepts and tools from probability theory are applied and developed in Biskup's lectures. These include Liouville Quantum Gravity, Gaussian multiplicative chaos, branching random walk, correlation inequalities, electric networks and random walk, concentration of measure, and more.

Despite their breadth of subject matter, multiple links exist between the three contributions to this volume. Their careful and clear presentation of topics of considerable contemporary interest in probability theory will be appreciated by beginning researchers and veterans alike.

Vancouver, Canada Gordon Slade

Preface

The 2017 PIMS-CRM Summer School in Probability was held at the Pacific Institute for the Mathematical Sciences (PIMS) at the University of British Columbia in Vancouver, Canada, during 5–30 June 2017. PIMS is proud to have sponsored nine Summer Schools in Probability. The 2017 Summer School was the ninth, following previous Summer Schools at the University of British Columbia in Vancouver in 2004, 2005, 2008, 2009, 2012, 2014, at the University of Washington in Seattle in 2010, and at the Centre de Recherches Mathématiques in Montréal in 2015.

The 2017 Summer School had 125 participants from 20 different countries. There were two main courses of 24 hours each, and three mini-courses of 3 hours each. Lecture notes from the main courses given by Marek Biskup and Hugo Duminil-Copin are included in this volume, as are the lecture notes from the mini-course by Christina Goldschmidt. The other two mini-courses were given by Martin Hairer (*A BPHZ theorem for stochastic PDEs*) and Sandra Cerrai (*SPDEs on graphs: an asymptotic approach*). In addition, twenty-nine 30-minute lectures were given by Summer School participants.

We thank PIMS for its support of the Summer Schools, and particularly Ruth Situma at PIMS who made sure everything ran smoothly. We also thank Saraí Hernández Torres for organizing the 30-minute lectures and generally helping to make the event a success. The Summer School was organized by Omer Angel, Mathav Murugan, Ed Perkins, and Gordon Slade.

The lecture notes contained in this volume provide introductory accounts of three of the most active and fascinating areas of research in modern probability theory:

- Scaling limits of random trees and random graphs (Christina Goldschmidt).
- Lectures on the Ising and Potts models on the hypercubic lattice (Hugo Duminil-Copin).
- Extrema of the two-dimensional discrete Gaussian free field (Marek Biskup).

These contributions, especially designed for graduate students entering research, provide thorough introductions which will be of value both to beginners and to experts.

Vancouver, Canada
May 2019

Martin T. Barlow
barlow@math.ubc.ca
Gordon Slade
slade@math.ubc.ca
Editors of the Proceedings
2017 PIMS-CRM Summer School in Probability

Sandra Cerrai "SPDEs on graphs: an asymptotic approach". Photo: Miguel Eichelberger (PIMS)

Martin Hairer "A BPHZ theorem for stochastic PDEs". Photo: Miguel Eichelberger (PIMS)

Contents

Scaling Limits of Random Trees and Random Graphs

Christina Goldschmidt

Abstract In the last 30 years, random combinatorial structures and their scaling limits have formed a flourishing area of research at the interface between probability and combinatorics. In this mini-course, I aim to show some of the beautiful theory that arises when considering scaling limits of random trees and graphs. Trees are fundamental objects in combinatorics and the enumeration of different classes of trees is a classical subject. In the first section, we will take as our basic object the genealogical tree of a critical Galton–Watson branching process. (As well as having nice probabilistic properties, this class turns out to include various natural types of random combinatorial tree in disguise.) In the same way as Brownian motion is the universal scaling limit for centred random walks of finite step-size variance, it turns out that all critical Galton–Watson trees with finite offspring variance have a universal scaling limit, Aldous' Brownian continuum random tree. The simplest model of a random network is the Erdős–Rényi random graph: we take n vertices, and include each possible edge independently with probability p. One of the most well-known features of this model is that it undergoes a phase transition. Take $p = c/n$. Then for $c < 1$, the components have size $O(\log n)$, whereas for $c > 1$, there is a giant component, comprising a positive fraction of the vertices, and a collection of components of size $O(\log n)$. (These statements hold with probability tending to 1 as $n \to \infty$.) In the second section, we will focus on the critical setting, $c = 1$, where the largest components have size of order $n^{2/3}$, and are "close" to being trees, in the sense that they have only finitely many more edges than a tree with the same number of vertices would have. We will see how to use a careful comparison with a branching process in order to derive the scaling limit of the critical Erdős–Rényi random graph. In the final section, we consider the setting of a critical random graph generated according to the configuration model with independent and identically distributed degrees. Here, under natural conditions we obtain the same scaling limit as in the Erdős–Rényi case (up to constants).

C. Goldschmidt (✉)
Department of Statistics and Lady Margaret Hall, University of Oxford,
Oxford, United Kingdom
e-mail: goldschm@stats.ox.ac.uk
URL: http://www.stats.ox.ac.uk/~goldschm

M. T. Barlow and G. Slade (eds.), *Random Graphs, Phase Transitions, and the Gaussian Free Field*, Springer Proceedings in Mathematics & Statistics 304,
https://doi.org/10.1007/978-3-030-32011-9_1

Keywords Scaling limits · Random graphs · Random trees · $\mathbb{R}$-trees · Brownian continuum random tree

Contents

These are (somewhat expanded) lecture notes for a 3-h long mini-course. The principal aim is to give an idea of the intuition behind the main results, rather than fully rigorous proofs. Some of the ideas are further explored in exercises.

Photo: Miguel Eichelberger (PIMS)

1 Galton–Watson Trees and the Brownian Continuum Random Tree

1.1 Uniform Random Trees

In order to build up some intuition, we start with perhaps the simplest model of a random tree. Let $\mathbb{T}_n$ be the set of (unordered) trees labelled by $[n] := \{1, 2, \ldots, n\}$, and let $\mathbb{T} = \cup_{n\geq 1}\mathbb{T}_n$. For example, $\mathbb{T}_3$ consists of the following trees.

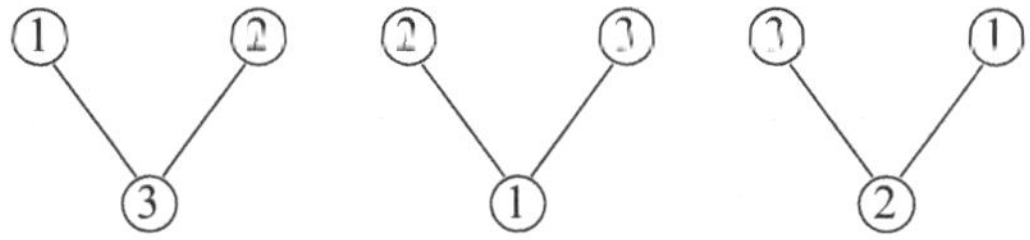

Cayley's formula says that $|\mathbb{T}_n| = n^{n-2}$. We let T_n be a tree chosen uniformly at random from the n^{n-2} elements of $\mathbb{T}_n$. Our first aim is to understand what T_n "looks like" for large n. In order to do this, it will be useful to have an algorithm for building T_n.

It will be technically easier to deal with $\mathbb{T}_n^\bullet$, the set of elements of $\mathbb{T}_n$ with a single distinguished vertex, called the *root*. Given a uniform element of $\mathbb{T}_n^\bullet$, we obtain a uniform element of $\mathbb{T}_n$ by simply forgetting the root.

The Aldous–Broder algorithm
Start from the complete graph on $[n]$.

- Pick a uniform vertex from which to start; this acts as a root.
- Run a simple random walk $(S_k)_{k\geq 0}$ on the graph (i.e. at each step, move to a vertex distinct from the current one, chosen uniformly at random).
- Whenever the walk visits a new vertex, keep the edge along which it was reached.
- Stop when all vertices have been visited.

Claim: the resulting rooted tree is uniform on $\mathbb{T}_n^\bullet$.

The random walk $(S_k)_{k\geq 0}$ has a uniform stationary distribution, and is reversible, so that we may talk about a stationary random walk $(S_k)_{k\in\mathbb{Z}}$. The dynamics of this random walk give rise to Markovian dynamics on $\mathbb{T}_n^\bullet$. In order to see this, let τ_k be the tree constructed from the random walk started at time k (which is rooted at S_k). For $1 \leq i \leq n$, let $\sigma_k(i) = \inf\{j \geq k : S_j = i\}$. Then the tree τ_k has edges

$$\{S_{\sigma_k(i)-1}, i\} \text{ for } 1 \leq i \leq n \text{ such that } \sigma_k(i) > k.$$

Now notice that $\sigma_{k+1}(i) \geq \sigma_k(i)$ for all $1 \leq i \leq n$. So conditionally on the tree τ_k, the tree τ_{k+1} must be independent of $\tau_{k-1}, \tau_{k-2}, \ldots$.

Since the random walk is stationary, the tree must be also. It remains to prove that its distribution is uniform on $\mathbb{T}_n^\bullet$.

Exercise 1 Consider the *time-reversed chain* (which must have the same stationary distribution). For $\tau, \tau' \in \mathbb{T}_n^\bullet$, write $q(\tau, \tau')$ for the transition probability from τ to τ' for the time-reversed chain.

1. Argue that the chain is irreducible on $\mathbb{T}_n^\bullet$.
2. Show that for fixed τ, $q(\tau, \tau') = 0$ or $1/(n-1)$.
3. Show that for fixed τ', $q(\tau, \tau') = 0$ or $1/(n-1)$.
4. It follows that $Q = (q(\tau, \tau'))_{\tau,\tau' \in \mathbb{T}_n^\bullet}$ is a doubly stochastic matrix. Deduce that the stationary distribution must be uniform.

A variant algorithm due to Aldous

Note that nothing changes if we permute all of the vertex-labels uniformly. So we may as well just do the labelling at the very end. Also, there are steps on which we do not add a new edge at all because the vertex to which the walk moves has already been visited. (Indeed, steps on which we add new edges are separated by geometrically-distributed numbers of steps on which we add no edges.) We may as well suppress this "wandering around" inside the structure we have already built.

- Start from a single vertex labelled 1.
- For $2 \le i \le n$, connect vertex i to vertex V_i by an edge, where

$$V_i = \begin{cases} i-1 & \text{with probability } 1 - \frac{i-2}{n-1} \\ k & \text{with probability } \frac{1}{n-1} \text{ for } 1 \le k \le i-2. \end{cases}$$

- Uniformly permute the vertex labels.

We may think of this algorithm as growing a sequence of paths with consecutive vertex-labels (of random lengths), with such a path ending whenever we reach a vertex labelled i which connects to $V_i \neq i-1$. The first such path has length

$$C_1^n := \inf\{i \ge 2 : V_i \neq i-1\}.$$

Our first glimpse into the scaling behaviour of T_n is given by the following simple proposition.

Proposition 1 *We have*

$$\frac{C_1^n}{\sqrt{n}} \xrightarrow{d} C_1$$

as $n \to \infty$, *where* $\mathbb{P}(C_1 > x) = \exp(-x^2/2)$, $x \ge 0$.

Proof We have

$$\mathbb{P}\left(n^{-1/2} C_1^n > x\right) = \mathbb{P}\left(C_1^n \ge \lfloor x\sqrt{n} \rfloor + 1\right) = \prod_{i=1}^{\lfloor x\sqrt{n} \rfloor - 2} \left(1 - \frac{i}{n-1}\right).$$

Taking logarithms and then Taylor expanding, we have

$$
\begin{aligned}
-\log \mathbb{P}\left(n^{-1/2}C_1^n > x\right) &= -\sum_{i=1}^{\lfloor x\sqrt{n}\rfloor-2} \log\left(1-\frac{i}{n-1}\right) \\
&= \sum_{i=1}^{\lfloor x\sqrt{n}\rfloor-2} \frac{i}{n-1} + o(1) \\
&= \frac{(\lfloor x\sqrt{n}\rfloor-2)(\lfloor x\sqrt{n}\rfloor-1)}{2(n-1)} + o(1) \to \frac{x^2}{2},
\end{aligned}
$$

as $n \to \infty$. □

Once we have built the first path of consecutive labels, we pick a uniform random point along it and start growing a second path of consecutive labels, etc.

Imagine now that edges in the tree have length 1. Formally, we do this by thinking of T_n as a metric space, where the points of the space are the vertices and the metric is given by the graph distance, for which we write d_n. The proposition suggests that, in order to get some sort of nice limit as $n \to \infty$, we should rescale the graph distance by $n^{-1/2}$.

Here is what turns out to be the limiting version of the algorithm.

Line-breaking construction

Let $C_1, C_2, \ldots$ be the points of an inhomogeneous Poisson process on $[0, \infty)$ of intensity measure $t dt$. (In particular, we have $\mathbb{P}(C_1 > x) = \exp\left(-\int_0^x t dt\right) = \exp(-x^2/2)$.) For each $i \geq 1$, conditionally on C_i, let $J_i \sim \mathrm{U}[0, C_i)$. Cut $[0, \infty)$ into intervals at the points given by the C_i's and, for $i \geq 1$, glue the line-segment $[C_i, C_{i+1})$ to the point J_i. (In particular, if we think of this as gradually building up a tree branch by branch, we glue $[C_i, C_{i+1})$ to a point chosen uniformly from the tree built so far.) Think of the union of all of these line-segments as a path metric space, and take its completion. This is (one somewhat informal definition of) the *Brownian continuum random tree (CRT)* $\mathcal{T}$. Write d for its metric.

Theorem 2 (Aldous [5], Le Gall [29]) *As* $n \to \infty$,

$$
\left(T_n, \frac{1}{\sqrt{n}} d_n\right) \xrightarrow{d} (\mathcal{T}, d).
$$

In order to make sense of this convergence, we need a topology on metric spaces, which we will discuss in Sect. 1.4. For the purposes of the present discussion, let us observe that one way to prove this theorem has at its heart the following joint convergence: if C_k^n is the kth element of the set $\{i \geq 2 : V_i \neq i-1\}$ and $J_k^n = V_{C_k^n}$ then

$$
\left(\frac{1}{\sqrt{n}}(C_1^n, J_1^n), \frac{1}{\sqrt{n}}(C_2^n, J_2^n), \ldots\right) \xrightarrow{d} ((C_1, J_1), (C_2, J_2), \ldots).
$$

(See Theorem 8 of Aldous [5].) We will later take a different approach in order to prove a more general version of Theorem 2.

1.2 Ordered Trees and Their Encodings

We will henceforth find it easier to work with rooted, ordered trees i.e. those with a distinguished vertex (the root) and such that the *children* of a vertex (its neighbours which are further away from the root) have a given planar order. We will use the standard *Ulam–Harris labelling* by elements of $\mathbb{U} := \cup_{n=0}^{\infty} \mathbb{N}^n$ where, by convention, $\mathbb{N}^0 := \{\emptyset\}$.

Write $\mathbf{T}$ for the set of finite rooted ordered trees. (To an element of $t \in \mathbb{T}$ we may associate a canonical element of $\mathbf{t} \in \mathbf{T}$ by rooting at the vertex labelled 1 in t and then embedding the children of a vertex in t in left-to-right order by increasing label.)

We will find it convenient to encode elements of $\mathbf{T}$ by discrete functions in two different ways. For $\mathbf{t} \in \mathbf{T}$ with n vertices, let $v_0, v_1, \ldots, v_{n-1}$ be the vertices listed in lexicographical order (so that, necessarily, $v_0 = \emptyset$). Let d denote the graph distance on $\mathbf{t}$. We define the *height function* of $\mathbf{t}$ to be $(H(i), 0 \le i \le n-1)$, where

$$H(i) := d(v_0, v_i), \quad 0 \le i \le n-1.$$

We imagine visiting the vertices of the tree in lexicographical order and simply recording the distance from the root at each step. It is straightforward to recover $\mathbf{t}$ from its height function.

Now let $K(i)$ be the number of children of v_i, for $i \ge 0$, and define the *depth-first walk* (or *Łukasiewicz path*) of $\mathbf{t}$ to be $(X(i), 0 \le i \le n)$, where $X(0) := 0$ and

$$X(i) := \sum_{j=0}^{i-1} (K(j) - 1), \quad 1 \le i \le n.$$

Again, we imagine visiting the vertices in lexicographical order, but this time we keep track of a *stack* of vertices which we "know about", but have not yet visited. At time 0, we are at the vertex v_0. Whenever we leave a vertex, we become aware of its children (if any) and add them to the stack. We add the children to the stack in reverse lexicographical order, so that the lexicographically smallest child of the vertex we have just left sits at the top at the stack. We also choose a new vertex to visit by taking the one from the top of the stack (it is then removed from the stack). Then for $0 \le i \le n-1$, the value $X(i)$ records the size of the stack when we visit vertex v_i.

See Fig. 1 for an example.

We observe straight away that $X(n) = \sum_{i=0}^{n-1} K(i) - n = -1$, since every vertex is the child of some other vertex, except v_0. On the other hand, for $i < n$, there is some non-negative number of vertices on the stack, so that $X(i) \ge 0$.

We shall now show that X also encodes $\mathbf{t}$ (see Proposition 1.2 of Le Gall [29] for a formal proof).

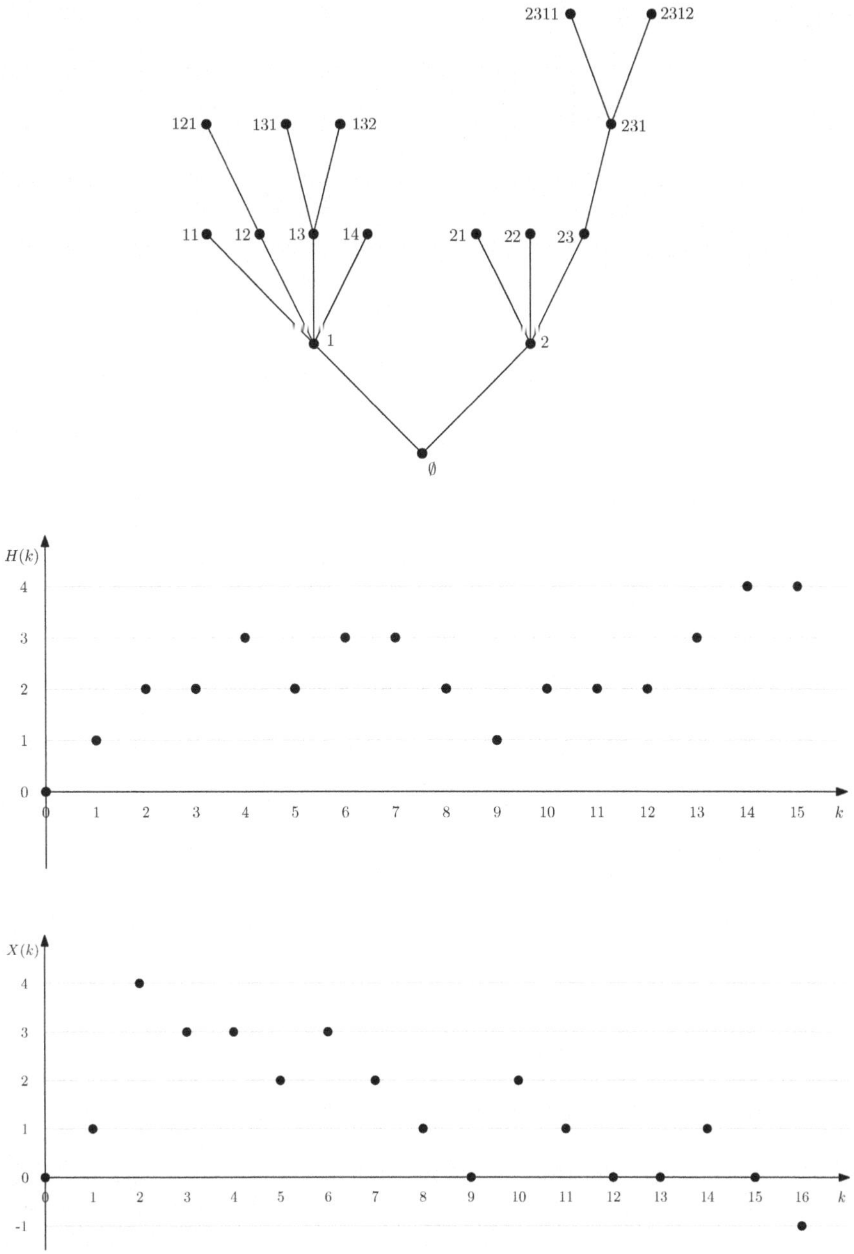

Fig. 1 A rooted ordered tree, its height function and its depth-first walk

Proposition 3 *For* $0 \le k \le n-1$,

$$H(k) = \#\left\{0 \le i \le k-1 : X(i) = \min_{i \le j \le k} X(j)\right\}.$$

Proof For any subtree of the original tree, the value of X once we have just finished exploring it is one less than its value when we visited the root of the subtree, whereas within the subtree, X takes at least its value at the root. Now, the height of v_k is equal to the number of subtrees we have begun but not completed exploring at the step before we reach v_k. The roots of these subtrees are times i before $k-1$ such that X has not yet gone lower by step k i.e.

$$H(k) = \#\left\{0 \le i \le k-1 : X(i) = \min_{i \le j \le k} X(j)\right\},$$

as desired. □

1.3 Galton–Watson Trees

Let us now take $T \in \mathbf{T}$ to be random by letting it be the family tree of a Galton–Watson branching process, with offspring distribution $(p_i)_{i \ge 0}$ i.e. each vertex gets a random number of children with distribution $(p_i)_{i \ge 0}$, independently of all other vertices. Let N be the total progeny (i.e. the number of vertices in the tree). We will impose the conditions $p_1 < 1$ and $\sum_{i \ge 0} i p_i \le 1$, under which $N < \infty$ almost surely. To avoid complicating the statements of our results, except where otherwise stated, we shall also assume that for all n sufficiently large, $\mathbb{P}(N = n) > 0$.

Proposition 4 *Let* $(R(k), k \ge 0)$ *be a random walk with* $R(0) = 0$ *and step distribution* $\nu(i) = p_{i+1}$, $i \ge -1$. *Set*

$$M = \inf\{k \ge 0 : R(k) = -1\}.$$

Then

$$(X(k), 0 \le k \le N) \stackrel{d}{=} (R(k), 0 \le k \le M).$$

See Proposition 1.5 and Corollary 1.6 of Le Gall [29] for a careful proof. So the depth-first walk of a (sub-critical or critical) Galton–Watson tree is a stopped random walk, which is a rather natural object from a probabilistic perspective. It turns out that many of the most natural combinatorial models of random trees are actually conditioned critical Galton–Watson trees.

Exercise 2 Let T be a Galton-Watson tree with Poisson(1) offspring distribution and total progeny N.

1. Fix a particular rooted ordered tree $\mathbf{t}$ with n vertices having numbers of children c_v, $v \in \mathbf{t}$. What is $\mathbb{P}(T = \mathbf{t})$?

2. Condition on the event $\{N = n\}$. Assign the vertices of T a uniformly random labelling by $[n]$, and let $\tilde{T}$ be the labelled tree obtained by forgetting the ordering and the root. Show that $\tilde{T}$ has the same distribution as T_n, a uniform random tree on n vertices.
   Hint: it suffices to show that the probability of obtaining a particular tree $\mathbf{t}$ *is a function of n only.*

Exercise 3 Let T be a Galton–Watson tree with offspring distribution $(p_k)_{k\geq 0}$ and total progeny N.

(a) Show that if $p_k = 2^{-k-1}$, $k \geq 0$ then, conditional on $N = n$, T is uniform on the set of ordered rooted trees with n vertices.
(b) Show that if $p_0 = 1/2$ and $p_2 = 1/2$ then, conditional on $N = 2n + 1$, T is uniform on the set of binary trees with n vertices of degree 3 and $n + 1$ leaves.

Suppose now that $\sum_{i=1}^{\infty} i p_i = 1$ and $\sigma^2 := \sum_{i=1}^{\infty} (i-1)^2 p_i \in (0, \infty)$. Write $(X^n(k), 0 \leq k \leq n)$ for the depth-first walk of our Galton–Watson tree conditioned on $N = n$.

Theorem 5 *As* $n \to \infty$,

$$\frac{1}{\sigma\sqrt{n}}(X^n(\lfloor nt \rfloor), 0 \leq t \leq 1) \xrightarrow{d} (e(t), 0 \leq t \leq 1),$$

where $(e(t), 0 \leq t \leq 1)$ *is a standard Brownian excursion.*

Using the fact that the depth-first walk of a Galton–Watson tree is a stopped random walk, this follows from a conditioned version of Donsker's invariance principle (Theorem 2.6 of Kaigh [27]). A highly non-trivial consequence of Theorem 5 is that, up to a scaling constant, the same is true for H^n, the conditioned height process.

Theorem 6 *As* $n \to \infty$,

$$\frac{\sigma}{\sqrt{n}}(H^n(\lfloor nt \rfloor), 0 \leq t \leq 1) \xrightarrow{d} 2(e(t), 0 \leq t \leq 1),$$

The Brownian CRT, which we encountered in Sect. 1.1 via the line-breaking construction, is the tree encoded (in a sense to be made precise in the next section) by $2(e(t), 0 \leq t \leq 1)$.

See Sects. 1.3 and 1.5 of Le Gall [29] for a complete proof of this theorem. We will give a sketch proof, not for the case of a single tree conditioned to have size n, but rather for a sequence of i.i.d. unconditioned critical Galton–Watson trees. (It is technically easier not to have to deal with the conditioning.) We encode this "forest" via the (shifted) concatenation of the depth-first walks of its trees: a new tree starts every time X reaches a new minimum. (We must be a little careful now with our interpretation of the quantity $X(k)$: it is the number of vertices on the stack, minus the number of components we have completely explored.) We take H to be defined, as before, via

$$H(k) = \#\left\{0 \le i \le k-1 : X(i) = \min_{i \le j \le k} X(j)\right\}. \tag{1.1}$$

Donsker's theorem easily gives

$$\frac{1}{\sigma\sqrt{n}}(X(\lfloor nt \rfloor), t \ge 0) \xrightarrow{d} (W(t), t \ge 0),$$

where W is a standard Brownian motion. The analogue of Theorem 6 is then as follows.

Theorem 7 *As* $n \to \infty$,

$$\frac{\sigma}{\sqrt{n}}(H^n(\lfloor nt \rfloor), t \ge 0) \xrightarrow{d} 2\left(W(t) - \inf_{0 \le s \le t} W(s), t \ge 0\right).$$

(The right-hand side has the same distribution as twice a reflecting Brownian motion $(|W(t)|, t \ge 0)$. We interpret this as encoding a forest of continuous trees, each corresponding to an excursion away from 0.)

The random walks which occur as depth-first walks of Galton–Watson trees have the special property that they are *skip-free to the left*, which means that they have step distribution concentrated on $\{-1, 0, 1, 2, \ldots\}$. It turns out that these random walks have particularly nice properties, some of which we explore in the next exercise.

Exercise 4 Let $(X(k), k \ge 0)$ be a random walk with step distribution $\nu(k), k \ge -1$. Assume that $\sum_{k \ge -1} k\nu(k) = 0$ and that $\sum_{k \ge -1} k^2 \nu(k) = \sigma^2 < \infty$. Suppose $X(0) = 0$ and let $T = \inf\{k \ge 1 : X(k) \ge 0\}$. This is called the *first weak (ascending) ladder time*. The random walk is recurrent, and so $T < \infty$ a.s. and it follows that the *first weak ladder height* $X(T)$ is finite a.s.

If the first step of the random walk is to 0 or above, then $T = 1$ and $X(T)$ is simply the new location of the random walk.

If the first step is to -1, on the other hand, things are more involved. In general, the random walk may now make several excursions which go below -1 and stay below it before returning to -1. Finally the walk leaves -1, perhaps initially going downwards, but eventually reaching $\{0, 1, 2, \ldots\}$ *without hitting* -1 *again*. Indeed, using the strong Markov property, we can see that the random walk makes a geometrically distributed number of excursions which return to -1 before it hits $\{0, 1, \ldots\}$, where the parameter of the geometric distribution is (by translation-invariance) $\mathbb{P}(X(T) > 0)$.

1. By conditioning on the first step of the random walk, and using the above considerations, show that for $k \ge 0$,

$$\mathbb{P}(X(T) = k) = \nu(k) + \nu(-1)\mathbb{P}(X(T) = k+1 | X(T) > 0).$$

2. Show that for $k \ge 0$,

$$\mathbb{P}(X(T) = k) = \sum_{j=0}^{\infty} \left(\frac{\nu(-1)}{\mathbb{P}(X(T) > 0)} \right)^j \nu(k+j).$$

3. Show directly that

$$\sum_{k=0}^{\infty} \bar{\nu}(k) = 1$$

where $\bar{\nu(k)} = \sum_{j=k}^{\infty} \nu(j)$.

4. Using the fact that $\sum_{k=0}^{\infty} \mathbb{P}(X(T) = k) = 1$, deduce carefully that we must have

$$\mathbb{P}(X(T) > 0) = \nu(-1),$$

and hence that $\mathbb{P}(X(T) = k) = \bar{\nu}(k)$ for $k \geq 0$.
Hint: you may want to use a probability generating function.

5. Finally, show that $\mathbb{E}[X(T)] = \sigma^2/2$.

This calculation is inspired by one by Jean-François Marckert and Abdelkader Mokkadem in [32]. They credit the argument to Feller.

We will use the result of the exercise to prove that the height process converges in the sense of finite-dimensional distributions.

Proposition 8 *For any $m \geq 1$ and any $0 \leq t_1 \leq t_2 \leq \ldots \leq t_m < \infty$,*

$$\frac{1}{\sqrt{n}} (H(\lfloor nt_1 \rfloor), \ldots, H(\lfloor nt_m \rfloor))$$
$$\xrightarrow{d} \frac{2}{\sigma} \left(W(t_1) - \inf_{0 \leq s \leq t_1} W(s), \ldots, W(t_m) - \inf_{0 \leq s \leq t_m} W(s) \right).$$

Proof Let $S(n) = \sup_{0 \leq k \leq n} X(k)$ and $I(n) = \inf_{0 \leq k \leq n} X(k)$. Let us introduce the time-reversed random walk, which takes the same jumps but in the opposite order:

$$\hat{X}^n(k) = X(n) - X(n-k).$$

Then

$$(\hat{X}^n(k), 0 \leq k \leq n) \stackrel{d}{=} (X(k), 0 \leq k \leq n).$$

Hence,

$$\begin{aligned}H(n) &= \#\left\{0 \le k \le n-1 : X(k) = \inf_{k \le j \le n} X(j)\right\} \\ &= \#\left\{1 \le i \le n : X(n-i) = \inf_{0 \le \ell \le i} X(n-\ell)\right\} \\ &= \#\left\{1 \le i \le n : \hat{X}^n(i) = \sup_{0 \le \ell \le i} \hat{X}^n(\ell)\right\}.\end{aligned}$$

By analogy, define

$$J(n) = \#\left\{1 \le i \le n : X(i) = \sup_{0 \le \ell \le i} X(\ell)\right\} = \#\{1 \le i \le n : X(i) = S(i)\}.$$

Note that

$$\sup_{0 \le k \le n} \hat{X}^n(k) = X(n) - \inf_{0 \le k \le n} X(k) = X(n) - I(n).$$

It follows that *for each fixed n*,

$$(S(n), J(n)) \stackrel{d}{=} (X(n) - I(n), H(n)).$$

Now define $T_0 = 0$ and $T_k = \inf\{i > T_{k-1} : X(i) = S(i)\}$, $k \ge 1$. Then the random variables $\{X(T_{k+1}) - X(T_k), k \ge 0\}$ are i.i.d. by the strong Markov property. By Exercise 4, they have mean $\sigma^2/2$.

We now claim that

$$\frac{H(n)}{X(n) - I(n)} \stackrel{p}{\to} \frac{2}{\sigma^2}, \tag{1.2}$$

as $n \to \infty$. To see this, write

$$\begin{aligned}S(n) = \sum_{k \ge 1 : T_k \le n} (S(T_k) - S(T_{k-1})) &= \sum_{k=1}^{J(n)} (S(T_k) - S(T_{k-1})) \\ &= \sum_{k=1}^{J(n)} (X(T_k) - X(T_{k-1}).\end{aligned}$$

Since $J(n) \to \infty$ as $n \to \infty$, by the Strong Law of Large Numbers we have

$$\frac{S(n)}{J(n)} \to \mathbb{E}[X(T_1)] = \frac{\sigma^2}{2} \quad \text{a.s.}$$

as $n \to \infty$. Since $(S(n), J(n)) \stackrel{d}{=} (X(n) - I(n), H(n))$ for each n, we deduce that

$$\frac{X(n) - I(n)}{H(n)} \stackrel{p}{\to} \frac{\sigma^2}{2}$$

as $n \to \infty$. Now, we know that

$$\frac{1}{\sqrt{n}}(X(\lfloor nt \rfloor), t \geq 0) \xrightarrow{d} \sigma(W(t), t \geq 0)$$

and so, by the continuous mapping theorem,

$$\frac{1}{\sqrt{n}}(X(\lfloor nt_1 \rfloor) - I(\lfloor nt_1 \rfloor), \ldots, X(\lfloor nt_m \rfloor) - I(\lfloor nt_m \rfloor))$$
$$\xrightarrow{d} \sigma\left(W(t_1) - \inf_{0 \leq s \leq t_1} W(s), \ldots, W(t_m) - \inf_{0 \leq s \leq t_m} W(s)\right).$$

The result then follows by using (1.2). □

1.4 $\mathbb{R}$-Trees Encoded by Continuous Excursions

We now turn to our notion of a continuous tree.

Definition 9 A compact metric space (T, d) is an $\mathbb{R}$-*tree* if the following conditions are fulfilled for every pair $x, y \in T$:

- There is a unique isometric map $f_{x,y} : [0, d(x, y)] \to T$ such that $f_{x,y}(0) = x$ and $f_{x,y}(d(x, y)) = y$. We write $[[x, y]] := f_{x,y}([0, d(x, y)])$.
- If g is a continuous injective map $[0, 1] \to T$ such that $g(0) = x$ and $g(1) = y$ then $g([0, 1]) = [[x, y]]$.

A *continuous excursion* is a continuous function $h : [0, \zeta] \to \mathbb{R}_+$ such that $h(0) = h(\zeta) = 0$ and $h(x) > 0$ for $x \in (0, \zeta)$, for some $0 < \zeta < \infty$. We will build all of our $\mathbb{R}$-trees from such functions. For a continuous excursion h, define first a pseudo-metric on $[0, \zeta]$ via

$$d_h(x, y) = h(x) + h(y) - 2 \inf_{x \wedge y \leq z \leq x \vee y} h(z).$$

Then define an equivalence relation by $x \sim y$ iff $d_h(x, y) = 0$, and let $\mathcal{T}_h$ be given by the quotient $[0, \zeta]/\sim$. Intuitively, we put glue on the underside of the function h and then imagine squashing the function from the right: whenever two parts of the function at the same height and with glue on them meet, they stick together. See below (distances in the tree should be interpreted vertically).

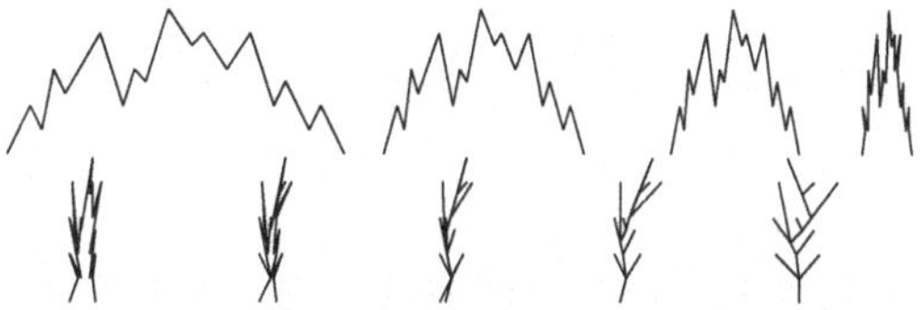

In particular, local minima of the function become branch-points of the tree.

Theorem 10 *For any continuous excursion h, $(\mathcal{T}_h, d_h)$ is an $\mathbb{R}$-tree.*

For a proof, see Theorem 2.2 of Le Gall [29]. For $t \in [0, \zeta]$, we write p_h for the canonical projection $[0, \zeta] \to \mathcal{T}_h$. It is usual to think of the tree as rooted at $\rho = p_h(0) = p_h(\zeta)$, the equivalence class of 0. It will be useful later to have a measure μ_h on $\mathcal{T}_h$ which is given by the push-forward of the Lebesgue measure on $[0, \zeta]$.

Now let $\mathcal{M}$ be the space of isometry classes of compact metric spaces. We endow $\mathcal{M}$ with the Gromov–Hausdorff distance. To define this, let (X, d) and (X', d') be (representatives of) elements of $\mathcal{M}$. A *correspondence* R is a subset of $X \times X'$ such that for all $x \in X$, there exists $x' \in X'$ such that $(x, x') \in R$ and vice versa. The *distortion* of R is

$$\mathrm{dis}(R) = \sup\{|d(x, y) - d'(x', y')| : (x, x'), (y, y') \in R\}.$$

The *Gromov–Hausdorff distance* between (X, d) and (X', d') is then given by

$$d_{\mathrm{GH}}((X, d), (X', d')) = \frac{1}{2} \inf_R \mathrm{dis}(R),$$

where the infimum is taken over all correspondences R between X and X'. Importantly, $(\mathcal{M}, d_{\mathrm{GH}})$ is a Polish space. (See Burago, Burago and Ivanov [17] for much more about the Gromov–Hausdorff distance.)

We define the Brownian CRT to be $(\mathcal{T}_{2e}, d_{2e})$, where $e = (e(t), 0 \leq t \leq 1)$ is a standard Brownian excursion.

1.5 Convergence to the Brownian CRT

Let T_n be a critical Galton–Watson tree with finite offspring variance $\sigma^2 > 0$, conditioned to have total size n, and let d_n be the graph distance on T_n.

Theorem 11 (Aldous [7], Le Gall [29]) *As $n \to \infty$,*

$$\left(T_n, \frac{\sigma}{\sqrt{n}} d_n\right) \xrightarrow{d} (\mathcal{T}_{2e}, d_{2e}).$$

Proof (I learnt this proof from Grégory Miermont.) By Skorokhod's representation theorem, we can find a probability space on which the convergence

$$\frac{\sigma}{\sqrt{n}}(H^n(\lfloor nt \rfloor), 0 \leq t \leq 1) \to 2(e(t), 0 \leq t \leq 1)$$

occurs almost surely (in the uniform norm). As usual, write $v_0, v_1, \ldots, v_{n-1}$ for the vertices of T_n in lexicographical order. Then $(T_n, \frac{\sigma}{\sqrt{n}} d_n)$ is isometric to $\{0, 1, \ldots, n -$

$1\}$ endowed with the distance

$$d^n(i,j) = \frac{\sigma}{\sqrt{n}} d_n(v_i, v_j).$$

Define a correspondence R_n between $\{0, 1, \ldots, n-1\}$ and $[0, 1]$ by setting $(i, s) \in R_n$ if $i = \lfloor ns \rfloor$; we also declare that $(n-1, 1) \in R_n$. Now endow $[0, 1]$ with the pseudo-metric d_{2e}. We will bound $\mathrm{dis}(R_n)$. Note first that if we write $u \wedge v$ for the most recent common ancestor of u and v, then

$$d_n(v_i, v_j) = d_n(v_0, v_i) + d_n(v_0, v_j) - 2d_n(v_0, v_i \wedge v_j).$$

By definition,

$$d_n(v_0, v_i) = H^n(i).$$

Moreover, it is not hard to see that

$$\left| d_n(v_0, v_i \wedge v_j) - \min_{i \le k \le j} H^n(k) \right| \le 1.$$

Now suppose that $(i, s), (j, t) \in R_n$ with $s \le t$. Then

$$\begin{aligned}
\left| d^n(i,j) - d_{2e}(s,t) \right| &= \left| \frac{\sigma}{\sqrt{n}} \left(H^n(\lfloor ns \rfloor) + H^n(\lfloor nt \rfloor) - 2d_n(v_0, v_i \wedge v_j) \right) \right. \\
&\qquad \left. - \left(2e(s) + 2e(t) - 4 \min_{s \le u \le t} e(u) \right) \right| \\
&\le \left| \frac{\sigma}{\sqrt{n}} \left(H^n(\lfloor ns \rfloor) + H^n(\lfloor nt \rfloor) - 2 \min_{s \le u \le t} H^n(\lfloor nu \rfloor) \right) \right. \\
&\qquad \left. - \left(2e(s) + 2e(t) - 4 \min_{s \le u \le t} e(u) \right) \right| + \frac{2\sigma}{\sqrt{n}}.
\end{aligned}$$

The right-hand side converges to 0 uniformly in $s, t \in [0, 1]$. Since

$$d_{\mathrm{GH}}\left(\left(T_n, \frac{\sigma}{\sqrt{n}} d_n \right), (\mathcal{T}_{2e}, d_{2e}) \right) \le \frac{1}{2} \mathrm{dis}(R_n),$$

the result follows. □

There are several steps along the way to the proof of Theorem 11 which, due to a lack of time, I have omitted. The following exercise is intended to lead you through a complete proof in one special case, assuming only Kaigh's theorem on the convergence of a random walk excursion.

Exercise 5 We have discussed the depth-first walk and the height function of a tree. A third encoding which is often used is the so-called *contour function* $(C(i), 0 \le i \le 2(n-1))$. For a tree $\mathbf{t} \in \mathbf{T}$, we imagine a particle tracing the outline of the tree from left to right at speed 1. (The picture below is for a labelled tree, with a planar embedding given by the labels.) Notice that we visit every vertex apart from the root ∅ a number of times given by its degree.

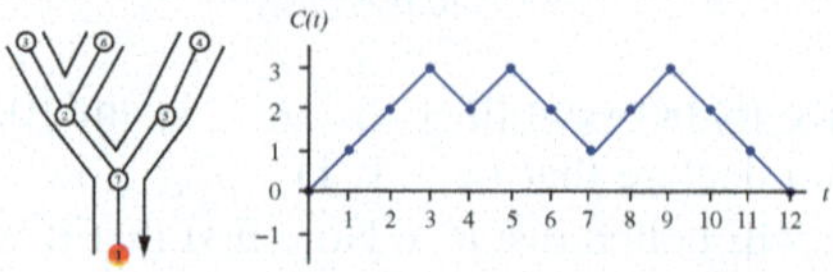

Let T_n be a Galton–Watson tree with offspring distribution $p(k) = 2^{-k-1}, k \geq 0$, conditioned to have total progeny $N = n$, as in Exercise 3(a). Let $(C^n(i), 0 \leq i \leq 2(n-1))$ be its contour function. It will be convenient to define a somewhat shifted version: let $\tilde{C}^n(0) = 0, \tilde{C}^n(2n) = 0$ and, for $1 \leq i \leq 2n-1, \tilde{C}^n(i) = 1 + C(i-1)$.

1. Show that $(\tilde{C}^n(i), 0 \leq i \leq 2n)$ has the same distribution as a simple symmetric random walk (i.e. a random walk which makes steps of $+1$ with probability $1/2$ and steps of -1 with probability $1/2$) conditioned to return to the origin for the first time at time $2n$.
Hint: first consider the unconditioned Galton–Watson tree with this offspring distribution.
2. It's straightforward to interpolate linearly to get a continuous function $\tilde{C}^n : [0, 2n] \to \mathbb{R}_+$. Let $\tilde{T}^n$ be the $\mathbb{R}$-tree encoded by this linear interpolation. Show that
$$d_{\mathrm{GH}}(T_n, \tilde{T}_n) \leq \frac{1}{2}.$$
Hint: notice that T_n considered as a metric space has only n points, whereas $\tilde{T}_n$ is an $\mathbb{R}$-tree and consists of uncountably many points. Draw a picture and find a correspondence.
3. Suppose that we have continuous excursions $f : [0, 1] \to \mathbb{R}_+$ and $g : [0, 1] \to \mathbb{R}_+$ which encode $\mathbb{R}$-trees $\mathcal{T}_f$ and $\mathcal{T}_g$. For $t \in [0, 1]$, let $p_f(t)$ be the image of t in the tree $\mathcal{T}_f$ and similarly for $p_g(t)$. Define a correspondence
$$R = \left\{(x, y) \in \mathcal{T}_f \times \mathcal{T}_g : x = p_f(t),\ y = p_g(t) \text{ for some } t \in [0, 1]\right\}.$$
Show that $\mathrm{dis}(R) \leq 4\|f - g\|_\infty$.
Hint: recall how the metric in an $\mathbb{R}$-tree is related to the function encoding it.
4. Observe that the variance of the step-size in a simple symmetric random walk is 1. Hence, by Theorem 5, we have
$$\frac{1}{\sqrt{2(n-1)}}(C^n(2(n-1)t), 0 \leq t \leq 1) \xrightarrow{d} (e(t), 0 \leq t \leq 1) \qquad (*)$$
as $n \to \infty$. Use this, (b) and (c) to prove directly that $(T_n, \frac{1}{\sqrt{n}}d_n)$ converges to a constant multiple of the Brownian CRT in the Gromov–Hausdorff sense.

Hint: you may want to use Skorokhod's representation theorem in order to work on a probability space where the convergence (*) occurs almost surely.

This approach is taken from Jean-François Le Gall and Grégory Miermont's lecture notes [31].

1.6 Properties of the Brownian CRT

A relatively straightforward extension of Theorem 11 shows that, in the appropriate topology (that generated by the Gromov–Hausdorff–Prokhorov distance; see Abraham, Hoscheit and Delmas [1] for a definition), the metric space $(T_n, \frac{\sigma}{\sqrt{n}} d_n)$ endowed additionally with the uniform measure on the vertices of T_n converges to $(\mathcal{T}_{2e}, d_{2e})$ endowed with the measure μ_{2e}, the push-forward of the Lebesgue measure on $[0, 1]$. In consequence, we refer to μ_{2e} as the *uniform measure* on $\mathcal{T}_{2e}$.

Consider picking points according to μ_{2e}. We may generate a sample from μ_{2e} simply by taking $p_{2e}(U)$ where $U \sim \mathrm{U}[0, 1]$ (recall that p_{2e} is the projection $[0, 1] \to \mathcal{T}_{2e}$). It turns out that $p_{2e}(U)$ is almost surely a leaf. (This may seem surprising at first sight. But we can think of it as saying, for example, that every vertex of a uniform random tree is at distance $o(\sqrt{n})$ from a leaf.) It is also the case that the rooted Brownian CRT $(\mathcal{T}_{2e}, d_{2e}, \rho)$ is invariant in distribution under random re-rooting at a point sampled from μ_{2e} (this follows because the same property is true for the uniform random tree T_n).

For fixed $k \geq 1$, let $X_1, X_2, \ldots, X_k$ be leaves of $\mathcal{T}_{2e}$ sampled according to μ_{2e}. Then the subtree of $\mathcal{T}_{2e}$ spanned by the set of points $\{\rho, X_1, \ldots, X_k\}$ has exactly the same distribution as the tree produced at step k in the line-breaking construction discussed in Sect. 1.1 [5].

The Brownian CRT has many other fascinating properties: for example, it is a random fractal, with Hausdorff and Minkowski dimension both equal to 2, almost surely [24, 30].

2 The Critical Erdős–Rényi Random Graph

We now turn to perhaps the simplest and best-known model of a random graph. Take n vertices labelled by $[n]$ and put an edge between any pair of them independently with probability p, for some fixed $p \in [0, 1]$. We write $G(n, p)$ for the resulting random graph. We will be interested in the connected components of $G(n, p)$ and, in particular, in their size and structure.

2.1 The Phase Transition and Component Sizes in the Critical Window

Let $p = c/n$ for some constant $c > 0$. The following statements hold with probability tending to 1 as $n \to \infty$:

- if $c < 1$, the largest connected component of $G(n, p)$ has size $\Theta(\log n)$;
- if $c > 1$, the largest connected component has size $\Theta(n)$ and the others are all of size $O(\log n)$.

In the latter case, we refer to the largest component as the *giant*.

Let us give an heuristic explanation for this phenomenon. We think about exploring the graph in a depth-first manner, which we will make more precise later. Firstly consider the vertex labelled 1. It has a $\text{Bin}(n-1, c/n) \approx \text{Po}(c)$ number of neighbours, say K. Consider the lowest-labelled of these neighbours. Conditionally on K, it itself has a $\text{Bin}(n-K-1, c/n)$ number of new neighbours. This distribution is still well-approximated by $\text{Po}(c)$, as long as $K = o(n)$. So we may think of exploring vertex by vertex and approximating the size of the component that we discover by the total progeny of a Galton–Watson branching process with $\text{Po}(c)$ offspring distribution, as long as the total number of vertices we have visited remains small relative to the size of the graph. If $c \leq 1$, such a branching process dies out with probability 1, which corresponds to obtaining a small component containing vertex 1. A similar argument will then work in subsequent components. If $c > 1$, there is positive probability that the branching process will survive. The branching process approximation holds good until we first explore a component which does not "die out"; this ends up being the giant component.

We will focus here on the critical case $c = 1$ or, more precisely, on the *critical window*: $p = \frac{1}{n} + \frac{\lambda}{n^{4/3}}$, $\lambda \in \mathbb{R}$. We will show in a moment that here the largest components have sizes on the order of $n^{2/3}$. With a view to later understanding the structure of these components, we will also track the *surplus* of each one, that is the number of edges it has more than a tree with the same number of vertices would: a component with m vertices and k edges has surplus $k - m + 1$.

Let us fix $\lambda \in \mathbb{R}$ and let $C_1^n, C_2^n, \ldots$ be the component sizes of $G\left(n, \frac{1}{n} + \frac{\lambda}{n^{4/3}}\right)$, listed in decreasing order, and let $S_1^n, S_2^n, \ldots$ be the corresponding surpluses.

Theorem 12 (Aldous [8]) *As $n \to \infty$,*

$$\left(\frac{1}{n^{2/3}}(C_1^n, C_2^n, \ldots), (S_1^n, S_2^n, \ldots)\right) \xrightarrow{d} ((C_1, C_2, \ldots), (S_1, S_2, \ldots))$$

where the limit has an explicit description to be given below. Convergence for the first sequence takes place in

$$\ell^2_{\downarrow} := \left\{ \mathbf{x} = (x_1, x_2, \ldots) : x_1 \geq x_2 \geq \ldots \geq 0, \sum_{i=1}^{\infty} x_i^2 < \infty \right\}$$

with the usual ℓ^2 distance; for the second sequence, it is in the product topology.

To describe the limit, let

$$W^\lambda(t) = W(t) + \lambda t - \frac{t^2}{2}, \quad t \geq 0,$$

where $(W(t), t \geq 0)$ is a standard Brownian motion. Let $B^\lambda(t) = W^\lambda(t) - \inf_{0 \leq s \leq t} W^\lambda(s)$ be the the process reflected at its minimum. Now draw the graph of the process B^λ and decorate it with the points of a rate 1 Poisson process in the plane, keeping only those points which fall between the x-axis and the function, as illustrated below. Conditionally on B^λ, we let $(P^\lambda(t), t \geq 0)$ be the inhomogeneous Poisson process of intensity $B^\lambda(t)$ at time t which describes the arrival of the points.

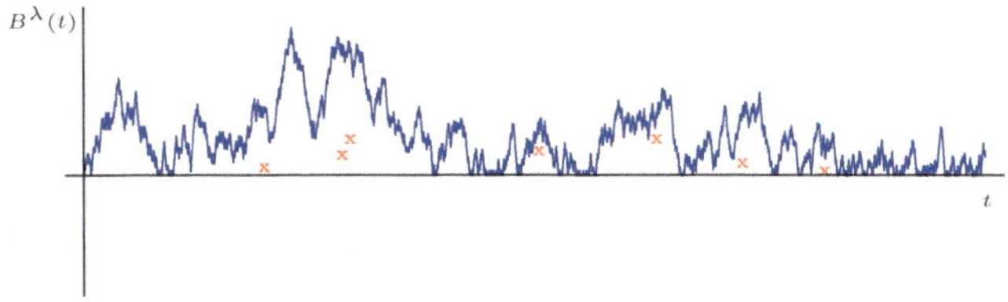

Then $(C_1, C_2, \ldots)$ is the ordered sequence of excursion-lengths above 0 of the process B^λ and $(S_1, S_2, \ldots)$ is the sequence of numbers of Poisson points falling in the corresponding excursions.

The key to the proof of this result is the depth-first walk (or a variant thereof: Aldous' original proof rather uses a *breadth-first* walk, but the details are essentially identical). Consider the component containing vertex 1. We explore as we would in a tree, simply ignoring any edges to vertices we have already seen (which make cycles), and using the labels to tell us which is the root of a component (the lowest-labelled vertex) and to provide a left-right ordering for the children of a vertex. When we reach the end of a component, we start exploring the component of the lowest-labelled vertex we have not yet seen, concatenating the depth-first walks. Note that, as in the case of a forest of Galton–Watson trees, we begin a new component every time the depth-first walk attains a new minimum. The following is essentially Theorem 3 of Aldous [8].

Theorem 13 (Aldous [8]) *Let X^n be the depth-first walk associated with $G\left(n, \frac{1}{n} + \frac{\lambda}{n^{4/3}}\right)$ and let N^n be the counting process of surplus edges (so that $N^n(k)$ is the number of surplus edges encountered by step k). Then the following convergences occur jointly:*

$$\frac{1}{n^{1/3}} \left(X^n(\lfloor n^{2/3} t \rfloor), t \geq 0\right) \xrightarrow{d} (W^\lambda(t), t \geq 0)$$

and

$$\left(N^n(\lfloor n^{2/3} t \rfloor), t \geq 0\right) \xrightarrow{d} (P^\lambda(t), t \geq 0)$$

as $n \to \infty$ (here both convergences are in the Skorokhod sense).

Proof (sketch) Write $v_0, v_1, \ldots$ for the vertices in the order we visit them, where $v_0 = 1$. Let $K^n(i)$ be the number of children of v_i, where children are *new* vertices we discover when we explore the neighbours of v_i. Then

$$X^n(k) = \sum_{i=0}^{k-1} (K^n(i) - 1).$$

Write $L^n(k) = -\inf_{0 \le i \le k} X^n(i)$. Then $L^n(k)$ is the number of components that we have fully explored before step k. When we visit v_i, there are $X^n(i) + L^n(i)$ vertices on the stack, and i vertices we have already visited. Hence, there are possible edges from v_i to any of the remaining $n - i - 1 - X^n(i) - L^n(i)$ vertices, each of which is present independently with probability $\frac{1}{n} + \frac{\lambda}{n^{4/3}}$. So, given $X^n(i)$ and $L^n(i)$,

$$K^n(i) \sim \text{Bin}\left(n - i - 1 - X^n(i) - L^n(i), \frac{1}{n} + \frac{\lambda}{n^{4/3}}\right),$$

with

$$\begin{aligned}
&\mathbb{E}\left[K^n(i) - 1 | X^n(0), \ldots, X^n(i)\right] \\
&\quad = \frac{\lambda}{n^{1/3}} - \frac{(i+1)}{n} - \frac{X^n(i) + L^n(i)}{n} - \frac{\lambda(i+1) + \lambda X^n(i) + \lambda L^n(i)}{n^{4/3}}.
\end{aligned}$$

In particular, (X^n, L^n) is a time-inhomogeneous Markov process. Suppose that $\sup_{0 \le i \le k} |X^n(i)| = o(n^{2/3})$ for $k = \Theta(n^{2/3})$. Then we may neglect the last two terms on the right-hand side to obtain that, for $i = \Theta(n^{2/3})$,

$$\mathbb{E}\left[K^n(i) - 1 | X^n(0), \ldots, X^n(i)\right] = \frac{\lambda}{n^{1/3}} - \frac{i}{n} + o(n^{-1/3}).$$

So, for $i = \Theta(n^{2/3})$, it is approximately the case that

$$X^n(i+1) - X^n(i) \sim \text{Po}\left(1 + \frac{\lambda}{n^{1/3}} - \frac{i}{n}\right) - 1.$$

Now let

$$M^n(k) = X^n(k) - \sum_{i=0}^{k-1} \left(\frac{\lambda}{n^{1/3}} - \frac{i}{n}\right) \approx X^n(k) - \frac{\lambda k}{n^{1/3}} + \frac{k^2}{2n}.$$

Then $(M^n(k), k \ge 0)$ is approximately a martingale. Since the step-sizes of X^n have variance approximately equal to 1, we may apply the martingale central limit theorem (Theorem 7.1.4 of [22]) to obtain

$$\left(\frac{1}{n^{1/3}}X^n(\lfloor tn^{2/3}\rfloor) - \lambda t + \frac{t^2}{2}, t \ge 0\right) \xrightarrow{d} (W(t), t \ge 0).$$

It follows by the continuous mapping theorem that

$$\frac{1}{n^{1/3}}\left(X^n(\lfloor tn^{2/3}\rfloor) + L^n(\lfloor tn^{2/3}\rfloor), t \ge 0\right) \xrightarrow{d} \left(W^\lambda(t) - \inf_{0\le s\le t} W^\lambda(s), t \ge 0\right).$$

Turning now to the surplus edges, there is potentially an edge between the vertex v_k we are currently visiting and any one of the $X^n(k) + L^n(k)$ vertices on the stack, which cannot be children of v_k, since they are already children of elements of $\{v_0, \ldots, v_{k-1}\}$. Any such edges which are present thus contribute to the surplus of a component. Since such edges occur independently and with probability $\frac{1}{n} + \frac{\lambda}{n^{4/3}}$, we get a Binomial$(X^n(k) + L^n(k), \frac{1}{n} + \frac{\lambda}{n^{4/3}})$ number of surplus edges from v_k. Now the conditional mean of this random variable, which is approximately $(X^n(\lfloor tn^{2/3}\rfloor) + L^n(\lfloor tn^{2/3}\rfloor))/n$, is on the order of $n^{-2/3}$, and so we see points only on time-intervals of order $n^{2/3}$ steps. But this is precisely the timescale in which we are interested and so taking limits as $n \to \infty$, we obtain a Poisson point process of intensity $W^\lambda(t) - \inf_{0\le s\le t} W^\lambda(s) = B^\lambda(t)$ at time $t \ge 0$. □

In the depth-first walk, we explore the components of the graph in size-biased random order. (Given the component sizes, vertex 1 lies in a component chosen with probability proportional to its size, and similarly for subsequent components.) In order to get from Theorem 13 to Theorem 12, one must show that in going from size-biased order to decreasing order, one doesn't lose track of any of the large components; see Sect. 2.3 of Aldous [8] for the details.

2.2 Component Structures

We saw in Proposition 3 that the depth-first walk of a tree can be used to recover its structure. The excursions of the depth-first walk in the graph setting thus encode a collection of *spanning trees* for the components. In the scaling limit, this depth-first walk becomes a Brownian motion with parabolic drift. Let us first look a little deeper into the properties of this limit process.

By the Cameron–Martin–Girsanov theorem (e.g. Theorem 38.5 in Sect. IV of [36]), for any bounded measurable test-function $f : C([0,t], \mathbb{R}) \to \mathbb{R}$, we have

$$\begin{aligned}&\mathbb{E}\left[f(W^\lambda(s), 0 \le s \le t)\right]\\ &\quad= \mathbb{E}\left[\exp\left(\int_0^t (\lambda - s)\mathrm{d}W(s) - \frac{1}{2}\int_0^t (\lambda - s)^2 \mathrm{d}s\right) f(W(s), 0 \le s \le t)\right].\end{aligned}$$

So W^λ is absolutely continuous with respect to a process which would encode a forest of Brownian CRT's. Consider the process of excursions of W^λ above its running minimum or, equivalently, the excursions of B^λ above 0. One can make sense of a sort of time-inhomogeneous excursion measure for B^λ, where the time-inhomogeneity manifests itself only in the *lengths* of the excursions. The shapes of the excursions themselves are absolutely continuous with respect to those of W. Indeed, let us write $\tilde{e}$ for an excursion of B^λ normalised to have length 1. Recall that e is a normalised Brownian excursion. Then for a bounded measurable test-function $g : \mathcal{C}([0,1], \mathbb{R}_+) \to \mathbb{R}$, we have

$$\mathbb{E}\left[g(\tilde{e})\right] = \frac{\mathbb{E}\left[g(e)\exp\left(\int_0^1 e(u)du\right)\right]}{\mathbb{E}\left[\exp\left(\int_0^1 e(u)du\right)\right]}.$$

More generally, for an excursion of length x (and an appropriately adjusted test-function), we have the usual Brownian scaling relation:

$$\mathbb{E}\left[g\left(\tilde{e}^{(x)}\right)\right] = \mathbb{E}\left[f\left(\sqrt{x}\tilde{e}(t/x), 0 \le t \le x\right)\right].$$

In summary, it is helpful to think of the reflected process B^λ in two parts: (a) it has excursions above 0 whose ordered lengths are the random variables $(C_1, C_2, \ldots)$ from Theorem 12 and (b) given these lengths, the excursions themselves are conditionally independent, with the ith longest having the same distribution as $\tilde{e}^{(C_i)}$. Part (a) of this decomposition is rather complicated but, taking that as an input, part (b) is comparatively simple.

The limiting picture for a single component corresponding to an excursion of length x is then as follows. We have an excursion $\tilde{e}^{(x)}$ (with law as above). Conditionally on $\tilde{e}^{(x)}$, the number of points falling under the excursion has Po $\left(\int_0^x \tilde{e}^{(x)}(u)du\right)$ distribution. Let $\tilde{\mathcal{T}}^{(x)}$ be the $\mathbb{R}$-tree encoded by $2\tilde{e}^{(x)}$ and write $p : [0, x] \to \tilde{\mathcal{T}}^{(x)}$ for the canonical projection. (We will also write $\tilde{\mathcal{T}}$ in place of $\tilde{\mathcal{T}}^{(1)}$.) A point $(t, y/2)$ under $\tilde{e}^{(x)}$ identifies the vertex $p(t)$ which is at height $2\tilde{e}^{(x)}(t)$ with the vertex at distance y along the path from the root to v in $\tilde{\mathcal{T}}^{(x)}$; we then endow the resulting space with the quotient metric. See Fig. 2 for an illustration.

Let $C_1^n, C_2^n, \ldots$ be the sequence of components of $G(n, \frac{1}{n} + \frac{\lambda}{n^{4/3}})$, listed in decreasing order of size. In order to keep the notation compact, for $a > 0$, write aC_i^n for the metric space formed by C_i^n endowed with the graph distance rescaled by a.

Theorem 14 (Addario-Berry, Broutin, G. [3]) *As $n \to \infty$,*

$$n^{-1/3}(C_1^n, C_2^n, \ldots) \xrightarrow{d} (\mathcal{C}_1, \mathcal{C}_2, \ldots),$$

where $\mathcal{C}_1, \mathcal{C}_2, \ldots$ is the sequence of random compact metric spaces corresponding to the excursions of Aldous' marked limit process B^λ in decreasing order of length.

The convergence here is with respect to the distance

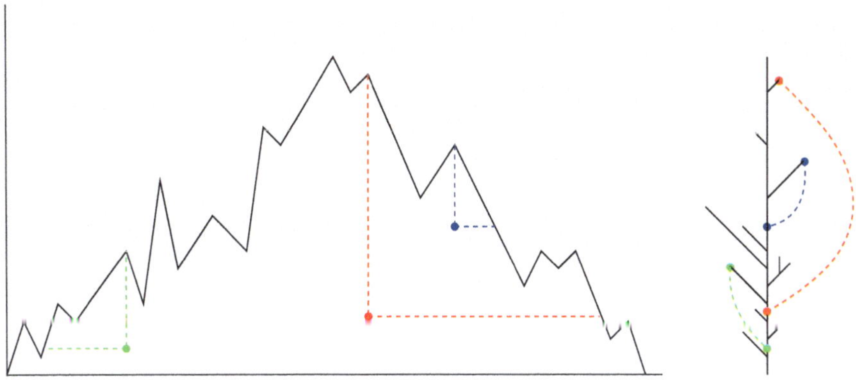

Fig. 2 Left: an excursion with three points. Right: the corresponding $\mathbb{R}$-tree with vertex-identifications (identify any pair of points joined by a dashed line)

$$\mathrm{dist}(\mathcal{A}, \mathcal{B}) = \left(\sum_{i=1}^{\infty} d_{\mathrm{GH}}(\mathcal{A}_i, \mathcal{B}_i)^4 \right)^{1/4},$$

where $\mathcal{A} = (\mathcal{A}_1, \mathcal{A}_2, \ldots)$ and $\mathcal{B} = (\mathcal{B}_1, \mathcal{B}_2, \ldots)$ are sequences of compact metric spaces.

Proof (sketch) Consider a component G of $G(n, p)$, conditioned to have a vertex set of size m (we take $[m]$ for simplicity). To any such component, we may associate a canonical spanning tree $T(G)$, called the *depth-first tree* of G: this is the tree we pick out when we do our depth-first walk, for which we write $(X(k), 0 \le k \le m)$.

Given a fixed tree $T \in \mathbb{T}_m$, which connected graphs G have $T(G) = T$? In other words, where might we put surplus edges into T such that we don't change the depth-first tree? Call any such edges *permitted*. It is straightforward to see that there are precisely $X(k)$ permitted edges at step k: one between v_k and each of the vertices which have been seen but not yet fully explored. So there are

$$a(T) := \sum_{k=0}^{m-1} X(k)$$

permitted edges in total. We call this the *area* of T. Now let

$$\mathbb{G}_T = \{ \text{ graphs } G \text{ such that } T(G) = T \}.$$

Then

$$\{\mathbb{G}_T : T \in \mathbb{T}_m\}$$

is a partition of the set of connected graphs on $[m]$. Moreover, $|\mathbb{G}_T| = 2^{a(T)}$, since each permitted edge may either be included or not.

Exercise 6 Let $\tilde{G}_m^p$ be a connected graph on vertices labelled by $[m]$ generated as follows:

– Pick a random tree $\tilde{T}_m^p$ such that

$$\mathbb{P}\left(\tilde{T}_m^p = T\right) \propto (1-p)^{-a(T)}, \quad T \in \mathbb{T}_m.$$

– Add each of the $a(\tilde{T}_m^p)$ permitted edges independently with probability p.

Show that $\tilde{G}_m^p$ has the same distribution as a component of $G(n, p)$ conditioned to have vertex set $[m]$.

It remains now to show that, for $m \sim xn^{2/3}$ and $p = \frac{1}{n} + \frac{\lambda}{n^{4/3}}$, we have

– $n^{-1/3}\tilde{T}_m^p \xrightarrow{d} \tilde{\mathcal{T}}^{(x)}$
– the locations of the surplus edges converge to the locations in the limiting picture.

For simplicity, let us take $x = 1$ and $\lambda = 0$, so that $p = m^{-3/2}$. (The general case is similar.) Write $\tilde{X}^m$ for the depth-first walk of $\tilde{T}_m^p$, and let $\tilde{H}^m$ be the corresponding height process defined, as usual, from $\tilde{X}^m$ via the relation (1.1). Then

$$a(\tilde{T}_m^p) = \int_0^m \tilde{X}^m(\lfloor s \rfloor)ds = m \int_0^1 \tilde{X}^m(\lfloor mt \rfloor)dt,$$

by a simple change of variables in the integral.

If T_m is a uniform random tree on $[m]$ and X^m is its depth-first walk, we know from Theorem 5 that

$$(m^{-1/2}X^m(\lfloor mt \rfloor), 0 \le t \le 1) \xrightarrow{d} (e(t), 0 \le t \le 1). \tag{2.1}$$

Moreover, by Exercise 6, for a bounded continuous test-function f,

$$\begin{aligned}
&\mathbb{E}\left[f(m^{-1/2}\tilde{X}^m(\lfloor mt \rfloor), 0 \le t \le 1)\right] \\
&= \frac{\mathbb{E}\left[f(m^{-1/2}X^m(\lfloor mt \rfloor), 0 \le t \le 1)(1-p)^{-m\int_0^1 X^m(\lfloor mu \rfloor)du}\right]}{\mathbb{E}\left[(1-p)^{-m\int_0^1 X^m(\lfloor mu \rfloor)du}\right]} \\
&= \frac{\mathbb{E}\left[f(m^{-1/2}X^m(\lfloor mt \rfloor), 0 \le t \le 1)(1-m^{-3/2})^{-m^{3/2}\int_0^1 m^{-1/2}X^m(\lfloor mu \rfloor)du}\right]}{\mathbb{E}\left[(1-m^{-3/2})^{-m^{3/2}\int_0^1 m^{-1/2}X^m(\lfloor mu \rfloor)du}\right]}.
\end{aligned}$$

We have

$$(1-m^{-3/2})^{-m^{3/2}\int_0^1 m^{-1/2}X^m(\lfloor mu \rfloor)du} \xrightarrow{d} \exp\left(\int_0^1 e(u)du\right)$$

as $m \to \infty$, by (2.1) and the continuous mapping theorem. The sequence of random variables on the left-hand side may be shown to be uniformly integrable (see Lemma

14 of [3]) and so

$$\mathbb{E}\left[f(m^{-1/2}\tilde{X}^m(\lfloor mt \rfloor), 0 \le t \le 1)\right] \to \mathbb{E}\left[f(\tilde{e})\right]$$

as $m \to \infty$. Similar reasoning then gives that

$$\mathbb{E}\left[f(m^{-1/2}\tilde{H}^m(\lfloor mt \rfloor), 0 \le t \le 1)\right] \to \mathbb{E}\left[f(2\tilde{e})\right],$$

which implies (by the same argument as in the proof of Theorem 11) that

$$\frac{1}{\sqrt{m}}\tilde{T}_m^p \xrightarrow{d} \tilde{\mathcal{T}}$$

as $m \to \infty$, in the Gromov–Hausdorff sense.

Now consider the surplus edges. It is straightforward to see that there is a bijection between permitted edges and integer points under the graph of the depth-first walk. A point at (k, ℓ) means "put an edge between v_k and the vertex at distance ℓ from the bottom of the stack". Since each permitted edge is present independently with probability p, the surplus edges form a point process, which converges on rescaling to our Poisson point process. Finally, surplus edges always join v_k and a younger child of some ancestor of v_k. In the limit, the distance between a vertex and its children vanishes, so that surplus edges are effectively to ancestors. □

3 Critical Random Graphs with i.i.d Random Degrees

We have seen that degrees in the Erdős–Rényi model are approximately Poisson distributed. In recent years, there has been much interest in modelling settings where this is certainly not the case. We will discuss one popular model which has arisen, with the principal aim of demonstrating that the results in the previous section are universal. We will restrict our attention to component sizes.

3.1 The Configuration Model

Suppose we wish to generate a graph G_n uniformly at random from those with vertex set $[n]$ and such that vertex i has degree vertex d_i, where $d_i \ge 1$ for $1 \le i \le n$ and $\ell_n = \sum_{i=1}^n d_i$ is even.

Assign d_i half-edges to vertex i. Label the half-edges in some arbitrary way by $1, 2, \ldots, \ell_n$. Then generate a uniformly random pairing of the half-edges to create full edges. This is known as the *configuration model*.

Clearly this may produce self-loops (edges whose endpoints are the same vertex) or multiple edges between the same pair of vertices, so in general the configuration model produces a multigraph, M_n. Assuming that there exists at least one simple graph with the given degrees then, conditionally on the event $\{M_n \text{ is a simple graph}\}$, M_n has the same law as G_n. This is a consequence of the following exercise.

Exercise 7 Fix a degree sequence $d_1, \ldots, d_n$. Show that the probability of generating a particular multigraph G with these degrees is

$$\frac{1}{(\ell_n - 1)!!} \frac{\prod_{i=1}^n d_i!}{2^{\mathrm{sl}(G)} \prod_{e \in E(G)} \mathrm{mult}(e)!},$$

where $\ell_n = \sum_{i=1}^n d_i$, $\mathrm{sl}(G)$ is the number of self-loops in G and $\mathrm{mult}(e)$ is the multiplicity of the edge $e \in E(G)$. (We recall the double factorial notation $n!! := \prod_{k=0}^{\lceil n/2 \rceil - 1} (n - 2k)$.)

We will take the degrees themselves to be random: let $D_1, D_2, \ldots, D_n$ be i.i.d. with the same law as some random variable D which has finite variance. We resolve the issue of $\sum_{i=1}^n D_i$ potentially being odd by simply throwing the last half-edge away when we generate the pairing in that case.

Let $\gamma = \mathbb{E}[D(D-1)] / \mathbb{E}[D]$. Then

$$\mathbb{P}(M_n \text{ is simple }) \to \exp\left(-\frac{\gamma}{2} - \frac{\gamma^2}{4}\right) > 0$$

(see Theorem 7.12 of van der Hofstad [25]), so that conditioning on simplicity will make sense for large n.

Theorem 15 (Molloy and Reed [33]) *If $\gamma < 1$ then, with probability tending to 1 as $n \to \infty$, there is no giant component; if $\gamma > 1$ then, with probability tending to 1 as $n \to \infty$, there is a unique giant component.*

Let us give an heuristic argument for why $\gamma = 1$ should be the critical point. An important point is that we may generate the pairing of the half-edges one by one, in any order that is convenient. So we will generate and explore the graph at the same time. Perform a depth-first exploration from an arbitrary vertex. Consider the first half-edge attached to the vertex (the one with the smallest half-edge label). It picks its pair uniformly from all those available, and so in particular it picks a half-edge belonging to a vertex chosen with probability proportional to its degree. The same will be true of subsequent half-edges. As long as we have not explored much of the graph, these degrees should have law close to the *size-biased distribution*, given by

$$\mathbb{P}\left(D^* = k\right) = \frac{k\mathbb{P}(D = k)}{\mathbb{E}[D]}, \quad k \geq 1.$$

Hence, we can compare to a branching process with offspring distribution $D^* - 1$, which has expectation

$$\mathbb{E}\left[D^* - 1\right] = \frac{\mathbb{E}\left[D^2\right]}{\mathbb{E}\left[D\right]} - 1 = \gamma.$$

The following exercise gives an idea of why Poisson degrees are particularly nice.

Exercise 8 Suppose that D is a non-negative integer-valued random variable with finite mean, and let D^* have the size-biased distribution

$$\mathbb{P}\left(D^* = k\right) = \frac{k\mathbb{P}\left(D = k\right)}{\mathbb{E}\left[D\right]}, \quad k \geq 1.$$

Show that $D^* - 1 \stackrel{d}{=} D$ if and only if D has a Poisson distribution.

3.2 *Scaling Limit for the Critical Component Sizes*

We will henceforth consider the configuration model with the following set-up: assume the degrees $D_1, D_2, \ldots, D_n$ are i.i.d. with the same distribution as D such that

- $\mathbb{P}\left(D \geq 1\right) = 1, \mathbb{P}\left(D = 2\right) < 1$;
- $\gamma = \frac{\mathbb{E}[D(D-1)]}{\mathbb{E}[D]} = 1$;
- $\mathbb{E}\left[D^3\right] < \infty$.

We write $\mu = \mathbb{E}\left[D\right]$ and $\beta = \mathbb{E}\left[D(D-1)(D-2)\right]$. We observe immediately that $\mathbb{E}\left[D^*\right] = 2$ and that $\mathrm{var}\left(D^*\right) = \beta/\mu$.

The analogue of Theorem 13 in this setting is as follows.

Theorem 16 (Riordan [35], Joseph [26]) *Let $C_1^n C_2^n, \ldots$ be the ordered component sizes of M_n or G_n. Then*

$$n^{-2/3}(C_1^n, C_2^n, \ldots) \xrightarrow{d} (C_1, C_2, \ldots)$$

in $\ell^2_\downarrow$, where the limit is given by the ordered sequence of excursion-lengths above past-minima of the process $(W^{\beta,\mu}(t), t \geq 0)$ defined by

$$W^{\beta,\mu}(t) := \sqrt{\frac{\beta}{\mu}}W(t) - \frac{\beta t^2}{2\mu^2}, \quad t \geq 0,$$

where W is a standard Brownian motion.

We give a sketch of a proof of this result, and refer the reader to Theorem 2.1 of [26] for a complete proof. To start with, we make precise the connection between a collection of i.i.d. random variables in size-biased random order and a collection of i.i.d. random variables with the size-biased distribution.

Exercise 9 Suppose that $D_1, D_2, \ldots, D_n$ are i.i.d. random variables with finite mean μ, and let $(\hat{D}_1^n, \hat{D}_2^n, \ldots, \hat{D}_n^n)$ be the same random variables in size-biased random order. That is, given the degree sequence $D_1, D_2, \ldots, D_n$, let Σ be a permutation of $[n]$ with conditional distribution

$$\mathbb{P}(\Sigma = \sigma | D_1, D_2, \ldots, D_n) = \frac{D_{\sigma(1)}}{\sum_{j=1}^n D_{\sigma(j)}} \frac{D_{\sigma(2)}}{\sum_{j=2}^n D_{\sigma(j)}} \cdots \frac{D_{\sigma(n)}}{D_{\sigma(n)}}, \quad \sigma \in \mathfrak{S}_n.$$

Then define

$$(\hat{D}_1^n, \hat{D}_2^n, \ldots, \hat{D}_n^n) = (D_{\Sigma(1)}, D_{\Sigma(2)}, \ldots, D_{\Sigma(n)}).$$

Now let $D_1^*, D_2^*, \ldots$ be i.i.d. with the (true) size-biased distribution. Show that for $m < n$ and $d_1, d_2, \ldots, d_m \geq 1$,

$$\mathbb{P}\left(\hat{D}_1^n = d_1, \hat{D}_2^n = d_2, \ldots, \hat{D}_m^n = d_m\right)$$
$$= \phi_m^n(d_1, d_2, \ldots, d_m)\mathbb{P}\left(D_1^* = d_1, D_2^* = d_2, \ldots, D_m^* = d_m\right),$$

where

$$\phi_m^n(d_1, d_2, \ldots, d_m) := \frac{n!\mu^m}{(n-m)!}\mathbb{E}\left[\prod_{i=1}^m \frac{1}{\sum_{j=i}^m d_j + \Delta_{n-m}}\right],$$

and $\Delta_{n-m} \stackrel{d}{=} D_1 + \ldots + D_{n-m}$.

Proof of Theorem 16 *(sketch)* We again use a depth-first walk but this time with a stack of unpaired half-edges. Start by picking a vertex with probability proportional to its degree. Declare one of its half-edges to be active and put the rest on the stack. Sample the active half-edge's pair (either on the stack or not) and remove both from further consideration. If we discovered a new vertex, add its remaining half-edges to the top of the stack. Then declare whichever half-edge is now on top of the stack to be active. If ever the stack becomes empty, pick a new vertex with probability proportional to its degree and continue.

In this procedure, we observe the vertex-degrees precisely in size-biased random order. Let $\tilde{X}(0) = 0$ and

$$\tilde{X}^n(k) := \sum_{i=1}^k (\hat{D}_i^n - 2), \quad k \geq 1.$$

Then $\tilde{X}^n$ behaves exactly like the depth-first walk except

- at the start of a component, where we should add $\hat{D}_i^n - 1$ rather than $\hat{D}_i^n - 2$ and
- whenever we pair the active half-edge with one on the stack.

Neither problem shows up in the limit (although showing this properly is somewhat technical). For the purposes of this sketch, we shall ignore the difference. We write

$X^*(0) = 0$ and let

$$X^*(i) = \sum_{j=1}^{i}(D_j^* - 2)$$

be a similar process built instead from i.i.d. size-biased random variables. Note that X^* is a centred random walk with step-variance equal to β/μ. In particular, by Donsker's theorem,

$$n^{-1/3}(X^*(\lfloor n^{2/3}s\rfloor), s \geq 0) \xrightarrow{d} \sqrt{\frac{\beta}{\mu}}(W(s), s \geq 0) \tag{3.1}$$

as $n \to \infty$. We aim to show that

$$n^{-1/3}(\tilde{X}(\lfloor n^{2/3}s\rfloor), s \geq 0) \xrightarrow{d} (W^{\beta,\mu}(s), s \geq 0).$$

By the Cameron–Martin–Girsanov theorem and integration by parts, for suitable test-functions f,

$$\begin{aligned}
&\mathbb{E}\left[f(W^{\beta,\mu}(s), 0 \leq s \leq t)\right] \\
&\mathbb{E}\left[\exp\left(-\sqrt{\frac{\beta}{\mu^3}}\int_0^t s\mathrm{d}W(s) - \frac{1}{2}\frac{\beta}{\mu^3}\int_0^t s^2\mathrm{d}s\right) f\left(\sqrt{\frac{\beta}{\mu}}(W(s), 0 \leq s \leq t)\right)\right] \\
&= \mathbb{E}\left[\exp\left(\sqrt{\frac{\beta}{\mu^3}}\int_0^t (W(s) - W(t))\mathrm{d}s - \frac{\beta t^3}{6\mu^3}\right) f\left(\sqrt{\frac{\beta}{\mu}}(W(s), 0 \leq s \leq t)\right)\right].
\end{aligned} \tag{3.2}$$

Exercise 9 gives us a way to obtain a discrete analogue of this change of measure. Write $x(i) = \sum_{j=1}^{i}(d_j - 2)$. Then we may rewrite

$$\begin{aligned}
&\phi_m^n(d_1, d_2, \ldots, d_m) \\
&= \frac{n!}{(n-m)!n^m}\mathbb{E}\left[\prod_{i=1}^{m}\frac{n\mu}{x(m) - x(i-1) + 2(m-i+1) + \Delta_{n-m}}\right] \\
&= \prod_{i=1}^{m-1}\left(1 - \frac{i}{n}\right)\mathbb{E}\left[\prod_{i=1}^{m}\frac{n\mu}{\Delta_{n-m} + 2(m-i+1) + x(m) - x(i-1)}\right] \\
&= \exp\left(\sum_{i=1}^{m-1}\log\left(1 - \frac{i}{n}\right)\right) \\
&\quad \times \mathbb{E}\left[\exp\left(-\sum_{i=1}^{m}\log\left(\frac{\Delta_{n-m} + 2(m-i+1) + x(m) - x(i-1)}{n\mu}\right)\right)\right].
\end{aligned}$$

Taylor expanding the logarithms, we get that this is approximately equal to

$$
\begin{aligned}
&\exp\left(-\sum_{i=1}^{m-1}\left[\frac{i}{n}+\frac{i^2}{2n^2}\right]-\sum_{i=1}^{m}\left[\frac{2(m-i+1)}{n\mu}-\frac{2(m-i+1)^2}{n^2\mu^2}\right]+m\right)\\
&\qquad\times\exp\left(-\frac{1}{n\mu}\sum_{i=1}^{m}(x(m)-x(i-1))\right)\mathbb{E}\left[\exp\left(-\frac{m}{n\mu}\Delta_{n-m}\right)\right]\\
&\approx\exp\left(\frac{1}{n\mu}\sum_{i=1}^{m-1}(x(i)-x(m))+m-\frac{(2+\mu)m^2}{2\mu n}+\frac{(2+\mu)(2-\mu)m^3}{6\mu^2n^2}\right)\\
&\qquad\times\mathbb{E}\left[\exp\left(-\frac{m}{n\mu}D_1\right)\right]^{n-m}.
\end{aligned}
$$

Using the moments of D_1, it is straightforward to show that its Laplace transform has the following asymptotic behaviour:

$$
\mathbb{E}\left[\exp(-\theta D_1)\right]=\exp\left(-\theta\mu+\frac{\theta^2\mu(2-\mu)}{2}-\frac{\theta^3}{6}(\beta+4\mu-6\mu^2+2\mu^3)+o(\theta^3)\right),
$$

as $\theta\downarrow 0$. Putting all of this together, almost everything cancels and we get that for $m=\lfloor tn^{2/3}\rfloor$,

$$
\phi_m^n(D_1^*,D_2^*,\ldots,D_m^*)\approx\exp\left(\frac{1}{n\mu}\sum_{i=1}^{m-1}(X^*(i)-X^*(m))-\frac{\beta t^3}{6\mu^3}\right).
$$

More work gets uniform integrability, and then we may conclude using (3.1) and the continuous mapping theorem that

$$
\begin{aligned}
&\mathbb{E}\left[f\left(n^{-1/3}\tilde{X}(\lfloor n^{2/3}s\rfloor),0\le s\le t\right)\right]\\
&=\mathbb{E}\left[\phi_m^n(D_1^*,D_2^*,\ldots,D_m^*)f\left(n^{-1/3}X^*(\lfloor n^{2/3}s\rfloor),0\le s\le t\right)\right]\\
&\to\mathbb{E}\left[\exp\left(\frac{1}{\mu}\int_0^t\sqrt{\frac{\beta}{\mu}}(W(s)-W(t))ds-\frac{\beta t^3}{6\mu^3}\right)f\left(\sqrt{\frac{\beta}{\mu}}(W(u),0\le s\le t)\right)\right]\\
&=\mathbb{E}\left[f(W^{\beta,\mu}(s),0\le s\le t)\right],
\end{aligned}
$$

where the last equality holds by (3.2).

Finally, it is possible to show that when exploring M_n, the first loop or multiple edge occurs at a time which is $\gg n^{2/3}$ and so the same distributional convergence holds if we condition on simplicity. □

4 Sources for These Notes and Suggested Further Reading

David Aldous' series of papers [5–7] and Jean-François Le Gall's paper [28] on the Brownian CRT remain an excellent source of inspiration. I have used Aldous' approach from [5, 6] in Sect. 1.1. The survey paper [6] gives what Aldous refers to as The Big Picture and is a great place to start reading about the Brownian CRT. I first learnt much of the material in Sect. 1 from a wonderful DEA lecture course at Paris VI in 2003 given by Jean-François Le Gall. These notes borrow heavily from his excellent survey of random trees [29].

To learn about generalisations of the Brownian CRT, in particular the so-called Lévy trees, see the monograph of Duquesne and Le Gall [21]. The *stable trees* are the scaling limits of critical Galton–Watson trees with offspring distributions which do not have finite variance, but instead lie in the domain of attraction of an α-stable law, for $\alpha \in (1, 2)$. They have particularly nice properties, including a line-breaking construction [23].

For those looking to learn about random graphs, I warmly recommend Remco van der Hofstad's recent book [25]. David Aldous' paper [8] on the critical Erdős–Rényi random graph and the multiplicative coalescent is essential reading for Sect. 2.1. Jim Pitman's St-Flour course [34] contains much complementary material. Section 2.2 is based on joint work with Louigi Addario-Berry and Nicolas Broutin [3]. To learn more about the properties of the metric space scaling limit of the Erdős–Rényi random graph, see the companion paper [2]. The results in these two papers played a key role in our proof, jointly with Grégory Miermont, of a scaling limit for the minimum-spanning tree of the complete graph endowed with i.i.d. random edge-weights from a continuous distribution [4].

The component sizes in the configuration model with critical i.i.d. degrees, as treated in Sect. 3, were studied by Joseph [26]; the component sizes and surpluses in a more general set-up were studied independently by Riordan [35]. See also Dhara, van der Hofstad, van Leeuwaarden and Sen [20]. The sketch proof of Theorem 16 presented here is based on joint work in progress with Guillaume Conchon-Kerjan [18]. The corresponding metric space scaling limit has been proved by Bhamidi and Sen [14]. In recent years, several other critical random graph models have been shown to possess the same scaling limit as the critical Erdős–Rényi random graph (either for the component sizes, or for the full component structures). See [9, 10, 12, 14, 15]. The case where the degree distribution does not have a third moment is more complicated and gives different scaling limits; see [11, 13, 16, 19, 26].

Acknowledgements My work is supported by EPSRC Fellowship EP/N004833/1. I am very grateful to two referees, whose careful reading of these notes considerably improved them. Thanks also to Louigi Addario-Berry for the picture of B^λ in Sect. 2.1. I would like to thank the organisers for doing a wonderful job of putting on the summer school and for making everyone feel so welcome, and to the participants for being such an interested and involved audience. Special thanks to Omer Angel for his hospitality and to Lior Silberman for the bike!

References

1. Abraham, R., Delmas, J.F., Hoscheit, P.: A note on the Gromov-Hausdorff-Prokhorov distance between (locally) compact metric measure spaces. Electron. J. Probab. **18**, paper no. 14, 21 pages (2013)
2. Addario-Berry, L., Broutin, N., Goldschmidt, C.: Critical random graphs: limiting constructions and distributional properties. Electron. J. Probab. **15**, 741–775, paper no. 25 (2010)
3. Addario-Berry, L., Broutin, N., Goldschmidt, C.: The continuum limit of critical random graphs. Probab. Theory Related Fields **152**(3–4), 367–406 (2012)
4. Addario-Berry, L., Broutin, N., Goldschmidt, C., Miermont, G.: The scaling limit of the minimum spanning tree of the complete graph. Ann. Probab. **45**(5), 3075–3144 (2017)
5. Aldous, D.: The continuum random tree. I. Ann. Probab. **19**(1), 1–28 (1991)
6. Aldous, D.: The continuum random tree. II. An overview. In: Stochastic analysis (Durham, 1990), *London Math. Soc. Lecture Note Ser.*, vol. 167, pp. 23–70. Cambridge Univ. Press, Cambridge (1991)
7. Aldous, D.: The continuum random tree. III. Ann. Probab. **21**(1), 248–289 (1993)
8. Aldous, D.: Brownian excursions, critical random graphs and the multiplicative coalescent. Ann. Probab. **25**(2), 812–854 (1997)
9. Bhamidi, S., Broutin, N., Sen, S., Wang, X.: Scaling limits of random graph models at criticality: Universality and the basin of attraction of the Erdős-Rényi random graph (2014+). arXiv:1411.3417 [math.PR]
10. Bhamidi, S., Budhiraja, A., Wang, X.: The augmented multiplicative coalescent, bounded size rules and critical dynamics of random graphs. Probab. Theory Related Fields **160**(3-4), 733–796 (2014)
11. Bhamidi, S., Dhara, S., van der Hofstad, R., Sen, S.: Universality for critical heavy-tailed network models: metric structure of maximal components (2017+). arXiv:1703.07145 [math.PR]
12. Bhamidi, S., van der Hofstad, R., van Leeuwaarden, J.S.H.: Scaling limits for critical inhomogeneous random graphs with finite third moments. Electron. J. Probab. **15**, no. 54, 1682–1703 (2010)
13. Bhamidi, S., van der Hofstad, R., van Leeuwaarden, J.S.H.: Novel scaling limits for critical inhomogeneous random graphs. Ann. Probab. **40**(6), 2299–2361 (2012)
14. Bhamidi, S., Sen, S.: Geometry of the vacant set left by random walk on random graphs, Wright's constants, and critical random graphs with prescribed degrees (2016+). Random Structures Algorithms, to appear. arXiv:1608.07153 [math.PR]
15. Bhamidi, S., Sen, S., Wang, X.: Continuum limit of critical inhomogeneous random graphs. Probab. Theory Related Fields **169**(1-2), 565–641 (2017)
16. Broutin, N., Duquesne, T., Wang, M.: Limits of multiplicative inhomogeneous random graphs and Lévy trees (2018+). arXiv:1804.05871 [math.PR]
17. Burago, D., Burago, Y., Ivanov, S.: A course in metric geometry, *Graduate Studies in Mathematics*, vol. 33. American Mathematical Society, Providence, RI (2001)
18. Conchon-Kerjan, G., Goldschmidt, C.: Stable graphs: the metric space scaling limits of critical random graphs with i.i.d. power-law degrees (2019+). In preparation.
19. Dhara, S., van der Hofstad, R., van Leeuwaarden, J., Sen, S.: Heavy-tailed configuration models at criticality (2016+). Ann. Inst. Henri Poincaré Probab. Stat. to appear. arXiv:1612.00650 [math.PR]
20. Dhara, S., van der Hofstad, R., van Leeuwaarden, J., Sen, S.: Critical window for the configuration model: finite third moment degrees. Electron. J. Probab. **22**, paper no. 16, 33 pages (2017)
21. Duquesne, T., Le Gall, J.F.: Random trees, Lévy processes and spatial branching processes. Astérisque **281**, vi+147 (2002)
22. Ethier, S.N., Kurtz, T.G.: Markov processes: Characterization and convergence. Wiley Series in Probability and Mathematical Statistics: Probability and Mathematical Statistics. John Wiley & Sons Inc., New York (1986)

23. Goldschmidt, C., Haas, B.: A line-breaking construction of the stable trees. Electron. J. Probab. **20**, paper no. 16, 24 pages (2015)
24. Haas, B., Miermont, G.: The genealogy of self-similar fragmentations with negative index as a continuum random tree. Electron. J. Probab. **9**, no. 4, 57–97 (2004)
25. van der Hofstad, R.: Random graphs and complex networks: volume I. Cambridge Series in Statistical and Probabilistic Mathematics. Cambridge University Press, Cambridge (2016)
26. Joseph, A.: The component sizes of a critical random graph with given degree sequence. Ann. Appl. Probab. **24**(6), 2560–2594 (2014)
27. Kaigh, W.: An invariance principle for random walk conditioned by a late return to zero. Ann. Probab. **4**(1), 115–121 (1976)
28. Le Gall, J.F.: The uniform random tree in a Brownian excursion. Probab. Theory Related Fields **96**(3), 369–383 (1993)
29. Le Gall, J.F.: Random trees and applications. Probab. Surv. **2**, 245–311 (2005)
30. Le Gall, J.F.: Random real trees. Ann. Fac. Sci. Toulouse Math. (6) **15**(1), 35–62 (2006)
31. Le Gall, J.F., Miermont, G.: Scaling limits of random trees and planar maps. In: Probability and statistical physics in two and more dimensions, *Clay Math. Proc.*, vol. 15, pp. 155–211. Amer. Math. Soc., Providence, RI (2012)
32. Marckert, J.F., Mokkadem, A.: The depth first processes of Galton-Watson trees converge to the same Brownian excursion. Ann. Probab. **31**(3), 1655–1678 (2003)
33. Molloy, M., Reed, B.: A critical point for random graphs with a given degree sequence. Random Structures Algorithms **6**(2-3), 161–180 (1995)
34. Pitman, J.: Combinatorial stochastic processes, *Lecture Notes in Mathematics*, vol. 1875. Springer-Verlag, Berlin (2006). Lectures from the 32nd Summer School on Probability Theory held in Saint-Flour, July 7–24, 2002, With a foreword by Jean Picard
35. Riordan, O.: The phase transition in the configuration model. Combin. Probab. Comput. **21**(1-2), 265–299 (2012)
36. Rogers, L.C.G., Williams, D.: Diffusions, Markov processes, and martingales. Vol. 2. Itô calculus. Cambridge Mathematical Library. Cambridge University Press, Cambridge (2000). Reprint of the second (1994) edition

Lectures on the Ising and Potts Models on the Hypercubic Lattice

Hugo Duminil-Copin

Abstract Phase transitions are a central theme of statistical mechanics, and of probability more generally. Lattice spin models represent a general paradigm for phase transitions in finite dimensions, describing ferromagnets and even some fluids (lattice gases). It has been understood since the 1980s that random geometric representations, such as the random walk and random current representations, are powerful tools to understand spin models. In addition to techniques intrinsic to spin models, such representations provide access to rich ideas from percolation theory. In recent years, for two-dimensional spin models, these ideas have been further combined with ideas from discrete complex analysis. Spectacular results obtained through these connections include the proofs that interfaces of the two-dimensional Ising model have conformally invariant scaling limits given by SLE curves and the fact that the connective constant of the self-avoiding walk on the hexagonal lattice is given by $\sqrt{2+\sqrt{2}}$. In higher dimensions, the understanding also progresses with the proof that the phase transition of Potts models is sharp, and that the magnetization of the three-dimensional Ising model vanishes at the critical point. These notes are largely inspired by [40, 42, 43].

Keywords Statistical physics · Phase transition · Ising and potts model · Percolation

Contents

H. Duminil-Copin (✉)
Institut des Hautes Études Scientifiques and Université de Genève, Geneva, Switzerland
e-mail: duminil@ihes.fr
URL: http://www.ihes.fr/~duminil

M. T. Barlow and G. Slade (eds.), *Random Graphs, Phase Transitions, and the Gaussian Free Field*, Springer Proceedings in Mathematics & Statistics 304,
https://doi.org/10.1007/978-3-030-32011-9_2

Photo: Miguel Eichelberger (PIMS)

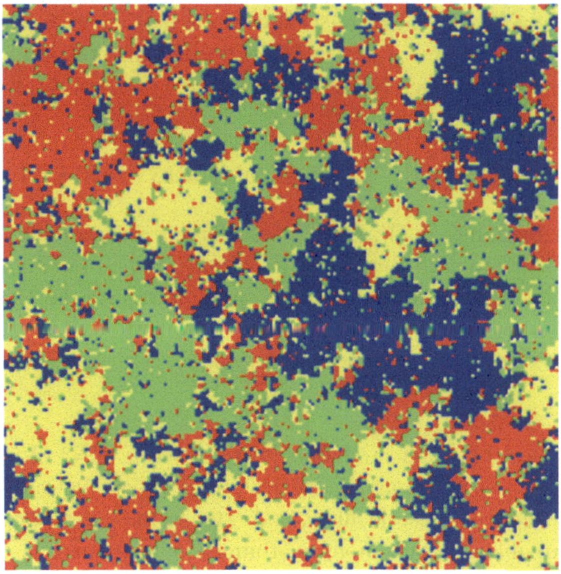

A simulation of the 4-state Potts model due to V. Beffara.

1 Graphical Representation of the Potts Model

1.1 Lattice Spin Models

Lattice models have been introduced as discrete models for real life experiments and were later on found useful to model a large variety of phenomena and systems ranging from ferroelectric materials to gas-liquid transitions. They also provide discretizations of Euclidean and Quantum Field Theories and are as such important from the point of view of theoretical physics. While the original motivation came from physics, they appeared as extremely complex and rich mathematical objects, whose study required the developments of important new tools that found applications in many other domains of mathematics.

The zoo of lattice models is very diverse: it includes models of spin glasses, quantum chains, random surfaces, spin systems, percolation models. Here, we focus on a smaller class of lattice models called spin systems. These systems are random collections of spin variables assigned to the vertices of a lattice. The archetypical example of such a model is provided by the Ising model, for which spins take value ± 1. We refer to [67] for an introduction on lattice models.

Definition of nearest-neighbor ferromagnetic lattice spin models In these notes, $\|\cdot\|$ denotes the Euclidean norm on $\mathbb{R}^d$. A graph $G = (V, E)$ is given by a vertex-set V and an edge set E which is a subset of pairs $\{x, y\} \subseteq V$. We will denote an (unoriented) edge with endpoints x and y by xy. While lattice models could be defined on very general lattices, we focus on the special case of the lattice given by the vertex-set $\mathbb{V} := \mathbb{Z}^d$ and the edge-set $\mathbb{E}$ composed of edges xy with endpoints x and y (in $\mathbb{Z}^d$) satisfying $\|x-y\| = 1$. Below, we use the notation $\mathbb{Z}^d$ to refer both to the lattice and its vertex-set. For a subgraph $G = (V, E)$ of $\mathbb{Z}^d$, we introduce the boundary of G defined by

$$\partial G := \{x \in V : \exists y \in \mathbb{Z}^d \text{ such that } xy \in \mathbb{E}\setminus E\}.$$

For a finite subgraph $G = (V, E)$ of $\mathbb{Z}^d$, attribute a *spin* variable σ_x belonging to a certain set $\Sigma \subseteq \mathbb{R}^r$ to each vertex $x \in V$. A *spin configuration* $\sigma = (\sigma_x : x \in V) \in \Sigma^V$ is given by the collection of all the spins. Introduce the Hamiltonian of σ defined by

$$H_G^{\mathrm{f}}(\sigma) := -\sum_{xy\in E} \sigma_x \cdot \sigma_y,$$

where $a \cdot b$ denotes the scalar product between a and b in $\mathbb{R}^d$. The above Hamiltonian corresponds to a ferromagnetic nearest-neighbor interaction. We will restrict ourselves to this case in these lectures, and refer to the corresponding papers for details on the possible generalizations to arbitrary interactions.

The *Gibbs measure on G at inverse temperature $\beta \geq 0$ with free boundary conditions* is defined by the formula

$$\mu_{G,\beta}^{\mathrm{f}}[f] := \frac{\displaystyle\int_{\Sigma^V} f(\sigma)\exp\big[-\beta H_G^{\mathrm{f}}(\sigma)\big]d\sigma}{\displaystyle\int_{\Sigma^V} \exp\big[-\beta H_G^{\mathrm{f}}(\sigma)\big]d\sigma} \tag{1.1}$$

for every $f : \Sigma^V \to \mathbb{R}$, where $d\sigma = \bigotimes_{x\in V} d\sigma_x$ is a product measure whose marginals $d\sigma_x$ are identical copies of a reference finite measure $d\sigma_0$ on Σ. Note that if $\beta = 0$, then spins are chosen independently according to the probability measure $d\sigma_0/\int_\Sigma d\sigma_0$.

Similarly, for $\mathrm{b} \in \Sigma$, introduce the *Gibbs measure $\mu_{G,\beta}^{\mathrm{b}}$ on G at inverse temperature β with boundary conditions* b defined as $\mu_{G,\beta}^{\mathrm{f}}[\,\cdot\,|\sigma_x = \mathrm{b}, \forall x \in \partial G]$.

A priori, Σ and $d\sigma_0$ can be chosen arbitrarily, thus leading to different examples of lattice spin models. The following (far from exhaustive) list of spin models already illustrates the vast variety of possibilities that such a formalism offers.

Ising model. $\Sigma = \{-1, 1\}$ and $d\sigma_0$ is the counting measure on Σ. This model was introduced by Lenz in [100] to model the temperature, called Curie's temperature, above which a magnet loses its ferromagnetic properties. It was studied by Ising in his PhD thesis [90].

Potts model. $\Sigma = \mathbb{T}_q$ ($q \geq 2$ is an integer), where $\mathbb{T}_q$ is the (unique) set in $\mathbb{R}^{q-1}$ (see Fig. 1) containing $\mathbb{1} := (1, 0, \dots, 0)$ such that for any $a, b \in \mathbb{T}_q$,

$$a \cdot b = \begin{cases} 1 & \text{if } a = b, \\ -\frac{1}{q-1} & \text{otherwise.} \end{cases}$$

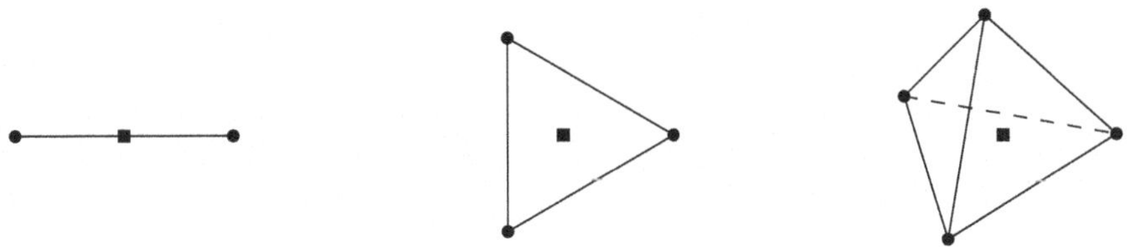

Fig. 1 From left to right, $\mathbb{T}_2$, $\mathbb{T}_3$ and $\mathbb{T}_4$

and $d\sigma_0$ is the counting measure on Σ. Note that $\mathbb{T}_q$ has exactly q elements.

This model was introduced as a generalization of the Ising model to more than two possible spins by Potts in [111] following a suggestion of his adviser Domb. While the model received little attention early on, it became an object of great interest in the last forty years. Since then, mathematicians and physicists have been studying it intensively, and a lot is known on its rich behavior.

Let us note that the Potts model is usually defined differently: the spins take values in $\{1, \ldots, q\}$, where each number is usually interpreted as a color. In this framework, the Hamiltonian is equal to the number of pairs of neighboring sites with different spins. The two formulations of the model are equivalent, up to a multiplication of the inverse-temperature β by a factor $\frac{q-1}{q}$.

Spin $O(n)$ model. Σ is the unit sphere in dimension n and $d\sigma_0$ is the surface measure. This model was introduced by Stanley in [122]. This is another generalization of the Ising model (the case $n = 1$ corresponds to the Ising model) to continuous spins. The $n = 2$ and $n = 3$ models were introduced before the general case and are called the XY and (classical) Heisenberg models respectively.

Discrete Gaussian Free Field (GFF). $\Sigma = \mathbb{R}$ and $d\sigma_0 = \exp(-\sigma_0^2/2)d\lambda(\sigma_0)$, where $d\lambda$ is the Lebesgue measure on $\mathbb{R}$. The discrete GFF is a natural model for random surfaces fluctuations. We refer to Biskup's lecture notes for details.

The ϕ_d^4 lattice model on $\mathbb{Z}^d$. $\Sigma = \mathbb{R}$ and $d\sigma_0 = \exp(-a\sigma_0^2 - b\sigma_0^4)d\lambda(\sigma_0)$, where $a \in \mathbb{R}$ and $b \geq 0$. This model interpolates between the GFF corresponding to $a = 1/2$ and $b = 0$, and the Ising model corresponding to the limit as $b = -a/2$ tends to $+\infty$.

Notation. The family of lattice models is so vast that it would be hopeless to discuss them in full generality. For this reason, we chose already (in the definition above) to focus on nearest-neighbor ferromagnetic interactions. Also, we will mostly discuss two generalizations of the Ising model, namely the Potts and $O(n)$ models.

Phase transition in Ising, Potts and $O(n)$ models We wish to illustrate that the theory of lattice spin models is both very challenging and very rich. For this, we wish to screen quickly through the possible behaviors of spin models. An important disclaimer: this section is not rigorous and most of the claims will not be justified before much later in the lectures. It is therefore not surprising if some of the claims of this section sound slightly bold at this time. We refer to [71] for a book on Gibbs measures and phase transitions.

Assume that the measures introduced above can be extended to infinite volume by taking weak limits of measures $\mu^{\rm f}_{G,\beta}$ and $\mu^{\rm b}_{G,\beta}$ as G tends to $\mathbb{Z}^d$ (sometimes called taking the thermodynamical limit), and denote the associated limiting measures by $\mu^{\rm f}_\beta$ and $\mu^{\rm b}_\beta$.

The behavior of the model in infinite volume can differ greatly depending on β. In order to describe the possible behaviors, introduce the following properties:

– The model exhibits *spontaneous magnetization* at β if

$$\mu^{\rm b}_\beta[\sigma_0\cdot {\rm b}]>0. \tag{MAG$_\beta$}$$

– The model exhibits *long-range ordering* at β if

$$\lim_{\|x\|\to\infty}\mu^{\rm f}_\beta[\sigma_0\cdot\sigma_x]>0. \tag{LRO$_\beta$}$$

– The model exhibits *exponential decay of correlations* at β if

$$\exists c_\beta>0 \text{ such that } \mu^{\rm f}_\beta[\sigma_0\cdot\sigma_x]\le e^{-c_\beta\|x\|} \text{ for all } x\in\mathbb{Z}^d. \tag{EXP$_\beta$}$$

(Note that, in the special case of Ising, Potts, and $O(n)$, the symmetries of Σ imply that $\mu^{\rm b}_\beta[\sigma_0\cdot{\rm b}]$ does not depend on the choice of b.) These three properties lead to three critical parameters separating phases in which they occur or not:

$$\begin{aligned}\beta_c^{\rm mag}&:=\inf\{\beta>0:({\rm MAG}_\beta)\},\\ \beta_c^{\rm lro}&:=\inf\{\beta>0:({\rm LRO}_\beta)\},\\ \beta_c^{\rm exp}&:=\sup\{\beta>0:({\rm EXP}_\beta)\}.\end{aligned}$$

The first parameter $\beta_c^{\rm mag}$ is usually called the *critical inverse temperature* and is simply denoted β_c. In the cases studied below, $\beta_c^{\rm lro}=\beta_c$ (see Sect. 1.4) and we therefore do not discuss when they are distinct in detail.

Models with Σ discrete for $d\ge 2$, or arbitrary Σ for $d\ge 3$, are expected to have spontaneous magnetization for $\beta\gg 1$ (thus proving that $\beta_c<\infty$). We will also see later that when $\beta_c<\infty$, one can often prove[1] that $\beta_c^{\rm exp}=\beta_c$. In such case, we say that the model undergoes a *sharp order/disorder phase transition*. If the model satisfies (MAG$_{\beta_c}$), the phase transition is said to be *discontinuous*; otherwise, it is *continuous*.

On the contrary, the Mermin-Wagner theorem [87, 105] states that a model on $\mathbb{Z}^2$ for which Σ is a compact continuous connected Lie group satisfies $\beta_c=+\infty$. Then, two cases are possible:

[1] One may also have $\beta_c^{\rm exp}<\beta_c<\infty$, as shown in [70] for the planar clock model with $q\gg 1$ states, but this situation is less common.

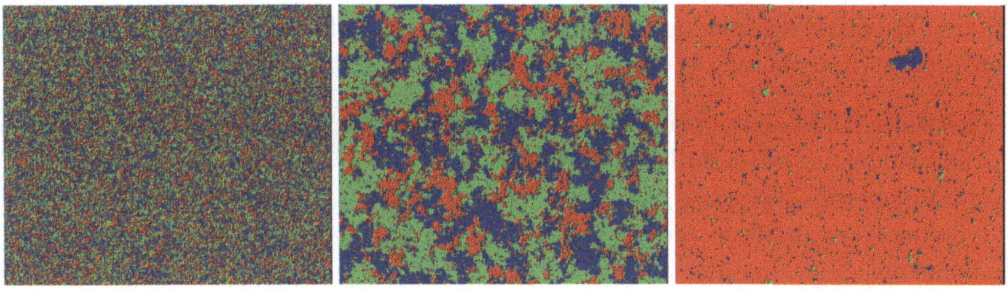

Fig. 2 Simulations of three-state planar Potts model at subcritical, critical and supercritical temperatures, due to V. Beffara

• $\beta^{\exp} = \infty$: the model does not undergo any phase transition. Polyakov [110] predicted this behavior for planar $O(n)$-models with $n \geq 3$. We refer to [51] and references therein for a more precise discussion.
• $\beta^{\exp} < \infty$: the model undergoes a *Berezinsky-Kosterlitz-Thouless* (BKT) phase transition. This type of phase transition is named after Berezinsky and Kosterlitz-Thouless,[2] who introduced it (non-rigorously) for the planar XY-model in two independent papers [17, 97]. Note that in such case, there is no spontaneous magnetization at any β (Fig. 2).

Exercise 1 Prove that for the Ising model on $\mathbb{Z}$, $\beta_c = \beta_c^{\exp} = \beta_c^{\mathrm{lro}} = +\infty$. Prove the same result for the Potts model with $q \geq 3$. What can be said for the spin $O(n)$ models?

To conclude this section, let us draw a panorama of questions. The table below gathers the behaviors that are expected for the Ising, Potts and spin $O(n)$ models.

<table>
<tr><td colspan="2"></td><td>$d = 2$</td><td>$d \geq 3$</td></tr>
<tr><td colspan="2">Ising</td><td colspan="2" rowspan="2">Continuous sharp order-disorder PT</td></tr>
<tr><td rowspan="2">Potts</td><td>$q \in \{3, 4\}$</td></tr>
<tr><td>$q \geq 5$</td><td colspan="2">Discontinuous sharp order-disorder PT</td></tr>
<tr><td rowspan="2">$O(n)$</td><td>$n = 2$</td><td>BKT PT</td><td rowspan="2">Continuous sharp order-disorder PT</td></tr>
<tr><td>$n \geq 3$</td><td>Absence of PT</td></tr>
</table>

The claims about Ising and Potts models will all be proved, except the discontinuity of the phase transition for $q \geq 3$ and $d \geq 3$, which is known only for $q \geq q_c(d) \gg 1$ [98] or $d \geq d_c(q) \gg 1$ [19]. We will not deal with continuous spins, but we mention that the understanding is more restricted there. In two dimensions, it is known that models with continuous spin symmetry cannot have an order-disorder phase transition [105]. The proof that the XY model (i.e. the $O(2)$ model) undergoes a BKT phase transition is due to Fröhlich and Spencer [70], while the existence of a phase transition in dimension $d \geq 3$ goes back to Fröhlich et al. [69]. To the best of our knowledge, there is no proof of sharpness or continuity in dimension $d \geq 3$.

[2]Kosterlitz and Thouless were awarded a Nobel prize in 2016 for their work on topological phase transitions.

Proving Polyakov's conjecture, i.e. that spin $O(n)$ models do not undergo any phase transition in dimension 2, is one of the biggest problems in mathematical physics.

Notation. The behavior of lattice models with a space of spins which is continuous is quite different from the one with discrete spins. For this reason, we choose to focus on typical examples of the second kind. From now on, we work with the Ising and Potts model only. We denote the measure of the q-state Potts model by $\mu^{\#}_{G,\beta,q}$. In order to lighten the notation, the measure of the Ising model is denoted by $\mu^{\#}_{G,\beta}$ rather than $\mu^{\#}_{G,\beta,2}$. Also, we will use $+$ and $-$ instead of $+1$ and -1.

1.2 Graphical Representation of Potts Models

We would like to have a more geometric grasp of correlations between spins of lattice models. In order to do so, we introduce another type of models, called percolation models.

A *percolation configuration* $\omega = (\omega_e : e \in E)$ on $G = (V, E)$ is an element of $\{0, 1\}^E$. If $\omega_e = 1$, the edge e is said to be *open*, otherwise e is said to be *closed*. A configuration ω can be seen as a subgraph of G with vertex-set V and edge-set $\{e \in E : \omega_e = 1\}$. A *percolation model* is given by a distribution on percolation configurations on G.

In order to study the connectivity properties of the (random) graph ω, we introduce some notation. A *cluster* is a maximal connected component of the graph ω (it may be an isolated vertex). Two vertices x and y are *connected in* ω if they are in the same cluster. We denote this event by $x \longleftrightarrow y$. For $A, B \subseteq \mathbb{Z}^d$, set $A \longleftrightarrow B$ if there exists a vertex of A connected to a vertex of B. We also allow ourselves to consider $B = \infty$, in which case we mean that a vertex in A is in an infinite cluster (Fig. 3).

The simplest example of percolation models is provided by *Bernoulli percolation*: each edge is open with probability p, and closed with probability $1 - p$, independently of the states of other edges. Below, the measure is denoted by $\mathbb{P}_p$ (its expectation is denoted by $\mathbb{E}_p$). This model was introduced by Broadbent and Hammersley in 1957 [28] and has been one of the most studied probabilistic model. We refer to [78] for a book on the subject.

Here, we will be interested in a slightly more complicated percolation model, named the *random-cluster model*, which is a percolation model in which the states (open or closed) of edges depend on each other. This model was introduced by Fortuin and Kasteleyn in [65] and is sometimes referred to as the *Fortuin-Kasteleyn percolation*. We refer to [77] for a very complete account on the subject.

Definition of the random-cluster model Let G be a finite subgraph of $\mathbb{Z}^d$. Let $o(\omega)$ and $c(\omega)$ denote the number of open and closed edges of ω. Define *boundary conditions* ξ to be a partition $P_1 \sqcup \cdots \sqcup P_k$ of ∂G. For boundary conditions ξ, define the graph ω^ξ obtained from ω by contracting, for each $1 \le i \le k$, all the vertices of P_i into one vertex. Also, let $k(\omega^\xi)$ be the number of clusters in the graph ω^ξ.

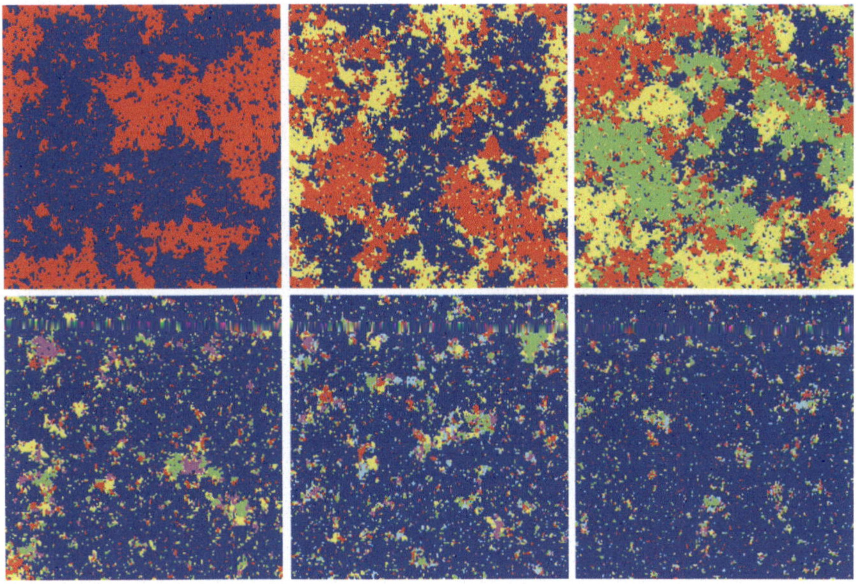

Fig. 3 Simulations of the critical planar Potts model with q equal to 2, 3, 4, 5, 6 and 9 respectively, due to V. Beffara. The behavior for $q \leq 4$ is clearly different from the behavior for $q > 4$. In the first three pictures, each color (corresponding to each element of $\mathbb{T}_q$) seems to play the same role, while in the last three, one color wins over the other ones

As an example, the *free boundary conditions* (denoted 0) correspond to the partition composed of singletons only: $\omega^0 = \omega$ and we prefer the lighter notation $k(\omega)$ to $k(\omega^0)$. The *wired boundary conditions* (denoted 1) correspond to the partition $\{\partial G\}$: $k(\omega^1)$ is the number of clusters obtained if all clusters touching the boundary are counted as 1. In general, a subgraph ξ of $\mathbb{Z}^d$ induces boundary conditions as follows: two vertices of ∂G are in the same P_i if they are in the same cluster of ξ. In this case, boundary conditions will often be identified with the graph ξ.

Exercise 2 Construct the random-cluster model on the torus as the random-cluster model on a finite box with a proper choice of boundary conditions.

Definition 1 The probability measure $\phi^{\xi}_{G,p,q}$ of the random-cluster model on G with *edge-weight* $p \in [0, 1]$, *cluster-weight* $q > 0$ and boundary conditions ξ is defined by

$$\phi^{\xi}_{G,p,q}[\omega] := \frac{p^{o(\omega)}(1-p)^{c(\omega)}q^{k(\omega^{\xi})}}{Z^{\xi}_{G,p,q}} \tag{1.2}$$

for every configuration $\omega \in \{0, 1\}^E$. The constant $Z^{\xi}_{G,p,q}$ is a normalizing constant, referred to as the *partition function*, defined in such a way that the sum over all configurations equals 1.

Fortuin and Kasteleyn introduced the random-cluster model as the unification of different models of statistical physics satisfying series/parallel laws when modifying the underlying graph:

- For $q = 1$, the random-cluster model corresponds to Bernoulli percolation. In this case, and to distinguish with the case $q \neq 1$, we prefer the notation $\mathbb{P}_p$ instead of the random-cluster notation.
- For integers $q \geq 2$, the model is related to Potts models; see Sect. 1.2.
- For $p \to 0$ and $q/p \to 0$, the model is connected to electrical networks via Uniform Spanning Trees; see Exercise 3.

Exercise 3 Consider a finite graph $G = (V, E)$. Prove that the limit of $\phi^0_{G,p,q}$ with $p \to 0$ and $q/p \to 0$ is the Uniform Spanning Tree on G, i.e. the uniform measure on connected subgraphs of the form $H = (V, F)$, with F not containing any cycle.

Let us mention two important properties of random-cluster models. For boundary conditions $\xi = P_1 \sqcup \cdots \sqcup P_k$ and $\psi \in \{0, 1\}^{E \setminus \{e\}}$, where $e = xy$, one may easily check that

$$\phi^\xi_{G,p,q}[\omega_e = 1 | \omega_{|E\setminus\{e\}} = \psi] = \phi^{\psi^\xi}_{\{e\},p,q}[\omega_e = 1]$$
$$= \begin{cases} p & \text{if } x \longleftrightarrow y \text{ in } \psi^\xi, \\ \dfrac{p}{p + q(1-p)} & \text{otherwise.} \end{cases} \tag{1.3}$$

Note that in particular the model satisfies the *finite-energy property*, meaning that there exists $c_{\mathrm{FE}} > 0$ such that for any e and ψ

$$\phi^\xi_{G,p,q}[\omega_e = 1 | \omega_{|E\setminus\{e\}} = \psi] \in [c_{\mathrm{FE}}, 1 - c_{\mathrm{FE}}]. \tag{FE}$$

Also, (1.3) can be extended by induction to any subgraph $G' = (V', E')$ of G, in the sense that for any boundary conditions ξ and any $\psi \in \{0, 1\}^{E \setminus E'}$ and $\psi' \in \{0, 1\}^{E'}$,

$$\phi^\xi_{G,p,q}[\omega_{|E'} = \psi' | \omega_{|E\setminus E'} = \psi] = \phi^{\psi^\xi}_{G',p,q}(\psi'). \tag{DMP}$$

(Recall the definition of the graph ψ^ξ from above.) This last property is called the *domain Markov property*.

Exercise 4 Prove carefully the finite-energy property (FE) and the domain Markov property (DMP).

The coupling between the random-cluster and Potts models The random-cluster model enables us to rephrase correlations in Potts models in terms of random subgraphs of $\mathbb{Z}^d$. This is the object of this section.

Consider an integer $q \geq 2$ and let G be a finite graph. Assume that a configuration $\omega \in \{0, 1\}^E$ is given. One can deduce a spin configuration $\sigma \in \mathbb{T}_q^V$ by assigning uniformly and independently to each cluster a spin. More precisely, consider an iid family of uniform random variables $\sigma_{\mathcal{C}}$ on $\mathbb{T}_q$ indexed by clusters $\mathcal{C}$ in ω. We then define σ_x to be equal to $\sigma_{\mathcal{C}}$ for every $x \in \mathcal{C}$. Note that all the vertices in the same cluster automatically receive the same spin.

Proposition 1 (Coupling for free boundary conditions). *Fix an integer $q \geq 2$, $p \in (0, 1)$ and G finite. If ω is distributed according to $\phi^0_{G,p,q}$ then σ constructed above is distributed according to the q-state Potts measure $\mu^{\mathrm{f}}_{G,\beta,q}$, where*

$$\beta := -\tfrac{q-1}{q} \ln(1-p). \tag{1.4}$$

Proof Consider the law $\mathbf{P}$ of the pair (ω, σ), where ω is a percolation configuration with free boundary conditions and σ is the corresponding spin configuration constructed as explained above. By definition, the first marginal of the distribution is sampled according to $\phi^0_{G,p,q}$. We wish to compute the law of the second marginal.

Say that the configurations $\sigma \in \mathbb{T}^V_q$ and $\omega \in \{0, 1\}^E$ are *compatible* if

$$\forall xy \in E\ :\ \omega_{xy} = 1 \Longrightarrow \sigma_x = \sigma_y.$$

Then, if ω and σ are not compatible, $\mathbf{P}[(\omega, \sigma)] = 0$, and if they are,

$$\mathbf{P}[(\omega, \sigma)] = \tfrac{1}{Z^0_{G,p,q}}\, p^{o(\omega)}(1-p)^{c(\omega)} q^{k(\omega)} \cdot q^{-k(\omega)} = \tfrac{1}{Z^0_{G,p,q}}\, p^{o(\omega)}(1-p)^{c(\omega)}.$$

For $\sigma \in \mathbb{T}^V_q$, introduce $E_\sigma := \{xy \in E : \sigma_x \neq \sigma_y\}$ and note that ω compatible with σ must satisfy $\omega_{xy} = 0$ for edges $xy \in E_\sigma$, and that there is no restriction on ω_{xy} for edges $xy \notin E_\sigma$. Summing $\mathbf{P}[(\omega, \sigma)]$ over configurations ω compatible with σ, we find

$$\mathbf{P}[\sigma] = \frac{1}{Z^0_{G,p,q}} (1-p)^{|E_\sigma|} \underbrace{\sum_{\omega' \in \{0,1\}^{E \setminus E_\sigma}} p^{o(\omega')}(1-p)^{c(\omega')}}_{=1} = \underbrace{\frac{\mathrm{e}^{-\beta|E|}}{Z^0_{G,p,q}}}_{C} \exp[-\beta H^{\mathrm{f}}_G(\sigma)].$$

In the second equality, we used that $1 - p = \exp(-\tfrac{q}{q-1}\beta)$ and

$$H^{\mathrm{f}}_G[\sigma] = \tfrac{1}{q-1}|E_\sigma| - |E \setminus E_\sigma| = \tfrac{q}{q-1}|E_\sigma| - |E|.$$

The proof follows readily since C does not depend on σ, hence is equal to $1/Z^{\mathrm{f}}_{G,\beta,q}$. □

Exercise 5 (*reverse procedure*) In the coupling above, what is the procedure to obtain the configuration ω from a configuration σ?

The same coloring procedure as above, except for the clusters $\mathcal{C}$ intersecting the boundary ∂G for which $\sigma_{\mathcal{C}}$ is automatically set to be equal to b, provides us with another coupling.

Proposition 2 (Coupling for monochromatic boundary conditions) *Fix an integer $q \geq 2$, $p \in (0,1)$ and G finite. If ω is distributed according to $\phi^1_{G,p,q}$, then σ constructed above is distributed according to the q-state Potts measure $\mu^{\mathrm{b}}_{G,\beta,q}$, where $\beta = -\frac{q-1}{q}\ln(1-p)$.*

Exercise 6 Write carefully the Proof of Proposition 2.

This coupling provides us with a dictionary between the properties of the random-cluster model and the Potts model. In order to illustrate this fact, let us mention two consequences.

Corollary 1 *Fix $d, q \geq 2$. Let G be a finite subgraph of $\mathbb{Z}^d$. Let $\beta > 0$ and $p \in [0,1]$ be connected by (1.4). For any $x \in V$,*

$$\mu^{\mathrm{f}}_{G,\beta,q}[\sigma_x \cdot \sigma_y] = \phi^0_{G,p,q}[x \longleftrightarrow y], \tag{1.5}$$

$$\mu^{\mathrm{b}}_{G,\beta,q}[\sigma_x \cdot \mathrm{b}] = \phi^1_{G,p,q}[x \longleftrightarrow \partial G]. \tag{1.6}$$

Proof We do the proof for $\mu^{\mathrm{f}}_{G,\beta,q}[\sigma_x \cdot \sigma_y]$. Consider the coupling $\mathbf{P}$ between ω and σ and denote its expectation by $\mathbf{E}$. If $x \leftrightarrow y$ denotes the event that x and y are connected in ω, we find that

$$\mu^{\mathrm{f}}_{G,\beta,q}[\sigma_x \cdot \sigma_y] = \mathbf{E}[\sigma_x \cdot \sigma_y \mathbb{1}_{x \longleftrightarrow y}] + \mathbf{E}[\sigma_x \cdot \sigma_y \mathbb{1}_{x \not\longleftrightarrow y}] = \phi^0_{G,p,q}[x \longleftrightarrow y],$$

where we used that $\sigma_x = \sigma_y$ if x is connected to y, and σ_x and σ_y are independent otherwise. The same reasoning holds for $\mu^{\mathrm{b}}_{G,\beta,q}[\sigma_x \cdot \mathrm{b}]$. □

As a side remark, note that we just proved that $\mu^{\mathrm{f}}_{G,\beta,q}[\sigma_x \cdot \sigma_y]$ and $\mu^{\mathrm{b}}_{G,\beta,q}[\sigma_x \cdot \mathrm{b}]$ are non-negative. In the case of the Ising model, one can extend the previous relation to the following: for any $A \subseteq V$,

$$\mu^{\mathrm{f}}_{G,\beta}[\sigma_A] = \phi^0_{G,p,2}[\mathcal{F}_A], \tag{1.7}$$

where $\sigma_A := \prod_{x \in A} \sigma_x$ and $\mathcal{F}_A$ is the event that every cluster of ω intersects A an even number of times. In particular, we deduce the first Griffiths inequality $\mu^{\mathrm{f}}_{G,\beta}[\sigma_A] \geq 0$.

Exercise 7 Prove (1.7).

1.3 The Mean-Field Model

In mathematical physics, it is often useful to approximate a model on a finite subgraph of $\mathbb{Z}^d$ by replacing it with the complete graph with the same number of vertices. Doing so, one avoids complications due to the special geometry of the graph. This approximation, called the *mean-field* approximation, provides insight on the model in large dimensions. In this section, which is somehow separated from the other sections, we wish to illustrate that the models defined above are quite elementary to study when working on the complete graph by discussing the case of Bernoulli percolation and the Ising model. We omit most of the proofs and recommend that the reader tries to show the different statements on his own.

Consider the complete graph G_n with vertex-set $\{1, \ldots, n\}$ and edge-set composed of all the unordered pairs of vertices. The random graph obtained by doing percolation of parameter p on G_n is called the *Erdös-Renyi graph* (denote the law of this graph by $\mathbb{P}_{n,p}$). The phase transition of the model is slightly different from the previous section, due to the fact that each one of the graphs G_n is finite, a fact which pushes us to consider the limit as n tends to infinity and $p = p(n)$ to zero at a certain rate. In [63], Erdös and Renyi proved the following result.

Theorem 1 *Fix $\lambda \in (0, \infty)$. If $\mathbf{C}$ is the cluster of 1, there exists $c = c(\lambda) > 0$ such that*

- *If $\lambda < 1$, for every $n, k \geq 1$, $\mathbb{P}_{n,p}[|\mathbf{C}| \geq k] \leq \exp(-ck)$.*
- *If $\lambda > 1$, $\liminf \mathbb{P}_{n,p}[|\mathbf{C}| \geq cn] =: \theta(\lambda) > 0$.*

The case of $\lambda < 1$ can be easily deduced by dominating $|\mathbf{C}|$ by the size of the cluster of a point in a regular tree of degree n with $p = p(n) = \lambda/n$, which itself can be studied using Galton-Watson trees (see Exercise 8). For $\lambda > 1$, the reasoning is slightly more subtle but not difficult. The study can be pushed much further, and for instance $\theta(\lambda)$ can be proved to behave as $(\lambda - 1)^{1+o(1)}$ as $\lambda \searrow 1$. We refer to [22] for a comprehensive study of the Erdös-Renyi graph.

Let us now turn to the Ising model on the complete graph. This model, called the *Curie-Weiss model*, has played an important role in the history of statistical physics as an example of a model undergoing a non-trivial phase transition for the average magnetization

$$m_n = \frac{1}{n} \sum_{x \in G_n} \sigma_x.$$

The symmetry of the model by global spin flip trivially implies that $\langle m_n \rangle_{G_n,\beta} = 0$, but a phase transition occurs as n tends to infinity and $\beta = \beta(n)$ to 0; see the next theorem and [67] for more detail.

Theorem 2 *Fix $\lambda \in (0, \infty)$. There exists $m(\lambda) \geq 0$ such that for every $\varepsilon > 0$, there exists $c = c(\varepsilon, \lambda) > 0$ such that for every $n \geq 1$,*

$$\langle ||m_n| - m(\lambda)| \rangle_{G_n,\beta} \leq \exp(-cn).$$

Furthermore, $m(\lambda) = 0$ if and only if $\lambda \leq 1$. Furthermore, $m(\lambda) = (\lambda - 1)^{1/2+o(1)}$ as $\lambda \searrow 0$.

Note that for $\lambda \leq 1$, the previous theorem implies that m_n concentrates around 0, and that deviations are exponentially unlikely. Concerning the case $\lambda > 1$, the m_n does not concentrate around a single value: it is necessarily close to $m(\lambda)$ or $-m(\lambda)$.

To conclude this section, one may wonder whether the mean-field approximation is truthful, for instance to guess critical exponents for the models on $\mathbb{Z}^d$, or whether it leads to wrong predictions. It appears that the answer is surprisingly non-trivial. In small dimensions (for instance dimension 2 or 3), the predictions emanating from the mean-field approximation are usually wrong. In particular, the exponent describing the behavior of $\theta(p)$ or $m(\beta)$ as p or β approaches the critical point is not the same as the one given above for the complete graph. Nonetheless, above a certain dimension, called the *critical upper dimension*, the exponents start taking their mean-field value, thus making the mean-field approximation truthful.

It is difficult to transform the mean-field approximation into a true mathematical theorem. For the nearest-neighbor Ising model on $\mathbb{Z}^d$, it was proved in [5] that the behavior is mean-field as soon as $d \geq 4$ (dimension 4 is predicted to be the critical upper dimension of the model) using the random current model and Reflection Positivity, two concepts which will be discussed in these lectures. For Bernoulli percolation, Hara and Slade used the *lace expansion technique*, first developed by [29] to study the Self-Avoiding Walk model, to prove that Bernoulli percolation on $\mathbb{Z}^d$ has a mean-field behavior for d sufficiently large; see [80]. The original argument has been refined: it is now known that Bernoulli percolation on $\mathbb{Z}^d$ has a mean-field behavior whenever $d > 10$; see [64]. Yet, the upper critical dimension is predicted to be equal to 6, thus leaving room for improvement. Other models, such as the spin $O(n)$ or the Potts model, also have a critical upper dimension, which varies depending on the models. Discussing their mean-field behavior would lead us too far from the main object of interest in these lecture notes, so that we propose to close the section here.

Exercise 8 Consider the d-ary tree $\mathbb{T}_d$, i.e. the infinite transitive graph of degree $d+1$ with no cycles.

1. For a vertex x in $\mathbb{T}_d$, compare the cluster of x for Bernoulli percolation of parameter p on $\mathbb{T}_d$ with the family tree of a Galton-Watson tree. What is the offspring distribution of this Galton-Watson tree?

2. Deduce from the theory of Galton-Watson trees that the critical point p_c of $\mathbb{T}_d$ is equal to $1/d$ and that there is no infinite cluster in $\mathbb{T}_d$ at p_c.

3. Study the behavior of the probability that x is in an infinite cluster as $p \searrow p_c$, and that x is connected to distance n at p_c.

1.4 The Percolation Phase Transition for the Random-Cluster Model

Positive association and monotonicity Up to now, we considered as granted the fact that spin-spin correlations of the Potts model were increasing in β, but this is not clear at all. One of the advantages of percolation configurations compared to spin configurations is that $\{0, 1\}^E$ is naturally ordered (simply say that $\omega \leq \omega'$ if $\omega_e \leq \omega'_e$ for any $e \in E$) so that we may define the notion of an *increasing* event:

$$\mathcal{A} \text{ is increasing} \quad \Longleftrightarrow \quad [(\omega \in \mathcal{A} \text{ and } \omega \leq \omega') \Longrightarrow \omega' \in \mathcal{A}]. \tag{1.8}$$

The random-cluster model with cluster-weight $q \geq 1$ enjoys some monotonicity properties regarding increasing events, and this special feature makes it more convenient to work with than Potts models.

> From now on, we always assume that the cluster-weight is larger or equal to 1, so that we will have the proper monotonicity properties (listed below).

We say that μ is *stochastically dominated* by ν if for any increasing event $\mathcal{A}$, $\mu[\mathcal{A}] \leq \nu[\mathcal{A}]$. Note that there is a natural way of checking that μ is stochastically dominated by ν. Assume that there exists a probability measure $\mathbf{P}$ on pairs $(\omega, \tilde{\omega}) \in \{0, 1\}^E \times \{0, 1\}^E$ such that

- the law of ω is μ,
- the law of $\tilde{\omega}$ is ν,
- $\mathbf{P}[\omega \leq \tilde{\omega}] = 1$.

Then, μ is automatically stochastically dominated by ν, since for any increasing event $\mathcal{A}$,

$$\mu[\mathcal{A}] = \mathbf{P}[\omega \in \mathcal{A}] = \mathbf{P}[\omega \in \mathcal{A} \text{ and } \omega \leq \tilde{\omega}] \leq \mathbf{P}[\tilde{\omega} \in \mathcal{A}] = \nu[\mathcal{A}].$$

When μ and ν are equal to two Bernoulli percolation measures $\mathbb{P}_p$ and $\mathbb{P}_{p'}$ with $p \leq p'$, it is quite simple to construct $\mathbf{P}$. Indeed, consider a collection of independent uniform $[0, 1]$ random variables $\mathbf{U}_e$ indexed by edges in E. Then, define ω and $\tilde{\omega}$ as follows

$$\omega_e = \begin{cases} 1 & \text{if } \mathbf{U}_e \geq 1 - p, \\ 0 & \text{otherwise} \end{cases} \qquad \text{and} \qquad \tilde{\omega}_e = \begin{cases} 1 & \text{if } \mathbf{U}_e \geq 1 - p', \\ 0 & \text{otherwise.} \end{cases}$$

By construction, ω and $\tilde{\omega}$ are respectively sampled according to $\mathbb{P}_p$ and $\mathbb{P}_{p'}$ (the states of different edges are independent, and the probability that an edge is open is respectively p and p') and $\omega \leq \tilde{\omega}$.

In general, it is more complicated to construct $\mathbf{P}$. The next lemma provides us with a convenient criterion to prove the existence of such a coupling. We say that a measure μ on $\{0, 1\}^E$ is *strictly positive* if $\mu(\omega) > 0$ for any $\omega \in \{0, 1\}^E$.

Lemma 1 *Consider two strictly positive measures μ and ν on $\{0, 1\}^E$ such that for any $e \in E$ and $\psi, \psi' \in \{0, 1\}^{E \setminus \{e\}}$ satisfying $\psi \leq \psi'$, one has*

$$\mu[\omega_e = 1 | \omega_{|E \setminus \{e\}} = \psi] \leq \nu[\omega_e = 1 | \omega_{|E \setminus \{e\}} = \psi']. \tag{1.9}$$

Then, there exists a measure $\mathbf{P}$ on pairs $(\omega, \tilde{\omega})$ with $\mathbf{P}[\omega \leq \tilde{\omega}] = 1$ such that ω and $\tilde{\omega}$ have laws μ and ν. In particular, μ is stochastically dominated by ν.

Proof In order to construct $\mathbf{P}$, we use a continuous-time Markov chain $(\omega^t, \tilde{\omega}^t)$ constructed as follows. Associate to each edge $e \in E$ an independent exponential clock and a collection of independent uniform $[0, 1]$ random variables $\mathbf{U}_{e,k}$.

At each time an exponential clock rings—say we are at time t and it is the k-th time the edge e rings—set (below ω^{t^-} and $\tilde{\omega}^{t^-}$ denote the configurations just before time t)

$$\omega_e^t = \begin{cases} 1 & \text{if } \mathbf{U}_{e,k} \geq \mu[\omega_e = 0 | \omega_{|E\setminus\{e\}} = \omega^{t^-}_{|E\setminus\{e\}}] \\ 0 & \text{otherwise,} \end{cases}$$

$$\tilde{\omega}_e^t = \begin{cases} 1 & \text{if } \mathbf{U}_{e,k} \geq \nu[\omega_e = 0 | \omega_{|E\setminus\{e\}} = \tilde{\omega}^{t^-}_{|E\setminus\{e\}}] \\ 0 & \text{otherwise.} \end{cases}$$

By definition, (ω^t) is an irreducible (because of strict positivity, one can go from any state to the state with all edges open, and back to any other configuration) continuous time Markov chain. The jump probabilities are such[3] that μ is its (unique) stationary measure. As a consequence, the law of ω^t converges to μ. Similarly, the law of $\tilde{\omega}^t$ converges to ν.

Finally, if the starting configurations ω^0 and $\tilde{\omega}^0$ are respectively the configurations with all edges closed, and all edges open, then $\omega^0 \leq \tilde{\omega}^0$ and the condition (1.9) implies that for all $t \geq 0$, $\omega^t \leq \tilde{\omega}^t$. Letting t tend to infinity provides us with a coupling of μ and ν with $\mathbf{P}[\omega \leq \tilde{\omega}] = 1$. □

Theorem 3 (Positive association) *Fix $q \geq 1$, $p \in [0, 1]$, ξ some boundary conditions and G finite. Then*

- (Comparison between boundary conditions) *For any increasing event $\mathcal{A}$ and $\xi' \geq \xi$ (meaning that the partition ξ' is coarser than the partition ξ),*

$$\phi^{\xi'}_{G,p,q}[\mathcal{A}] \geq \phi^{\xi}_{G,p,q}[\mathcal{A}]. \tag{CBC}$$

- (Monotonicity) *For any increasing event A and any $p' \geq p$,*

$$\phi^{\xi}_{G,p',q}[\mathcal{A}] \geq \phi^{\xi}_{G,p,q}[\mathcal{A}]. \tag{MON}$$

- (Fortuin-Kasteleyn-Ginibre inequality) *For any increasing events $\mathcal{A}$ and $\mathcal{B}$,*

$$\phi^{\xi}_{G,p,q}[\mathcal{A} \cap \mathcal{B}] \geq \phi^{\xi}_{G,p,q}[\mathcal{A}]\phi^{\xi}_{G,p,q}[\mathcal{B}]. \tag{FKG}$$

The assumption $q \geq 1$ is not simply technical: these properties above fail when $q < 1$. For instance, a short computation on a small graph shows that the random-cluster model with $q < 1$ does not satisfy the FKG inequality. Also recall that as $p \to 0$ and $q/p \to 0$, one may obtain the Uniform Spanning Tree, which is known to be edge-negatively correlated. It is natural to expect some form of negative correlation for random-cluster models with $q < 1$, but no general result is known as of today.

---

[3]The probability that $\omega_e^t = 1$ is exactly the probability that $\omega_e = 1$ knowing the state of all the other edges.

One important feature of the comparison between boundary conditions is that the free and wired boundary conditions are extremal in the following sense: for any increasing event $\mathcal{A}$ and any boundary conditions ξ,

$$\phi^0_{G,p,q}[\mathcal{A}] \le \phi^\xi_{G,p,q}[\mathcal{A}] \le \phi^1_{G,p,q}[\mathcal{A}]. \tag{1.10}$$

For more applications of (CBC), we refer to Exercises 10 and 11.

Proof We wish to apply the previous lemma. Consider an edge $e = xy \in E$ and $\psi \le \psi'$ two configurations in $\{0,1\}^{E\setminus\{e\}}$. Recall that (1.3) is stating that

$$\phi^\xi_{G,p,q}[\omega_e = 1|\omega_{|E\setminus\{e\}} = \psi] = \begin{cases} p & \text{if } x \text{ and } y \text{ are connected in } \psi^\xi, \\ \dfrac{p}{p+q(1-p)} & \text{otherwise.} \end{cases}$$

Observe that if x and y are connected in ψ^ξ, they also are in $(\psi')^\xi$ (and a fortiori in $(\psi')^{\xi'}$), and that $p \ge \frac{p}{p+q(1-p)}$ (since $q \ge 1$). With the previous observations, (MON) and (CBC) follow readily from the previous lemma.

For (FKG), we need to be slightly more careful. Without loss of generality, we may assume that $\mathcal{B}$ has positive probability. Define the measures $\mu = \phi^\xi_{G,p,q}$ and $\nu = \mu[\cdot|\mathcal{B}]$. One may easily check that (1.9) is satisfied. The measure ν is not strictly positive, but this played a role only in proving that the Markov chains had unique invariant measures. The fact that $\tilde\omega^0$ is in $\mathcal{B}$ (since $\mathcal{B}$ is non-empty and increasing, and all the edges are open in $\tilde\omega^0$) implies that the stationary measure of $(\tilde\omega^t)$ is ν, so that the conclusions of the previous lemma are still valid and ν stochastically dominates μ. As a consequence,

$$\phi^\xi_{G,p,q}[\mathcal{A}] = \mu[\mathcal{A}] \le \nu[\mathcal{A}] = \frac{\phi^\xi_{G,p,q}[\mathcal{A}\cap\mathcal{B}]}{\phi^\xi_{G,p,q}[\mathcal{B}]},$$

which proves (FKG). □

The coupling between random-cluster and Potts models implies the following nice consequence of monotonicity.

Corollary 2 *Fix G finite and $q \ge 2$ an integer. The functions $\beta \longmapsto \mu^{\mathrm{f}}_{G,\beta,q}[\sigma_x \cdot \sigma_y]$ and $\beta \longmapsto \mu^{\mathrm{b}}_{G,\beta,q}[\sigma_x \cdot \mathrm{b}]$ are non-decreasing.*

Exercise 9 (*Second Griffiths inequality*) Using the coupling with the random-cluster model, prove that for any set of vertices A and B,

$$\mu^{\mathrm{f}}_{G,\beta}[\sigma_A\sigma_B] \geq \mu^{\mathrm{f}}_{G,\beta}[\sigma_A]\mu^{\mathrm{f}}_{G,\beta}[\sigma_B]. \qquad \text{(2ndGriffiths)}$$

(This inequality is called the *second Griffiths inequality*, see the next sections for more detail.)

Exercise 10 (*Comparison with boundary conditions 1*) Fix $p \in [0, 1]$, $q \geq 1$, a finite graph $G = (V, E)$ and ξ some boundary conditions. Let F be a subset of E and H be the graph with edge-set F and vertex-set given by the endpoints of edges in F. Then, for any increasing events $\mathcal{A}$ and $\mathcal{B}$ depending only on edges in F and $E\setminus F$ respectively, show that

$$\phi^0_{H,p,q}[\mathcal{A}] \leq \phi^\xi_{G,p,q}[\mathcal{A}|\mathcal{B}] \leq \phi^1_{H,p,q}[\mathcal{A}].$$

Exercise 11 (*Comparison with boundary conditions 2*) Consider a graph $G = (V, E)$ and $F \subseteq E$. Let $G' = (W, F)$ be the graph with edge-set F and vertex-set given by the endpoints of the edges in F. Let $\mathcal{A}$ be an increasing event depending on edges in F only. Let ξ be some boundary conditions on ∂G.

1. Define the set $\mathrm{S} = \mathrm{S}(\omega)$ of vertices in V not connected in ω to a vertex in ∂G. Show that for any $S \subseteq V$, the event $\{\mathrm{S} = S\}$ is measurable in terms of edges with at least one endpoint outside S.
2. Fix $S \subseteq V$. Consider the graph H with vertex-set S and edge-set composed of edges in E with both endpoints in S. Use the previous observation to prove that

$$\phi^\xi_{G,p,q}[\mathcal{A}, \mathrm{S} = S \,|\, \partial G' \not\longleftrightarrow \partial G] \leq \phi^0_{H,p,q}[\mathcal{A}]\phi^\xi_{G,p,q}[\mathrm{S} = S \,|\, \partial G' \not\longleftrightarrow \partial G].$$

3. Prove that $\phi^\xi_{G,p,q}[\mathcal{A} \,|\, \partial G' \not\longleftrightarrow \partial G] \leq \phi^0_{G,p,q}[\mathcal{A}]$.
4. We now restrict ourselves to two dimensions. A circuit is a path starting and ending at the same vertex. Let $\mathcal{B}$ be the event that there exists an open circuit in $E\setminus F$ disconnecting W from ∂G. Prove that $\phi^\xi_{G,p,q}[\mathcal{A} \,|\, \mathcal{B}] \geq \phi^1_{G,p,q}[\mathcal{A}]$.

Exercise 12 (*Holley and FKG lattice conditions*) 1. Show that for strictly positive measures, (1.9) is equivalent to the Holley criterion: for any ω and ω',

$$\nu[\omega \vee \omega']\mu[\omega \wedge \omega'] \geq \nu[\omega]\mu[\omega'], \qquad \text{(Holley)}$$

where $\vee$ and $\wedge$ are the min and max of two configurations.
2. Show that for a strictly positive measure μ, (FKG) holds if the FKG lattice condition holds: for any ω and any edges e and f,

$$\mu[\omega^{ef}]\mu[\omega_{ef}] \geq \mu[\omega^e_f]\mu[\omega^f_e], \qquad \text{(FKGlatticecondition)}$$

where ω^{ef}, ω_{ef}, ω^e_f and ω^f_e denote the configurations ω' coinciding with ω except at e and f, where (ω'_e, ω'_f) are equal respectively to $(1, 1)$, $(0, 0)$, $(1, 0)$ and $(0, 1)$.

Exercise 13 Is there a monotonicity in q at fixed p?

Phase transition in the random-cluster and Potts models When discussing phase transitions, we implicitly considered infinite-volume Potts measures to define β_c. Their definition is not a priori clear since the Hamiltonian would then be an infinite sum of terms equal to 1 or $-1/(q-1)$. One can always consider sub-sequential limits of measures $\mu^{\mathrm{f}}_{G,\beta,q}$, but one can in fact do much better using the random-cluster model: monotonicity properties of the previous section enable us to prove convergence of certain sequences of measures.

Below and in the rest of this document, set for every $n \geq 0$, $\Lambda_n := [-n, n]^d \cap \mathbb{Z}^d$. Also, E_n will denote the set of edges between two vertices of Λ_n.

Proposition 3 *Fix $q \geq 1$. There exist two (possibly equal) measures $\phi^0_{p,q}$ and $\phi^1_{p,q}$ on $\{0,1\}^{\mathbb{E}}$, called the infinite-volume random-cluster measures with free and wired boundary conditions respectively, such that for any event $\mathcal{A}$ depending on a finite number of edges,*

$$\lim_{n\to\infty} \phi^1_{\Lambda_n,p,q}[\mathcal{A}] = \phi^1_{p,q}[\mathcal{A}] \quad \text{and} \quad \lim_{n\to\infty} \phi^0_{\Lambda_n,p,q}[\mathcal{A}] = \phi^0_{p,q}[\mathcal{A}].$$

One warning: while boundary conditions cannot be defined as a partition of the boundary in infinite volume, one still needs to keep track of the dependency on boundary conditions for finite-volume measures when constructing the measure. Therefore, the measures $\phi^1_{p,q}$ and $\phi^0_{p,q}$ have no reason to be the same and we will see examples of values of p and q for which they are in fact different. In addition to this, one may imagine other infinite-volume measures obtained via limits of measures on finite graphs with arbitrary (and possibly random) boundary conditions.

Proof We deal with the case of free boundary conditions. Wired boundary conditions are treated similarly. Fix an increasing event $\mathcal{A}$ depending on edges in Λ_N only. We find that for any $n \geq N$,

$$\phi^0_{\Lambda_{n+1},p,q}[\mathcal{A}] \overset{\text{(DMP)}}{=} \phi^0_{\Lambda_{n+1},p,q}[\phi^{\xi}_{\Lambda_n,p,q}[\mathcal{A}]] \overset{\text{(CBC)}}{\geq} \phi^0_{\Lambda_n,p,q}[\mathcal{A}],$$

where ξ is the random boundary conditions induced by the configuration $\omega_{|E_{n+1}\setminus E_n}$. We deduce that $(\phi^0_{\Lambda_n,p,q}[\mathcal{A}])_{n\geq 0}$ is increasing, and therefore converges to a certain value $P[\mathcal{A}]$ as n tends to infinity.

Since the probability of an event $\mathcal{B}$ depending on finitely many edges can be written by inclusion-exclusion (see Exercise 14) as a combination of the probability of increasing events, taking the same combination defines a natural value $P(\mathcal{B})$ for which $\phi^0_{\Lambda_n,p,q}[\mathcal{B}]$ converges to $P(\mathcal{B})$.

The fact that $(\phi^0_{\Lambda_n,p,q})_{n\geq 0}$ are probability measures implies that the function P (which is a priori defined on the set of events depending on finitely many edges) can be extended into a probability measure on $\mathcal{F}_{\mathbb{E}}$. We denote this measure by $\phi^0_{p,q}$. □

Exercise 14 For $\psi \in \{0,1\}^E$, write $\{\omega \in \{0,1\}^{\mathbb{E}} : \omega_e = \psi_e, \forall e \in E\}$ as $A\setminus B$ with $B \subseteq A$ two increasing events. Deduce that any event depending on finitely many edges can be written by inclusion-exclusion using increasing events.

The properties of finite-volume measures (FKG inequality, monotonicity, ordering between boundary conditions) extend to infinite volume in a straightforward fashion. In particular, one may define a *critical parameter* $p_c \in [0,1]$ such that

$$\begin{aligned} p_c = p_c(q,d) &:= \inf\{p > 0 : \phi^1_{p,q}[0 \leftrightarrow \infty] > 0\} \\ &= \sup\{p > 0 : \phi^1_{p,q}[0 \leftrightarrow \infty] = 0\}. \end{aligned}$$

Let us conclude by explaining what this implies for Potts models. One can extend the coupling between random-cluster models and Potts models in order to construct $q+1$ measures $\mu^{\mathrm{f}}_{\beta,q}$ and $\mu^{\mathrm{b}}_{\beta,q}$ with $\mathrm{b} \in \mathbb{T}_q$ on $\mathbb{Z}^d$ by doing the same couplings as in finite volume, except that clusters intersecting the boundary are replaced by infinite clusters.

Corollary 3 *The measures $\mu^{\mathrm{f}}_{\beta,q}$ and $\mu^{\mathrm{b}}_{\beta,q}$ with $\mathrm{b} \in \mathbb{T}_q$ are the limits of the measures $\mu^{\mathrm{f}}_{\Lambda_n,\beta,q}$ and $\mu^{\mathrm{b}}_{\Lambda_n,\beta,q}$. Furthermore, if β and p satisfy (1.4), then*

$$m^*(\beta,q) := \mu^{\mathrm{b}}_{\beta,q}[\sigma_0 \cdot \mathrm{b}] = \phi^1_{p,q}[0 \longleftrightarrow \infty].$$

The proof of the corollary is immediate from the convergence of the random-cluster measures and the coupling. Note that this enables us to define rigorously

$$\beta_c = \beta_c(q,d) := \inf\{\beta > 0 : m^*(\beta,q) > 0\} = \sup\{\beta > 0 : m^*(\beta,q) = 0\},$$

which is related to $p_c(q,d)$ by the formula

$$\beta_c(q,d) := -\tfrac{q-1}{q} \log\big[1 - p_c(q,d)\big]. \tag{1.11}$$

Exercise 15 Is there some ordering between the measures $\phi^1_{p,q}$ in $q \geq 1$ at fixed p? Deduce from the study of Bernoulli percolation that $p_c(q,d) > 0$.

Exercise 16 1. Prove that for $p < p_c$, $\phi^1_{p,q} = \phi^0_{p,q}$.
2. (To do after reading Sect. 2.4) Prove that on $\mathbb{Z}^2$, $\phi^1_{p,q} = \phi^0_{p,q}$ for $p > p_c$.

Exercise 17 A probability measure ϕ on $\{0,1\}^{\mathbb{E}}$ is called an *infinite-volume random-cluster* measure with parameters p and q if for every finite graph $G = (V, E)$,

$$\phi[\omega_{|E} = \eta \,|\, \mathcal{F}_E] = \phi^\xi_{G,p,q}[\eta] \quad , \quad \forall \eta \in \{0,1\}^E,$$

where ξ are the boundary conditions induced by the configuration outside G and $\mathcal{F}_E$ is the σ-algebra induced by $(\omega_e : e \notin E)$. Prove that $\phi^0_{p,q} \leq \phi \leq \phi^1_{p,q}$ for any infinite-volume measure with parameters p and $q \geq 1$. Deduce that there exists a unique infinite-volume measure if and only if $\phi^0_{p,q} = \phi^1_{p,q}$.

Exercise 18 Using (FE), prove that $p_c(q,d) > 0$.

Long-range ordering and spontaneous magnetization Let us now focus on the following question: is (LRO$_\beta$) equivalent to (MAG$_\beta$)? In terms of random-cluster model, this gets rephrased as follows: is $\phi^1_{p,q}[0 \leftrightarrow \infty] = 0$ equivalent to $\phi^0_{p,q}[0 \leftrightarrow x]$ tends to 0 as $\|x\|$ tends to infinity? Two things could prevent this from happening. First, $\phi^1_{p,q}$ and $\phi^0_{p,q}$ could be different. Second, it may be that, when an infinite cluster exists, then automatically infinitely many of them do, so that the probability that two vertices are connected tends to zero.

Let us first turn to the second problem and prove that the infinite cluster, when it exists, is unique.

Theorem 4 *Fix $p \in [0,1]$ and $q \geq 1$. For # equal to 0 or 1, either $\phi^{\#}_{p,q}[0 \leftrightarrow \infty] = 0$ or*

$$\phi^{\#}_{p,q}[\exists \text{ a unique infinite cluster}] = 1.$$

This result was first proved in [6] for Bernoulli percolation. It was later obtained via different types of arguments. The beautiful argument presented here is due to Burton and Keane [31].

We begin by studying ergodic properties of $\phi^1_{p,q}$ and $\phi^0_{p,q}$. Let τ_x be a translation of the lattice by $x \in \mathbb{Z}^d$. This translation induces a shift on the space of configurations $\{0,1\}^{\mathbb{E}}$. Define $\tau_x\mathcal{A} := \{\omega \in \{0,1\}^{\mathbb{E}} : \tau_x^{-1}\omega \in \mathcal{A}\}$. An event $\mathcal{A}$ is *invariant under translations* if for any $x \in \mathbb{Z}^d$, $\tau_x\mathcal{A} = \mathcal{A}$. A measure μ is *invariant under translations* if $\mu[\tau_x\mathcal{A}] = \mu[\mathcal{A}]$ for any event $\mathcal{A}$ and any $x \in \mathbb{Z}^d$. The measure is said to be *ergodic* if any event invariant under translation has probability 0 or 1.

Lemma 2 *The measures $\phi^1_{p,q}$ and $\phi^0_{p,q}$ are invariant under translations and ergodic.*

Proof Let us treat the case of $\phi^1_{p,q}$, the case of $\phi^0_{p,q}$ is left to the reader (Exercise 20). Let $\mathcal{A}$ be an increasing event depending on finitely many edges, and $x \in \mathbb{Z}^d$. Choose k such that $x \in \Lambda_k$. Since $\Lambda_{n-k} \subseteq \tau_x\Lambda_n \subseteq \Lambda_{n+k}$, the comparison between boundary conditions (CBC) gives

$$\phi^1_{\Lambda_{n+k},p,q}[\tau_x\mathcal{A}] \leq \phi^1_{\tau_x\Lambda_n,p,q}[\tau_x\mathcal{A}] \leq \phi^1_{\Lambda_{n-k},p,q}[\tau_x\mathcal{A}].$$

We deduce that

$$\phi^1_{p,q}[\mathcal{A}] = \lim_{n\to\infty} \phi^1_{\Lambda_n,p,q}[\mathcal{A}] = \lim_{n\to\infty} \phi^1_{\tau_x\Lambda_n,p,q}[\tau_x\mathcal{A}] = \phi^1_{p,q}[\tau_x\mathcal{A}].$$

Since the increasing events depending on finitely many edges span the σ-algebra of measurable events, we obtain that $\phi^1_{p,q}$ is invariant under translations.

Any event can be approximated by events depending on finitely many edges, hence the ergodicity follows from mixing (see Exercise 19), i.e. from the property that for any events $\mathcal{A}$ and $\mathcal{B}$ depending on finitely many edges,

$$\lim_{\|x\|\to\infty} \phi^1_{p,q}[\mathcal{A} \cap \tau_x\mathcal{B}] = \phi^1_{p,q}[\mathcal{A}]\phi^1_{p,q}[\mathcal{B}]. \qquad \text{(Mixing)}$$

Observe that by inclusion-exclusion, it is sufficient to prove the equivalent result for $\mathcal{A}$ and $\mathcal{B}$ increasing and depending on finitely many edges. Let us give ourselves these two increasing events $\mathcal{A}$ and $\mathcal{B}$ depending on edges in Λ_k only, and $x \in \mathbb{Z}^d$. The FKG inequality and the invariance under translations of $\phi^1_{p,q}$ imply that

$$\phi^1_{p,q}[\mathcal{A} \cap \tau_x\mathcal{B}] \geq \phi^1_{p,q}[\mathcal{A}]\phi^1_{p,q}[\tau_x\mathcal{B}] = \phi^1_{p,q}[\mathcal{A}]\phi^1_{p,q}[\mathcal{B}].$$

In the other direction, for any $n \geq 2k$, if x is far enough from the origin, then Λ_n and $\tau_x \Lambda_n$ do not intersect. Thus, the comparison between boundary conditions (more precisely Exercise 10 for $H = \Lambda_N$ with $N \geq n + k$, and then a limit as N tends to infinity) gives

$$\phi^1_{p,q}[\mathcal{A} \cap \tau_x \mathcal{B}] \leq \phi^1_{\Lambda_n,p,q}[\mathcal{A}]\phi^1_{\tau_x \Lambda_n,p,q}[\tau_x \mathcal{B}] = \phi^1_{\Lambda_n,p,q}[\mathcal{A}]\phi^1_{\Lambda_n,p,q}[\mathcal{B}].$$

The result follows by taking x to infinity. □

Exercise 19 Prove that the mixing property (Mixing) implies ergodicity. *Hint.* Consider an event $\mathcal{A}$ which is invariant by translation and approximate it by an event $\mathcal{B}$ depending on finitely many edges. Then, use that the probability that $\mathcal{B} \cap \tau_x \mathcal{B}$ tends to the square of the probability of $\mathcal{B}$ together with the fact that $\mathcal{A} = \mathcal{A} \cap \tau_x \mathcal{A}$.

Exercise 20 Prove that $\phi^0_{p,q}$ is invariant under translations and ergodic.

Proof (*Theorem* 4) We present the proof in the case of wired boundary conditions and for $p \in (0, 1)$ (the result is obvious for p equal to 0 or 1). Let $\mathcal{E}_{\leq 1}$, $\mathcal{E}_{<\infty}$ and $\mathcal{E}_\infty$ be the events that there are no more than one, finitely many and infinitely many infinite clusters respectively. Since having no infinite cluster is an event which is invariant under translations, it has probability 0 or 1 by ergodicity, and it is therefore sufficient to prove that $\phi^1_{p,q}[\mathcal{E}_{\leq 1}] = 1$ (Fig. 4).

Let us start by showing that $\phi^1_{p,q}[\mathcal{E}_{<\infty} \setminus \mathcal{E}_{\leq 1}] = 0$. By ergodicity, $\mathcal{E}_{<\infty}$ and $\mathcal{E}_{\leq 1}$ both have probability equal to 0 or 1. Since $\mathcal{E}_{\leq 1} \subseteq \mathcal{E}_{<\infty}$, we only need to prove that $\phi^1_{p,q}[\mathcal{E}_{<\infty}] > 0$ implies $\phi^1_{p,q}[\mathcal{E}_{\leq 1}] > 0$. Let $\mathcal{F}$ be the event that all (there may be none) the infinite clusters intersect Λ_n. Since $\mathcal{F}$ is independent of E_n, (DMP) together with (FE) imply that

$$\phi^1_{p,q}[\mathcal{F} \cap \{\omega_e = 1, \forall e \in E_n\}] \geq \phi^1_{p,q}[\mathcal{F}]\, c_{\mathrm{FE}}^{|E_n|}.$$

Now, assume that $\phi^1_{p,q}[\mathcal{E}_{<\infty}] > 0$. Since any configuration in the event on the left contains zero or one infinite cluster (all the vertices in Λ_n are connected), choosing n large enough that $\phi^1_{p,q}[\mathcal{F}] \geq \frac{1}{2}\phi^1_{p,q}[\mathcal{E}_{<\infty}] > 0$ implies that $\phi^1_{p,q}[\mathcal{E}_{\leq 1}] > 0$.

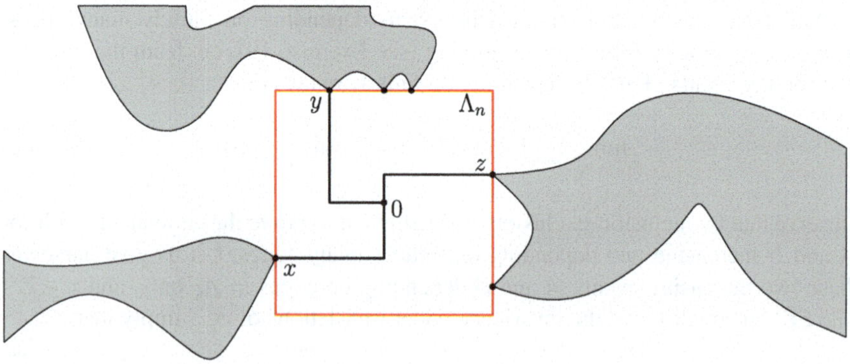

Fig. 4 Construction of a trifurcation at the origin starting from three disjoint infinite clusters (in gray) intersecting Λ_n. The three paths inside Λ_n are vertex-disjoint, except at the origin

We now exclude the possibility of an infinite number of infinite clusters. Consider $n > 0$ large enough that

$$\phi^1_{p,q}[K \text{ infinite clusters intersect the box} \Lambda_n] \geq \tfrac{1}{2}\phi^1_{p,q}[\mathcal{E}_\infty], \tag{1.12}$$

where $K = K(d)$ is large enough that three vertices x, y, z of $\partial\Lambda_n$ at a distance at least three of each other that are connected to infinity in $\omega_{|\mathbb{E}\setminus E_n}$. Using these three vertices, one may modify[4] the configuration in E_n as follows:

1. Choose three paths in Λ_n intersecting each other only at the origin, and intersecting $\partial\Lambda_n$ only at one point, which is respectively x, y and z.
2. Open all the edges of these paths, and close all the other edges in E_n.

We deduce from this construction that

$$\phi^1_{p,q}[\mathcal{T}_0] \geq c_{\mathrm{FE}}^{|E_n|} \cdot \tfrac{1}{2}\phi^1_{p,q}[\mathcal{E}_\infty], \tag{1.13}$$

where $\mathcal{T}_0$ is the following event: $\mathbb{Z}^d\setminus\{0\}$ contains three distinct infinite clusters which are connected to 0 by an open edge. A vertex $x \in \mathbb{Z}^d$ is called a *trifurcation* if $\tau_x\mathcal{T}_0 =: \mathcal{T}_x$ occurs.

Fix $n \geq 1$ and denote the number of trifurcations in Λ_n by T. By invariance under translation, $\phi^1_{p,q}[\mathcal{T}_x] = \phi^1_{p,q}[\mathcal{T}_0]$ and therefore

$$\phi^1_{p,q}[\mathsf{T}] = \phi^1_{p,q}[\mathcal{T}_0] \times |\Lambda_n|. \tag{1.14}$$

Let us now bound deterministically T. In order to do this, first perform the following two "peelings" of the set $F_0 := \{e_1, \ldots, e_r\}$ of edges in E_n that are open in ω.

- For each $1 \leq i \leq r$, if e_i is on a cycle formed by edges in F_{i-1}, set $F_i = F_{i-1}\setminus\{e_i\}$, otherwise, set $F_i = F_{i-1}$. In the end, the set $\tilde{F}_0 := F_r = \{f_1, \ldots, f_s\}$ is a forest.
- For each $1 \leq j \leq s$, if $\tilde{F}_{j-1}\setminus\{f_j\}$ contains a cluster not intersecting $\partial\Lambda_n$, then set $\tilde{F}_j$ to be $\tilde{F}_{j-1}\setminus\{f_j\}$ and the cluster in question. Otherwise, set $\tilde{F}_j = \tilde{F}_{j-1}$. At the end, $\tilde{F}_s$ is a forest whose leaves belong to $\partial\Lambda_n$.

Since the trifurcations are vertices of degree at least three in this forest, we deduce that T is smaller than the number of leaves in the forest, i.e. $\mathsf{T} \leq |\partial\Lambda_n|$. This gives

$$\phi^1_{p,q}[\mathcal{T}_0] \overset{(1.14)}{=} \frac{\phi^1_{p,q}[\mathsf{T}]}{|\Lambda_n|} \leq \frac{|\partial\Lambda_n|}{|\Lambda_n|} \longrightarrow 0 \quad \text{as } n \to \infty.$$

Combined with (1.13), this implies that $\phi^1_{p,q}[\mathcal{E}_\infty] = 0$. The claim follows. □

[4]Note that one may wish to pick $K = 3$ in (1.12) instead of a (a priori) larger K, but that this choice would make the construction of the trifurcations described below more difficult due to the fact that the three clusters may arrive very close to each other on the corner of Λ_n, and therefore prevent us from "rewiring them" to construct a trifurcation at the origin.

Exercise 21 We say that an (countable) infinite locally finite transitive graph $\mathbb{G}$ is amenable if

$$\inf_{G \subseteq \mathbb{G}} \frac{|\partial G|}{|G|} = 0.$$

Show that Theorem 4 still holds in this context. What about graphs which are not amenable, do we always have uniqueness of the infinite cluster?

We now turn to the first problem and prove the following.

Theorem 5 *For $q \geq 1$, the set of edge-weights p for which $\phi^1_{p,q} \neq \phi^0_{p,q}$ is at most countable.*

The theorem implies that for any $p > p_c$, there exists $p' \in (p_c, p)$ such that $\phi^1_{p',q} = \phi^0_{p',q}$. As a consequence,

$$\phi^0_{p,q}[0 \longleftrightarrow \infty] \geq \phi^0_{p',q}[0 \longleftrightarrow \infty] = \phi^1_{p',q}[0 \longleftrightarrow \infty] > 0. \tag{1.15}$$

In other words, $\phi^0_{p,q}[0 \leftrightarrow \infty] > 0$ for any $p > p_c$ and we could have defined the critical point using the free boundary conditions instead of the wired ones. We will use this fact quite often.

The Proof of Theorem 5 goes back to Lebowitz and Martin-Löf [92] in the case of the Ising model. The very elegant argument harvests the convexity of the free energy (see Exercise 22). Here, we present a slightly rephrased version of this argument, which relies on the fact that the probability of an edge to be open is increasing.

Proof Before diving into the proof, let us remark that

$$\phi^0_{p,q} = \phi^1_{p,q} \qquad \Longleftrightarrow \qquad \phi^0_{p,q}[\omega_e] = \phi^1_{p,q}[\omega_e], \ \forall e \in \mathbb{E}.$$

The direct implication being obvious, we assume the assertion on the right and try to prove the one on the left. Consider an increasing event $\mathcal{A}$ depending on a finite set E of edges, then if $\mathbf{P}_n$ denotes the increasing coupling between $\omega \sim \phi^0_{\Lambda_n,p,q}$ and $\tilde{\omega} \in \phi^1_{\Lambda_n,p,q}$ constructed in the Proof of Lemma 1, we find that

$$\begin{aligned} 0 \leq \phi^1_{\Lambda_n,p,q}[\mathcal{A}] - \phi^0_{\Lambda_n,p,q}[\mathcal{A}] &= \mathbf{P}_n[\tilde{\omega} \in \mathcal{A}, \omega \notin \mathcal{A}] \\ &\leq \sum_{e \in E} \mathbf{P}_n[\tilde{\omega}_e = 1, \omega_e = 0] \\ &= \sum_{e \in E} \phi^1_{\Lambda_n,p,q}[\omega_e] - \phi^0_{\Lambda_n,p,q}[\omega_e]. \end{aligned}$$

Letting n go to infinity implies that $\phi^1_{p,q}[\mathcal{A}] = \phi^0_{p,q}[\mathcal{A}]$. Since increasing events depending on finitely many edges generate the σ-algebra, this gives that $\phi^1_{p,q} = \phi^0_{p,q}$.

Our goal is to prove that $\phi^1_{p,q}[\omega_0] = \phi^0_{p,q}[\omega_e]$ at any point of continuity of $p \mapsto \phi^1_{p,q}[\omega_e]$. Since this function is increasing, it has at most countably many points of discontinuity and the theorem will follow. Below, we fix such a point of continuity p. We also consider $p' < p$ and set $a := \phi^0_{p,q}[\omega_e]$ and $b := \phi^1_{p',q}[\omega_e]$.

Consider $0 < \varepsilon < \min\{1-a, b\}$ and $n \geq 1$. The comparison between boundary conditions gives that

$$\phi^0_{\Lambda_n,p,q}[o(\omega)] \leq a|E_n| \quad \text{so that} \quad \phi^0_{\Lambda_n,p,q}[o(\omega) \leq (a+\varepsilon)|E_n|] \geq \varepsilon, \tag{1.16}$$

$$\phi^1_{\Lambda_n,p',q}[o(\omega)] \geq b|E_n| \quad \text{so that} \quad \phi^1_{\Lambda_n,p',q}[o(\omega) \geq (b-\varepsilon)|E_n|] \geq \varepsilon. \tag{1.17}$$

(For the inequalities on the right, we also used that $0 \leq o(\omega) \leq |E_n|$.) Now, observe that for every random variable X,

$$\phi^\xi_{G,p',q}[X] = \frac{\phi^\xi_{G,p,q}[X\lambda^{o(\omega)}]}{\phi^\xi_{G,p,q}[\lambda^{o(\omega)}]}$$

since

$$\begin{aligned}
&\sum_{\omega\in\{0,1\}^E} X(\omega)p'^{o(\omega)}(1-p')^{c(\omega)}q^{k(\omega)} \\
&\quad = (1-p')^{|E|} \sum_{\omega\in\{0,1\}^E} X(\omega)\big(\tfrac{p'}{1-p'}\big)^{o(\omega)} q^{k_\xi(\omega)} \\
&\quad = (1-p')^{|E|} \sum_{\omega\in\{0,1\}^E} \lambda^{o(\omega)} X(\omega)\big(\tfrac{p}{1-p}\big)^{o(\omega)} q^{k(\omega)} \\
&\quad = \big(\tfrac{1-p'}{1-p}\big)^{|E|} \sum_{\omega\in\{0,1\}^E} \lambda^{o(\omega)} X(\omega) p^{o(\omega)}(1-p)^{c(\omega)} q^{k(\omega)}.
\end{aligned}$$

Using this, the fact that $k(\omega^1) \leq k(\omega) \leq k(\omega^1) + |\partial\Lambda_n|$ and setting $\lambda := \frac{p'(1-p)}{(1-p')p} < 1$, we find that

$$\begin{aligned}
\varepsilon \overset{(1.17)}{\leq} \phi^1_{\Lambda_n,p',q}[o(\omega) > (b-\varepsilon)|E_n|] &\leq q^{|\partial\Lambda_n|}\,\phi^0_{\Lambda_n,p',q}[o(\omega) > (b-\varepsilon)|E_n|] \\
&\leq q^{|\partial\Lambda_n|}\frac{\phi^0_{\Lambda_n,p,q}[\lambda^{o(\omega)}\mathbb{1}_{o(\omega)>(b-\varepsilon)|E_n|}]}{\phi^0_{\Lambda_n,p,q}[\lambda^{o(\omega)}\mathbb{1}_{o(\omega)\leq(a+\varepsilon)|E_n|}]} \\
&\overset{(1.16)}{\leq} \frac{q^{|\partial\Lambda_n|}\lambda^{(b-a-2\varepsilon)|E_n|}}{\varepsilon}.
\end{aligned}$$

The fact that $|E_n|/|\partial\Lambda_n|$ tends to infinity as n tends to infinity implies that $b \leq a + 2\varepsilon$. Since this is true for any $\varepsilon > 0$, we deduce $b \leq a$. Letting p' tend to p and using the continuity of $p' \mapsto \phi^1_{p',q}[\omega_e]$ at p gives that $\phi^1_{p,q}[\omega_e] \leq \phi^0_{p,q}[\omega_e]$. Since we already have $\phi^1_{p,q}[\omega_0] \geq \phi^0_{p,q}[\omega_e]$, this concludes the proof. □

Exercise 22 1. Show that $Z^1_{\Lambda_{2n},p,q} \geq (Z^1_{\Lambda_n,p,q})^{2^d}$.
2. Deduce that $f^1_n(p,q) := \frac{1}{|E_{2^n}|}\log(Z^1_{\Lambda_{2^n},p,q})$ converges to a quantity $f(p,q)$ (called the *free energy*).
3. Show that $f^0_n(p,q) := \frac{1}{|E_{2^n}|}\log(Z^0_{\Lambda_{2^n},p,q})$ converges to $f(p,q)$ as well.
4. Show that the right and left derivatives of

$$g : t \mapsto f\Big(\frac{e^t}{1+e^t}, q\Big) + \log(1+e^t)$$

are respectively $\phi^1_{p,q}[\omega_e]$ and $\phi^0_{p,q}[\omega_e]$.
5. Show that g is convex and therefore not differentiable in at most countably many points. Conclude.

Let us conclude this section by stating the following corollary for the Potts model.

Corollary 4 *Consider the Potts model on $\mathbb{Z}^d$. For any $\beta > \beta_c$,* (LRO$_\beta$) *holds true, while for any $\beta < \beta_c$,* (LRO$_\beta$) *does not hold.*

Note that we do not claim that the property is equivalent to (MAG$_\beta$) since at β_c, one may have (MAG$_{\beta_c}$) but not (LRO$_{\beta_c}$).

Proof By the coupling with the random-cluster model, we need to prove that $\phi^0_{p,q}[0 \leftrightarrow x]$ tends to 0 when $p < p_c$, which is obvious, and that $\phi^0_{p,q}[0 \leftrightarrow x]$ does not tend to 0 when $p > p_c$, which follows from

$$\phi^0_{p,q}[0 \longleftrightarrow x] \geq \phi^0_{p,q}[0 \longleftrightarrow \infty, x \longleftrightarrow \infty] \geq \phi^0_{p,q}[0 \longleftrightarrow \infty]^2 \stackrel{(1.15)}{>} 0,$$

where the first inequality is due to the uniqueness of the infinite cluster, and the second to the FKG inequality and the invariance under translations. □

2 Computation of Critical Points and Sharp Phase Transitions

We would now like to discuss how the critical point of a planar percolation model can sometimes be computed, and how fast correlations decay when $p < p_c$. We start by studying Bernoulli percolation, and then focus on the random-cluster model.

2.1 *Kesten's Theorem*

In this section, we focus on the case $d = 2$. We begin by discussing the duality relation for Bernoulli percolation. Consider the dual lattice $(\mathbb{Z}^2)^* := (\frac{1}{2}, \frac{1}{2}) + \mathbb{Z}^2$ of the lattice $\mathbb{Z}^2$ defined by putting a vertex in the middle of each face, and edges between nearest neighbors. Each edge $e \in \mathbb{E}$ is in direct correspondence with an edge

e^* of the dual lattice crossing it in its middle. For a finite graph $G = (V, E)$, let G^* be the graph with edge-set $E^* = \{e^*, e \in E\}$ and vertex-set given by the endpoints of the edges in E^*.

A configuration ω is naturally associated with a dual configuration ω^*: every edge e which is closed (resp. open) in ω corresponds to an open (resp. closed) edge e^* in ω^*. More formally,

$$\omega^*_{e^*} := 1 - \omega_e \qquad \forall e \in E.$$

Note that if ω is sampled according to $\mathbb{P}_p$, then ω^* is sampled according to $\mathbb{P}_{1-p}$. This duality relation suggests that the critical point of Bernoulli percolation on $\mathbb{Z}^2$ is equal to 1/2. We discuss different levels of heuristic leading to this prediction.

Heuristic level 0 The simplest non-rigorous justification of the fact that $p_c = 1/2$ invokes the uniqueness of the phase transition, i.e. the observation that the model should undergo a single change of macroscopic behavior as p varies. This implies that p_c must be equal to $1 - p_c$, since otherwise the model will change at p_c (with the appearance of an infinite cluster in ω), and at $1 - p_c$ (with the disappearance of an infinite cluster in ω^*). Of course, it seems difficult to justify why there should be a unique phase transition. This encourages us to try to improve our heuristic argument.

Heuristic level 1 One may invoke a slightly more subtle argument. On the one hand, assume for a moment that $p_c < 1/2$. In such case, for any $p \in (p_c, 1 - p_c)$, there (almost surely) exist infinite clusters in both ω and ω^*. Since the infinite cluster is unique almost surely, this seems to be difficult to have coexistence of an infinite cluster in ω and an infinite cluster in ω^*, and it therefore leads us to believe that $p_c \geq 1/2$. On the other hand, assume that $p_c > 1/2$. In such case, for any $p \in (1 - p_c, p_c)$, there (almost surely) exist no infinite cluster in both ω and ω^*. This seems to contradict the intuition that if clusters are all finite in ω, then ω^* should contain an infinite cluster. This reasoning is wrong in general (there may be no infinite cluster in both ω and ω^*), but it seems still believable that this should not occur for a whole range of values of p. Again, the argument is fairly weak here and we should improve it.

Heuristic level 2 Consider the event, called $\mathcal{H}_n$, corresponding to the existence of a path of open edges of ω in $R_n := [0, n] \times [0, n - 1]$ going from the left to the right side of R_n. Observe that the complement of the event $\mathcal{H}_n$ is the event that there exists a path of open edges in ω^* going from top to bottom in the graph R_n^*; see Fig. 5. Using the rotation by $\pi/2$, one sees that at $p = 1/2$, these two events have the same probability, so that

$$\mathbb{P}_{1/2}[\mathcal{H}_n] = \tfrac{1}{2} \quad \forall n \geq 1. \tag{2.1}$$

Now, one may believe that for $p < p_c$, the clusters are so small that the probability that one of them contains a path crossing R_n from left to right tends to 0, which would imply that the probability of $\mathcal{H}_n$ would tend to 0, and therefore that $p_c \leq 1/2$. On the other hand, one may believe that for $p > p_c$, the infinite cluster is so omnipresent that it contains with very high probability a path crossing R_n from left to right, thus

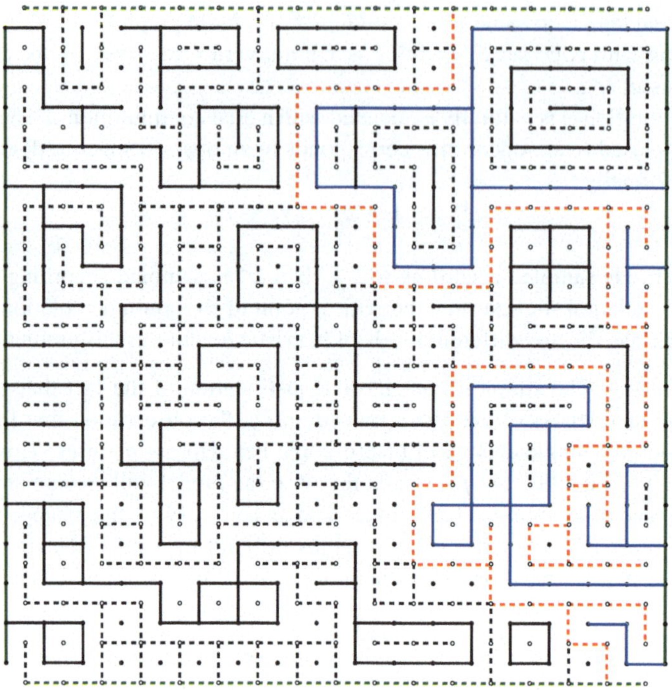

Fig. 5 The rectangle R_n together with its dual R_n^* (the green edges on the boundary are irrelevant for the crossing, so that we may consider only the black edges, for which the dual graph is isomorphic to the graph itself (by rotating it). The dual edges (in red) of the edge-boundary of the cluster of the right boundary in ω (in blue) is a cluster in ω^* crossing from top to bottom in R_n^*

implying that the probability of $\mathcal{H}_n$ would tend to 1. This would give $p_c \geq 1/2$. Unfortunately, the first of these two claims is difficult to justify. Nevertheless, the second one can be proved as follows.

Proposition 4 *Assume that* $\mathbb{P}_p[0 \leftrightarrow \infty] > 0$, *then* $\lim_{n\to\infty} \mathbb{P}_p[\mathcal{H}_n] = 1$.

Proof Fix $n \geq k \geq 1$. Since a path from Λ_k to Λ_n ends up either on the top, bottom, left or right side of Λ_n, the square root trick using the FKG inequality (See Exercise 23) implies that

$$\mathbb{P}_p[\Lambda_k \text{ is connected in } \Lambda_n \text{ to the left of } \Lambda_n] \geq 1 - \mathbb{P}_p[\Lambda_k \not\longleftrightarrow \infty]^{1/4}.$$

Set $n' = \lfloor (n-1)/2 \rfloor$. Consider the event $\mathcal{A}_n$ that $(n', n') + \Lambda_k$ is connected in R_n to the left of R_n, and $(n'+2, n') + \Lambda_k$ is connected in R_n to the right of R_n. We deduce that

$$\mathbb{P}_p[\mathcal{A}_n] \geq 1 - 2\mathbb{P}_p[\Lambda_k \not\longleftrightarrow \infty]^{1/4}.$$

Fig. 6 Construction in the Proof of Proposition 4. One path connects the left side of R_n (in blue) to the blue hatched area. The other one on the right side of R_n (in red) to the red hatched area. The two paths must be in the same cluster (of R_n) by uniqueness, which therefore must contain a path from left to right

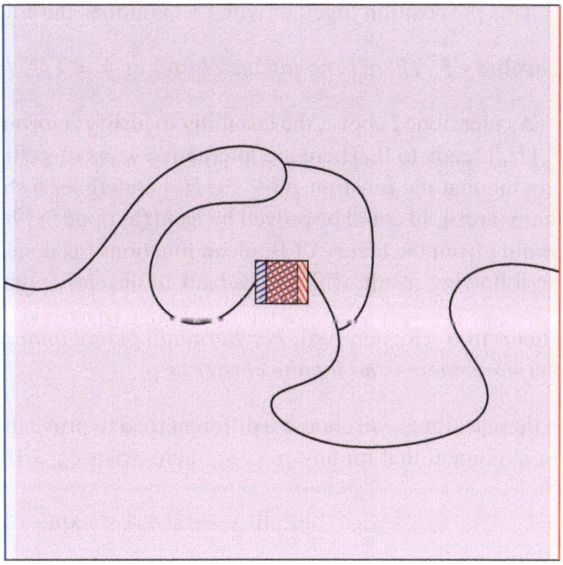

The uniqueness of the infinite cluster implies[5] that

$$\liminf_{n\to\infty} \mathbb{P}_p[\mathcal{H}_n] = \liminf_{n\to\infty} \mathbb{P}_p[\mathcal{A}_n] \geq 1 - 2\mathbb{P}_p[\Lambda_k \not\longleftrightarrow \infty]^{1/4}.$$

Letting k tend to infinity and using that the infinite cluster exists almost surely, we deduce that $\mathbb{P}_p[\mathcal{H}_n]$ tends to 1 (Fig. 6). □

Exercise 23 (*Square root trick*) Prove, using (FKG), that for any increasing events $\mathcal{A}_1, \ldots, \mathcal{A}_r$,

$$\max\{\mathbb{P}_p[\mathcal{A}_i] : 1 \leq i \leq r\} \geq 1 - \Big(1 - \mathbb{P}_p\Big[\bigcup_{i=1}^{r} \mathcal{A}_i\Big]\Big)^{1/r}.$$

Exercise 24 (*Zhang argument*) 1. Show that

$$\mathbb{P}_{1/2}[\text{top of } \Lambda_n \text{ is connected to infinity outside } \Lambda_n] \geq 1 - \mathbb{P}_{1/2}[\Lambda_n \not\longleftrightarrow \infty]^{1/4}.$$

2. Deduce that the probability of the event $\mathcal{B}_n$ that there exist infinite paths in ω from the top and bottom of Λ_n to infinity in $\mathbb{Z}^2 \setminus \Lambda_n$, and infinite paths in ω^* from the left and right sides to infinity satisfies

$$\mathbb{P}_{1/2}[\mathcal{B}_n] \geq 1 - 4\mathbb{P}_{1/2}[\Lambda_n \not\longleftrightarrow \infty]^{1/4}.$$

3. Using (FE) and the uniqueness of the infinite cluster, prove that $\mathbb{P}_{1/2}[\Lambda_n \not\leftrightarrow \infty]$ cannot tend to 0.

[5]The event $\mathcal{A}_n \setminus \mathcal{H}_n$ is included in the event that there are two distinct clusters in R_n going from Λ_k to ∂R_n. The intersection of the latter events for $n \geq 1$ is included in the event that there are two distinct infinite clusters, which has zero probability. Thus, the probability of $\mathcal{A}_n \setminus \mathcal{H}_n$ goes to 0 as n tends to infinity.

This proposition together with (2.1) implies the following corollary

Corollary 5 *There is no infinite cluster at $p = 1/2$. In particular, $p_c \geq 1/2$.*

As mentioned above, the last thing to justify rigorously is the fact that for $p < p_c$, $\mathbb{P}_p[\mathcal{H}_n]$ tends to 0. There are alternative ways of getting the result, in particular by proving that the function $p \mapsto \mathbb{P}_p[\mathcal{H}_n]$ undergoes a sharp threshold[6] near $1/2$. This sharp threshold could be proved by hand (as done in [96]), or using abstract theorems coming from the theory of Boolean functions (as done in [24]). Overall, one obtains the following result, which goes back to the early eighties.

Theorem 6 (Kesten [96]) *For Bernoulli percolation on $\mathbb{Z}^2$, p_c is equal to 1/2. Furthermore, there is no infinite cluster at p_c.*

In these lectures, we choose a different road to prove that $\mathbb{P}_p[\mathcal{H}_n]$ tends to 0. Assume for a moment that for any $p < p_c$, there exists $c_p > 0$ such that for all $n \geq 1$,

$$\mathbb{P}_p[0 \longleftrightarrow \partial\Lambda_n] \leq \exp(-c_p n).$$

Then, $\mathbb{P}_p[\mathcal{H}_n]$ tends to 0 as n tends to infinity since

$$\begin{aligned}\mathbb{P}_p[\mathcal{H}_n] &\leq \sum_{k=0}^{n-1} \mathbb{P}_p[(0,k) \text{ is connected to the right of } R_n] \\ &\leq n\, \mathbb{P}_p[0 \longleftrightarrow \partial\Lambda_n] \leq n \exp(-c_p n).\end{aligned}$$

Overall, Kesten's theorem thus follows from the following result.

Theorem 7 *Consider Bernoulli percolation on $\mathbb{Z}^d$,*

1. *For $p < p_c$, there exists $c_p > 0$ such that for all $n \geq 1$, $\mathbb{P}_p[0 \leftrightarrow \partial\Lambda_n] \leq \exp(-c_p n)$.*
2. *There exists $c > 0$ such that for $p > p_c$, $\mathbb{P}_p[0 \leftrightarrow \infty] \geq c(p - p_c)$.*

Note that the second item, called the *mean-field lower bound* is not relevant for the proof of Kesten's Theorem. Also note that Theorem 7 is a priori way too strong compared to what is needed since it holds in arbitrary dimensions.

---

[6]A sequence (f_n) of continuous homeomorphisms from $[0, 1]$ onto itself satisfies a sharp threshold if for any $\varepsilon > 0$, $\Delta_n(\varepsilon) := f_n^{-1}(1-\varepsilon) - f_n^{-1}(\varepsilon)$ tends to 0.

Exercise 25 $(p_c(\mathbb{G}) + p_c(\mathbb{G}^*) = 1)$ In this exercise, we use the notation $A \overset{B}{\longleftrightarrow} C$ the event that A and C are connected by a path using vertices in B only. Consider Bernoulli percolation on a planar lattice $\mathbb{G}$ embedded in such a way that $\mathbb{Z}^2$ acts transitively on $\mathbb{G}$. We do not assume any symmetry of the lattice. We call the left, right, top and bottom parts of a rectangle Left, Right, Top and Bottom. Also, $\mathcal{H}(n,k)$ and $\mathcal{V}(n,k)$ are the events that $[0,n] \times [0,k]$ is crossed horizontally and vertically by paths of open edges.

1. Use the Borel-Cantelli lemma and Theorem 7 (one may admit the fact that the theorem extends to this context) to prove that for $p < p_c(\mathbb{G})$, there exists finitely many open circuits surrounding a given vertex of $\mathbb{G}^*$. Deduce that $p_c(\mathbb{G}) + p_c(\mathbb{G}^*) \le 1$.
We want to prove the converse inequality by contradiction. From now on, we assume that both $p > p_c(\mathbb{G})$ and $p^* > p_c(\mathbb{G}^*)$.
2. For $s > 0$ and $x \in \mathbb{Z}^2$, define $S_x = x + [0,s]^2$. Prove that for any rectangle R, there exists $x = x(R) \in R \cap \mathbb{Z}^2$ such that there exists x' and x'' neighbors of x in $\mathbb{Z}^2$ satisfying

$$\mathbb{P}_p[S_x \overset{R}{\longleftrightarrow} \text{Bottom}] \ge \mathbb{P}_p[S_x \overset{R}{\longleftrightarrow} \text{Top}] \quad \mathbb{P}_p[S_x \overset{R}{\longleftrightarrow} \text{Left}] \ge \mathbb{P}_p[S_x \overset{R}{\longleftrightarrow} \text{Right}], \tag{2.2}$$

$$\mathbb{P}_p[S_{x'} \overset{R}{\longleftrightarrow} \text{Top}] \ge \mathbb{P}_p[S_{x'} \overset{R}{\longleftrightarrow} \text{Bottom}] \quad \mathbb{P}_p[S_{x''} \overset{R}{\longleftrightarrow} \text{Right}] \ge \mathbb{P}_p[S_{x''} \overset{R}{\longleftrightarrow} \text{Left}]. \tag{2.3}$$

3. Set $\mathbb{H} := \mathbb{R}_+ \times \mathbb{R}$, $\ell_+ := \{0\} \times \mathbb{R}_+$, $\ell_- := \{0\} \times \mathbb{R}_-$ and $\ell = \ell_- \cup \ell_+$. Prove that there exists $x = x(m)$ with first coordinate equal to m satisfying

$$\mathbb{P}_p[S_x \overset{\mathbb{H}}{\longleftrightarrow} \ell_-] \ge \mathbb{P}_p[S_x \overset{\mathbb{H}}{\longleftrightarrow} \ell_+] \quad \text{and} \quad \mathbb{P}_p[S_{x+(0,1)} \overset{\mathbb{H}}{\longleftrightarrow} \ell_-] \le \mathbb{P}_p[S_{x+(0,1)} \overset{\mathbb{H}}{\longleftrightarrow} \ell_+].$$

4. Using the square root trick, deduce that

$$\mathbb{P}_p[S_x \overset{\mathbb{H}}{\longleftrightarrow} \ell_-] \ge 1 - \sqrt{\mathbb{P}_p[S_x \not\longleftrightarrow \ell]} \quad \text{and} \quad \mathbb{P}_p[S_{x+(0,1)} \overset{\mathbb{H}}{\longleftrightarrow} \ell_+] \ge 1 - \sqrt{\mathbb{P}_p[S_{x+(0,1)} \not\longleftrightarrow \ell]}.$$

5. Using the fact that there exists a unique infinite cluster in ω almost surely, prove that the probability that $\{0\} \times [0,1]$ is connected in $\omega^* \cap \mathbb{H}$ to infinity is tending to 0.
6. Prove that the distance between $x(R)$ and the boundary of R is necessarily tending to infinity as $\min\{n,k\}$ tends to infinity.
7. Using $x(R)$, prove that $\max\{\mathbb{P}_p[\mathcal{V}(n,k)], \mathbb{P}_p[\mathcal{H}(n,k+1)]\}$ tends to 1 and $\min\{\mathbb{P}_p[\mathcal{V}(n,k)], \mathbb{P}_p[\mathcal{H}(n,k)]\}$ tends to 0 as $\min\{k,n\}$ tends to infinity. *Hint.* Use the square root trick and the uniqueness criterion like in the previous questions.
8. By considering the largest integer k such that $\mathbb{P}_p[\mathcal{V}(n,k)] \ge \mathbb{P}_p[\mathcal{H}(n,k)]$, reach a contradiction. Deduce that $p_c(\mathbb{G}) + p_c(\mathbb{G}^*) \ge 1$.
9. (to do after Sect. 2.4) How does this argument extend to random-cluster models with $q \ge 1$?

2.2 *Two Proofs of Sharpness for Bernoulli Percolation*

Theorem 7 was first proved by Aizenman and Barsky [2] and Menshikov [104] (these two proofs are presented in [76]). Here, we choose to present two new arguments from [54, 61, 62].

Before diving into the proofs, note that for any function $X : \{0,1\}^E \longrightarrow \mathbb{R}$ where E is finite,

$$\frac{\mathrm{d}\mathbb{E}_p[X]}{\mathrm{d}p} = \tfrac{1}{p(1-p)} \sum_{e \in E} \mathrm{Cov}_p[X, \omega_e], \tag{DF}$$

(where Cov_p is the covariance for $\mathbb{P}_p$) which is obtained readily by differentiating the quantity $\mathbb{E}_p[X] = \sum_{\omega \in \{0,1\}^E} X(\omega) p^{\sum_e \omega_e} (1-p)^{\sum_e 1-\omega_e}$. We insist on the fact that we are considering functions X depending on finitely many edges only (in particular it is clear that $\mathbb{E}_p[X]$ is analytic).

Proof using the $\varphi_p(S)$ quantity Let S be a finite set of vertices containing the origin. We say that $0 \xleftrightarrow{S} x$ if 0 is connected to x using only edges between vertices of S. We denote the edge-boundary of S by

$$\Delta S = \big\{xy \subseteq \mathbb{E} : x \in S, y \notin S\big\}.$$

For $p \in [0, 1]$ and $0 \in S \subseteq \mathbb{Z}^d$, define

$$\varphi_p(S) := p \sum_{xy \in \Delta S} \mathbb{P}_p[0 \xleftrightarrow{S} x]. \tag{2.4}$$

Set

$$\tilde{p}_c := \sup\big\{p \in [0, 1] : \exists S \ni 0 \text{ finite with } \varphi_p(S) < 1\big\}. \tag{2.5}$$

Step 1: for $p < \tilde{p}_c$, (EXP$_p$) holds true. By definition, one can fix a finite set S containing the origin, such that $\varphi_p(S) < 1$. Choose $L > 0$ such that $S \subseteq \Lambda_{L-1}$. Consider $k \geq 1$ and assume that the event $0 \leftrightarrow \partial\Lambda_{kL}$ holds. Introduce the random variable $\mathsf{C} := \{x \in S : x \xleftrightarrow{S} 0\}$ corresponding to the cluster of 0 in S. Since $S \cap \partial\Lambda_{kL} = \emptyset$, one can find an open edge $xy \in \Delta S$ such that $0 \xleftrightarrow{S} x$ and $y \xleftrightarrow{\mathsf{C}^c} \partial\Lambda_{kL}$. Using the union bound, and then decomposition on the possible realizations of C, we find

$$\begin{aligned}
&\mathbb{P}_p[0 \longleftrightarrow \partial\Lambda_{kL}]\\
&\quad\leq \sum_{xy \in \Delta S} \sum_{C \subseteq S} \mathbb{P}_p\big[\{0 \xleftrightarrow{S} x\} \cap \{\mathsf{C} = C\} \cap \{\omega_{xy} = 1\} \cap \{y \xleftrightarrow{C^c} \partial\Lambda_{kL}\}\big]\\
&\quad\leq \sum_{xy \in \Delta S} \sum_{C \subseteq S} \mathbb{P}_p\big[\{0 \xleftrightarrow{S} x\} \cap \{\mathsf{C} = C\}\big] \cdot p \cdot \mathbb{P}_p\big[y \xleftrightarrow{C^c} \partial\Lambda_{kL}\big]\\
&\quad\leq p\Big(\sum_{xy \in \Delta S} \sum_{C \subseteq S} \mathbb{P}_p\big[\{0 \xleftrightarrow{S} x\} \cap \{\mathsf{C} = C\}\big]\Big)\mathbb{P}_p\big[0 \longleftrightarrow \partial\Lambda_{(k-1)L}\big]\\
&\quad\leq p\Big(\sum_{xy \in \Delta S} \mathbb{P}_p\big[0 \xleftrightarrow{S} x\big]\Big)\mathbb{P}_p\big[0 \longleftrightarrow \partial\Lambda_{(k-1)L}\big]\\
&\quad= \varphi_p(S)\mathbb{P}_p\big[0 \longleftrightarrow \partial\Lambda_{(k-1)L}\big].
\end{aligned}$$

In the second line, we used that $\{y \xleftrightarrow{C^c} \partial\Lambda_{kL}\}$, $\{\omega_{xy} = 1\}$ and $\{0 \xleftrightarrow{S} x\} \cap \{\mathsf{C} = C\}$ are independent. Indeed, these events depend on disjoint sets of edges: the first one on edges with both endpoints outside of C, the second one on xy only, and the third one on edges between vertices of S with at least one endpoint in C. In the third line, we used $y \in \Lambda_L$ implies

$$\mathbb{P}_p[y \overset{C^c}{\longleftrightarrow} \partial\Lambda_{kL}] \le \mathbb{P}_p[0 \longleftrightarrow \partial\Lambda_{(k-1)L}].$$

In the fourth line, we used that the events $\{0 \overset{S}{\longleftrightarrow} x\} \cap \{\mathsf{C} = C\}$ partition the event $0 \overset{S}{\longleftrightarrow} x$. Induction on k gives $\mathbb{P}_p[0 \leftrightarrow \partial\Lambda_{kL}] \le \varphi_p(S)^k$, thus proving the claim.

Step 2: For $p > \tilde{p}_c$, $\mathbb{P}_p[0 \leftrightarrow \infty] \ge \frac{p-\tilde{p}_c}{p(1-\tilde{p}_c)}$. Let us start by the following lemma providing a differential inequality valid for every p. Define $\theta_n(p) := \mathbb{P}_p[0 \leftrightarrow \partial\Lambda_n]$.

Lemma 3 *Let $p \in (0, 1)$ and $n \ge 1$,*

$$\theta_n'(p) \ge \tfrac{1}{p(1-p)} \cdot \inf_{\substack{S \subseteq \Lambda_n \\ 0 \in S}} \varphi_p(S) \cdot \big(1 - \theta_n(p)\big). \tag{2.6}$$

Let us first see how the second step follows from Lemma 3. Above $\tilde{p}_c$, (2.6) becomes $\theta_n' \ge \frac{1}{p(1-p)}(1 - \theta_n)$ which can be rewritten as

$$\big[\log\big(\tfrac{1}{1-\theta_n}\big)\big]' \ge \big[\log\big(\tfrac{p}{1-p}\big)\big]'.$$

Integrating between $\tilde{p}_c$ and p implies that for every $n \ge 1$,

$$\theta_n(p) \ge \frac{p - \tilde{p}_c}{p(1 - \tilde{p}_c)}.$$

By letting n tend to infinity, we obtain the desired lower bound on $\mathbb{P}_p[0 \leftrightarrow \infty]$.

Proof (*Lemma* 3) Apply (DF) to $X := -\mathbb{1}_{0 \not\longleftrightarrow \partial\Lambda_n}$ to get

$$\theta_n'(p) = \tfrac{1}{p(1-p)} \sum_{e \in E_n} \mathbb{E}_p\big[\mathbb{1}_{0 \not\longleftrightarrow \partial\Lambda_n}(p - \omega_e)\big]. \tag{2.7}$$

Fix an edge e and consider the event A that $\omega_{|E_n \setminus \{e\}}$ satisfies the following three properties

- P1 one of the endpoints of e is connected to 0,
- P2 the other one is connected to $\partial\Lambda_n$,
- P3 0 is not connected to $\partial\Lambda_n$.

(This event corresponds in the standard terminology to the fact that the edge e is *pivotal* for $0 \not\longleftrightarrow \partial\Lambda_n$ but this is irrelevant here.) By definition, ω_e is independent of $\{0 \not\longleftrightarrow \partial\Lambda_n\} \cap A^c$. Since ω_e is a Bernoulli random variable of parameter p, we deduce that

$$\mathbb{E}_p[\mathbb{1}_{A^c}\mathbb{1}_{0 \not\longleftrightarrow \partial\Lambda_n}(p - \omega_e)] = 0.$$

Also, for $\omega \in A$, 0 is not connected to $\partial\Lambda_n$ if and only if the edge e is closed, and in this case ω itself (not only its restriction to $E_n \setminus \{e\}$) satisfies P1, P2 and P3. Therefore, we can write

$$\mathbb{E}_p[\mathbb{1}_A \mathbb{1}_{0 \not\leftrightarrow \partial\Lambda_n}(p-\omega_e)] = p\,\mathbb{P}_p[\omega \text{ satisfies P1, P2 and P3}].$$

Overall, the previous discussion implies that (2.7) can be rewritten as

$$\theta_n'(p) = \frac{1}{p(1-p)} \sum_{\substack{x,y\in\Lambda_n \\ xy\in E_n}} p\,\mathbb{P}_p\big[0 \longleftrightarrow x,\, y \longleftrightarrow \partial\Lambda_n,\, 0 \not\longleftrightarrow \partial\Lambda_n\big]. \tag{2.8}$$

Introduce $\mathsf{S} := \{z \in \Lambda_n : z \not\longleftrightarrow \partial\Lambda_n\}$ and a fixed set S. The intersection of $\{\mathsf{S} = S\}$ with the event on the right-hand side of (2.8) can be rewritten nicely. The fact that $0 \not\longleftrightarrow \partial\Lambda_n$ becomes the condition that S contains 0. Furthermore, the conditions $0 \longleftrightarrow x$ and $y \longleftrightarrow \partial\Lambda_n$ get rephrased as $xy \in \Delta S$ and 0 is connected to x in S. Thus, partitioning the event on the right of (2.8) into the possible values of S gives

$$\begin{aligned}\theta_n'(p) &= \frac{1}{p(1-p)} \sum_{0\in S\subseteq\Lambda_n} \sum_{xy\in\Delta S} p\,\mathbb{P}_p\big[0 \overset{S}{\longleftrightarrow} x,\, \mathsf{S} = S\big] \\ &= \frac{1}{p(1-p)} \sum_{0\in S\subseteq\Lambda_n} \sum_{xy\in\Delta S} p\,\mathbb{P}_p\big[0 \overset{S}{\longleftrightarrow} x\big]\mathbb{P}_p[\mathsf{S} = S\big] \\ &\geq \frac{1}{p(1-p)} \cdot \inf_{0\in S\subseteq\Lambda_n} \varphi_p(S)\cdot(1-\theta_n(p)),\end{aligned}$$

where in the second line we used that $0 \overset{S}{\longleftrightarrow} x$ is measurable in terms of edges with both endpoints in S, and $\mathsf{S} = S$ is measurable in terms of the other edges. In the last line, we used that the family of events $\{\mathsf{S} = S\}$ with $S \ni 0$ partition the event that 0 is not connected to $\partial\Lambda_n$. □

Steps 1 and 2 conclude the proof since $\tilde{p}_c$ must be equal to p_c, and therefore the proof of the theorem.

Exercise 26 (*Percolation with long-range interactions*) Consider a family $(J_{x,y})_{x,y\in\mathbb{Z}^d}$ of non-negative coupling constants which is invariant under translations, meaning that $J_{x,y} = J(x-y)$ for some function J. Let $\mathbf{P}_\beta$ be the bond percolation measure on $\mathbb{Z}^d$ defined as follows: for $x, y \in \mathbb{Z}^d$, $\{x, y\}$ is *open* with probability $1 - \exp(-\beta J_{x,y})$, and *closed* with probability $\exp(-\beta J_{x,y})$.
1. Define the analogues $\tilde{\beta}_c$ and $\varphi_\beta(S)$ of $\tilde{p}_c$ and $\varphi_p(S)$ in this context.
2. Show that there exists $c > 0$ such that for any $\beta \geq \tilde{\beta}_c$, $\mathbf{P}_\beta[0 \longleftrightarrow \infty] \geq c(\beta - \tilde{\beta}_c)$.
3. Show that if the interaction is finite range (i.e. that there exists $R > 0$ such that $J(x) = 0$ for $\|x\| \geq R$), then for any $\beta < \tilde{\beta}_c$, there exists $c_\beta > 0$ such that $\mathbf{P}_\beta[0 \longleftrightarrow \partial\Lambda_n] \leq \exp(-c_\beta n)$ for all n.
4. In the general case, show that for any $\beta < \tilde{\beta}_c$, $\sum_{x\in\mathbb{Z}^d} \mathbf{P}_\beta[0 \longleftrightarrow x] < \infty$.

Hint. Consider S such that $\varphi_\beta(S) < 1$ and show that for $n \geq 1$ and $x \in \Lambda_n$, $\sum_{y\in\Lambda_n} \mathbf{P}_\beta[x \overset{\Lambda_n}{\longleftrightarrow} y] \leq \frac{|S|}{1-\varphi_\beta(S)}$.

Remark 1 Since $\varphi_p(\{0\}) = 2dp$, we find $p_c(d) \geq 1/2d$. Also, $p_c(d) \leq p_c(2) = \frac{1}{2}$.

Remark 2 The set of parameters p such that there exists a finite set $0 \in S \subseteq \mathbb{Z}^d$ with $\varphi_p(S) < 1$ is an open subset of $[0, 1]$. Since this set is coinciding with $[0, p_c)$, we deduce that $\varphi_{p_c}(\Lambda_n) \geq 1$ for any $n \geq 1$. As a consequence, the expected size of the cluster of the origin satisfies at p_c,

$$\sum_{x\in\mathbb{Z}^d} \mathbb{P}_{p_c}[0 \longleftrightarrow x] \geq \tfrac{1}{dp_c} \sum_{n\geq 0} \varphi_{p_c}(\Lambda_n) = +\infty.$$

In particular, $\mathbb{P}_{p_c}[0 \leftrightarrow x]$ cannot decay faster than algebraically (see Exercise 27 for more detail).

Exercise 27 (*Definition of the correlation length*) Fix $d \geq 2$ and set $e_1 = (1, 0, \dots, 0)$.
1. Prove that, for any $p \in [0, 1]$ and $n, m \geq 0$, $\mathbb{P}_p[x_0 \longleftrightarrow (m+n)e_1] \geq \mathbb{P}_p[x_0 \longleftrightarrow me_1] \cdot \mathbb{P}_p[x_0 \longleftrightarrow ne_1]$
2. Deduce that $\xi(p) = \left(\lim_{n\to\infty} -\frac{1}{n} \log \mathbb{P}_p[0 \longleftrightarrow ne_1]\right)^{-1}$ and that $\mathbb{P}_p[0 \longleftrightarrow ne_1] \leq \exp(-n/\xi(p))$.
3. Prove that $\xi(p)$ tends to infinity as p tends to p_c.
4. Prove that for any $x \ni \partial\Lambda_n$,
$$\mathbb{P}_{p_c}[0 \longleftrightarrow 2ne_1] \geq \mathbb{P}_{p_c}[0 \longleftrightarrow x]^2.$$
5. Using that $\varphi_{p_c}(\Lambda_n) \geq 1$ for every n, prove that there exists $c > 0$ such that for any $x \in \mathbb{Z}^d$, $\mathbb{P}_{p_c}[0 \leftrightarrow x] \geq \frac{c}{\|x\|^{2d(d-1)}}$.

Proof using randomized algorithms The second proof uses the notion of random decision trees or equivalently randomized algorithms (the two terms will be used interchangeably). In theoretical science, determining the computational complexity of tasks is a difficult problem (think of P against NP). To simplify the problem, computer scientists came up with computational problems involving so-called *decision trees*. Informally speaking, a decision tree associated with a Boolean function f takes $\omega \in \{0, 1\}^n$ as an input, and reveals algorithmically the value of ω at different coordinates one by one. At each step, which coordinate will be revealed next depends on the values of ω revealed so far. The algorithm stops as soon as the value of f is the same no matter the values of ω on the remaining coordinates. The question is then to determine how many bits of information must be revealed before the algorithm stops.

Formally, a decision tree is defined as follows. Consider a finite set E of cardinality n. For a n-tuple $x = (x_1, \dots, x_n)$ and $t \leq n$, write $x_{[t]} = (x_1, \dots, x_t)$ and $\omega_{x_{[t]}} = (\omega_{x_1}, \dots, \omega_{x_t})$. A *decision tree* $T = (e_1, \psi_t, t < n)$ takes $\omega \in \{0, 1\}^E$ as an input and gives back an ordered sequence $e = (e_1, \dots, e_n)$ constructed inductively as follows: for any $2 \leq t \leq n$,

$$e_t = \psi_t(e_{[t-1]}, \omega_{e_{[t-1]}}) \in E\setminus\{e_1, \dots, e_{t-1}\},$$

where ψ_t is a function interpreted as the decision rule at time t (ψ_t takes the location and the value of the bits for the first $t-1$ steps of the induction, and decides of the next bit to query). For $f : \{0, 1\}^E \to \mathbb{R}$, define

$$\tau(\omega) = \tau_{f,T}(\omega) := \min\left\{t \geq 1 : \forall\omega' \in \{0, 1\}^E,\quad \omega'_{e_{[t]}} = \omega_{e_{[t]}} \Longrightarrow f(\omega) = f(\omega')\right\}.$$

Remark 3 In computer science, a decision tree is usually associated directly to a boolean function f and defined as a rooted directed tree in which each internal nodes are labeled by elements of E, leaves by possible outputs, and edges are in correspondence with the possible values of the bits at vertices (see [109] for a formal definition). In particular, the decision trees are usually defined up to τ, and not later on.

The OSSS inequality, originally introduced in [109] as a step toward a conjecture of Yao [128], relates the variance of a Boolean function to the influence of the variables and the computational complexity of a random decision tree for this function.

Theorem 8 (OSSS for Bernoulli percolation) *Consider $p \in [0,1]$ and a finite set of edges E. Fix an increasing function $f : \{0,1\}^E \longrightarrow [0,1]$ and an algorithm T. We have*

$$\mathrm{Var}_p(f) \ \le\ 2\sum_{e\in E} \delta_e(f,T)\,\mathrm{Cov}_p[f,\omega_e], \tag{2.9}$$

where $\delta_e(f,T) := \mathbb{P}_p\big[\exists t \le \tau(\omega)\ :\ e_t = e\big]$ is the revealment (of f) for the decision tree T.

The general inequality does not require f to be increasing, but we will only use it in this context.

Proof Our goal is to apply a Linderberg-type argument. Consider two independent sequences ω and $\tilde\omega$ of iid Bernoulli random variables of parameter p. Write $\mathbb{P}$ for the coupling between these variables (and $\mathbb{E}$ for its expectation). Construct $\mathbf{e}$ by setting $\mathbf{e}_1 = e_1$ and for $t \ge 1$, $\mathbf{e}_{t+1} := \psi_t(\mathbf{e}_{[t]}, \omega_{\mathbf{e}_{[t]}})$. Similarly, define

$$\tau := \min\big\{t \ge 1 : \forall x \in \{0,1\}^E,\ x_{\mathbf{e}_{[t]}} = \omega_{\mathbf{e}_{[t]}} \Rightarrow f(x) = f(\omega)\big\}.$$

Finally, for $0 \le t \le n$, define

$$\omega^t := (\tilde\omega_{\mathbf{e}_1}, \dots, \tilde\omega_{\mathbf{e}_t}, \omega_{\mathbf{e}_{t+1}}, \dots, \omega_{\mathbf{e}_{\tau-1}}, \tilde\omega_{\mathbf{e}_\tau}, \tilde\omega_{\mathbf{e}_{\tau+1}}, \dots, \tilde\omega_{\mathbf{e}_n}),$$

where it is understood that the n-tuple under parentheses is equal to $\tilde\omega$ if $t \ge \tau$. (We used a slight abuse of notation, the order here is shuffled to match the order in which the edges are revealed by the algorithm.)

Since ω^0 and ω coincide on $\mathbf{e}_t$ for any $t \le \tau$, we deduce that $f(\omega^0) = f(\omega)$. Also, $f(\omega^n) = f(\tilde\omega)$ since $\omega^n = \tilde\omega$. As a consequence, conditioning on ω gives

$$\begin{aligned}\mathrm{Var}_p(f) &\le \mathbb{E}_p\big[|f - \mathbb{E}_p[f]|\big] = \mathbb{E}\big[\big|\,\mathbb{E}[f(\omega^0)|\omega] - \mathbb{E}[f(\omega^n)|\omega]\,\big|\big]\\ &\le \mathbb{E}\big[|f(\omega^0) - f(\omega^n)|\big].\end{aligned}$$

The triangular inequality and the observation that $\omega^t = \omega^{t-1}$ for any $t > \tau$ gives that

$$\mathrm{Var}_p(f) \le \sum_{t=1}^{n} \mathbb{E}\big[|f(\omega^t) - f(\omega^{t-1})|\big] = \sum_{t=1}^{n} \mathbb{E}\big[|f(\omega^t) - f(\omega^{t-1})|\mathbb{1}_{t\le\tau}\big].$$

Let us now decomposed this expression according to the possible values for $\mathbf{e}_t$. Note that $\mathbf{e}_t$ is measurable in terms of $\omega_{[t-1]}$, and that τ is a stopping time, so that $\{t \le \tau\} = \{\tau \le t-1\}^c$ is also measurable in terms of $\omega_{[t-1]}$. Overall, we get that

$$\mathrm{Var}_p(f) \le \sum_{e \in E} \sum_{t=1}^{n} \mathbb{E}\big[\mathbb{E}\big[|f(\omega^t) - f(\omega^{t-1})| \mid \omega_{[t-1]}\big] \mathbb{1}_{t \le \tau, \mathbf{e}_t = e}\big].$$

Let $f^1(\omega)$ and $f^0(\omega)$ denote the function f applied to the configuration equal to ω except at e where it is equal to 1 or to 0 respectively. Note that since f is increasing, we find that $f^1 \ge f^0$. Now, conditionally on $\omega_{[t-1]}$ and $\{t \le \tau, \mathbf{e}_t = e\}$, both ω^t and ω^{t-1} are sequences of iid Bernoulli random variables of parameter p, differing (potentially) exactly at e (since $\omega^t_e = \tilde{\omega}_e$ and $\omega^{t-1}_e = \omega_e$). We deduce that

$$\mathbb{E}\big[\,|f(\omega^t) - f(\omega^{t-1})| \mid \omega_{[t-1]}\big] = 2p(1-p)\mathbb{E}_p[f^1(\omega) - f^0(\omega)] = 2\mathrm{Cov}_p[f, \omega_e].$$

Recalling that $\sum_{t=1}^{n} \mathbb{P}[t \le \tau, \mathbf{e}_t = e] = \delta_e(f, T)$ concludes the proof. □

Let us start by the proof of a general lemma (Fig. 7).

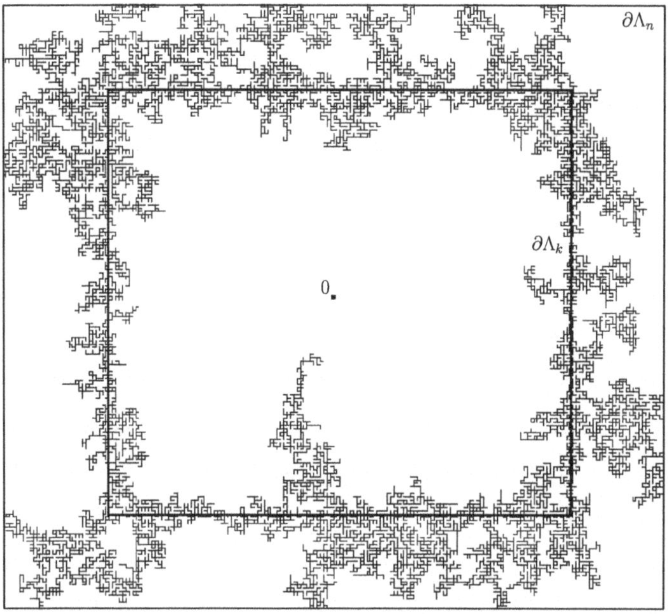

Fig. 7 A realization of the clusters intersecting $\partial\Lambda_k$. Every edge having one endpoint in this set has been revealed by the decision tree. Furthermore in this specific case, we know that 0 is not connected to the boundary of Λ_n

Lemma 4 *Consider a converging sequence of increasing differentiable functions $f_n : [0, x_0] \longrightarrow [0, M]$ satisfying*

$$f_n' \geq \frac{n}{\Sigma_n} f_n \tag{2.10}$$

for all $n \geq 1$, where $\Sigma_n = \sum_{k=0}^{n-1} f_k$. Then, there exists $x_1 \in [0, x_0]$ such that

P1 *For any $x < x_1$, there exists $c_x > 0$ such that for any n large enough, $f_n(x) \leq \exp(-c_x n)$.*

P2 *For any $x > x_1$, $f = \lim_{n\to\infty} f_n$ satisfies $f(x) \geq x - x_1$.*

Proof Define

$$x_1 := \inf\Big\{x \,:\, \limsup_{n\to\infty} \frac{\log \Sigma_n(x)}{\log n} \geq 1\Big\}.$$

Assume $x < x_1$. Fix $\delta > 0$ and set $x' = x - \delta$ and $x'' = x - 2\delta$. We will prove that there is exponential decay at x'' in two steps.

First, there exists an integer N and $\alpha > 0$ such that $\Sigma_n(x) \leq n^{1-\alpha}$ for all $n \geq N$. For such an integer n, integrating $f_n' \geq n^\alpha f_n$ between x' and x—this differential inequality follows from (2.10), the monotonicity of the functions f_n (and therefore Σ_n) and the previous bound on $\Sigma_n(x)$ – implies that

$$f_n(x') \leq M \exp(-\delta n^\alpha), \quad \forall n \geq N.$$

Second, this implies that there exists $\Sigma < \infty$ such that $\Sigma_n(x') \leq \Sigma$ for all n. Integrating $f_n' \geq \frac{n}{\Sigma} f_n$ for all n between x'' and x'—this differential inequality is again due to (2.10), the monotonicity of Σ_n, and the bound on $\Sigma_n(x')$—leads to

$$f_n(x'') \leq M \exp(-\frac{\delta}{\Sigma} n), \quad \forall n \geq 0.$$

Assume $x > x_1$. For $n \geq 1$, define the function $T_n := \frac{1}{\log n} \sum_{i=1}^n \frac{f_i}{i}$. Differentiating T_n and using (2.10), we obtain

$$T_n' = \frac{1}{\log n} \sum_{i=1}^n \frac{f_i'}{i} \overset{(2.10)}{\geq} \frac{1}{\log n} \sum_{i=1}^n \frac{f_i}{\Sigma_i} \geq \frac{\log \Sigma_{n+1} - \log \Sigma_1}{\log n},$$

where in the last inequality we used that for every $i \geq 1$,

$$\frac{f_i}{\Sigma_i} \geq \int_{\Sigma_i}^{\Sigma_{i+1}} \frac{dt}{t} = \log \Sigma_{i+1} - \log \Sigma_i.$$

For $x' \in (x_1, x)$, using that $\Sigma_{n+1} \geq \Sigma_n$ is increasing and integrating the previous differential inequality between x' and x gives

$$T_n(x) - T_n(x') \geq (x - x') \frac{\log \Sigma_n(x') - \log M}{\log n}.$$

Hence, the fact that $T_n(x)$ converges to $f(x)$ as n tends to infinity implies

$$f(x) - f(x') \;\geq\; (x - x')\Big[\limsup_{n\to\infty} \frac{\log \Sigma_n(x')}{\log n}\Big] \;\geq\; x - x'.$$

Letting x' tend to x_1 from above, we obtain $f(x) \geq x - x_1$. □

We now present the Proof of Theorem 7. We keep the notation introduced in the previous section

$$\theta_n(p) = \mathbb{P}_p[0 \longleftrightarrow \partial\Lambda_n] \quad\text{and}\quad S_n := \sum_{k=0}^{n-1} \theta_k.$$

Lemma 5 *For any $n \geq 1$, one has*

$$\sum_{xy\in E_n} \mathrm{Cov}_p[\mathbb{1}_{0\leftrightarrow\partial\Lambda_n}, \omega_e] \geq \frac{n}{8S_n} \cdot \theta_n(1 - \theta_n).$$

The proof is based on Theorem 8 applied to a well-chosen decision tree determining $\mathbb{1}_{0\leftrightarrow\partial\Lambda_n}$. One may simply choose the trivial decision tree checking every edge of the box Λ_n. Unfortunately, the revealment of the decision tree being 1 for every edge, the OSSS inequality will not bring us much information. A slightly better decision tree would be provided by the decision tree discovering the cluster of the origin "from inside". Edges far from the origin would then be revealed by the decision tree if (and only if) one of their endpoints is connected to the origin. This provides a good bound for the revealment of edges far from the origin, but edges close to the origin are still revealed with large probability. In order to avoid this last fact, we will rather choose a family of decision trees discovering the clusters of $\partial\Lambda_k$ for $1 \leq k \leq n$ and observe that the average of their revealment for a fixed edge will always be small.

Proof For any $k \in [\![1, n]\!]$, we wish to construct a decision tree T determining $\mathbb{1}_{0\leftrightarrow\partial\Lambda_n}$ such that for each $e = uv$,

$$\delta_e(T) \leq \mathbb{P}_p[u \longleftrightarrow \partial\Lambda_k] + \mathbb{P}_p[v \longleftrightarrow \partial\Lambda_k]. \tag{2.11}$$

Note that this would conclude the proof since we obtain the target inequality by applying Theorem 8 for each k and then summing on k. As a key, we use that for $u \in \Lambda_n$,

$$\sum_{k=1}^{n} \mathbb{P}_p[u \longleftrightarrow \partial\Lambda_k] \;\leq\; \sum_{k=1}^{n} \mathbb{P}_p[u \longleftrightarrow \partial\Lambda_{|k-d(u,0)|}(u)] \;\leq\; 2S_n.$$

We describe the decision tree T, which corresponds first to an exploration of the clusters in Λ_n intersecting $\partial\Lambda_k$ that does not reveal any edge with both endpoints outside these clusters, and then to a simple exploration of the remaining edges.

More formally, we define $\mathbf{e}$ (instead of the collection of decision rules ϕ_t) using two growing sequences $\partial\Lambda_k = V_0 \subseteq V_1 \subseteq \cdots \subseteq V$ and $\emptyset = F_0 \subseteq F_1 \subseteq \cdots \subseteq F$ (where F is the set of edges between two vertices within distance n of the origin) that should be understood as follows: at step t, V_t represents the set of vertices that the decision tree found to be connected to $\partial\Lambda_k$, and F_t is the set of explored edges discovered by the decision tree until time t.

Fix an ordering of the edges in F. Set $V_0 = \partial\Lambda_k$ and $F_0 = \emptyset$. Now, assume that $V_t \subseteq V$ and $F_t \subseteq F$ have been constructed and distinguish between two cases:

- If there exists an edge $e = xy \in F\setminus F_t$ with $x \in V_t$ and $y \notin V_t$ (if more than one exists, pick the smallest one for the ordering), then set $\mathbf{e}_{t+1} = e$, $F_{t+1} = F_t \cup \{e\}$ and set
$$V_{t+1} := \begin{cases} V_t \cup \{x\} & \text{if } \omega_e = 1 \\ V_t & \text{otherwise.} \end{cases}$$
- If e does not exist, set $\mathbf{e}_{t+1}$ to be the smallest $e \in F\setminus F_t$ (for the ordering) and set $V_{t+1} = V_t$ and $F_{t+1} = F_t \cup \{e\}$.

As long as we are in the first case, we are still discovering the clusters of $\partial\Lambda_k$. Also, as soon as we are in the second case, we remain in it. The fact that τ is not greater than the last time we are in the first case gives us (2.11).

Note that τ may a priori be strictly smaller than the last time we are in the first case (since the decision tree may discover a path of open edges from 0 to $\partial\Lambda_n$ or a family of closed edges disconnecting the origin from $\partial\Lambda_n$ before discovering the whole clusters of $\partial\Lambda_k$). □

We are now in a position to provide our alternative proof of exponential decay. Fix $n \geq 1$. Lemma 5 together with the different formula gives

$$\theta_n' = \tfrac{1}{p(1-p)} \sum_{e\in E_n} \mathrm{Cov}(\mathbb{1}_{0\leftrightarrow\partial\Lambda_n}, \omega_e) \geq \frac{n}{2S_n} \cdot \theta_n(1-\theta_n).$$

To conclude, fix $p_0 \in (p_c, 1)$ and observe that for $p \leq p_0$, $1-\theta_n(p) \geq 1-\theta_1(p_0) > 0$. Then, apply Lemma 4 to $f_n = \frac{2}{(1-\theta_1(p_0))}\theta_n$.

Other models can be treated using the OSSS inequality (to mention only two, Voronoi percolation [53] and Boolean percolation [55]) but the study of the random-cluster model requires a generalization of the OSSS inequality, which we present below.

Let us make a small detour, analyze what we did in the previous proof, and discuss the study of averages of boolean functions. We proved an inequality of the form

$$\theta_n' \geq C_n\theta_n \tag{2.12}$$

for a constant C_n that was large as soon as θ_n was small. In particular, when θ_n was decaying polynomially fast, C_n was polynomially large, a statement which allowed us to prove that θ_n was decaying stretched exponentially fast and then exponentially fast for smaller values of p (see the proof of **P1** of Lemma 4).

Historically, differential inequalities like (2.12) were obtained using abstract sharp threshold theorems. The general theory of sharp thresholds for discrete product spaces was initiated by Kahn, Kalai and Linial in [93] in the case of the uniform measure on $\{0, 1\}^n$, i.e. in the case of $\mathbb{P}_p$ with $p = 1/2$. There, Kahn, Kalai and Linial used the Bonami-Beckner inequality [10, 26] to deduce inequalities between the variance of a boolean function and so-called influences of this function. Bourgain et al. [27] extended these inequalities to product spaces $\{0, 1\}^n$ and to $\mathbb{P}_p$ with arbitrary $p \in [0, 1]$. For completeness, let us state a version of this result due to Talagrand [123]: there exists a constant $c > 0$ such that for any $p \in [0, 1]$ and any increasing event A,

$$\mathbb{P}_p[A](1 - \mathbb{P}_p[A]) \le c \log \tfrac{1}{p(1-p)} \sum_{e \in E} \frac{\mathrm{Cov}_p[\mathbb{1}_A, \omega_e]}{\log(1/\mathrm{Cov}_p[\mathbb{1}_A, \omega_e])}.$$

Notice that as soon as all covariances are small, the sum of covariances is large. This result can seem counterintuitive at first but it is definitely very efficient to prove differential inequalities like (2.12). In particular, $\mathrm{Cov}_p[\mathbb{1}_A, \omega_e] \le \mathbb{P}_p[A]$ so that applying the previous displayed equation to $A = \{0 \leftrightarrow \partial\Lambda_n\}$ gives

$$\theta_n(1 - \theta_n) \le \frac{c_p}{\log(1/\theta_n)} \theta_n'.$$

In order to compare this inequality to what we got with the OSSS inequality, let us look at the case where θ_n is decaying polynomially fast. In this case, the value of C_n is of order $\log n$. This is not a priori sufficient to prove that θ_n decays exponentially fast for smaller values of p since it only improves the decay of θ_n by small polynomials. From this point of view, the logarithm in the expression $\log(1/\theta_n)$ is catastrophic.

Mathematicians succeeded in going around this difficulty by considering crossing events (see Sect. 5 for more detail). A beautiful example of the application of sharp threshold results to percolation theory is the result of Bollobás and Riordan about critical points of planar percolation models [23, 24].

Recently, Graham and Grimmett [74] managed to extend the BKKKL/Talagrand result to random-cluster models. Combined with ideas from [23], this led to a computation of the critical point of the random-cluster model (see below). Nonetheless, these proofs involving crossing probabilities are pretty specific to planar models and, to the best of our knowledge, fail to apply in higher dimensions. In particular, it seems necessary to use a generalization of the OSSS inequality rather than a generalization of the BKKKL/Talagrand result, which is what we propose to do in the next section.

Exercise 28 *(A k-dependent percolation model)* Consider a family of iid Bernoulli random variables $(\eta_x)_{x\in\mathbb{Z}^d}$ of parameter $1-p$ and say that an edge $e \in \mathbb{Z}^d$ is open if both endpoints are at a distance less than or equal to R from any $x \in \mathbb{Z}^d$ with $\eta_x = 1$ (it corresponds to taking the vacant set of balls of radius R centered around the vertices $x \in \mathbb{Z}^d$ with $\eta_x = 1$). Adapt the previous proof to show that the model undergoes a sharp phase transition, and that (EXP$_p$) holds for any $p < p_c$.

2.3 Sharpness for Random-Cluster Models

We now turn to the proof of the following generalization of Theorem 7.

Theorem 9 (Duminil-Copin et al. [54]) *Consider the random-cluster model on $\mathbb{Z}^d$ with $q \geq 1$.*

1. *There exists $c > 0$ such that for $p > p_c$, $\phi^1_{p,q}[0 \leftrightarrow \infty] \geq c(p-p_c)$.*
2. *For $p < p_c$, there exists $c_p > 0$ such that for all $n \geq 1$,*

$$\phi^1_{\Lambda_n,p,q}[0 \longleftrightarrow \partial\Lambda_n] \leq \exp(-c_p n).$$

The result extends to any infinite locally-finite quasi-transitive graph $\mathbb{G}$. The proof will be based on the following improvement of the OSSS inequality (2.9). Below, $\mathrm{Var}_{G,p,q}$ and $\mathrm{Cov}_{G,p,q}$ are respectively the variance and the covariance for $\phi^1_{G,p,q}$.

Theorem 10 *Consider $q \geq 1$, $p \in [0,1]$, and a finite graph G. Fix an increasing function $f : \{0,1\}^E \longrightarrow [0,1]$ and an algorithm T. We have*

$$\mathrm{Var}_{G,p,q}(f) \leq C_{G,p,q} \sum_{e\in E} \delta_e(f,T)\,\mathrm{Cov}_{G,p,q}[f,\omega_e], \tag{2.13}$$

where $\delta_e(f,T) := \mathbb{P}_p\big[\exists t \leq \tau(\omega) \,:\, e_t = e\big]$ is the revealment (of f) for the decision tree T, and $c_{G,p,q}$ is defined by

$$C_{G,p,q} := \frac{1}{\inf_{e\in E} \mathrm{Var}_{G,p,q}(\omega_e)}.$$

Before proving this statement, let us remark that it implies the theorem in the same way as in Bernoulli percolation.

Proof (Theorem 9) Set $\theta_n(p) = \phi^1_{\Lambda_{2n},p,q}[0 \leftrightarrow \partial\Lambda_n]$ and $S_n = \sum_{k=0}^{n-1} \theta_k$. Following the same reasoning as in Lemma 5, we find

$$\sum_{e\in E_{2n}} \mathrm{Cov}_{\Lambda_{2n},p,q}(\mathbb{1}_{0\leftrightarrow\partial\Lambda_n},\omega_e) \geq \mathrm{Var}_{\Lambda_{2n},p,q}(\omega_e) \frac{n\,\theta_n(1-\theta_n)}{2C_{\Lambda_{2n},p,q} \max_{x\in\Lambda_n} \sum_{k=0}^{n-1} \phi^1_{\Lambda_{2n},p,q}[x \leftrightarrow \partial\Lambda_k(x)]},$$

where $\Lambda_k(x)$ is the box of size k around x. Since $\Lambda_{2k}(x) \subseteq \Lambda_{2n}$ for any $x \in \Lambda_n$ and $2k \leq n$, we deduce

$$\sum_{k=0}^{n-1} \phi^1_{\Lambda_{2n},p,q}[x \leftrightarrow \partial\Lambda_k(x)] \leq 2 \sum_{k=0}^{(n-1)/2} \phi^1_{\Lambda_{2n},p,q}[x \leftrightarrow \partial\Lambda_k(x)]$$
$$\overset{(CBC)}{\leq} 2 \sum_{k=0}^{(n-1)/2} \theta_k(p) \leq 2S_n(p).$$

Overall, we find

$$\sum_{e\in E_{2n}} \mathrm{Cov}_{\Lambda_{2n},p,q}(\mathbb{1}_{0\leftrightarrow\partial\Lambda_n}, \omega_e) \geq \mathrm{Var}_{\Lambda_{2n},p,q}(\omega_e)\frac{n}{4C_{\Lambda_{2n},p,q} \cdot S_n} \cdot \theta_n(1-\theta_n).$$

Now, (DF) trivially extends to random-cluster models with $q > 0$ so that

$$\frac{\mathrm{d}}{\mathrm{d}p}\phi^1_{\Lambda_{2n},p,q}[0 \leftrightarrow \partial\Lambda_n] = \tfrac{1}{p(1-p)} \sum_{e\in E_{2n}} \mathrm{Cov}_{\Lambda_{2n},p,q}(\mathbb{1}_{0\leftrightarrow\partial\Lambda_n}, \omega_e).$$

We deduce that for $p \in [p_0, p_1]$,

$$\theta'_n \geq c\frac{n}{S_n}\theta_n,$$

where

$$c := \tfrac{1}{C}\phi^1_{\Lambda_{2n},p_0,q}(\omega_e)(1 - \phi^1_{\Lambda_{2n},p_1,p}[\omega_e])(1 - \theta_1(p_1)) > 0$$

(the constant C is the maximum over $p \in [p_0, p_1]$ and n of $C_{\Lambda_{2n},p,q}$).

To conclude, observe that measurability and the comparison between boundary conditions imply that

$$\liminf \theta_n \geq \liminf \phi^1_{p,q}[0 \longleftrightarrow \partial\Lambda_n] = \phi^1_{p,q}[0 \longleftrightarrow \infty]$$

and that for any $k \geq 1$,

$$\limsup \theta_n \leq \limsup \phi^1_{\Lambda_{2n},p,q}[0 \longleftrightarrow \partial\Lambda_k] = \phi^1_{p,q}[0 \longleftrightarrow \partial\Lambda_k].$$

Letting k tend to infinity implies that θ_n tends to $\phi^1_{p,q}[0 \leftrightarrow \infty]$. We are therefore in a position to apply Lemma 4, which implies the first item of Theorem 9 and the fact that for $p < p_c$, there exists $c_p > 0$ such that for any $n \geq 0$, $\theta_n(p) \leq \exp(-c_p n)$. It remains to observe that

$$\phi^1_{\Lambda_{2n},p,q}[0 \longleftrightarrow \partial\Lambda_{2n}] \le \phi^1_{\Lambda_{2n},p,q}[0 \longleftrightarrow \partial\Lambda_n] = \theta_n(p)$$

to obtain the second item of the theorem.[7] □

We now turn to the Proof of Theorem 10. The strategy is a combination of the original proof of the OSSS inequality for product measures (which is a Efron-Stein type reasoning), together with an encoding of random-cluster measures in terms of iid random variables.

We start by a useful lemma explaining how to construct ω with a certain law μ on $\{0, 1\}^E$ from iid uniform random variables. Recall the notation $\mathbf{E}$ and $e_{[t]}$. For $u \in [0, 1]^n$ and $e \in \mathbf{E}$, define $F^\mu_e(u) = x$ inductively for $1 \le t \le n$ by

$$x_{e_t} := \begin{cases} 1 & \text{if } u_t \ge \mu[\omega_{e_t} = 0 \mid \omega_{e_{[t-1]}} = x_{e_{[t-1]}}], \\ 0 & \text{otherwise.} \end{cases} \tag{2.14}$$

Lemma 6 *Let* $\mathbf{U}$ *be an iid sequence of uniform* $[0, 1]$ *random variables, and* $\mathbf{e}$ *a random variable taking values in* $\mathbf{E}$*. Assume that for every* $1 \le t \le n$*,* $\mathbf{U}_t$ *is independent of* $(\mathbf{e}_1, \dots, \mathbf{e}_t)$*, then* $\mathbf{X} = F^\mu_{\mathbf{e}}(\mathbf{U})$ *has law* μ*.*

Proof Let $x \in \{0, 1\}^E$ and $e \in \mathbf{E}$ such that $\mathbb{P}[\mathbf{X} = x, \mathbf{e} = e] > 0$. The probability $\mathbb{P}[\mathbf{X} = x, \mathbf{e} = e]$ can be written as

$$\prod_{t=1}^n \mathbb{P}[\mathbf{X}_{e_t} = x_{e_t} \mid \mathbf{e}_{[t]} = e_{[t]}, \mathbf{X}_{e_{[t-1]}} = x_{e_{[t-1]}}]$$
$$\times \prod_{t=1}^n \mathbb{P}[\mathbf{e}_t = e_t \mid \mathbf{e}_{[t-1]} = e_{[t-1]}, \mathbf{X}_{e_{[t-1]}} = x_{e_{[t-1]}}].$$

(All the conditionings are well defined, since we assumed $\mathbb{P}[\mathbf{X} = x, \mathbf{e} = e] > 0$.) Since $\mathbf{U}_t$ is independent of $\mathbf{e}_{[t]}$ and $\mathbf{U}_{[t-1]}$ (and thus $\mathbf{X}_{e_{[t-1]}}$), the definition (2.14) gives

$$\mathbb{P}[\mathbf{X}_{e_t} = x_{e_t} \mid \mathbf{e}_{[t]} = e_{[t]}, \mathbf{X}_{e_{[t-1]}} = x_{e_{[t-1]}}] = \mu[\omega_{e_t} = x_{e_t} \mid \omega_{e_{[t-1]}} = x_{e_{[t-1]}}]$$

so that the first product is equal to $\mu[\omega = x]$ independently of e. Fixing $x \in \{0, 1\}^E$, and summing on $e \in \mathbf{E}$ satisfying $\mathbb{P}[\mathbf{X} = x, \mathbf{e} = e] > 0$ gives

$$\mathbb{P}[\mathbf{X} = x] = \sum_e \mathbb{P}[\mathbf{X} = x, \mathbf{e} = e]$$
$$= \mu[\omega = x] \sum_e \prod_{t=1}^n \mathbb{P}[\mathbf{e}_t = e_t | \mathbf{e}_{[t-1]} = e_{[t-1]}, \mathbf{X}_{e_{[t-1]}} = x_{e_{[t-1]}}] = \mu[\omega = x].$$

□

[7] Formally, we only obtained the result for n even, but the result for n odd can be obtained similarly.

Proof (*Theorem* 8) Consider two independent sequences of iid uniform [0, 1] random variables $\mathbf{U}$ and $\mathbf{V}$. Write $\mathbb{P}$ for the coupling between these variables (and $\mathbb{E}$ for its expectation). Construct $(\mathbf{e}, \mathbf{X}, \tau)$ inductively as follows: set $\mathbf{e}_1 = e_1$, and for $t \geq 1$,

$$\mathbf{X}_{\mathbf{e}_t} = \begin{cases} 1 & \text{if } \mathbf{U}_t \geq \phi^1_{G,p,q}[\omega_{\mathbf{e}_t} = 0 \,|\, \omega_{\mathbf{e}_{[t-1]}} = \mathbf{X}_{\mathbf{e}_{[t-1]}}] \\ 0 & \text{otherwise} \end{cases} \quad \text{and} \quad \mathbf{e}_{t+1} := \psi_{t+1}(\mathbf{e}_{[t]}, \mathbf{X}_{\mathbf{e}_{[t]}}),$$

and $\tau := \min\big\{t \geq 1 : \forall x \in \{0,1\}^E, x_{\mathbf{e}_{[t]}} = \mathbf{X}_{\mathbf{e}_{[t]}} \Rightarrow f(x) = f(\mathbf{X})\big\}$. Finally, for $0 \leq t \leq n$, define $\mathbf{Y}^t := F_{\mathbf{e}}(\mathbf{W}^t)$, where

$$\mathbf{W}^t := \mathbf{W}^t(\mathbf{U}, \mathbf{V}) = (\mathbf{V}_1, \dots, \mathbf{V}_t, \mathbf{U}_{t+1}, \dots, \mathbf{U}_\tau, \mathbf{V}_{\tau+1}, \dots, \mathbf{V}_n)$$

(in particular $\mathbf{W}^t$ is equal to $\mathbf{V}$ if $t \geq \tau$).

Lemma 6 applied to $(\mathbf{U}, \mathbf{e})$ gives that $\mathbf{X}$ has law μ and is $\mathbf{U}$-measurable. Lemma 6 applied to $(\mathbf{V}, \mathbf{e})$ implies that $\mathbf{Y}^n$ has law μ and is independent of $\mathbf{U}$. Therefore,

$$\phi^1_{G,p,q}[|f - \phi^1_{G,p,q}(f)|] \leq \mathbb{E}\big[\,\big|\,\mathbb{E}[f(\mathbf{X})|\mathbf{U}] - \mathbb{E}[f(\mathbf{Y}^n)|\mathbf{U}]\,\big|\,\big] \leq \mathbb{E}\big[|f(\mathbf{X}) - f(\mathbf{Y}^n)|\big].$$

Exactly as for iid random variables, $f(\mathbf{X}) = f(\mathbf{Y}^0)$. Following the same lines as in the iid case, we obtain (recall that f takes values in [0, 1])

$$\mathrm{Var}_{G,p,q}(f) \leq \sum_{e \in E} \sum_{t=1}^n \mathbb{E}\Big[\mathbb{E}\big[|f(\mathbf{Y}^t) - f(\mathbf{Y}^{t-1})| \,\big|\, \mathbf{U}_{[t-1]}\big]\, \mathbb{1}_{t \leq \tau, \mathbf{e}_t = e}\Big]$$

so that the proof of the theorem follows from the fact that on $\{t \leq \tau, \mathbf{e}_t = e\}$,

$$\mathbb{E}\big[\,|f(\mathbf{Y}^t) - f(\mathbf{Y}^{t-1})| \,\big|\, \mathbf{U}_{[t-1]}\big] \;\leq\; \tfrac{1}{\mathrm{Var}_{G,p,q}(\omega_e)} \mathrm{Cov}_{G,p,q}(f, \omega_e). \tag{2.15}$$

Note that $F^\mu_e(u)$ is both increasing in u and in μ (for stochastic domination). We deduce that both $\mathbf{Y}^{t-1}$ and $\mathbf{Y}^t$ are sandwiched between

$$\mathbf{Z} := F_{\mathbf{e}}^{\phi^1_{G,p,q}[\cdot|\omega_e=0]}(\mathbf{W}^{t-1}) = F_{\mathbf{e}}^{\phi^1_{G,p,q}[\cdot|\omega_e=0]}(\mathbf{W}^t)$$

and

$$\mathbf{Z}' := F_{\mathbf{e}}^{\phi^1_{G,p,q}[\cdot|\omega_e=1]}(\mathbf{W}^{t-1}) = F_{\mathbf{e}}^{\phi^1_{G,p,q}[\cdot|\omega_e=1]}(\mathbf{W}^t).$$

Since $\mathbf{W}^t$ is independent of $\mathbf{U}_{[t-1]}$, Lemma 6 and the fact that f is increasing give us

$$\begin{aligned}\mathbb{E}\big[\,|f(\mathbf{Y}^t) - f(\mathbf{Y}^{t-1})| \,\big|\, \mathbf{U}_{[t-1]}\big] &\leq \mathbb{E}[f(\mathbf{Z}')] - \mathbb{E}[f(\mathbf{Z})] \\ &= \phi^1_{G,p,q}[f(\omega)|\omega_e = 1] - \phi^1_{G,p,q}[f(\omega)|\omega_e = 0] \\ &= \frac{\mathrm{Cov}_{G,p,q}(f, \omega_e)}{\phi^1_{G,p,q}[\omega_e](1 - \phi^1_{G,p,q}[\omega_e])}.\end{aligned}$$

□

2.4 Computation of the Critical Point for Random-Cluster Models on $\mathbb{Z}^2$

The goal of this section is to explain how one can compute the critical point of the random-cluster model on $\mathbb{Z}^2$ using Theorem 9. As mentioned in the end of Sect. 2.2, the following theorem was first proved using a sharp threshold theorem, and we refer to [11, 50, 52] for alternative proofs.

Theorem 11 (Beffara and Duminil-Copin [11]) *For the random-cluster model on $\mathbb{Z}^2$ with cluster-weight $q \geq 1$,*

$$p_c = \frac{\sqrt{q}}{1+\sqrt{q}}.$$

Also, for $p < p_c$, there exists $c_p > 0$ such that $\phi^1_{p,q}[0 \leftrightarrow \partial\Lambda_n] \leq \exp(-c_p n)$ for all $n \geq 0$.

This theorem has the following corollary.

Corollary 6 (Beffara and Duminil-Copin [11]) *The critical inverse-temperature of the Potts model on $\mathbb{Z}^2$ satisfies*

$$\beta_c(q) := \tfrac{q-1}{q} \log(1+\sqrt{q}).$$

We start by discussing duality for random-cluster models. The boundary conditions on a finite subgraph $G = (V, E)$ of $\mathbb{Z}^2$ are called *planar* if they are induced by some configuration $\xi \in \{0, 1\}^{\mathbb{E}\setminus E}$. For any planar boundary conditions ξ, one can associate dual boundary conditions ξ^* on G^* induced by the configuration $\xi^*_{e^*} = 1 - \xi_e$ for any $e \notin E$.

As an example, the free boundary conditions correspond to $\xi_e = 0$ for all $e \in \mathbb{E}\setminus E$. Similarly, when G is connected and has connected complement, the wired boundary conditions correspond to $\xi_e = 1$ for all $e \in \mathbb{E}\setminus E$. (This explains the notations 0 and 1 for the free and wired boundary conditions.) In this case, the dual of wired boundary conditions is the free ones, and vice-versa.

A typical example of non-planar boundary conditions is given by "periodic" boundary conditions on Λ_n, for which (k, n) and $(k, -n)$ (resp. (n, k) and $(-n, k)$) are paired for every $k \in [\![-n, n]\!]$. Another (slightly less interesting) example is given by the wired boundary conditions when G has non-connected complement in $\mathbb{Z}^2$.

Proposition 5 (Duality). *Consider a finite graph G and planar boundary conditions ξ. If ω has law $\phi^\xi_{G,p,q}$, then ω^* has law $\phi^{\xi^*}_{G^*,p^*,q}$, where p^* is the solution of*

$$\frac{pp^*}{(1-p)(1-p^*)} = q.$$

There is a specific value of p for which $p = p^*$. This value will be denoted $p_{\rm sd}$, and satisfies

$$p_{\rm sd}(q) = \frac{\sqrt{q}}{1+\sqrt{q}}.$$

Proof Let us start with G connected with connected complement, and free boundary conditions. Let v, e, f and c be the number of vertices, edges, faces and clusters of the graph $(\omega^*)^1$ embedded in the plane.[8] We wish to interpret Euler's formula in terms of $k(\omega)$, $o(\omega^*)$ and $k((\omega^*)^1)$. First, v is a constant not depending on ω and e is equal to $o(\omega^*)$. Also, the bounded faces of the graph are in direct correspondence with the clusters of ω, and therefore $f = k(\omega) + 1$ (note that there is exactly one unbounded face). Overall, Euler's formula ($f = c + e - v + 1$) gives

$$k(\omega) = k((\omega^*)^1) + o(\omega^*) - v.$$

Set $Z := Z^0_{G,p,q}$ and recall that $\frac{q(1-p)}{p} = \frac{p^*}{1-p^*}$. Since $c(\omega) = o(\omega^*)$, we get

$$\begin{aligned}\phi^0_{G,p,q}[\omega] &= \tfrac{p^{|E|}}{Z}\big(\tfrac{1-p}{p}\big)^{c(\omega)} q^{k(\omega)}\\ &= \tfrac{p^{|E|}q^{-v}}{Z}\big(\tfrac{1-p}{p}\big)^{o(\omega^*)} q^{k((\omega^*)^1)+o(\omega^*)}\\ &= \tfrac{p^{|E|}q^{-v}}{Z}\big(\tfrac{p^*}{1-p^*}\big)^{o(\omega^*)} q^{k((\omega^*)^1)}\\ &= \phi^1_{G^*,p^*,q}[\omega^*].\end{aligned}$$

(Note that we also proved that $Z^1_{G^*,p^*,q} = Z^0_{G,p,q} q^v p^{-|E|}(1-p^*)^{|E|}$.)

For arbitrary planar boundary conditions, the proof follows from the domain Markov property. Indeed, pick n large enough so that there exists $\psi \in \{0,1\}^{E_n \setminus E}$ inducing the boundary conditions ξ (such an n always exists), and introduce

$$\omega^\psi_e = \begin{cases} \omega_e & \text{if } e \in E,\\ \psi_e & \text{if } e \in E_n \setminus E,\end{cases}$$

Since the boundary conditions induced by ψ^* on $E_n^* \setminus E^*$ with the boundary of Λ_n wired are exactly ξ^*, we deduce that

$$\phi^\xi_{G,p,q}[\omega] \overset{(DMP)}{=} c\,\phi^0_{\Lambda_n,p,q}[\omega^\psi] = c\,\phi^1_{\Lambda_n^*,p^*,q}[(\omega^\psi)^*] \overset{(DMP)}{=} \phi^{\xi^*}_{G^*,p^*,q}[\omega^*],$$

where c is a constant not depending on ω. This concludes the proof. □

[8] Recall that $(\omega^*)^1$ is the graph ω^* where all vertices of ∂G^* are identified together. This graph can clearly be embedded in the plane by "moving" the vertices of ∂G^* to a single point chosen in the exterior face of ω, and drawing the edges incident to ∂G^* by "extending" the corresponding edges of ω^* by continuous curves not intersecting each other or edges of ω, and going to this chosen point.

Exercise 29 (*Duality for the random-cluster model on the torus*) Let $\mathbb{T}_n = [0,n]^2$ and consider the boundary conditions where $(0,k)$ and (n,k) are identified for any $0 \le k \le n$, and $(j,0)$ and (j,n) are identified for any $0 \le j \le n$. We write the measure $\phi_{\mathbb{T}_n,p,q}$.

1. A configuration ω is said to have a *net* if ω^* does not contain any non-retractable loop. Let $t(\omega)$ be equal to 0 if ω^* has a net, 2 if ω has a net, and 1 otherwise. Prove that

$$|V| + f(\omega) + t(\omega) = k(\omega) + o(\omega) + 1,$$

where $f(\omega)$ is the number of faces in the configuration.
2. Show that $\widetilde{\phi}_{\mathbb{T}_n,p,q}(\omega) = \widetilde{\phi}_{\mathbb{T}_n^*,p^*,q}(\omega^*)$, where

$$\widetilde{\phi}_{\mathbb{T}_n,p,q}(\omega) = \sqrt{q}^{t(\omega)} \cdot \frac{p^{o(\omega)}(1-p)^{c(\omega)}q^{k(\omega)}}{\widetilde{Z}^{per}_{\mathbb{T}_n,p,q}}.$$

3. Deduce that the probability of $\mathcal{H}_n$ is exactly $1/2$ for the measure $\widetilde{\phi}_{\mathbb{T}_n,p_{sd},q}$.

We are now in a position to prove Theorem 11.

Proof (*Theorem* 11) The previous duality relation enables us to generalize the duality argument for crossing events. Indeed, considering the limit (as $G \nearrow \mathbb{Z}^2$) of the duality relation between wired and free boundary conditions, we get that the dual measure of $\phi^1_{p,q}$ is $\phi^0_{p^*,q}$. Recall that $\mathcal{H}_n$ is the event that the rectangle of size $n+1$ times n is crossed horizontally. For $q \ge 1$, the self-duality at p_{sd} implies that

$$\phi^1_{p_{sd},q}[\mathcal{H}_n] + \phi^0_{p_{sd},q}[\mathcal{H}_n] = 1.$$

The comparison between boundary conditions thus implies

$$\phi^1_{p_{sd},q}[\mathcal{H}_n] \ge \tfrac{1}{2} \ge \phi^0_{p_{sd},q}[\mathcal{H}_n]. \tag{2.16}$$

Note that the $\phi^1_{p_{sd},q}[\mathcal{H}_n]$ is no longer equal to $1/2$. Indeed, the complement event is still a rotated version of $\mathcal{H}_n$, but the law of ω^* is not the same as the one of ω, since the boundary conditions are free instead of wired.

We are ready to conclude. The fact that $\phi^1_{p_{sd},q}[\mathcal{H}_n] \ge 1/2$ implies that $\phi^1_{p_{sd},q}[0 \leftrightarrow \partial\Lambda_n] \ge 1/(2n)$ (exactly as for Bernoulli percolation). Since this quantity is not decaying exponentially fast, Theorem 9 gives that $p_c \le p_{sd}$.

Also, if $\phi^0_{p,q}[0 \leftrightarrow \infty] > 0$, then $\lim \phi^0_{p,q}[\mathcal{H}_n] = 1$. Indeed, the measure is ergodic (Lemma 2) and satisfies the almost sure uniqueness of the infinite cluster (Theorem 4). Since it also satisfies the FKG inequality, the proof of Proposition 4 works the same for random-cluster models with $q \ge 1$. Together with (2.16), this implies that $\phi^0_{p_{sd},q}[0 \leftrightarrow \infty] = 0$ and therefore that $p_{sd} \le p_c$. □

Remark 4 Note that we just proved that $\phi^0_{p_c,q}[0 \leftrightarrow \infty] = 0$.

Exercise 30 (*Critical points of the triangular and hexagonal lattices*) Define p such that $p^3 + 1 = 3p$ and set p_c for the critical parameter of the triangular lattice.

1. Consider a graph G and add a vertex x inside the triangle u, v, w. Modify the graph F by removing edges uv, vw and wu, and adding xu, xv and xw. The new graph is denoted G'. Show that the Bernoulli percolation of parameter p on G can be coupled to the Bernoulli percolation of parameter p on G' in such a way that connections between different vertices of G are the same.

2. Using exponential decay in subcritical for the triangular lattice, show that if $p < p_c$, the percolation of parameter $1 - p$ on the hexagonal lattice contains an infinite cluster almost surely. Using the transformation above, reach a contradiction.
3. Prove similarly that $p \leq p_c(\mathbb{T})$.
4. Find a degree three polynomial equation for the critical parameter of the hexagonal lattice.
5. What happens for the random-cluster model?

3 Where Are We Standing? And a Nice Conjecture...

Up to now, we proved that the critical inverse-temperature of the Potts model exists, and that it corresponds to the point where long-range ordering emerges. We also proved monotonicity of correlations. In the specific case of $\mathbb{Z}^2$, we computed the critical point exactly. Last but not least, we proved that correlations decay exponentially fast when $\beta < \beta_c$. Overall, we gathered a pretty good understanding of the off-critical phase, but we have little information on the critical one. In particular, we would like to determine whether the phase transition of Potts models is continuous or not. In terms of random-cluster model, it corresponds to deciding whether $\phi^1_{p_c,q}[0 \leftrightarrow \infty]$ is equal to 0 or not.

We proved in the previous section that for critical Bernoulli percolation on $\mathbb{Z}^2$, there was no infinite cluster almost surely. For $q > 1$, we only managed to prove this result for the free boundary conditions. This is therefore not sufficient to discriminate between a continuous and a discontinuous phase transition for planar Potts models.

Before focusing on this question in the next sections, let us briefly mention that even for Bernoulli percolation, knowing whether there exists an infinite cluster at criticality is a very difficult question in general. For $\mathbb{Z}^d$ with $d \geq 3$, the absence of an infinite cluster at criticality was proved using lace expansion for $d \geq 19$ [80] (it was recently improved to $d \geq 11$ [64]). The technique involved in the proof is expected to work until $d \geq 6$. For $d \in \{3, 4, 5\}$, the strategy will not work and the following conjecture remains one of the major open questions in our field.

Conjecture 1 *For any* $d \geq 2$, $\mathbb{P}_{p_c}[0 \longleftrightarrow \infty] = 0$.

Some partial results were obtained in $\mathbb{Z}^3$ in the past decades. For instance, it is known that the probability, at p_c of an infinite cluster in $\mathbb{N} \times \mathbb{Z}^2$ is zero [7]. Let us also mention that $\mathbb{P}_{p_c(\mathbb{Z}^2 \times G)}[0 \longleftrightarrow \infty]$ was proved to be equal to 0 on graphs of the form $\mathbb{Z}^2 \times G$, where G is finite; see [56], and on graphs with exponential growth in [14, 84] (see also the following exercise).

This exercise presents the beautiful proof due to Tom Hutchcroft of absence of percolation at criticality for amenable locally-finite transitive graphs with exponential growth. We say that $\mathbb{G}$ has exponential growth if there exists $c_{\text{vg}} > 0$ such that $|\Lambda_n| \geq \exp(c_{\text{vg}} n)$.

Exercise 31 ($\mathbb{P}_{p_c}[0 \leftrightarrow \infty] = 0$ *for amenable Cayley graphs with exponential growth*) Let $\mathbb{G}$ be an amenable infinite locally-finite transitive graphs with exponential growth.

1. Use amenability to prove that $\mathbb{P}_{p_c}[0 \leftrightarrow \infty] > 0 \Longrightarrow \inf\{\mathbb{P}_{p_c}[x \leftrightarrow y], x, y \in \mathbb{G}\} > 0$. *Hint*: use Exercise 21.
2. Use the FKG inequality to prove that $u_n(p) = \inf\{\mathbb{P}_p[x \leftrightarrow 0], x \in \partial\Lambda_n\}$ satisfies that for every n and m,

$$u_{n+m}(p) \geq u_n(p)u_m(p).$$

3. Adapt Step 1 of the Proof of Theorem 7 (see also Question 4 of Exercise 26) to get that for any $p < p_c$, $\sum_{x \in \mathbb{G}} \mathbb{P}_p[0 \longleftrightarrow x] < \infty$.
4. Use the two previous questions to deduce that for any $p < p_c$, $u_n(p) \leq \exp(-c_{\text{vg}} n)$ for every $n \geq 1$.
5. Conclude.

4 Continuity of the Phase Transition for the Ising Model

Many aspects of the Ising model are simpler to treat than in other models (including Bernoulli percolation). We therefore focus on this model first. We will prove that the phase transition of the model is always continuous for the nearest neighbor model on $\mathbb{Z}^d$ with $d \geq 2$. Before proceeding further, let us mention that the Ising model does not always undergo a continuous phase transition: the long-range model on $\mathbb{Z}$ with coupling constants $J_{x,y} = 1/|x - y|^2$ undergoes a discontinuous phase transition (we refer to [3] for details).

The section is organized as follows. We start by providing a simple argument proving that in two dimensions, the phase transition is continuous. We then introduce a new object, called the random current representation, and study its basic properties. Finally, we use the properties of this model to prove that the phase transition is continuous in dimension $d \geq 3$.

4.1 An Elementary Argument in Dimension d = 2

We present a very elegant argument, due to Wendelin Werner, of the following.

Proposition 6 *On* $\mathbb{Z}^2$, $\mu^+_{\beta_c}[\sigma_0] = 0$.

Proof The crucial observation is the following: the measure $\mu^{\text{f}}_{\beta_c}$ is mixing, and therefore ergodic. Indeed, recall that $\sigma \sim \mu^{\text{f}}_{\beta_c}$ can be obtained from a percolation configuration $\omega \sim \phi^0_{p_c,2}$ by coloring independently different clusters. The absence of an infinite cluster for $\phi^0_{p_c,2}$ (Remark 4) enables us to deduce the mixing property of $\mu^{\text{f}}_{\beta_c}$ from the one of $\phi^0_{p_c,2}$ (see Exercise 32).

The Burton-Keane argument implies that when existing, the infinite cluster of minuses is unique. Consider the event $\widetilde{\mathcal{H}}_n$ that there exists a path of minuses in $[0,n]^2$ crossing from left to right. The complement of this event contains the event that there exists a path of pluses in $[0,n]^2$ crossing from top to bottom. We deduce that $\mu^{\rm f}_{\beta_c}[\widetilde{\mathcal{H}}_n] \le \frac{1}{2}$ for every $n \ge 1$. The proof of Proposition 4 works the same here and we deduce that the probability that there is an infinite cluster of minuses is zero, since otherwise the probability of $\widetilde{\mathcal{H}}_n$ would tend to 1.

We now prove that $\mu^+_{\beta_c}[\sigma_0]$ is smaller than or equal to 0, which immediately implies that it is equal to zero since we already know that it is larger than or equal to 0. Consider the set C of $x \in \Lambda_n$ which are not connected to $\partial\Lambda_n$ by a path of minuses. Conditionally on $\{\mathsf{C} = C\}$, the law of the configuration in C is equal to μ^+_{C,β_c} since $\{\mathsf{C} = C\}$ is measurable in terms of spins outside C or on ∂C, and that spins on ∂C are all pluses (we use the Gibbs property for lattice models, which is obtained similarly to the domain Markov property for random-cluster models). Also note that

$$\mu^+_{C,\beta_c}[\sigma_0] = \phi^1_{C,p_c,2}[0 \longleftrightarrow \partial C] \ge \phi^1_{p_c,2}[0 \longleftrightarrow \infty] = \mu^+_{\beta_c}[\sigma_0].$$

Note that if $0 \notin \mathsf{C}$, then $\sigma_0 = -1$. We deduce that

$$\begin{aligned} 0 = \mu^{\rm f}_{\beta_c}[\sigma_0] &= \mu^{\rm f}_{\beta_c}[\sigma_0 \mathbb{1}_{0\notin\mathsf{C}}] + \sum_{0\in C\subseteq\Lambda_n} \mu^+_{C,\beta_c}[\sigma_0]\mu^{\rm f}_{\beta_c}[\mathsf{C} = C] \\ &\ge -\mu^{\rm f}_{\beta_c}[0 \notin \mathsf{C}] + \mu^+_{\beta_c}[\sigma_0] \sum_{0\in C\subseteq\Lambda_n} \mu^{\rm f}_{\beta_c}[\mathsf{C} = C] \\ &= -\mu^{\rm f}_{\beta_c}[0 \notin \mathsf{C}] + \mu^+_{\beta_c}[\sigma_0]\mu^{\rm f}_{\beta_c}[0 \in \mathsf{C}]. \end{aligned}$$

Letting n tend to infinity and using that $\mu^{\rm f}_{\beta_c}[0 \notin \mathsf{C}_n]$ tends to zero (since there is no infinite cluster of minuses) gives the result. □

Exercise 32 Prove the mixing property of $\mu^{\rm f}_{\beta_c}$.

Exercise 33 Prove that the Ising model satisfies (CBC) and (FKG) for the natural order on $\{\pm 1\}^V$.

4.2 *High-Temperature Expansion, Random Current Representation and Percolation Interpretation of Truncated Correlations*

For many reasons, the Ising model is special among Potts models. One of these reasons is the $+/-$ gauge symmetry: flipping all the spins leaves the measure invariant (for free boundary conditions). We will harvest this special feature in the following.

The *high temperature expansion* of the Ising model is a graphical representation introduced by van der Waerden [126]. It relies on the following identity based on the fact that $\sigma_x\sigma_y \in \{-1, +1\}$:

$$e^{\beta\sigma_x\sigma_y} = \cosh(\beta) + \sigma_x\sigma_y \sinh(\beta) = \cosh(\beta)\left[1 + \tanh(\beta)\sigma_x\sigma_y\right]. \qquad (4.1)$$

For a finite graph G, the notation η will always refer to a percolation configuration in $\{0, 1\}^E$ (we will still use the notation $o(\eta)$ for the number of edges in η). We prefer the notation η instead of ω to highlight the fact that η will have *source constraints*, i.e. that the parity of its degree at every vertex will be fixed. More precisely, write $\partial\eta$ for the set of vertices of η with odd degree. Note that $\partial\eta = \emptyset$ is equivalent to saying that η is an even subgraph of G, i.e. that the degree at each vertex is even.

For $A \subseteq V$, set

$$\sigma_A := \prod_{x\in A} \sigma_x.$$

Proposition 7 *Let G be a finite graph, $\beta > 0$, and $A \subseteq V$. We find*

$$\sum_{\sigma\in\{\pm1\}^V} \sigma_A \exp[-\beta H_G^{\mathrm{f}}(\sigma)] = 2^{|V|}\cosh(\beta)^{|E|} \sum_{\partial\eta=A} \tanh(\beta)^{o(\eta)}. \qquad (4.2)$$

Proof Using (4.1) for every $xy \in E$ gives

$$\begin{aligned}\sum_{\sigma\in\{\pm1\}^V} \sigma_A \exp[-\beta H_G^{\mathrm{f}}(\sigma)] &= \sum_{\sigma\in\{\pm1\}^V} \sigma_A \prod_{xy\in E} e^{\beta\sigma_x\sigma_y}\\ &= \cosh(\beta)^{|E|} \sum_{\sigma\in\{\pm1\}^V} \sigma_A \prod_{xy\in E}\left[1 + \tanh(\beta)\sigma_x\sigma_y\right]\\ &= \cosh(\beta)^{|E|} \sum_{\sigma\in\{\pm1\}^V}\sum_{\eta\in\{0,1\}^E} \tanh(\beta)^{o(\eta)}\sigma_A \prod_{\substack{xy\in E\\ \eta_{xy}=1}} \sigma_x\sigma_y\\ &= \cosh(\beta)^{|E|} \sum_{\eta\in\{0,1\}^E} \tanh(\beta)^{o(\eta)} \sum_{\sigma\in\{\pm1\}^V} \sigma_A \prod_{\substack{xy\in E\\ \eta_{xy}=1}} \sigma_x\sigma_y.\end{aligned}$$

Using the involution of $\{0, 1\}^V$ sending σ to the configuration coinciding with σ except at x where the spin is flipped, one sees that if any of the terms σ_x appears with an odd power in the previous sum over σ, then the sum equals 0. Since the power corresponds to the degree of x in η if $x \notin A$, and is equal to the degree minus 1 if $x \in A$, we deduce that

$$\sum_{\sigma\in\{\pm1\}^V} \sigma_A \prod_{\substack{xy\in E\\ \eta_{xy}=1}} \sigma_x\sigma_y = \begin{cases} 2^{|V|} & \text{if } \partial\eta = A,\\ 0 & \text{otherwise}\end{cases}$$

and the formula therefore follows. □

Exercise 34 (*Kramers-Wannier duality*) 1. Show that there exists a correspondence between even subgraphs of G and spin configurations for the Ising model on G^*, with $+$ boundary conditions on the exterior face.
2. Express the partition function of the Ising model at inverse-temperature β^* on G^* with $+$ boundary conditions in terms of even subgraphs of G.
3. For which value of β^* do we obtain the same expression (up to a multiplicative constant) as (4.2).

The previous expansion of the partition function is called the *high-temperature expansion*. We deduce

$$\mu^{\mathrm{f}}_{G,\beta}[\sigma_A] = \frac{\sum_{\partial\eta=A} \tanh(\beta)^{o(\eta)}}{\sum_{\partial\eta=\emptyset} \tanh(\beta)^{o(\eta)}} \geq 0. \tag{4.3}$$

(The inequality is called Griffiths's first inequality). Notice two things about the high-temperature expansion of spin-spin correlations:

- the sums in the numerator and denominator of (4.3) are running on different types of graphs (the source constraints are not the same), which illustrates a failure of this representation: we cannot a priori rewrite this quantity as a probability.
- when squaring this expression, we end up considering, in the numerator and denominator, two sums over pairs of configurations η_1 and η_2 with $\partial\eta_1 = \partial\eta_2$. This means that $\eta_1 + \eta_2$ has even degree at each vertex, both in the numerator and denominator.

In order to harvest this second observation, we introduce a system of currents. This introduction is only a small detour, since we will quickly get back to percolation configurations.

A *current* $\mathbf{n}$ on G is a function from E to $\mathbb{N} := \{0, 1, 2, ...\}$ (the notation $\mathbf{n}$ will be reserved to currents). A *source* of $\mathbf{n} = (\mathbf{n}_{xy} : xy \in E)$ is a vertex x for which $\sum_{y\sim x} \mathbf{n}_{xy}$ is odd. The set of sources of $\mathbf{n}$ is denoted by $\partial\mathbf{n}$. Also set

$$w_\beta(\mathbf{n}) = \prod_{xy\in E} \frac{\beta^{\mathbf{n}_{xy}}}{\mathbf{n}_{xy}!}.$$

One may follow the Proof of Proposition 7 with the Taylor expansion

$$\exp(\beta\sigma_x\sigma_y) = \sum_{\mathbf{n}_{xy}=0}^{\infty} \frac{(\beta\sigma_x\sigma_y)^{\mathbf{n}_{xy}}}{\mathbf{n}_{xy}!}$$

replacing (4.1) to get

$$\sum_{\sigma\in\{\pm1\}^V} \sigma_A \exp[-\beta H^{\mathrm{f}}_G(\sigma)] = 2^{|V|} \sum_{\partial\mathbf{n}=A} w_\beta(\mathbf{n}), \tag{4.4}$$

(this expression is called the *random current expansion* of the partition function) from which we deduce an expression for correlations which is very close to (4.3)

$$\mu^{\mathrm{f}}_{G,\beta}[\sigma_A] = \frac{\displaystyle\sum_{\partial\mathbf{n}=A} w_\beta(\mathbf{n})}{\displaystyle\sum_{\partial\mathbf{n}=\emptyset} w_\beta(\mathbf{n})}. \tag{4.5}$$

The random current perspective on the Ising model's phase transition is driven by the hope that the onset of long range order coincides with a percolation transition in a system of duplicated currents (this point of view was used first in [1, 75], see also [45] and references therein for a recent account). While we managed to rewrite the spin-spin correlations of the Ising model in terms of the random-cluster model or the high-temperature expansion, the representations fail to apply to truncated correlations.[9] From this point of view, the expression (4.5) is slightly better than (4.3) when considering the product of two spin-spin correlations since weighted sums over two "independent" currents $\mathbf{n}_1$ and $\mathbf{n}_2$ can be rewritten in terms of the sum over a single current $\mathbf{m}$ (see below). This seemingly tiny difference enables us to switch the sources from one current to another one and to recover a probabilistic interpretation in terms of a percolation model.

More precisely, recall that $\mathcal{F}_A$ is the event that every cluster of the percolation configuration is intersecting A an even number of times.[10] We will prove below that for any $A \subseteq V$,

$$\mu^{\mathrm{f}}_{G,\beta}[\sigma_A]^2 = \mathbf{P}^{\emptyset}_{G,\beta}[\mathcal{F}_A], \tag{4.6}$$

where $\mathbf{P}^{\emptyset}_{G,\beta}$ is a percolation model defined as follows (we define a slightly more general percolation model which will be used later). For $B \subseteq V$,

$$\mathbf{P}^{B}_{G,\beta}[\omega] = \frac{\displaystyle\sum_{\substack{\partial\mathbf{n}_1=B\\ \partial\mathbf{n}_2=\emptyset}} w_\beta(\mathbf{n}_1)w_\beta(\mathbf{n}_2)\mathbb{1}_{\widehat{\mathbf{n}_1+\mathbf{n}_2}=\omega}}{\displaystyle\sum_{\substack{\partial\mathbf{n}_1=B\\ \partial\mathbf{n}_2=\emptyset}} w_\beta(\mathbf{n}_1)w_\beta(\mathbf{n}_2)} \tag{4.7}$$

for any $\omega \in \{0,1\}^E$, where to each current $\mathbf{n}$, we associate a percolation configuration $\widehat{\mathbf{n}}$ on E by setting $\widehat{\mathbf{n}}_{xy} = 1$ if $\mathbf{n}_{xy} > 0$, and 0 otherwise.

This yields an alternative graphical representation for spin-spin correlations, which can be compared to the expression $\mu^{\mathrm{f}}_{G,\beta}[\sigma_A] = \phi^0_{G,p,q}[\mathcal{F}_A]$ obtained using the random-cluster model. It involves the same increasing event $\mathcal{F}_A$ (see Exercise 7), but for a different percolation model, and for the square of spin-spin correlations this time.

---

[9] Truncated correlations is a vague term referring to differences of correlation functions (for instance $\mu^+_\beta[\sigma_x\sigma_y] - \mu^+_\beta[\sigma_x]\mu^+_\beta[\sigma_y]$ or $\mu^+_\beta[\sigma_x\sigma_y] - \mu^{\mathrm{f}}_\beta[\sigma_x\sigma_y]$ or $U_4(x_1, x_2, x_3, x_4)$ defined later in this section).

[10] When $A = \{x, y\}$, the event $\mathcal{F}_A$ is simply the event that x and y are connected to each other.

Let us now prove the following lemma, which leads immediately to (4.6).

Lemma 7 (Switching lemma [1, 75]). *For any $A, B \subseteq V$ and any $F : \mathbb{N}^E \to \mathbb{R}$,*

$$\sum_{\substack{\partial \mathbf{n}_1 = A \\ \partial \mathbf{n}_2 = B}} F(\mathbf{n}_1 + \mathbf{n}_2) w_\beta(\mathbf{n}_1) w_\beta(\mathbf{n}_2) = \sum_{\substack{\partial \mathbf{n}_1 = A\Delta B \\ \partial \mathbf{n}_2 = \emptyset}} F(\mathbf{n}_1 + \mathbf{n}_2) w_\beta(\mathbf{n}_1) w_\beta(\mathbf{n}_2) \mathbb{1}_{\widehat{\mathbf{n}_1+\mathbf{n}_2} \in \mathcal{F}_B}, \tag{switch}$$

where $A\Delta B := (A\backslash B) \cup (B\backslash A)$ is the symmetric difference between the sets A and B.

Before proving this lemma, let us mention a few implications. First, (4.6) follows directly from this lemma since

$$\mu^{\mathrm{f}}_{G,\beta}[\sigma_A]^2 = \frac{\displaystyle\sum_{\substack{\partial \mathbf{n}_1 = A \\ \partial \mathbf{n}_2 = A}} w_\beta(\mathbf{n}_1) w_\beta(\mathbf{n}_2)}{\displaystyle\sum_{\substack{\partial \mathbf{n}_1 = \emptyset \\ \partial \mathbf{n}_2 = \emptyset}} w_\beta(\mathbf{n}_1) w_\beta(\mathbf{n}_2)} \overset{\text{(switch)}}{=} \mathbf{P}^{\emptyset}_{G,\beta}[\mathcal{F}_A],$$

We can go further and try to rewrite more complicated expressions. For instance,

$$\mu^{\mathrm{f}}_{G,\beta}[\sigma_A]\mu^{\mathrm{f}}_{G,\beta}[\sigma_B] = \frac{\displaystyle\sum_{\substack{\partial \mathbf{n}_1 = A \\ \partial \mathbf{n}_2 = B}} w_\beta(\mathbf{n}_1) w_\beta(\mathbf{n}_2)}{\displaystyle\sum_{\substack{\partial \mathbf{n}_1 = \emptyset \\ \partial \mathbf{n}_2 = \emptyset}} w_\beta(\mathbf{n}_1) w_\beta(\mathbf{n}_2)} \overset{(switch)}{=} \mu^{\mathrm{f}}_{G,\beta}[\sigma_A\sigma_B] \cdot \mathbf{P}^{A\Delta B}_{G,\beta}[\mathcal{F}_B].$$

In particular, the fact that the probability on the right is smaller or equal to 1 gives the second Griffiths inequality

$$\mu^{\mathrm{f}}_{G,\beta}[\sigma_A\sigma_B] \geq \mu^{\mathrm{f}}_{G,\beta}[\sigma_A]\mu^{\mathrm{f}}_{G,\beta}[\sigma_B]. \tag{G2}$$

Note that here we have an explicit formula for the difference between the average of $\sigma_A\sigma_B$ and the product of the averages, which was not the case in the proof presented in Exercise 9 (which was using the FKG inequality and the random-cluster model). This will be the main advantage of the previous representation: it will enable us to rewrite truncated correlations in terms of connectivity properties of this new percolation model.

Let us conclude this section by mentioning that we saw two representations of the Ising model partition function in this section: the high-temperature expansion in terms of even subgraphs, and the random current expansion. The coupling between the random-cluster model with $q = 2$ and the Ising model provides us with a third expansion (we leave it to the reader to write it properly). There exist several other representations: the low-temperature expansion, the representation in terms of dimers, the Kac-Ward expansion (see e.g. [33, 39, 101, 102]).

Exercise 35 (*Lebowitz's inequality*) Set $\mu := \mu^{\mathrm{f}}_{G,\beta}$ and σ_i for the spin at a vertex x_i. Define

$$U_4(x_1, x_2, x_3, x_4) = \mu[\sigma_1\sigma_2\sigma_3\sigma_4] - \mu[\sigma_1\sigma_2]\mu[\sigma_3\sigma_4] - \mu[\sigma_1\sigma_3]\mu[\sigma_2\sigma_4] - \mu[\sigma_1\sigma_4]\mu[\sigma_2\sigma_3]$$

for $x_1, x_2, x_3, x_4 \in G$. Using the switching lemma, show that

$$U_4(x_1, x_2, x_3, x_4) = -2\mu[\sigma_1\sigma_2\sigma_3\sigma_4]\mathbf{P}^{\{x_1,x_2,x_3,x_4\}}_{G,\beta}[x_1, x_2, x_3, x_4 \text{ all connected}].$$

Note that in particular $U_4(x_1, x_2, x_3, x_4) \le 0$, which is known as Lebowitz's inequality.

Proof (*the switching lemma*) We make the change of variables $\mathbf{m} = \mathbf{n}_1 + \mathbf{n}_2$ and $\mathbf{n} = \mathbf{n}_1$. Since

$$w_\beta(\mathbf{n})w_\beta(\mathbf{m}-\mathbf{n}) = \prod_{xy\in E} \frac{\beta^{(\mathbf{m}-\mathbf{n})_{xy}}}{(\mathbf{m}-\mathbf{n})_{xy}!}\frac{\beta^{\mathbf{n}_{xy}}}{\mathbf{n}_{xy}!} = w_\beta(\mathbf{m})\binom{\mathbf{m}}{\mathbf{n}},$$

where $\binom{\mathbf{m}}{\mathbf{n}} := \prod_{xy\in E}\binom{\mathbf{m}_{xy}}{\mathbf{n}_{xy}}$, we deduce that

$$\sum_{\substack{\partial\mathbf{n}_1=A\\ \partial\mathbf{n}_2=B}} F(\mathbf{n}_1+\mathbf{n}_2)w_\beta(\mathbf{n}_1)w_\beta(\mathbf{n}_2) = \sum_{\partial\mathbf{m}=A\Delta B} F(\mathbf{m})w_\beta(\mathbf{m}) \sum_{\mathbf{n}\le\mathbf{m},\partial\mathbf{n}=B}\binom{\mathbf{m}}{\mathbf{n}}. \quad (4.8)$$

Now, consider the multigraph $\mathcal{M}$ obtained from $\mathbf{m}$ as follows: the vertex set is V and x and y in V are connected by $\mathbf{m}_{xy}$ edges. Then, $\binom{\mathbf{m}}{\mathbf{n}}$ can be interpreted as the number of subgraphs of $\mathcal{M}$ with exactly $\mathbf{n}_{xy}$ edges between x and y. As a consequence,

$$\sum_{\mathbf{n}\le\mathbf{m},\partial\mathbf{n}=B}\binom{\mathbf{m}}{\mathbf{n}} = |\{\mathcal{N}\subseteq\mathcal{M} : \partial\mathcal{N} = B\}|,$$

where $\partial\mathcal{N} = B$ means that $\mathcal{N}$ has odd degree on vertices of B, and even degree everywhere else.

Note that this number is 0 if $\widehat{\mathbf{m}} \notin \mathcal{F}_B$. Indeed, any subgraph $\mathcal{N}$ with $\partial\mathcal{N} = B$ contains disjoint paths pairing the vertices of B. In particular, any cluster of $\mathcal{M}$ intersecting an element x of B must also intersect the element of B paired to x by $\mathcal{N}$.

On the other hand, if $\widehat{\mathbf{m}} \in \mathcal{F}_B$, then any cluster of $\mathcal{M}$ intersects an even number of vertices in B. We claim that in this case there exists $\mathcal{K} \subseteq \mathcal{M}$ with $\partial\mathcal{K} = B$. The fact that $\widehat{\mathbf{m}} \in \mathcal{F}_B$ clearly implies the existence of a collection of paths in $\mathcal{M}$ pairing the vertices of B.[11] A priori, these paths may self-intersect or intersect each other. We now prove that this is not the case if the collection has minimal total length among all the possible choices for such collections of paths. Assume for instance that there exist an edge $e = xy$ and two paths $\gamma = \gamma_1 \circ xy \circ \gamma_2$ and $\gamma' = \gamma_1' \circ yx \circ \gamma_2'$, where we use the intuitive notation that γ is the concatenation of a path γ_1 going to x, then using the edge e, and a path γ_2 from y to the end, and similarly for γ' (note that we may reverse γ', so that we can assume it first goes through y and then through x). But in this case the paths $\gamma_1 \circ \gamma_2'$ and $\gamma_1' \circ \gamma_2$ also pair the same vertices, and have shorter length. The same argument shows that the paths must be self-avoiding. To conclude, simply set $\mathcal{K}$ to be the graph with edge set composed of edges in the paths constructed above.

The map $\mathcal{N} \mapsto \mathcal{N} \Delta \mathcal{K}$ is a bijection (in fact an involution) mapping subgraphs of $\mathcal{M}$ with $\partial\mathcal{N} = B$ to subgraphs of $\mathcal{M}$ with $\partial\mathcal{N} = \emptyset$. As a consequence, in this case

$$|\{\mathcal{N} \subseteq \mathcal{M} : \partial\mathcal{N} = B\}| = |\{\mathcal{N} \subseteq \mathcal{M} : \partial\mathcal{N} = \emptyset\}|.$$

Overall,

$$\sum_{\mathbf{n} \leq \mathbf{m}, \partial\mathbf{n} = B} \binom{\mathbf{m}}{\mathbf{n}} = \mathbb{1}_{\widehat{\mathbf{m}} \in \mathcal{F}_B} \sum_{\mathbf{n} \leq \mathbf{m}, \partial\mathbf{n} = \emptyset} \binom{\mathbf{m}}{\mathbf{n}}.$$

Inserting this in (4.8) and making back the change of variables $\mathbf{n}_1 = \mathbf{n}$ and $\mathbf{n}_2 = \mathbf{m} - \mathbf{n}_1$ gives the result. □

Exercise 36 How should currents be defined in order to rewrite correlations for the Ising model with Hamiltonian $H_G^{\mathrm{f}}(\sigma) = -\sum_{x,y \in V} J_{x,y} \sigma_x \sigma_y$, where $J_{x,y}$ are coupling constants? What is $w_\beta(\mathbf{n})$ in this context? Is the switching lemma still true?

To conclude this section, note that the currents enter in the definition of $\mathbf{P}_{G,\beta}^B$ only through their sources and their traces (i.e. whether they are positive or 0), so that we could have replaced currents taking values in $\mathbb{N}$ by objects taking values in $\{0, 1, 2\}$ with 0 if the current is 0, 1 if it is odd and 2 if it is positive and even. But also note that they do not rely only on the degree of the percolation configuration at every vertex (or equivalently on sources), so that the high-temperature expansion would not have been enough to define $\mathbf{P}_{G,\beta}^A$. The random current representation is crucial to express truncated correlation functions. This is the end of the detour and we will now try to use the percolation representation coming from random currents to prove continuity of correlations.

[11] Meaning that these paths start and end in B, and each element in B appears exactly once in the set of beginning and ends of these paths.

Exercise 37 In this exercise, we consider three measures on G:

- The first one, denoted $\mathrm{P}^{\emptyset}_{G,\beta}$, is attributing a weight to configurations $\eta \in \{0,1\}^E$ proportional to $\tanh(\beta)^{o(\eta)} \mathbb{1}_{\partial\eta=\emptyset}$ (this measure is sometimes known as the loop $O(1)$ model).
- The second one, denoted by $\mathbb{P}^{\emptyset}_{G,\beta}$, is attributing a weight to configuration $n \in \{0,1\}^E$ proportional to

$$\sum_{\mathbf{n}\in\mathbb{N}^E} w_\beta(\mathbf{n}) \mathbb{1}_{\widehat{\mathbf{n}}=n}.$$

- The last one is given by $\phi^0_{G,\beta}$, where $\beta = -\frac{1}{2}\log(1-p)$.

1. Prove that $n \sim \mathbb{P}^{\emptyset}_{G,\beta}$ is obtained from $\eta \sim \mathrm{P}^{\emptyset}_{G,\beta}$ by opening independently additional edges with parameter $p_1 = 1 - \frac{1}{\cosh\beta}$.
2. Consider a graph ω. How many even subgraphs does it contain? Deduce from this formula that if one picks uniformly at random an even subgraph η from $\omega \sim \phi^{\mathrm{f}}_{G,\beta}$, one obtains a random even subgraph of law $\mathrm{P}^{\emptyset}_{G,\beta}$.
3. Prove that by opening independently additional edges from $\eta \sim \mathrm{P}^{\emptyset}_{G,\beta}$ with probability $p_2 = \tanh(\beta)$, one recovers $\omega \sim \phi^0_{G,\beta}$.
4. What is the procedure to go from $n \sim \mathbb{P}^{\emptyset}_{G,\beta}$ to $\omega \sim \phi^0_{G,\beta}$?
5. Use Kramers-Wannier duality (Exercise 34) to prove that $\mathrm{P}^{\emptyset}_{G,\beta}$ is the law of the interfaces between pluses and minuses in an Ising model with $+$ boundary conditions on G^* with inverse-temperature β^*.

4.3 Continuity of the Phase Transition for Ising Models on $\mathbb{Z}^d$ for $d \geq 3$

In this section, we prove that the phase transition of the Ising model is continuous for any $d \geq 3$. Let us start by saying that, as in the case of $\mathbb{Z}^2$, the critical Ising model with free boundary conditions on $\mathbb{Z}^d$ does not have long-range ordering. This can easily be seen from a classical result, called the *infrared bound*: for any $\beta < \beta_c$,

$$\mu^{\mathrm{f}}_\beta[\sigma_x\sigma_y] \leq \tfrac{C}{\beta} G(x,y), \tag{IR}$$

where $G(x,y)$ is the Green function of simple random walk, or equivalently the spin-spin correlations for the discrete GFF. The proof of this inequality is based on the so-called reflection-positivity (RP) technique introduced by Fröhlich et al. [69]; see e.g. [20] for a review. This technique has many applications in different fields of mathematical physics. We added the constant $C > 0$ factor compared to the standard statement (where $C = 1/2$) since the infrared bound is proved in Fourier space, and involves an averaging over y. One may then use the Messager-Miracle inequality (see Exercise 38 or the original reference [106]) to get a bound for any fixed x and y.

By letting $\beta \nearrow \beta_c$, (IR) implies that

$$\mu^{\mathrm{f}}_{\beta_c}[\sigma_x\sigma_y] \leq \tfrac{C}{\beta_c} G(x,y). \tag{4.9}$$

Here, it is important to understand what we did: we took the limit as $\beta \nearrow \beta_c$ of $\mu^{\mathrm{f}}_{\beta_c}[\sigma_x\sigma_y]$. This left-continuity is not true for $\mu^+_{\beta_c}[\sigma_x\sigma_y]$ since we use here the following exchange of two supremums:

$$\begin{aligned}\mu^{\mathrm{f}}_{\beta_c}[\sigma_x\sigma_y] &= \sup_n \mu^{\mathrm{f}}_{\Lambda_n,\beta_c}[\sigma_x\sigma_y]\\ &= \sup_n \sup_{\beta<\beta_c} \mu^{\mathrm{f}}_{\Lambda_n,\beta}[\sigma_x\sigma_y]\\ &= \sup_{\beta<\beta_c} \sup_n \mu^{\mathrm{f}}_{\Lambda_n,\beta}[\sigma_x\sigma_y] = \sup_{\beta<\beta_c} \mu^{\mathrm{f}}_{\beta}[\sigma_x\sigma_y].\end{aligned}$$

For $\mu^+_\beta[\sigma_x\sigma_y]$, one of the supremums would be an infimum and the previous argument would not hold. In fact, $\beta \mapsto \mu^+_\beta[\sigma_x\sigma_y]$ is right continuous (since then it involves only infimums). Similarly, $p \mapsto \phi^0_{p,q}[A]$ and $p \mapsto \phi^1_{p,q}[A]$ are respectively left and right continuous for increasing events depending on finitely many edges (and therefore for any event depending on finitely many edges). Also, by taking an increasing sequence of increasing events A_n with limit A, then $p \mapsto \phi^0_{p,q}[A]$ is left-continuous. Note that this is not true for any measurable event, an archetypical example being $A := \{0 \leftrightarrow \infty\}$, which is the limit of a decreasing sequence of increasing events.

Let us go back to the consequence of (4.9). Since the simple random walk is transient on $\mathbb{Z}^d$ for $d \geq 3$, the right-hand side tends to 0 as $\|x - y\|$ tends to infinity. This claim implies that (LRO$_{\beta_c}$) does not hold.

Exercise 38 We wish to prove that

$$\mu^{\mathrm{f}}_\beta[\sigma_0\sigma_x] \leq \mu^{\mathrm{f}}_\beta[\sigma_0\sigma_y] \qquad \text{(Mes-Mir)}$$

if $x = (x_1, \dots, x_d)$ and $y = (y_1, \dots, y_d)$ satisfy either of the following two conditions

C1 $0 \leq x_1 \leq y_1$ and $x_i = y_i$ for every $i \geq 2$;
C2 $x_1 + x_2 = y_1 + y_2$ and $0 \leq x_1 - x_2 \leq y_1 - y_2$ and $x_i = y_i$ for every $i \geq 3$.

1. Define the graph $\mathbb{L}$ obtained from $\mathbb{Z}^d$ by adding another edge between u and $u + (1, 0, \dots, 0)$, for each u with first coordinate u_1 equal to x_1 (let E be the set of new edges). How should we set coupling constants $J_{x,y}$ (in the sense of Exercise 36) on edges of $\mathbb{E} \cup E$ to have a model which is equivalent to the original model on $\mathbb{Z}^d$?
2. Define $E_1 := \{uv \subseteq \mathbb{Z}^d : u_1, v_1 \leq x_1\} \cup E$ and $E_2 := \mathbb{E}\setminus E_1$. Show that

$$\sum_{\partial\mathbf{n}=\{0,y\}} w(\mathbf{n}) = \sum_{\substack{\mathbf{n}_1\in\mathbb{N}^{E_1},\mathbf{n}_2\in\mathbb{N}^{E_2}\\ \partial(\mathbf{n}_1+\mathbf{n}_2)=\{0,y\}}} w(\mathbf{n}_1)w(\mathbf{n}_2) \quad\text{and}\quad \sum_{\partial\mathbf{n}=\{0,x\}} w(\mathbf{n}) = \sum_{\substack{\mathbf{n}_1\in\mathbb{N}^{E_1},\mathbf{n}_2\in\mathbb{N}^{E_2}\\ \partial(\mathbf{n}_1+\mathbf{n}_2)=\{0,x\}}} w(\mathbf{n}_1)w(\mathbf{n}_2).$$

2. Consider $y = x + (1, 0, \dots, 0)$. Using the current $\mathbf{n}_2' \in \mathbb{N}^{E_1}$ obtained from $\mathbf{n}_2$ by taking the reflection with respect to $\{z \in \mathbb{R}^d : z_1 = x_1 + 1/2\}$ (with the convention that an edge of E is sent to the corresponding edge of $\mathbb{Z}^d$ with the same endpoints), show that

$$\sum_{\partial\mathbf{n}=\{0,y\}} w(\mathbf{n}) \leq \sum_{\partial\mathbf{n}=\{0,x\}} w(\mathbf{n}).$$

Hint. Use the multi-graph $\mathcal{M}$ obtained from $\mathbf{n}_1 + \mathbf{n}_2'$ and observe that for terms involved in the left-hand side, $\mathcal{M}$ necessarily contains a path from x to $\{z \in \mathbb{R}^d : z_1 = y_1\}$, and that in such case one may use a "switching lemma".
3. Prove (Mes-Mir) under condition **C1**. Adapt the proof to show (Mes-Mir) under condition **C2**.
4. Show that (Mes-Mir) implies that if $e_1 = (1, 0, \dots, 0)$, then for every $x \in \partial\Lambda_n$,

$$\mu^{\mathrm{f}}_\beta[\sigma_0\sigma_{ne_1}] \geq \mu^{\mathrm{f}}_\beta[\sigma_0\sigma_x] \geq \mu^{\mathrm{f}}_\beta[\sigma_0\sigma_{dne_1}]. \qquad (4.10)$$

It is unclear whether the absence of long-range ordering for $\mu^{\mathrm{f}}_{\beta_c}$ implies that $\mu^{+}_{\beta_c}[\sigma_0] = 0$ since $\mu^{\mathrm{f}}_{\beta_c}$ and $\mu^{+}_{\beta_c}$ are a priori different. We will see that $\mu^{\mathrm{b}}_{\beta_c}$ is different from $\mu^{\mathrm{f}}_{\beta_c}$ for q-state Potts model with $q \geq 4$ in two dimensions. Nonetheless, the following result tells us that this is never the case for the Ising model on $\mathbb{Z}^d$.

Theorem 12 (Aizenman et al. [4]) *For the Ising model at inverse temperature β on $\mathbb{Z}^d$,* (MAG$_\beta$) *implies* (LRO$_\beta$). *As a consequence, the spontaneous magnetization is equal to* 0 *at criticality.*

The proof is based on the percolation representation obtained using random currents. We will prove that the difference between spin-spin correlations with plus and free boundary conditions (which can be understood as truncated correlations) can be expressed in terms of a percolation model. We will then study the ergodic properties of this model to derive that (MAG$_\beta$) implies (LRO$_\beta$). The strategy is somewhat similar to Sect. 1.4.

Step 1: expressing truncated correlations using a percolation model based on random currents. Let us start by expressing spin-spin correlations with + boundary conditions in terms of currents. Let $\mathrm{int}(V) := V\backslash\partial G$ be the set of interior vertices. In order to do so, perform the same expansion (with Taylor series) as for the free boundary conditions to obtain

$$\sum_{\sigma\in\{\pm1\}^V} \sigma_A \exp[-\beta H^{\mathrm{f}}_G(\sigma)]\mathbb{1}_{\sigma_{|\partial G}=+} = 2^{|\mathrm{int}(V)|} \sum_{\partial\mathbf{n}\cap\mathrm{int}(V)=A} w_\beta(\mathbf{n}), \tag{4.11}$$

where we use that

$$\sum_{\substack{\sigma\in\{\pm1\}^V\\ \sigma_{|\partial G}=+}} \sigma_A \prod_{xy\in E} (\sigma_x\sigma_y)^{\mathbf{n}_{xy}} = \begin{cases} 2^{|\mathrm{int}(V)|} & \text{if } \partial\mathbf{n}\cap\mathrm{int}(V) = A \\ 0 & \text{otherwise.} \end{cases}$$

Introduce, as in the case of free boundary conditions, the measure

$$\mathbf{P}^{+}_{G,\beta}[\omega] = \frac{\displaystyle\sum_{\substack{\partial\mathbf{n}_1\cap\mathrm{int}(V)=\emptyset\\ \partial\mathbf{n}_2=\emptyset}} w_\beta(\mathbf{n}_1)w_\beta(\mathbf{n}_2)\mathbb{1}_{\widehat{\mathbf{n}_1+\mathbf{n}_2}=\omega}}{\displaystyle\sum_{\substack{\partial\mathbf{n}_1\cap\mathrm{int}(V)=\emptyset\\ \partial\mathbf{n}_2=\emptyset}} w_\beta(\mathbf{n}_1)w_\beta(\mathbf{n}_2)}.$$

The key observation in the proof below is the following lemma.

Lemma 8 *For G finite and $\beta > 0$, we have that*

$$\mu^{+}_{G,\beta}[\sigma_x\sigma_y]\mu^{\mathrm{f}}_{G,\beta}[\sigma_x\sigma_y] = \mathbf{P}^{+}_{G,\beta}[x \longleftrightarrow y] \qquad \forall x, y \in V\backslash\partial G, \tag{4.12}$$

$$\mu^{+}_{G,\beta}[\sigma_x\sigma_y] - \mu^{\mathrm{f}}_{G,\beta}[\sigma_x\sigma_y] \leq \tfrac{1}{\sinh(\beta)}\mathbf{P}^{+}_{G,\beta}[\mathcal{A}(xy)] \qquad \forall xy \in E, \tag{4.13}$$

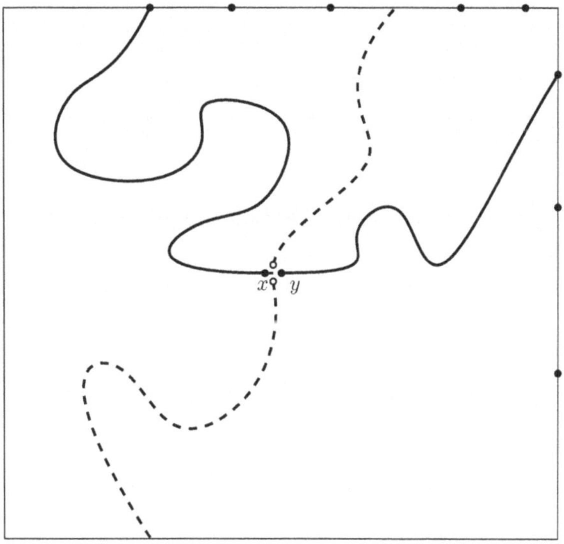

Fig. 8 Event $\mathcal{A}(xy)$. The bullets in the boundary are representing the sources of $\mathbf{n}_1$

where $\mathcal{A}(xy)$ *is the event that* $\omega_{xy} = 1$ *and that in* $\omega \setminus \{xy\}$*,* x *and* y *are connected to* ∂G *but not to each other (see Fig. 8).*

Proof Equation 4.12 follows easily from (4.11) and the switching lemma by decomposing into possible values of $\partial\mathbf{n}_1 \cap \partial G$:

$$\mu^+_{G,\beta}[\sigma_x\sigma_y]\mu^{\mathrm{f}}_{G,\beta}[\sigma_x\sigma_y] = \frac{\displaystyle\sum_{A\subseteq\partial G}\sum_{\substack{\partial\mathbf{n}_1=A\Delta\{x,y\}\\ \partial\mathbf{n}_2=\{x,y\}}} w_\beta(\mathbf{n}_1)w_\beta(\mathbf{n}_2)}{\displaystyle\sum_{\substack{\partial\mathbf{n}_1\cap\mathrm{int}(V)=\emptyset\\ \partial\mathbf{n}_2=\emptyset}} w_\beta(\mathbf{n}_1)w_\beta(\mathbf{n}_2)}$$

$$= \frac{\displaystyle\sum_{A\subseteq\partial G}\sum_{\substack{\partial\mathbf{n}_1=A\\ \partial\mathbf{n}_2=\emptyset}} w_\beta(\mathbf{n}_1)w_\beta(\mathbf{n}_2)\mathbb{1}_{\widehat{x\overset{\mathbf{n}_1+\mathbf{n}_2}{\longleftrightarrow}y}}}{\displaystyle\sum_{\substack{\partial\mathbf{n}_1\cap\mathrm{int}(V)=\emptyset\\ \partial\mathbf{n}_2=\emptyset}} w_\beta(\mathbf{n}_1)w_\beta(\mathbf{n}_2)} = \mathbf{P}^+_{G,\beta}[x \longrightarrow y].$$

Let us now turn to (4.13), which is slightly more subtle. We may use the switching lemma (in a similar fashion to what we did above) to get

$$\mu^+_{G,\beta}[\sigma_x\sigma_y] - \mu^{\mathrm{f}}_{G,\beta}[\sigma_x\sigma_y] = \frac{\displaystyle\sum_{\partial\mathbf{n}_1\cap\mathrm{int}(V)=\{x,y\}} w_\beta(\mathbf{n}_1)}{\displaystyle\sum_{\partial\mathbf{n}_1\cap\mathrm{int}(V)=\emptyset} w_\beta(\mathbf{n}_1)} - \frac{\displaystyle\sum_{\partial\mathbf{n}_2=\{x,y\}} w_\beta(\mathbf{n}_2)}{\displaystyle\sum_{\partial\mathbf{n}_2=\emptyset} w_\beta(\mathbf{n}_2)}$$

$$= \frac{\displaystyle\sum_{\substack{\partial \mathbf{n}_1 \cap \mathrm{int}(V)=\{x,y\} \\ \partial \mathbf{n}_2=\emptyset}} w_\beta(\mathbf{n}_1)w_\beta(\mathbf{n}_2)\mathbb{1}_{(\mathbf{n}_1,\mathbf{n}_2)\in\mathcal{B}(xy)}}{\displaystyle\sum_{\substack{\partial \mathbf{n}_1 \cap \mathrm{int}(V)=\emptyset \\ \partial \mathbf{n}_2=\emptyset}} w_\beta(\mathbf{n}_1)w_\beta(\mathbf{n}_2)}, \qquad (4.14)$$

where $\mathcal{B}(xy)$ is the event that x and y are not connected in $\widehat{\mathbf{n}_1 + \mathbf{n}_2}$. Note that on $\mathcal{B}(xy)$, the source constraints in $\mathbf{n}_1$ are forcing x and y to be connected to sources of $\mathbf{n}_1$, i.e. to ∂G. Also note that $\mathbf{n}_1$ is equal to 0 on xy.

Consider $(\mathbf{n}_1, \mathbf{n}_2) \in \mathcal{B}(xy)$ and define the set of currents $C(\mathbf{n}_1)$ coinciding with $\mathbf{n}_1$ except at xy, where they have an odd value. Note that currents in $C(\mathbf{n}_1)$ have no sources in $\mathrm{int}(V)$ (since $\mathbf{n}_1$ was equal to 0 at xy, and that now the current is odd) and that the trace of $\tilde{\mathbf{n}}_1 + \mathbf{n}_2$ is in $\mathcal{A}(xy)$ for any $\tilde{\mathbf{n}}_1 \in C(\mathbf{n}_1)$. Finally, note that the sets $C(\mathbf{n}_1)$ are disjoint for different $\mathbf{n}_1$ and that the sum of the weights of currents in $C(\mathbf{n}_1)$ is equal to $\sinh(\beta)w_\beta(\mathbf{n}_1)$. Overall, changing $\mathbf{n}_1$ to the sum over $\tilde{\mathbf{n}}_1$ in $C(\mathbf{n}_1)$ in the previous sum gives the result. □

Step 2: ergodic properties of the infinite volume version of $\mathbf{P}^+_{G,\beta}$. We now prove that one may define an infinite-volume version of $\mathbf{P}^+_{G,\beta}$, denoted $\mathbf{P}^+_\beta$, which is invariant under translations and ergodic. We further prove that $\mathbf{P}^+_\beta$ contains at most one infinite cluster almost surely.

Lemma 9 *The sequence of measures* $\mathbf{P}^+_{\Lambda_n,\beta}$ *converges to a measure* $\mathbf{P}^+_\beta$ *on* $\{0, 1\}^{\mathbb{E}}$ *which is invariant under translations and ergodic.*

Proof Note that ω with law $\mathbf{P}^+_{G,\beta}$ is obtained as the union of two independent configurations obtained by taking the trace of only one current, either sourceless or with sources located on ∂G. Furthermore, conditioned on the parity of the currents (which we recall is simply the high-temperature expansion η), whether a current is positive or not is decided independently for each edge (if the current is odd, it must be positive, otherwise it is positive with probability $1 - \cosh(\beta)^{-1}$). Therefore, convergence, invariance under translation and ergodicity of the limit of the laws of the parity of each current implies the claim. Since the proof is the same in both cases, we focus on the current defined on G.

Let $\mathrm{P}^{\mathrm{f}}_{G,\beta}$ be the measure defined for $\eta \in \{0, 1\}^E$ by

$$\mathrm{P}^{\mathrm{f}}_{G,\beta}[\eta] = \frac{\tanh(\beta)^{o(\eta)}}{\displaystyle\sum_{\partial\eta=\emptyset} \tanh(\beta)^{o(\eta)}}.$$

For a set of edges F, define the event $\mathcal{C}_F$ that $\eta_e = 0$ for each $e \in F$. We deduce from Lemma 7 that

$$\mathrm{P}^{\mathrm{f}}_{G,\beta}[\mathcal{C}_F] = \mu^{\mathrm{f}}_{G,\beta}\big[e^{-\beta K_F}\big]\cosh(\beta)^{|F|} \qquad (4.15)$$

with $K_F(\sigma) := \sum_{xy\in F} \sigma_x\sigma_y$. The convergence of the above expression follows now directly from the convergence of Ising measures as $G \nearrow \mathbb{Z}^d$. Since the events $\mathcal{C}_F$ with F finite generate the σ-algebra, we obtain the convergence of measures. The invariance under translations of μ^{f}_β implies immediately the invariance under translations of the limiting measure $\mathrm{P}^{\mathrm{f}}_\beta$.

To prove ergodicity, we prove that the measure is mixing, which we only need to prove for events of the form $\mathcal{C}_F$. Fix two finite sets F and F' of edges. Using the expression (4.15) for x large enough so that $F \cap (x + F') = \emptyset$, we find

$$\frac{\mathrm{P}^{\mathrm{f}}_\beta[\mathcal{C}_{F\cup(x+F')}]}{\mathrm{P}^{\mathrm{f}}_\beta[\mathcal{C}_F]\mathrm{P}^{\mathrm{f}}_\beta[\mathcal{C}'_F]} = \frac{\mu^{\mathrm{f}}_\beta\big[e^{-\beta K_F}e^{-\beta K_{x+F'}}\big]}{\mu^{\mathrm{f}}_\beta\big[e^{-\beta K_F}\big]\mu^{\mathrm{f}}_\beta\big[e^{-\beta K_{F'}}\big]}. \tag{4.16}$$

Ergodicity can therefore be presented as an implication of the statement that this ratio tends to 1. To see this convergence, observe that μ^{f}_β is itself mixing for functions of the spin-space that are even. Indeed, consider A and B two sets of even cardinality and x large enough that $A \cap (x + B) = \emptyset$. The coupling with the random-cluster model gives

$$\frac{\mu^{\mathrm{f}}_\beta[\sigma_A\sigma_{x+B}]}{\mu^{\mathrm{f}}_\beta[\sigma_A]\mu^{\mathrm{f}}_\beta[\sigma_B]} = \frac{\phi^0_{p,2}[\mathcal{F}_{A\cup(x+B)}]}{\phi^0_{p,2}[\mathcal{F}_A]\phi^0_{p,2}[\mathcal{F}_B]} \longrightarrow 1,$$

with $p = 1 - e^{-2\beta}$, where the last convergence is due to the mixing property of $\phi^{\mathrm{f}}_{p,2}$ and the fact that

$$\phi^0_{p,2}[\mathcal{F}_{A\cup(x+B)}\setminus\mathcal{F}_A \cap \mathcal{F}_{x+B}] \longrightarrow 0.$$

(This last fact is due to the fact that this event is included in the event that there are two disjoint clusters going from A to distance $\|x\|/2$, which by uniqueness of the infinite cluster, has a probability going to zero as x goes to infinity.) Since the random-variables σ_A for $|A|$ even generate the σ-algebra of even functions of the spin-space, the result follows. □

The ergodicity of $\mathbf{P}^+_\beta$ implies that an infinite cluster exists with probability either 0 or 1. We now prove that when it exists, it is unique almost surely.

Lemma 10 *For any $\beta > 0$, there exists at most one infinite cluster $\mathbf{P}_\beta$-almost surely.*

Proof The proof of this theorem follows from the Burton-Keane argument presented in the Proof of Theorem 4. The only difference lies in the fact that we do not have the finite-energy property (FE) anymore. Nevertheless, we have the following insertion tolerance claim, which we leave as an exercise (see Exercise 39): for any event $\mathcal{A}$ depending on edges different from xy,

$$\mathbf{P}^+_\beta[\omega_{xy} = 1|\mathcal{A}] \geq c_{\mathrm{IT}}. \tag{IT}$$

Note that the probability on the left may a priori be arbitrarily close to 1, but the previous lower bound is sufficient for our purpose.

Recall the notation from the Proof of Theorem 4. First, one may check that the proof that $\mathbf{P}^+_\beta[\mathcal{E}_{<\infty}\setminus\mathcal{E}_{\leq 1}] = 0$ is the same. To exclude the possibility of an infinite number of infinite clusters, one cannot really work with trifurcations anymore since constructing them would require the finite-energy rather than the insertion tolerance. Nevertheless, one can work with a notion of coarse trifurcation, where 0 is a *coarse trifurcation* if edges in E_N are open, and $\omega\setminus E_N$ contains at least three infinite clusters intersecting Λ_N. Similarly, one defines the fact that x is a coarse trifurcation. Note that if there is an infinite number of infinite clusters, then the construction of Theorem 4 and (IT) imply that coarse trifurcations occur with positive probability. The end of the proof works the same, except that up to $|\Lambda_N|$ coarse trifurcations can intersect a fixed coarse trifurcation, so that the deterministic bound for the number of trifurcations is now $|\Lambda_N| \cdot |\partial\Lambda_n|$. □

Step 3: conclusion of the proof. The proof now follows readily. We work by contraposition and assume that (LRO_β) does not hold. Then, taking the limit of (4.12) (as $G \nearrow \mathbb{Z}^d$) together with ergodicity and the uniqueness of the infinite cluster (when it exists) implies that there is no infinite cluster for $\mathbf{P}^+_\beta$ almost surely. Now, the limit as G tends to $\mathbb{Z}^d$ of $\mathcal{A}(xy)$ is included in the event that there exists an infinite cluster. Therefore, by taking the limit as G tends to $\mathbb{Z}^d$ in (4.13), we obtain that $\mu^{\mathrm{f}}_\beta[\sigma_x\sigma_y] = \mu^+_\beta[\sigma_x\sigma_y]$ for any $xy \in \mathbb{E}$.

To conclude, note that this translates into the fact that $\phi^0_{p,2}[x \leftrightarrow y] = \phi^1_{p,2}[x \leftrightarrow y]$. Yet, a simple computation[12] gives that

$$\phi^0_{p,2}[\omega_{xy}] = \tfrac{p}{2}(1 + \phi^0_{p,2}[x \longleftrightarrow y]).$$

and similarly for $\phi^1_{p,2}$ so that $\phi^0_{p,2}[\omega_{xy}] = \phi^1_{p,2}[\omega_{xy}]$. We already argued in the Proof of Theorem 5 that this implies $\phi^0_{p,2} = \phi^1_{p,2}$. In particular, for any $x, y \in \mathbb{Z}^d$,

$$\mu^+_\beta[\sigma_x\sigma_y] = \phi^1_{p,2}[x \longleftrightarrow y] = \phi^0_{p,2}[x \longleftrightarrow y] = \mu^{\mathrm{f}}_\beta[\sigma_x\sigma_y]$$

tends to 0 as $\|x - y\|$ tends to infinity (by the infrared bound). By uniqueness of the infinite connected component, we deduce that $\phi^1_{p,2}[0 \longleftrightarrow \infty] = 0$, which is the claim.

[12]Or more elegantly the use of Exercise 5, which states that $\phi^0_{p,2}[\omega_{xy}] = p \cdot \mu^{\mathrm{f}}_\beta[\sigma_x = \sigma_y]$, which combined with $\mu^{\mathrm{f}}_\beta[\sigma_x = \sigma_y] = 2\mu^{\mathrm{f}}_\beta[\sigma_x\sigma_y] - 1 = 2\phi^0_{p,2}[x \leftrightarrow y] - 1$ gives the requested equality.

Exercise 39 Prove (IT) using an argument similar to the proof of (4.13).

Exercise 40 Consider the Ising model with a magnetic field h, i.e. the model with Hamiltonian

$$H_{G,h}(\sigma) = H_G(\sigma) - \sum_{x \in V} h\sigma_x.$$

We denote the infinite-volume measure by $\mu^{\mathrm{f}}_{\beta,h}$.

1. Interpret the correlations of the model in terms of random currents on the graph G^h with vertex-set $V \cup \{\mathfrak{g}\}$and edge-set given by $E \cup \{x\mathfrak{g} : x \in V\}$. What is the weight $w_{\beta,h}(\mathbf{n})$ of a current?
2. Following a reasoning similar to the proof of (4.13), show that $\mu^{\mathrm{f}}_{\beta,h}[\sigma_x\sigma_0] - \mu^{\mathrm{f}}_{\beta,h}[\sigma_x]\mu^{\mathrm{f}}_{\beta,h}[\sigma_0]$ can be reinterpreted as the probability under a system of duplicated currents that x is connected to 0 but not to $\mathfrak{g}$.
3. Using insertion tolerance, prove that this probability decays exponentially fast in $\|x\|$

4.4 Polynomial Decay at Criticality for d ≥ 3

It is natural to ask how fast the spin-spin correlations decay at criticality. We will prove the following

Theorem 13 *For $d \geq 3$, there exists $c, C \in (0, \infty)$ such that for every $x \in \mathbb{Z}^d$,*

$$\frac{c}{\|x\|^{d-1}} \leq \mu^{\mathrm{f}}_{\beta_c}[\sigma_0\sigma_x] \leq \frac{C}{\|x\|^{d-2}}. \tag{4.17}$$

Proof The upper bound is provided by the infrared bound (IR). For the lower bound, we invoke Simon's inequality (see Exercise 41), stating that for every $n \geq 1$,

$$\mu^{\mathrm{f}}_{\beta}[\sigma_0\sigma_x] \leq \sum_{y \in \partial\Lambda_n} \mu^{\mathrm{f}}_{\beta}[\sigma_0\sigma_y]\,\mu^{\mathrm{f}}_{\beta}[\sigma_y\sigma_x].$$

Assume that $\varphi_\beta(\Lambda_n) := \sum_{y \in \partial\Lambda_n} \mu^{\mathrm{f}}_{\beta}[\sigma_0\sigma_y] < 1$ for some n, and observe that by a reasoning similar to the first step of Sect. 2.2, we find that if $x \in \Lambda_{kn}$,

$$\mu^{\mathrm{f}}_{\beta}[\sigma_0\sigma_x] \leq \varphi_\beta(\Lambda_n)^k.$$

Now, $\beta \mapsto \mu^{+}_{\beta}[\sigma_0\sigma_x]$ is continuous from the right since it is the infimum of the continuous increasing functions $\beta \mapsto \mu^{+}_{G,\beta}[\sigma_0\sigma_x]$. We deduce that

$$\lim_{\beta \searrow \beta_c} \varphi_\beta(\Lambda_n) \leq \lim_{\beta \searrow \beta_c} \sum_{y \in \partial\Lambda_n} \mu^{+}_{\beta}[\sigma_0\sigma_y] = \sum_{y \in \partial\Lambda_n} \mu^{+}_{\beta_c}[\sigma_0\sigma_y] = \varphi_{\beta_c}(\Lambda_n),$$

where in the last equality we used that $\mu^{\mathrm{f}}_{\beta_c} = \mu^{+}_{\beta_c}$.

Therefore, if $\varphi_{\beta_c}(\Lambda_n) < 1$, then $\varphi_\beta(\Lambda_n) < 1$ for some $\beta > \beta_c$. By the reasoning above, this would imply that correlations decay exponentially fast for $\beta > \beta_c$, which is absurd. In conclusion, $\varphi_{\beta_c}(\Lambda_n) \geq 1$ for every $n \geq 1$.

The Messager-Miracle inequality (Mes-Mir) used twice (more precisely (4.10)) implies that for any $y \in \partial\Lambda_n$,

$$\mu^{\rm f}_{\beta_c}[\sigma_0\sigma_{ne_1}] \ge \mu^{\rm f}_{\beta_c}[\sigma_0\sigma_y] \ge \mu^{\rm f}_{\beta_c}[\sigma_0\sigma_{dne_1}] \tag{4.18}$$

where $e_1 = (1, 0, \ldots, 0)$. The left inequality together with $\varphi_{\beta_c}(\Lambda_n) \ge 1$ imply that

$$\mu^{\rm f}_{\beta_c}[\sigma_0\sigma_{ne_1}] \ge \frac{1}{|\partial\Lambda_n|}$$

for every n. The proof follows readily from the right inequality of (4.18). □

Exercise 41 (*Simon's inequality*) Using the switching lemma, prove Simon's inequality: for any set S disconnecting x from y (in the sense that any path from x to y intersects S),

$$\mu^{\rm f}_{G,\beta}[\sigma_x\sigma_z] \le \sum_{y\in S} \mu^{\rm f}_{G,\beta}[\sigma_x\sigma_y]\mu^{\rm f}_{G,\beta}[\sigma_y\sigma_z]. \tag{Simon}$$

A slightly stronger inequality, called Lieb's inequality, can also be obtained using random currents (the proof is more difficult). The improvement lies in the fact that $\mu^{\rm f}_{G,\beta}[\sigma_x\sigma_y]$ can be replaced by $\mu^{\rm f}_{S,\beta}[\sigma_x\sigma_y]$:

$$\mu^{\rm f}_{G,\beta}[\sigma_x\sigma_z] \le \sum_{y\in S} \mu^{\rm f}_{S,\beta}[\sigma_x\sigma_y]\mu^{\rm f}_{G,\beta}[\sigma_y\sigma_z]. \tag{Lieb}$$

5 Continuity/Discontinuity of the Phase Transition for the Planar Random-Cluster Model

We now turn to the case of the random-cluster model in two dimensions. We will discuss the following result.

Theorem 14 *Consider the random-cluster model with cluster-weight $q \ge 1$ on $\mathbb{Z}^2$. Then $\phi^1_{p_c,q}[0 \leftrightarrow \infty] = 0$ if and only if $q \le 4$.*

As an immediate corollary, we obtain the following result.

Corollary 7 *The phase transition of the Potts model is continuous for $q \in \{2, 3, 4\}$ and discontinuous for $q \ge 5$.*

The section is organized as follows. We first study crossing probabilities for planar random-cluster models by building a Russo-Seymour-Welsh type theory for these models. This part enables us to discriminate between two types of behavior:

- the continuous one in which crossing probabilities do not go to zero, even when boundary conditions are free (which correspond to the worse ones for increasing events). In this case, the infinite-volume measures with free and wired boundary conditions are equal and correlations decay polynomially fast.

– the discontinuous one in which crossing probabilities with free boundary conditions go to zero exponentially fast. In this case, the infinite-volume measure with free boundary conditions looks subcritical in the sense that the probability that 0 is connected to distance n is decaying exponentially fast, while the infinite-volume measure with wired boundary conditions contains an infinite cluster almost surely.

We then prove that for $q \leq 4$, the probability of being connected to distance n for the free boundary conditions goes to zero at most polynomially fast, thus proving that we are in the continuous case. In order to do that, we introduce parafermionic observables. Finally, we discuss the $q > 4$ case, in which we sketch the proof that the probability of being connected to distance n decays exponentially fast, thus proving that we are in the discontinuous phase.

5.1 *Crossing Probabilities in Planar Random-Cluster Models*

We saw that the probability of crossing squares was equal to 1/2 for Bernoulli percolation, and that it was either bounded from above or below by $1/2$ for random-cluster models depending on the boundary conditions. This raises the question of probabilities of crossing more complicated shapes, such as rectangle with aspect ratio $\rho \neq 1$. While this could look like a technical question, we will see that studying crossing probabilities is instrumental in the study of critical random cluster models.

We begin with some general notation. For a rectangle $R := [a, b] \times [c, d]$ (when a, b, c or d are not integers, an implicit rounding operation is performed), introduce the event $\mathcal{H}(R)$ that R is crossed horizontally, i.e. that the left side $\{a\} \times [c, d]$ is connected by a path in $\omega \cap R$ to the right side $\{b\} \times [c, d]$. Similarly, define $\mathcal{V}(R)$ be the event that R is crossed vertically, i.e. that the bottom side $[a, b] \times \{c\}$ is connected by a path in $\omega \cap R$ to the top side $[a, b] \times \{d\}$. When $R = [0, n] \times [0, k]$, we rather write $\mathcal{V}(n, k)$ and $\mathcal{H}(n, k)$.

Exercise 42 Consider Bernoulli percolation (of parameter p) on a planar transitive locally finite infinite graph with $\pi/2$ symmetry.

1. Using the rectangles $R_1 = [0, n] \times [0, 2n]$, $R_2 = [0, n] \times [n, 3n]$, $R_3 = [0, n] \times [2n, 4n]$, $R_4 = [0, 2n] \times [n, 2n]$ and $R_5 = [0, 2n] \times [2n, 3n]$, show that

$$\mathbb{P}_p[\mathcal{H}(n, 4n)] \leq 5\mathbb{P}[\mathcal{H}(n, 2n)].$$

2. Deduce that $u_{2n} \leq 25u_n^2$ where $u_n = \mathbb{P}_p[\mathcal{H}(n, 2n)]$. Show that (u_n) decays exponentially fast as soon as there exists n such that $u_n < \frac{1}{25}$.
3. Deduce that $u_n \geq \frac{1}{25}$ for every n or (EXP$_p$). What did we prove at p_c?

The RSW theory for infinite-volume measures Recall from (2.16) that we know that

$$\phi^1_{p_c,q}[\mathcal{H}(n, n)] \geq \phi^1_{p_c,q}[\mathcal{H}(n + 1, n)] \geq \tfrac{1}{2}.$$

It is natural to wish to improve this result by studying crossing probabilities for wired boundary conditions for rectangles of fixed aspect ratio remain bounded away from 0 when n tends to infinity. This is the object of the following theorem.

Theorem 15 (Beffara and Duminil-Copin [11]) *Let* $\rho>0$*, there exists* $c = c(\rho)>0$ *such that for every* $n \geq 1$*,*

$$\phi^1_{p_c,q}[\mathcal{H}(\rho n, n)] \geq c.$$

For Bernoulli percolation, a uniform upper bound follows easily from the uniform lower bound and duality since the complement of the event that a rectangle is crossed vertically is the event that the dual rectangle is crossed horizontally in the dual configuration. This is not the case for general random-cluster models since the dual measure is the measure with free boundary conditions. In fact, we will see in the next sections that a uniform upper bound is not necessarily true: crossing probabilities could go to 1 for wired boundary conditions, and to 0 for free ones. It was therefore crucial to state this theorem for "favorable" boundary conditions at infinity.

Also, as soon as a uniform lower bound (in n) for $\rho = 2$ is proved, then one can easily combine crossings in different rectangles to obtain a uniform lower bound for any $\rho > 1$. Indeed, define (for integers $i \geq 0$) the rectangles $R_i := [in, (i+2)n] \times [0, n]$ and the squares $S_i := R_i \cap R_{i+1}$. Then,

$$\phi^1_{p_c,q}[\mathcal{H}(\rho n, n)] \geq \phi^1_{p_c,q}\Big[\bigcap_{i \leq \rho}(\mathcal{H}(R_i) \cap \mathcal{V}(S_i))\Big] \overset{\text{(FKG)}}{\geq} c(2)^{2\lfloor \rho \rfloor}.$$

One may even prove lower bounds for crossing probabilities in arbitrary topological rectangles (see Exercise 43 below).

Exercise 43 Consider a simply connected domain with a smooth boundary Ω with four distinct points a, b, c and d on the boundary. Let $(\Omega_\delta, a_\delta, b_\delta, c_\delta, d_\delta)$ be the finite graph with four marked points on the boundary defined as follows: Ω_δ is equal to $\Omega \cap \delta\mathbb{Z}^2$ (we assume here that it is connected and of connected complement, so that the boundary is a simple path) and $a_\delta, b_\delta, c_\delta, d_\delta$ be the four points of $\partial\Omega_\delta$ closest to a, b, c and d.
Prove that there exists $c = c(q, \Omega, a, b, c, d) > 0$ such that for any $\delta > 0$,

$$\phi^1_{p_c,q}[(a_\delta b_\delta) \overset{\Omega_\delta}{\longleftrightarrow} (c_\delta d_\delta)] \geq c,$$

where $(a_\delta b_\delta)$ and $(c_\delta d_\delta)$ are the portions of $\partial\Omega_\delta$ from a_δ to b_δ, and from c_δ to d_δ, when going counterclockwise around $\partial\Omega_\delta$.

On the other hand, it is a priori not completely clear how to obtain a lower bound for $\rho = 2$ (or any $\rho > 1$) from a lower bound for $\rho = 1$. In fact, the main difficulty of Theorem 15 lies in passing from crossing squares with probabilities bounded uniformly from below to crossing rectangles in the hard direction with probabilities bounded uniformly from below. In other words, the main step is the following proposition.

Proposition 8 *For every* $n \geq 1$, $\phi^1_{p_c,q}[\mathcal{H}(2n, n)] \geq \frac{1}{16(1+q^2)}\phi^1_{p_c,q}[\mathcal{H}(n, n)]^6$.

A statement claiming that crossing a rectangle in the hard direction can be expressed in terms of the probability of crossing squares is called a Russo-Seymour-Welsh (RSW) type theorem. For Bernoulli percolation on $\mathbb{Z}^2$, this RSW result was first proved in [115, 118]. Since then, many proofs have been produced (for Bernoulli), among which [23–25, 124, 125]. We refer to [60] for a review of recent progress in this field. Here, we provide a proof for random-cluster models.

Proof We treat the case of n even, the case n odd can be done similarly. Let us introduce the two rectangles

$$R := [-2n, 2n] \times [-n, n] \qquad S := [0, 2n] \times [-n, n] \qquad S' := [-2n, 0] \times [-n, n].$$

Also introduce the notation $\alpha := \phi^1_{p_c,q}[\mathcal{H}(S)]$. Also, define the sets

$$\begin{aligned} A^+ &:= \{-2n\} \times [0, n] & B^+ &:= \{0\} \times [0, n] & C^+ &:= \{2n\} \times [0, n] \\ A^- &:= \{-2n\} \times [-n, 0] & B^- &:= \{0\} \times [-n, 0] & C^- &:= \{2n\} \times [-n, 0]. \end{aligned}$$

By symmetry with respect to the x-axis, the probability that there is a path in $\omega \cap S$ from B to C^+ is larger than or equal to $\alpha/2$. Similarly, the probability that there is a path in $\omega \cap S$ from B^- to C is larger than or equal to $\alpha/2$. Since the probability of $\mathcal{V}(S)$ is also α. The combination of these three events implies the event $\mathcal{E}$ that there exists a path in $\omega \cap S$ from B^- to C^+. Thus, the FKG inequality gives

$$\phi^1_{p_c,q}[\mathcal{E}] \geq \tfrac{\alpha^3}{4}.$$

Let $\mathcal{E}'$ be the event that there exists a path in $\omega \cap S'$ from A^- to B^+. By symmetry with respect to the origin, we have

$$\phi^1_{p_c,q}[\mathcal{E}'] \geq \tfrac{\alpha^3}{4}.$$

On the event $\mathcal{E} \cap \mathcal{E}'$, consider the paths of edges Γ and Γ' defined by:

- Γ is the bottom-most open crossing of S from B^- to C^+,
- Γ' is the top-most open crossing of S' from A^- to B^+,

Construct the graph $G = G(\Gamma, \Gamma')$ with edge-set composed of edges with at least one endpoint in the cluster of the origin in $\mathbb{R}^2 \setminus (\Gamma \cup \Gamma' \cup \sigma\Gamma \cup \sigma\Gamma')$ (here the paths are considered as subsets of $\mathbb{R}^2$), where $\sigma\Gamma$ and $\sigma\Gamma'$ are the reflections of Γ and Γ' with respect to the y-axis; see Fig. 9.

Let us assume for a moment that we have the following bound: for any two possible realizations γ and γ' of Γ and Γ',

$$\phi^{\mathrm{mix}}_{G,p_c,q}[\gamma \stackrel{G}{\longleftrightarrow} \gamma'] \geq \tfrac{1}{1+q}, \tag{5.1}$$

where the mix boundary conditions correspond to wired on γ and γ', and free elsewhere (i.e. the partition is given by $P_1 = \gamma$, $P_2 = \gamma'$ and singletons). Then,

$$\phi^1_{p_c,q}[\mathcal{H}(4n, 2n)] = \phi^1_{p_c,q}[\mathcal{H}(R)] \tag{5.2}$$

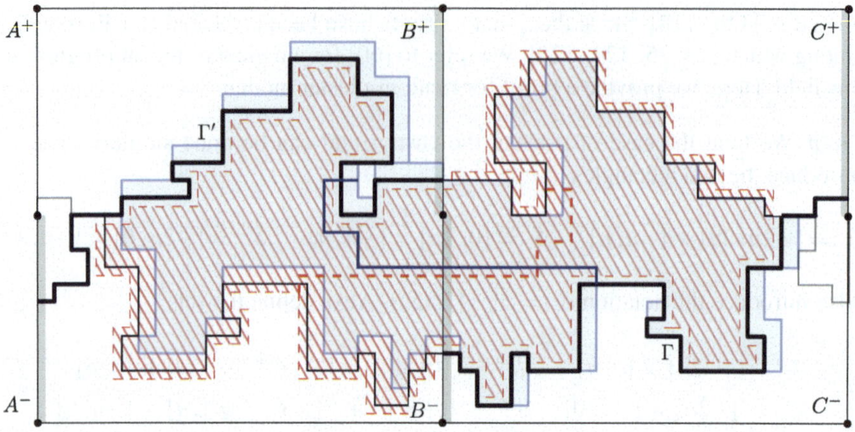

Fig. 9 The sets $A^\pm$, $B^\pm$ and $C^\pm$. We depicted Γ, Γ' and their symmetric with respect to the y-axis. In gray, the set G. Hatched in red, the dual graph of G together with a path in ω^* preventing the existence of a path from Γ to Γ' in G. In blue, the translation by $(1/2, 1/2)$ of the symmetric of G with respect to the y-axis, as well as the image by the same transformation of the dashed path. This path of ω' crossed G from Γ to Γ'

$$
\begin{aligned}
&\geq \phi^1_{p_c,q}[\{\Gamma \stackrel{G}{\longleftrightarrow} \Gamma'\} \cap \mathcal{E} \cap \mathcal{E}']\\
&= \sum_{\gamma,\gamma'} \phi^1_{p_c,q}[\{\gamma \stackrel{G}{\longleftrightarrow} \gamma'\} \cap \{\Gamma = \gamma, \Gamma' = \gamma'\}]\\
&\geq \sum_{\gamma,\gamma'} \phi^{\mathrm{mix}}_{G,p_c,q}[\gamma \stackrel{G}{\longleftrightarrow} \gamma'] \cdot \phi^1_{p_c,q}[\Gamma = \gamma, \Gamma' = \gamma']\\
&\stackrel{(5.1)}{\geq} \tfrac{1}{1+q} \sum_{\gamma,\gamma'} \phi^1_{p_c,q}[\Gamma = \gamma, \Gamma' = \gamma']\\
&= \tfrac{1}{1+q}\, \phi^1_{p_c,q}[\mathcal{E} \cap \mathcal{E}'] \stackrel{(\mathrm{FKG})}{\geq} \tfrac{\alpha^6}{16(1+q)},
\end{aligned}
$$

where in the fourth line we used the fact that $\Gamma = \gamma$ and $\Gamma' = \gamma'$ are measurable events of edges not in G, and that the boundary conditions induced on ∂G always dominate the mixed boundary conditions. In the last line, we used that the events $\{\Gamma = \gamma, \Gamma' = \gamma'\}$ partition $\mathcal{E} \cap \mathcal{E}'$, and the lower bounds on the probability of the events $\mathcal{E}$ and $\mathcal{E}'$ proved above.

We now turn to the proof of (5.1). We wish to use a symmetry argument (similar to the proof that crossing a square has probability larger or equal to 1/2). We believe the argument to be more transparent on Fig. 9 and we refer to its caption.

Fix $G = (V, E)$. Since the mix boundary conditions are planar boundary conditions, it will be simpler to consider a configuration $\xi \in \{0, 1\}^{\mathbb{E}\setminus E}$ inducing them. We choose the following one: $\xi_e = 1$ for all edges $e \in \gamma \cup \gamma'$ and $\xi_e = 0$ for all other edges. Set ω^ξ to be the configuration coinciding with ω on E, and with ξ on $\mathbb{E}\setminus E$.

Consider ω' to be the translation by $(1/2, 1/2)$ and then reflection with respect to the y axis of $(\omega^\xi)^*$. By duality, the law of ω' on G is dominated by the mix$'$ boundary conditions defined to be wired on $\gamma \cup \gamma'$, and free elsewhere (i.e. $P_1 = \gamma \cup \gamma'$ and then singletons[13]). The absence of path in ω from γ to γ' is included in the event that there is a path in $\omega'_{|E}$ from γ to γ', so that

$$1 - \phi^{\rm mix}_{G,p_c,q}[\gamma \overset{G}{\longleftrightarrow} \gamma'] \le \phi^{\rm mix'}_{G,p_c,q}[\gamma \overset{G}{\longleftrightarrow} \gamma'] \le q\,\phi^{\rm mix}_{G,p_c,q}[\gamma \overset{G}{\longleftrightarrow} \gamma'],$$

where in the second inequality we used that the Radon-Nikodym derivative is smaller or equal to q since $k_{\rm mix}(\omega) - k_{\rm mix'}(\omega) \in \{0, 1\}$. The inequality (5.1) follows readily. This concludes the proof. □

Let us conclude this section by recalling that crossing probabilities in rectangles are expected to converge to explicit functions of ρ as n tends to infinity. More generally, crossing probabilities in topological rectangles should be conformally invariant; see [119] for the case of site percolation (see also [13, 127] for reviews) and [15, 38, 91] for the case of the Ising model.

Here, we present a beautiful argument due to Vincent Tassion proving some weak form of crossing property for general FKG measures (with sufficient symmetry). We refer to [125] for more detail.

Exercise 44 (*Weak RSW for FKG measures*) Consider a measure μ on $\{0, 1\}^{\mathbb{E}}$ which is invariant under the graph isomorphisms of $\mathbb{Z}^2$ onto itself. We further assume that μ satisfies the FKG inequality. We assume that $\inf_n \mu[\mathcal{H}(n, n)] > 0$. The goal of this exercise is to prove that

$$\limsup_n \mu[\mathcal{H}(3n, n)] > 0. \tag{5.3}$$

1. Let $\mathcal{E}_n$ be the event that the left side of $[-n, n]^2$ is connected to the top-right corner (n, n). Use the FKG inequality to prove that $\limsup_n \mu[\mathcal{E}_n] > 0$ implies (5.3).
2. Assume the limit superior above is zero. Now, for any $-n \le \alpha < \beta \le n$, define the event $\mathcal{F}_n(\alpha, \beta)$ to be the existence of a crossing from the left side of $[-n, n]^2$ to the segment $\{n\} \times [\alpha, \beta]$. We consider the function

$$h_n(\alpha) = \mu[\mathcal{F}_n(0, \alpha)] - \mu[\mathcal{F}_n(\alpha, n)].$$

Show that h_n is an increasing function, and that there exists $c_0 > 0$ such that $h_n(n) > c_0$ for all n.
3. Assume that $h_n(n/2) < c_0/2$. Use (FKG) to prove that (5.3).
4. Assume that $h_n(n/2) > c_0/2$, and let $\alpha_n = \inf\{\alpha : h(\alpha) > c_0/2\}$. Define the event $\mathcal{X}_n(\alpha)$ by the existence of a cluster in $[-n, n]^2$ connecting the four segments $\{-n\} \times [-n, -\alpha]$, $\{-n\} \times [\alpha, n]$, $\{n\} \times [-n, -\alpha]$, and $\{n\} \times [\alpha, n]$. Prove that there exists a constant $c_1 > 0$ independent of n such that $\mu[\mathcal{X}_n(\alpha)] \ge c_1$.
5. Prove that, for infinitely many n's, $\alpha_n < 2\alpha_{2n/3}$.
6. Prove that, whenever $\alpha_n < 2\alpha_{2n/3}$, there exists a constant c_2 such that $\mu[\mathcal{H}(8/3n, 2n)] > c_2$. Conclude.

A dichotomy for random-cluster models Physicists work with several definitions of continuous phase transitions. For instance, a continuous phase transition may refer to the divergence of the correlation length, the continuity of the order parameter (here the spontaneous magnetization or the density of the infinite cluster), the uniqueness of the Gibbs states at criticality, the divergence of the susceptibility, the scale invariance at criticality, etc. From a mathematical point of view, these properties are not clearly

[13]Note that they are not equal to the mix boundary conditions since γ and γ' are wired together.

equivalent (there are examples of models for which they are not), and they therefore refer to a priori different notions of continuous phase transition.

In the following result, we use the study of crossing probabilities to prove that all these properties are equivalent for the planar random-cluster model.

Theorem 16 (Duminil-Copin et al. [57]). *Let $q \geq 1$, the following assertions are equivalent at criticality:*

- **P1** *(Absence of an infinite cluster)* $\phi^1_{p_c,q}[0 \longleftrightarrow \infty] = 0$.
- **P2** *(Uniqueness of the infinite-volume measure)* $\phi^0_{p_c,q} = \phi^1_{p_c,q}$.
- **P3** *(Infinite susceptibility)* $\sum_{x \in \mathbb{Z}^2} \phi^0_{p_c,q}[0 \longleftrightarrow x] = \infty$.
- **P4a** *(Slow decay with free boundary conditions)* $\lim_{n\to\infty} \frac{1}{n^{1/3}} \log \phi^0_{p_c,q}[0 \longleftrightarrow \partial\Lambda_n] = 0$.
- **P4b** *(Sub-exponential decay for free boundary conditions)* $\lim_{n\to\infty} \frac{1}{n} \log \phi^0_{p_c,q}[0 \longleftrightarrow \partial\Lambda_n] = 0$.
- **P5** *(Uniform crossing probabilities) There exists $c = c(\rho) > 0$ such that for all $n \geq 1$ and all boundary conditions ξ, if R denotes the rectangle $[-n, (\rho+1)n] \times [-n, 2n]$, then*

$$c \leq \phi^\xi_{R,p_c,q}[\mathcal{H}(\rho n, n)] \leq 1 - c. \tag{5.4}$$

The previous theorem does not show that these conditions are all satisfied, only that they are equivalent. In fact, whether the conditions are satisfied or not depend on the value of q, as we will see in the next two sections.

While Properties **P1**–**P4b** are quite straightforward to interpret, **P5** is maybe more mysterious. One may wonder why having bounds that are uniform in boundary conditions is so relevant. The answer will become clear in the next sections: uniformity in boundary conditions is crucial to handle quantitatively dependencies between events in different parts of the graph. Note that the lower bound in **P5** is a priori much stronger than the result of Theorem 15 since the study of the previous section provided no information for free boundary conditions, even for crossing squares.

Let us conclude this discussion by noticing that property **P5** is not equivalent to the stronger statement **P5'** where boundary conditions are put on the boundary of the rectangle $R' := [0, \rho n] \times [0, n]$ instead of R. In fact, the probability of crossing R' with free boundary conditions on $\partial R'$ tends to 0 for the random-cluster model with cluster-weight $q = 4$, while **P5** is still true there. One may show that **P5'** is true for $q < 4$, but the proof is more complicated (see [49, 57] for proofs for $q = 2$ and $q \in [1, 4)$ respectively).

Last but not least, observe that the upper bound in (5.4) follows from the lower bound by duality.

The Proof of Theorem 16 can be divided in several steps. First, one can see that several implications are essentially trivial.

Proposition 9 *We have that* **P5** $\Rightarrow$ **P1** $\Rightarrow$ **P2** $\Rightarrow$ **P3** $\Rightarrow$ **P4a** $\Rightarrow$ **P4b**.

The last implication **P4b** ⇒ **P5** is the most difficult and is postponed to the next section. In fact we will only prove **P4a** ⇒ **P5** since this will be sufficient for the applications we have in mind. We refer to [57] for the proof of **P4b** ⇒ **P5**.

Proof The implications **P3** ⇒ **P4a** ⇒ **P4b** are completely obvious, and **P1** ⇒ **P2** is the object of Exercise 17. For **P5** ⇒ **P1**, introduce the event $\mathcal{A} := \mathcal{V}([-3n, 3n] \times [2n, 3n])$. If $\partial\Lambda_n$ is connected to $\partial\Lambda_{4n}$, then one of the four rotated versions of the event $\mathcal{A}$ must also occur (where the angles of the rotation are $\frac{\pi}{2}k$ with $0 \le k \le 3$). Therefore,

$$\phi^1_{\Lambda_{4n}\setminus\Lambda_n,p_c,q}[\partial\Lambda_n \longleftrightarrow \partial\Lambda_{4n}] \stackrel{\text{(FKG)}}{\le} 1 - \phi^1_{\Lambda_{4n}\setminus\Lambda_n,p_c,q}[\mathcal{A}^c]^4 \stackrel{\mathbf{P5}}{\le} 1 - c,$$

where $c := c(6)^4$ (we also used the comparison between boundary conditions in the second inequality). By successive applications of the domain Markov property and the comparison between boundary conditions (Exercise 10), we deduce the existence of $\alpha > 0$ such that

$$\begin{aligned}\phi^1_{\Lambda_n,p_c,q}[0 \longleftrightarrow \partial\Lambda_n] &\le \prod_{4^k \le n} \phi^1_{\Lambda_{4^k}\setminus\Lambda_{4^{k-1}},p_c,q}[\partial\Lambda_{4^{k-1}} \longleftrightarrow \partial\Lambda_{4^k}] \\ &\le (1-c)^{\lfloor \log_4 n \rfloor} \le n^{-\alpha} \end{aligned} \tag{5.5}$$

which gives **P1** by passing to the limit. For **P2** ⇒ **P3**, recall the definition of $\mathcal{H}_n$ so that

$$n\,\phi^0_{p_c,q}[0 \longleftrightarrow \partial\Lambda_n] \ge \phi^0_{p_c,q}[\mathcal{H}_n] \stackrel{\mathbf{P2}}{=} \phi^1_{p_c,q}[\mathcal{H}_n] \stackrel{(2.16)}{=} 1/2. \tag{5.6}$$

We deduce that

$$\begin{aligned}\sum_{x\in\mathbb{Z}^2} \phi^0_{p_c,q}[0 \longleftrightarrow x] &= \sum_{n=0}^{\infty} \sum_{x\in\partial\Lambda_n} \phi^0_{p_c,q}[0 \longleftrightarrow x] \\ &\ge \sum_{n\ge1} \phi^0_{p_c,q}[0 \longleftrightarrow \partial\Lambda_n] \ge \sum_{n\ge1} \tfrac{1}{2n} = +\infty. \end{aligned}$$

□

Note that (5.5) and (5.6) show that under **P5**, for all $n \ge 1$,

$$\frac{1}{2n} \le \phi^0_{p_c,q}[0 \longleftrightarrow \partial\Lambda_n] \le \frac{1}{n^\alpha}. \tag{PD}$$

This is one among a long list of properties implied by **P5**. Let us mention a few others: mixing properties (Exercise 46, the existence of sub-sequential scaling limits for interfaces, the value of certain critical exponents called *universal* critical exponents (it has nothing to do with the universality of the model itself), the fractal nature of large clusters (with some explicit bounds on the Hausdorff dimension). It is also an important step toward the understanding of conformal invariance of the model, scaling relations between several critical exponents, etc.

In the next two exercises, we assume **P5**.

Exercise 45 1. Prove that there exists $c>0$ such that $\phi^0_{p_c,q}[0\longleftrightarrow\partial\Lambda_n]\le c\phi^0_{p_c,q}[0\longleftrightarrow\partial\Lambda_{2n}]$.
2. Prove that there exist $c_1,c_2>0$ such that for any $x\in\partial\Lambda_n$,

$$c_1\phi^0_{p_c,q}[0\longleftrightarrow\partial\Lambda_n]^2\le\phi^0_{p_c,q}[0\longleftrightarrow x]\le c_2\phi^0_{p_c,q}[0\longleftrightarrow\partial\Lambda_n]^2.$$

Exercise 46 (*Polynomial mixing*) 1. Show that there exists a constant $c>0$ such that for any $n\ge 2k$ and any event $\mathcal{A}$ depending on edges in Λ_k only, $\phi^\xi_{\Lambda_k,p_c,q}[\Lambda_k\not\longleftrightarrow\partial\Lambda_n|\mathcal{A}]\ge 1-(\frac{k}{n})^c$.

2. Construct a coupling between $\omega\sim\phi^\xi_{\Lambda_{2n},p_c,q}$ and $\tilde\omega\sim\phi^1_{\Lambda_{2n},p_c,q}$ in such a way that ω and $\tilde\omega$ coincide on Λ_k when Λ_k is not connected to $\partial\Lambda_n$ in $\tilde\omega$. *Hint.* Construct the coupling step by step using an exploration of the cluster connected to the boundary. Deduce that

$$\phi^\xi_{\Lambda_k,p_c,q}[\mathcal{A}]\ge(1-(\tfrac{k}{n})^c)\phi^1_{\Lambda_k,p_c,q}[\mathcal{A}].$$

3. Construct a coupling between $\omega\sim\phi^\xi_{\Lambda_{2n},p_c,q}$ and $\tilde\omega\sim\phi^1_{\Lambda_{2n},p_c,q}$ in such a way that ω and $\tilde\omega$ coincide on Λ_k when there exists an open circuit in ω surrounding Λ_k. Deduce that

$$\phi^1_{\Lambda_k,p_c,q}[\mathcal{A}]\ge(1-(\tfrac{k}{n})^c)\phi^\xi_{\Lambda_k,p_c,q}[\mathcal{A}].$$

4. Deduce that for any event $\mathcal{B}$ depending on edges outside Λ_n only,

$$\left|\phi^1_{p_c,q}[\mathcal{A}\cap\mathcal{B}]-\phi^1_{p_c,q}[\mathcal{A}]\phi^1_{p_c,q}[\mathcal{B}]\right|\le 2\left(\tfrac{k}{n}\right)^c\phi^1_{p_c,q}[\mathcal{A}]\phi^1_{p_c,q}[\mathcal{B}].$$

Proof of P4a$\Rightarrow$P5 of Theorem 16 We drop the dependency on $q\ge 1$ and p_c in the subscripts of the measures. In the next proofs, we omit certain details of reasoning concerning comparison with respect to boundary conditions. We already encountered such arguments several times (for instance in Exercise 11 and in the proof of Proposition 8). We encourage the reader to try to fill up the details of each one of these omissions (Exercise 49).

In order to prove **P4a$\Rightarrow$P5**, we developed a geometric renormalization for crossing probabilities: crossing probabilities at scale $2n$ are expressed in terms of crossing probabilities at scale n. The renormalization scheme is built in such a way that as soon as the crossing probability passes below a certain threshold, they start decaying stretched exponentially fast. As a consequence, either crossing probabilities remain bounded away from 0, or they decay to 0 stretched exponentially fast. Let $\mathcal{A}_n$ be the event that there exists a circuit[14] of open edges in $\Lambda_{2n}\setminus\Lambda_n$ surrounding the origin and set

$$u_n:=\phi^0_{\Lambda_{8n}}[\mathcal{A}_n].$$

The proof articulates around Proposition 10 below, which relates u_n to u_{7n}.

Proposition 10 *There exists a constant $C<\infty$ such that $u_{7n}\le C\,u_n^2$ for all $n\ge 1$.*

[14]i.e. a path of edges starting and ending at the same point.

This statement allows us to prove recursively that $Cu_{7^k n} \leq (Cu_n)^{2^k}$. In particular, if there exists n such that $Cu_n < 1$, then $u_{7^k n}$ decays stretched exponentially fast. Therefore, the proof of **P4a** $\Rightarrow$ **P5** follows trivially from the previous proposition and the following fairly elementary facts:

- (u_n) bounded away from 0 implies **P5** (Exercise 47).
- Stretched exponential decay of $u_{7^k n}$ implies stretched exponential decay of $\phi^0[0 \leftrightarrow \partial\Lambda_n]$ (Exercise 48). Note that this last fact is very intuitive since in order to have a circuit in $\Lambda_{2n} \setminus \Lambda_n$ surrounding the origin, one must have fairly big clusters.

Exercise 47 1. Fix $\varepsilon > 0$ and $\rho > 0$. Combine circuits in annuli to prove the existence of $c = c(\rho, \varepsilon) > 0$ such that for all n, if $R = [-\varepsilon n, (\rho + \varepsilon)n] \times [-\varepsilon n, (1 + \varepsilon)n]$ then $\phi^0_R[\mathcal{H}(\rho n, n)] \geq c$.

2. Deduce that $\phi^\xi_R[\mathcal{H}(\rho n, n)] \geq c$. for every boundary conditions ξ.

3. Use duality to prove that for any ξ and n, $\phi^\xi_R[\mathcal{H}(\rho n, n)] \leq 1 - c'$ for some $c' = c'(\rho, \varepsilon) > 0$.

Exercise 48 In this exercise, we assume that $\limsup \frac{1}{n^\alpha} \log \phi^0[0 \longleftrightarrow \partial\Lambda_n] = 0$ for some constant $\alpha > 0$.

1. Prove that $\phi^0[0 \longleftrightarrow \partial\Lambda_n] \leq \sum_{k \geq n} \sum_{x \notin \Lambda_k} \phi^0_{\Lambda_k}[0 \longleftrightarrow x]$. *Hint*. Use the farthest point on the cluster of 0 and an argument similar to Exercise 16.

2. Deduce that $\limsup_{x \to \infty} \frac{1}{k^\alpha} \log \phi^0_{\Lambda_k}[0 \longleftrightarrow x] = 0$, where k is defined in such a way that $x \in \partial\Lambda_k$.

3. Prove that $\limsup_{x \to \infty} \frac{1}{k^\alpha} \log \phi^0_{\Lambda_{3k}}[0 \longleftrightarrow 2ke_1] = 0$, where $e_1 = (1, 0, \ldots, 0)$.

4. Prove that for every n, $\limsup_{k \to \infty} \frac{1}{k^\alpha} \log u_{7^k n} = 0$.

To prove Proposition 10, first consider the strip $\mathbb{S} = \mathbb{Z} \times [-n, 2n]$, and the random-cluster measure $\phi^{1/0}_{\mathbb{S}}$ with free boundary conditions on $\mathbb{Z} \times \{-n\}$ and wired everywhere else. We refer to Exercise 50 for the (slightly technical) proof of this lemma.

Lemma 11 *For all $\rho > 0$, there exists a constant $c > 0$ such that for all $n \geq 1$,*

$$\phi^{1/0}_{\mathbb{S}}[\mathcal{H}(\rho n, n)] \geq c. \tag{5.7}$$

Even though we do not provide a proof of this statement, the intuition is fairly convincing: the boundary conditions are still somehow "balanced" between primal and dual configurations, and it is therefore not so surprising that crossing probabilities are bounded away from above.

In the next lemma, we consider horizontal crossings in rectangular shaped domains with free boundary conditions on the bottom and wired elsewhere.

Lemma 12 *For all $\rho > 0$ and $\ell \geq 2$, there exists $c = c(\rho, \ell) > 0$ such that for all $n > 0$,*

$$\phi^{1/0}_D[\mathcal{H}(\rho n, n)] \geq c \tag{5.8}$$

with $D = [0, \rho n] \times [-n, \ell n]$, and $\phi^{1/0}_D$ is the random-cluster measure with free boundary conditions on the bottom side, and wired on the three other sides.

Proof For $\ell = 2$, Lemma 11 and the comparison between boundary conditions (used on the sides) imply the result readily. Now, assume that the result holds for ℓ and let us prove it for $\ell + 1$. The comparison between boundary conditions in $[0, \rho n] \times [0, (\ell + 1)n]$ implies that

$$\phi_D^{1/0}[\mathcal{H}(R)] \geq c(\rho, \ell),$$

where $R = [0, \rho n] \times [n, 2n]$. The comparison between boundary conditions implies that conditioned on $\mathcal{H}(R)$, the measure restricted to edges in $R' := [0, \rho n] \times [0, n]$ dominates the restriction (to R') of the measure on $D' := \mathbb{S} \cap D$ with free boundary conditions on the bottom of D', and wired on the other sides. We deduce that

$$\phi_D^{1/0}[\mathcal{H}(\rho n, n) \cap \mathcal{H}(R)] \geq c(\rho, 2)\phi_D^{1/0}[\mathcal{H}(R)] \geq c(\rho, 2)c(\rho, \ell). \qquad \square$$

Proof (*Proposition* 10) Fix $n \geq 1$ and set $N := 56n$. Below, the constants c_i are independent of n. Define $\mathcal{A}_n^{\pm}$ to be the translates of the event $\mathcal{A}_n$ by $z_{\pm} := (\pm 5n, 0)$.

Conditioned on $\mathcal{A}_{7n}$, the restriction of the measure to Λ_{7n} dominates the restriction of the measure with wired boundary conditions at infinity. Using this in the second inequality, we find

$$\phi_{\Lambda_N}^0[\mathcal{A}_n^+ \cap \mathcal{A}_n^-] \geq \phi_{\Lambda_N}^0[\mathcal{A}_n^+ \cap \mathcal{A}_n^- \cap \mathcal{A}_{7n}] \geq \phi^1[\mathcal{A}_n^+ \cap \mathcal{A}_n^-]\, u_{7n} \geq c_1 u_{7n}, \tag{5.9}$$

where in the last inequality we combined crossings in rectangles of aspect ratio 4 to create circuits, and then used Theorem 15 to bound the probability from below (which is justified since the boundary conditions are wired at infinity).

Let $\mathcal{B}_n$ be the event that $R := [-N, N] \times [-2n, 2n]$ is not connected to $R' := [-N, N] \times [-3n, 3n]$. Under $\phi_N^0[\,\cdot\,|\mathcal{A}_n^+ \cap \mathcal{A}_n^-]$, the boundary conditions outside of R are dominated by wired boundary conditions on R and free boundary conditions on the boundary of Λ_N. As a consequence, Lemma 12 applied to the dual measure in the two rectangles $[-N, N] \times [2n, N]$ and $[-N, N] \times [-N, -2n]$ implies that

$$\phi_{\Lambda_N}^0[\mathcal{B}_n | \mathcal{A}_n^+ \cap \mathcal{A}_n^-] \geq c_2. \tag{5.10}$$

Altogether, (5.9) and (5.10) lead to the estimate

$$\phi_{\Lambda_N}^0[\mathcal{A}_n^+ \cap \mathcal{A}_n^- \cap \mathcal{B}_n] \geq c_3\, u_{7n}. \tag{5.11}$$

Define C to be the set of points in R' which are not connected to the top or the bottom sides of R'. Let $\mathcal{C}_n$ be the event that the left and right sides of $S := [-3n, 3n]^2$ are not connected together in C. Conditionally on $A_n^+ \cap \mathcal{A}_n^- \cap \mathcal{B}_n \cap \{\mathsf{C} = C\}$, the boundary conditions in C are dominated by the boundary conditions of the restriction (to C) of the measure in S with free on the top and bottom sides of S, and wired on the left and right. A duality argument in S implies that the probability to have a top to bottom dual crossing is bounded from below by $\frac{1}{1+q}$. This implies that in C, the probability of a dual crossing from top to bottom is a fortiori bounded from below by $\frac{1}{1+q}$. Averaging on the possible values of C, this implies (Fig. 10)

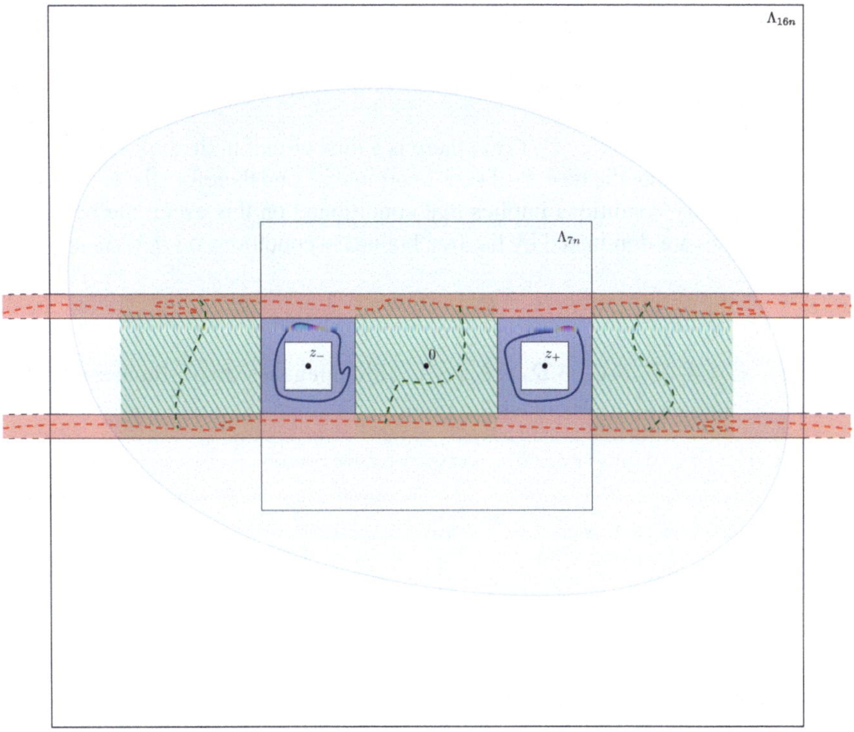

Fig. 10 The different events involved in the construction. In light blue, a circuit in ω implying the occurrence of the event $\mathcal{A}_{7n}$. Inside the circuit, the measure dominates a random-cluster measure with wired boundary conditions at infinity. Therefore, conditionally on $\mathcal{A}_{7n}$, one can construct the two blue circuits (corresponding to $\mathcal{A}_n^{\pm}$) with positive probability. For the rest of the construction, the event $\mathcal{A}_{7n}$ is not taken into account anymore. The dashed paths in the red areas correspond to two paths in ω^*, which imply the occurrence of the event $\mathcal{B}_n$. In the top red rectangle (which goes further left and right, but could not be drawn on the picture), conditionally on $\mathcal{A}_n^+ \cap \mathcal{A}_n^-$, the boundary conditions are dominated by wired boundary conditions on the bottom and free on the boundary of $\partial \Lambda_N$, hence one can apply Lemma 12. The same reasoning is valid for the bottom red rectangle. The dashed green paths correspond to events in ω^* implying the occurrence of $\mathcal{C}_n$, $\mathcal{D}_n$ and $\mathcal{E}_n$. The dashed green areas correspond to the intersection of C with $[-3n, 3n]^2$, $[-13n, -7n] \times [-3n, 3n]$ and $[7n, 13n] \times [-3n, 3n]$. In these areas, the boundary conditions are dominated by the wired boundary conditions on left and right, and free on top and bottom

$$\phi^0_{\Lambda_N}[\mathcal{C}_n \mid \mathcal{A}_n^+ \cap \mathcal{A}_n^- \cap \mathcal{B}_n] \geq \tfrac{1}{1+q}. \tag{5.12}$$

A similar reasoning gives that if $\mathcal{D}_n$ and $\mathcal{E}_n$ denote respectively the events that the left and right sides of $[-13n, 7n] \times [-3n, 3n]$ and $[7n, 13n] \times [-3n, 3n]$ are not connected in S, we have

$$\phi^0_{\Lambda_N}[\mathcal{C}_n \cap \mathcal{D}_n \cap \mathcal{E}_n \mid \mathcal{A}_n^+ \cap \mathcal{A}_n^- \cap \mathcal{B}_n] \geq \tfrac{1}{(1+q)^3}$$

which, together with (5.11), leads to

$$\phi^0_{\Lambda_N}[\mathcal{A}_n^+ \cap \mathcal{A}_n^- \cap \mathcal{B}_n \cap \mathcal{C}_n \cap \mathcal{D}_n \cap \mathcal{E}_n] \ge c_4\, u_{7n}. \tag{5.13}$$

Now, on $\mathcal{A}_n^- \cap \mathcal{B}_n \cap \mathcal{C}_n \cap \mathcal{D}_n \cap \mathcal{E}_n$, there is a dual circuit in the box Λ of size $8n$ around z_+ surrounding the box Λ' of size $2n$ around z_+ and therefore the comparison between boundary conditions implies that conditioned on this event, the boundary conditions in Λ' are dominated by the free boundary conditions on $\partial\Lambda$. As a consequence,

$$\phi^0_{\Lambda_N}[\mathcal{A}_n^+ | \mathcal{A}_n^- \cap \mathcal{B}_n \cap \mathcal{C}_n \cap \mathcal{D}_n \cap \mathcal{E}_n] \le \phi^0_{\Lambda_{8n}}[\mathcal{A}_n] = u_n.$$

Similarly, $\phi^0_{\Lambda_N}[\mathcal{A}_n^- | \mathcal{B}_n \cap \mathcal{C}_n \cap \mathcal{D}_n \cap \mathcal{E}_n] \le u_n$. Plugging these two estimates in (5.13) gives $u_n^2 \ge c_4 u_{7n}$, which concludes the proof. □

Exercise 49 Fill in the details of the different comparison between boundary conditions used in the last two proofs.

Exercise 50 Below, we use the notation $A \overset{S}{\longleftrightarrow} B$ to denote the existence of a path from A to B staying in S. We assume that $\phi_{\mathbb{S}}^{1/0}[\mathcal{H}(\rho n, n)] \le \frac{1}{2}$. Set

$$\mathbb{S}' := \mathbb{Z} \times [0, n] \quad R := [0, 9\lambda] \times [0, n] \quad R' := [4\lambda, 9\lambda] \times [0, n].$$

Set $\lambda = n/11$ and $\ell_i = [i\lambda, (i+1)\lambda] \times \{0\}$.

1. Show that if $\phi_{\mathbb{S}}^{1/0}[\ell_i \overset{\mathbb{S}'}{\longleftrightarrow} \ell_{i+2}] \ge c$, then $\phi_{\mathbb{S}}^{1/0}[\mathcal{H}(\rho n, n)] \ge c^{11\rho}$.
2. Show that $\phi_{\mathbb{S}}^{1/0}[\mathcal{V}(\rho n, n)] \ge \frac{1}{2}$.
3. Deduce that one of the following two conditions occur:

C1 $\phi_{\mathbb{S}}^{1/0}[\ell_4 \overset{R}{\longleftrightarrow} \mathbb{Z} \times \{n\}] \ge \frac{1}{44\rho}$.

C2 $\phi_{\mathbb{S}}^{1/0}[\ell_4 \overset{R}{\longleftrightarrow} \{0\} \times \mathbb{Z}] \ge \frac{1}{88\rho}$.

4. Assume that C1 holds true. Show that $\phi_{\mathbb{S}}^{1/0}[\ell_2 \overset{\mathbb{S}'}{\longleftrightarrow} \ell_4] \ge \frac{1}{1+q}(\frac{1}{36\rho})^2$. *Hint*. Reason as in the proof of (5.12).
5. Assume that C2 holds true. Show that

$$\phi_{\mathbb{S}}^{1/0}[\{\ell_4 \overset{R}{\longleftrightarrow} \{7\lambda\} \times \mathbb{Z}\} \cap \{\ell_6 \overset{R'}{\longleftrightarrow} \{4\lambda\} \times \mathbb{Z}\}] \ge (\tfrac{1}{88\rho})^2.$$

* Construct a symmetric domain to prove that

$$\phi_{\mathbb{S}}^{1/0}[\ell_4 \overset{\mathbb{S}'}{\longleftrightarrow} \ell_6] \ge \tfrac{1}{1+q}(\tfrac{1}{88\rho})^2.$$

6. Conclude.

5.2 *Proving Continuity for q ≤ 4: The Parafermionic Observables*

In this section, we prove that for $q \in [1, 4]$, **P1–5** are satisfied by proving that **P4a** is satisfied. In order to do so, we introduce the so-called parafermionic observables.

The next section is intended to offer an elementary application of the parafermionic observable by studying a slightly different problem, namely the question of computing the connective constant of the hexagonal lattice. We will then go back to the random-cluster model later on.

Computing the connective constant of the hexagonal lattice Let $\mathbb{H} = (\mathbb{V}, \mathbb{E})$ be the hexagonal lattice (for now, we assume that 0 is a vertex of $\mathbb{H}$ and we assume that the edge on the right of 0 is horizontal). Points in the plane are considered as complex numbers. A *walk* γ *of length* n is a path $\gamma : \{0, \ldots, n\} \mapsto \mathbb{V}$ such that $\gamma_0 = 0$ and $\gamma_i\gamma_{i+1} \in \mathbb{E}$ for any $i < n$. The walk is *self-avoiding* if $\gamma_i = \gamma_j$ implies $i = j$. Let c_n be the number of self-avoiding walks of length n.

A self-avoiding walk of length $n + m$ can be uniquely cut into a self-avoiding walk of length n and a translation of a self-avoiding walk of length m. Hence,

$$c_{n+m} \le c_n c_m,$$

from which it follows (by Fekete's lemma on sub-multiplicative sequences of real numbers) that there exists $\mu_c \in [1, +\infty)$, called the *connective constant*, such that

$$\mu_c := \lim_{n\to\infty} c_n^{1/n}.$$

On the hexagonal lattice, Nienhuis [107, 108] used the Coulomb gas formalism to conjecture non-rigorously what μ_c should be. In this section, we present a mathematical proof of this prediction.

Theorem 17 (Duminil-Copin and Smirnov [59]) *We have* $\mu_c = \sqrt{2 + \sqrt{2}}$.

Before diving into the argument, let us recall the following classical fact. We choose to leave the proof of this statement as an exercise (Exercise 51) since the argument is instructive. A *self-avoiding bridge* is a self-avoiding walk $\gamma : \{0, \ldots, n\} \mapsto \mathbb{H}$ satisfying that $0 < \mathrm{Re}(\gamma_i) \le \mathrm{Re}(\gamma_n)$ for every $1 \le i \le n$. Let b_n be the number of bridges of length n.

Proposition 11 (Hammersley-Welsh *[79]). We have that* $\lim_{n\to\infty} b_n^{1/n} = \mu_c$.

Exercise 51 1. Prove that $b_n^{1/n}$ converges to a value μ and that $b_n \le \mu^n$ for all n.
2. Let h_n be the number of (half-space) self-avoiding walks with $\mathrm{Re}(\gamma_i) > 0$ for all $i \ge 1$. Prove that

$$c_n \le \sum_{k=0}^{n} h_{k+1} h_{n+1-k}.$$

Hint. Cut the walk at a point of maximal first coordinate and add horizontal edges. Deduce that $\lim_{n\to\infty} h_n^{1/n} = \mu_c$.
3. By decomposing with respect to the last point with maximal first coordinate, show that

$$h_n \le \sum_{k=0}^{n} b_k h_{n-k}.$$

4. Let p_n be the number of partitions of n into integers, i.e. the number of $h_1 \geq h_2 \geq \cdots \geq h_\ell$ such that $h_1 + \cdots + h_\ell = n$. Let $P_n = \sum_{k=0}^{n} p_k$. By iterating the decomposition above, and observing that the width of the different half-space walks is decreasing, deduce that

$$h_n \leq P_n \mu^n.$$

5. Prove that the generation function $P$of the number p_n of partitions of an integer satisfies

$$P(t) = \sum_{n=0}^{\infty} P_n t^n = \prod_{n=1}^{\infty} \frac{1}{1-t^n}.$$

6. Deduce that $\mu = \mu_c$. Remark: One may also invoke a result of Hardy-Ramanujan stating that $p_n \leq \exp(O(\sqrt{n}))$ to make the previous result quantitative.

Assume that the lattice has mesh size 1 and is shifted by $(-\frac{1}{2}, 0)$ so that the origin is now a *mid-edge*, i.e. the middle of an edge, which we call a. We also assume that this edge is horizontal (as in Fig. 11). We now consider that self-avoiding walks are in fact starting at a and ending at mid-edges. Their length, denoted by $|\gamma|$, is still the

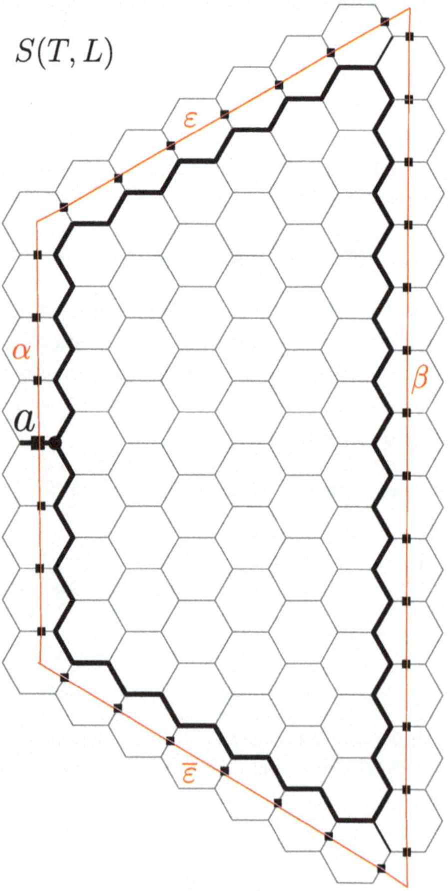

Fig. 11 The graph $S(T, L)$ and its boundary parts α, β, ε and $\bar{\varepsilon}$

number of vertices on it. We consider a truncated vertical strip $S(T, L)$ of width T cut at height L at an angle of $\pi/3$ (see Fig. 11), i.e.

$$S(T, L) := \{z \in \mathbb{C} : 0 \leq \text{Re}(z) \leq \tfrac{3}{2}T \text{ and } \sqrt{3}|\text{Im}(z)| \leq 3L + \text{Re}(z)\}.$$

Denote by α the left boundary of $S(T, L)$ and by β the right one. Symbols ε and $\bar{\varepsilon}$ denote the top and bottom boundaries of $S(T, L)$. For $x > 0$, introduce the following quantities:

$$A_{T,L} := \sum_{\substack{\gamma \subseteq S(T,L) \\ \gamma \text{ ends on } \alpha \setminus \{a\}}} x^{|\gamma|} \qquad B_{T,L} := \sum_{\substack{\gamma \subseteq S(T,L) \\ \gamma \text{ ends on } \beta}} x^{|\gamma|} \qquad E_{T,L} := \sum_{\substack{\gamma \subseteq S(T,L) \\ \gamma \text{ ends on } \varepsilon \cup \bar{\varepsilon}}} x^{|\gamma|}.$$

We will prove the following lemma.

Lemma 13 *If $x := 1/\sqrt{2 + \sqrt{2}}$, then for any $T, L \geq 0$,*

$$1 = \cos\left(\tfrac{3\pi}{8}\right) A_{T,L} + B_{T,L} + \cos\left(\tfrac{\pi}{4}\right) E_{T,L}. \tag{5.14}$$

Before proving this statement, let us show how it implies the claim. Observe that sequences $(A_{T,L})_{L>0}$ and $(B_{T,L})_{L>0}$ are increasing in L and are bounded. They therefore converge. We immediately deduce that $(E_{T,L})_{L>0}$ also does. Let A_T, B_T and E_T be the corresponding limits.

Upper bound on the connective constant Observe that $B_T \leq 1$ for any T (since $B_{T,L} \leq 1$) so that for any $y < x$,

$$\sum_{n=0}^{\infty} b_n y^n \leq \sum_{T \geq 0} B_T \left(\tfrac{y}{x}\right)^T < \infty$$

(we use that a bridge of width T has length at least T). Proposition 11 thus implies

$$\mu_c = \lim_{n \to \infty} b_n^{1/n} \leq \sqrt{2 + \sqrt{2}}. \tag{5.15}$$

Lower bound on the connective constant Assume first that $E_T > 0$ for some T. Then,

$$\sum_{n=0}^{n} c_n x^n \geq \sum_{L=0}^{\infty} E_{T,L} = +\infty,$$

which implies $\mu_c \geq \sqrt{2 + \sqrt{2}}$. Assume on the contrary that $E_T = 0$ for all T. Taking the limit in (5.14) implies

$$1 = \cos\left(\tfrac{3\pi}{8}\right) A_T + B_T. \tag{5.16}$$

Observe that self-avoiding walks entering into account for A_T and not for A_{T-1} have to visit a vertex $x \in \mathbb{V}$ on the right of the strip of width T, i.e. satisfying

$\mathrm{Re}(x) = \frac{3}{2}T - \frac{1}{2}$. Cutting such a walk at the first such point (and adding half-edges to the two halves), we obtain two bridges. We conclude that

$$A_T - A_{T-1} \le \tfrac{1}{x} B_T^2. \tag{5.17}$$

Combining (5.16) for $T - 1$ and T with (5.17) gives

$$0 = 1 - 1 = \cos\left(\tfrac{3\pi}{8}\right)(A_T - A_{T-1}) + B_T - B_{T-1} \le \cos\left(\tfrac{3\pi}{8}\right)\tfrac{1}{x}B_T^2 + B_T - B_{T-1},$$

so

$$\cos\left(\tfrac{3\pi}{8}\right)\tfrac{1}{x}B_T^2 + B_T \ge B_{T-1}.$$

By induction, it is easy to check that

$$B_T \ge \frac{\min[B_1, x/\cos\left(\frac{3\pi}{8}\right)]}{T}$$

for every $T \ge 1$. This implies that $\mu_c \ge \sqrt{2+\sqrt{2}}$ in this case as well since

$$\sum_{n=0}^{\infty} b_n x^n = \sum_{T=0}^{\infty} B_T = +\infty.$$

At the light of the previous discussion, we shall now prove Lemma 13. Fix T and L. Introduce the *parafermionic observable*[15] defined as follows: for a mid-edge z in $S(T, L)$, set

$$F(z) := \sum_{\substack{\gamma \subseteq S(T,L) \\ \gamma \text{ ends at } z}} \mathrm{e}^{-\mathrm{i}\sigma W_\gamma(a,z)} x^{|\gamma|},$$

where $\sigma := \frac{5}{8}$ and $W_\gamma(u, v)$ is equal to $\frac{\pi}{3}$ times the number of left turns minus the number of right turns made by the walk γ when going from u to v.

Lemma 14 *For any $v \in \mathbb{V} \cap S(T, L)$,*

$$(p - v)F(p) + (q - v)F(q) + (r - v)F(r) = 0, \tag{5.18}$$

where p, q, r are the mid-edges of the three edges incident to v.

Proof In this proof, we further assume that the mid-edges p, q and r are oriented counterclockwise around v. Note that $(p - v)F(p) + (q - v)F(q) + (r - v)F(r)$ is a sum of "contributions"

[15] Let us mention that there are other instances of parafermionic observables for the self-avoiding walk, see [9, 72]. We do not discuss this further here since our goal is to quickly move back to the random-cluster model.

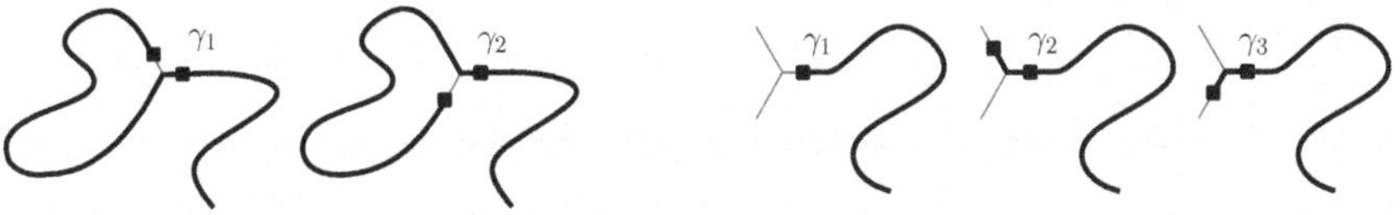

Fig. 12 **Left**: a pair of walks visiting the three mid-edges and matched together. **Right**: a triplet of walks, one visiting one mid-edge, the other two visiting two mid-edges, which are matched together

$$c(\gamma) = (z - v)\mathrm{e}^{-\mathrm{i}\sigma W_\gamma(a,z)} x^{|\gamma|}$$

over all possible walks γ finishing at $z \in \{p, q, r\}$. The set of such walks can be partitioned into pairs and triplets of walks in the following way, see Fig. 12:

Walks visiting the three mid-edges p, q and r can be grouped in pairs: If a walk γ_1 visits all three mid-edges, it means that the edges belonging to γ_1 form a self-avoiding path up to v plus (up to a half-edge) a self-avoiding loop from v to v. One can associate to γ_1 the walk passing through the same edges, but exploring the loop from v to v in the other direction.

Walks not visiting the three mid-edges p, q and r can be grouped in triplets: If a walk γ_1 visits only one mid-edge, it can be grouped with two walks γ_2 and γ_3 that visit exactly two mid-edges by prolonging the walk one step further (there are two possible choices). The reverse is true: a walk visiting exactly two mid-edges belongs to the group of a walk visiting only one mid-edge (this walk is obtained by erasing the last step).

If the sum of contributions for each pair and each triplet described above vanishes, then the total sum is zero. We now intend to show that this is the case.

Let γ_1 and γ_2 be two walks that are grouped as in the first case. Without loss of generality, we assume that γ_1 ends at q and γ_2 ends at r. Since γ_1 and γ_2 coincide up to the mid-edge p (they are matched together), we deduce that $|\gamma_1| = |\gamma_2|$ and

$$\begin{aligned} W_{\gamma_1}(a, q) &= W_{\gamma_1}(a, p) + W_{\gamma_1}(p, q) = W_{\gamma_1}(a, p) - \tfrac{4\pi}{3}, \\ W_{\gamma_2}(a, r) &= W_{\gamma_2}(a, p) + W_{\gamma_2}(p, r) = W_{\gamma_1}(a, p) + \tfrac{4\pi}{3}. \end{aligned}$$

In order to evaluate the winding of γ_1 between p and q, we used the fact that a is on the boundary of $S(T, L)$ so that the walk does necessarily four more turns on the right than turns on the left between p and q. Altogether,

$$\begin{aligned} c(\gamma_1) + c(\gamma_2) &= (q - v)\mathrm{e}^{-\mathrm{i}\sigma W_{\gamma_1}(a,q)} x^{|\gamma_1|} + (r - v)\mathrm{e}^{-\mathrm{i}\sigma W_{\gamma_2}(a,r)} x^{|\gamma_2|} \\ &= (p - v)\mathrm{e}^{-\mathrm{i}\sigma W_{\gamma_1}(a,p)} x^{|\gamma_1|} \left(j\bar{\lambda}^4 + \bar{j}\lambda^4\right) = 0 \end{aligned}$$

where $j = \mathrm{e}^{\mathrm{i}2\pi/3}$ and $\lambda = \exp(-\mathrm{i}5\pi/24)$ (here we use the crucial choice of $\sigma = \frac{5}{8}$).

Let $\gamma_1, \gamma_2, \gamma_3$ be three walks matched as in the second case. Without loss of generality, we assume that γ_1 ends at p and that γ_2 and γ_3 extend γ_1 to q and r

respectively. As before, we easily find that $|\gamma_2| = |\gamma_3| = |\gamma_1| + 1$ and

$$W_{\gamma_2}(a,q) = W_{\gamma_2}(a,p) + W_{\gamma_2}(p,q) = W_{\gamma_1}(a,p) - \tfrac{\pi}{3},$$
$$W_{\gamma_3}(a,r) = W_{\gamma_3}(a,p) + W_{\gamma_3}(p,r) = W_{\gamma_1}(a,p) + \tfrac{\pi}{3}.$$

Following the same steps as above, we obtain

$$c(\gamma_1) + c(\gamma_2) + c(\gamma_3) = (p - v)e^{-i\sigma W_{\gamma_1}(a,p)} x^{|\gamma_1|} \left(1 + xj\bar{\lambda} + x\bar{j}\lambda\right) = 0.$$

Here is the *only* place where we use the crucial fact that $x^{-1} = \sqrt{2+\sqrt{2}} = 2\cos\frac{\pi}{8}$. The claim follows readily by summing over all pairs and triplets. □

Exercise 52 (*Parafermionic observable for the loop $O(n)$-model*) Consider the loop $O(n)$ model defined as follows. Let $E(\Omega)$ be the set of even subgraphs of $\Omega \subseteq \mathbb{H}$ (equivalently, these are the families of non-intersecting loops). Also, let $E_{a,z}(\Omega)$ be the family of loops, plus one self-avoiding walk $\gamma(\omega)$ going from a to z not intersecting any of the loops. Define the *parafermionic observable*

$$F(z) = \sum_{\omega \in E_{a,z}(\Omega)} \mathrm{e}^{-i\sigma W_\gamma(a,z)} x^{|\omega|} n^{\ell(\omega)},$$

where $|\omega|$ is the total length of the loops and the self-avoiding walk, and $\ell(\omega)$ is the number of loops. Note that this model generalizes both the self-avoiding walk ($n = 0$) and the Ising model on the hexagonal lattice ($n = 1$) via the high-temperature expansion.
Show that for $n \in [0, 2]$, there exist two values of σ, and for each one a single value of x such that F satisfies (5.18). *The smallest of the two values of x is conjectured by Nienhuis to be the critical point of the model.*

Proof (*Lemma* 13) Sum the relation (5.18) over all $v \in \mathbb{V} \cap S(T, L)$. Values at interior mid-edges cancel and we end up with

$$0 = -\sum_{z\in\alpha} F(z) + \sum_{z\in\beta} F(z) + j\sum_{z\in\varepsilon} F(z) + \bar{j}\sum_{z\in\bar{\varepsilon}} F(z), \tag{5.19}$$

where $j = \mathrm{e}^{2i\pi/3}$. Using the symmetry of the domain with respect to the x axis, we deduce that $F(\bar{z}) = \bar{F}(z)$. Observe that the winding of any self-avoiding walk from a to the bottom part of α is $-\pi$ while the winding to the top part is π. We conclude

$$\sum_{z\in\alpha} F(z) = F(a) + \sum_{z\in\alpha\setminus\{a\}} F(z) = 1 + \frac{\mathrm{e}^{-i5\pi/8} + \mathrm{e}^{i5\pi/8}}{2} A_{T,L} = 1 - \cos\left(\tfrac{3\pi}{8}\right) A_{T,L}.$$

Above, we have used the fact that the only walk from a to a is of length 0. Similarly, the winding from a to any half-edge in β (resp. ε and $\bar{\varepsilon}$) is 0 (resp. $\frac{2\pi}{3}$ and $-\frac{2\pi}{3}$), therefore

$$\sum_{z\in\beta} F(z) = B_{T,L} \quad \text{and} \quad j\sum_{z\in\varepsilon} F(z) + \bar{j}\sum_{z\in\bar{\varepsilon}} F(z) = \cos\left(\tfrac{\pi}{4}\right) E_{T,L}.$$

The lemma follows readily by plugging these three formulæ in (5.19). □

The Proof of Lemma 13 can be understood in the following way. Coefficients in (5.18) are three cubic roots of unity multiplied by $p - v$, so that the left-hand side can be seen as a discrete integral along an elementary contour on the dual lattice in the following sense. For a closed path $c = (z_i)_{i \leq n}$ of vertices in the triangular lattice $\mathbb{T}$ dual to $\mathbb{H}$, define the discrete integral of a function F on mid-edges by

$$\oint_c F(z)dz := \sum_{i=0}^{n-1} F\left(\tfrac{z_i + z_{i+1}}{2}\right)(z_{i+1} - z_i). \tag{5.20}$$

Equation (5.18) at $v \in \mathbb{V}$ implies that the discrete contour integral going around the face of $\mathbb{T}$ corresponding to v is zero. Decomposing a closed path into a sum of elementary triangles gives that the discrete integral along any closed path vanishes.

The fact that the integral of the parafermionic observable along closed path vanishes is a glimpse of conformal invariance of the model in the sense that the observable satisfies a weak notion of discrete holomorphicity. Nevertheless, these relations do not uniquely determine F. Indeed, the number of mid-edges (and therefore of unknown variables) exceeds the number of linear relations (5.18) (which corresponds to the number of vertices). Nonetheless, one can combine the fact that the discrete integral along the exterior boundary of $S(T, L)$ vanishes with the fact that the winding of self-avoiding walks ending at boundary mid-edges is deterministic and explicit. This extra information is sufficient to derive some non-trivial information on the model. In the next section, we will use a similar idea in the case of random-cluster models.

The loop representation and the parafermionic observable In order to define parafermionic observables for random-cluster models, we first discuss the loop representation of the model.

In the definitions below, we recommend looking at Figs. 13, 14 *and* 15.

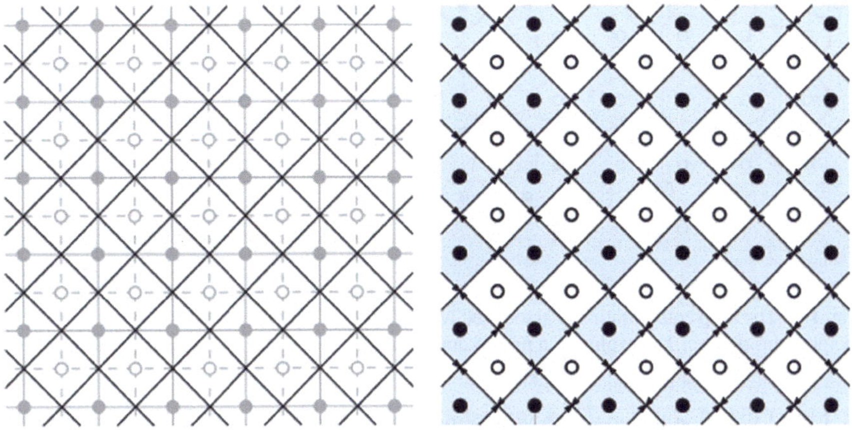

Fig. 13 On the left, the lattice $\mathbb{Z}^2$, its dual lattice $(\mathbb{Z}^2)^*$ and medial lattice $(\mathbb{Z}^2)^\diamond$. On the right, a natural orientation on the medial lattice

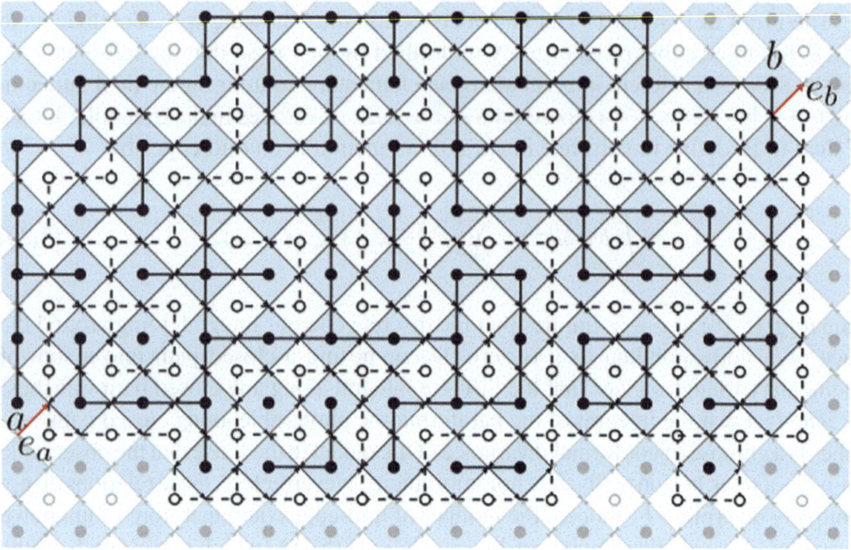

Fig. 14 The configuration ω (in bold lines) with its dual configuration ω^* (in dashed lines). Notice that the edges of ω are open on (ba), and that those of ω^* are open on $(ab)^*$

Fig. 15 The loop configuration $\overline{\omega}$ associated with the primal and dual configurations ω and ω^* in the previous picture. The exploration path is drawn in bold. It starts at e_a and finishes at e_b

Let Ω be a connected graph with connected complement in $\mathbb{Z}^2$, and a and b two vertices on its boundary. The triplet (Ω, a, b) is called a Dobrushin domain. The set $\partial\Omega$ is divided into two boundary arcs denoted by (ab) and (ba): the first one goes from a to b when going counterclockwise around $\partial\Omega$, while the second goes from b to a. The *Dobrushin boundary conditions* are defined to be free on (ab) and wired on (ba). In other words, the partition is composed of (ba) together with singletons. Note that the state of edges on (ba) is now irrelevant since the vertices of (ba) are wired together anyway. We will therefore consider that edges on (ba) are not in Ω (this will be relevant when defining Ω^*). Also, the Dobrushin boundary conditions are planar, and it is therefore convenient to choose a configuration ξ inducing them. We set $\xi_e = 0$ for all $e \in \mathbb{E}\setminus E$ except for edges on (ba), for which $\xi_e = 1$. Below, the measure on (Ω, a, b) with Dobrushin boundary conditions is denoted by $\phi^{a,b}_{\Omega,p,q}$.

Let Ω^* be the dual of the graph Ω (recall that edges in (ba) are not part of Ω anymore). We draw the dual configuration ω^* with the additional condition that edges between vertices of $\partial\Omega^*$ that are bordering (ab) are open in ω^* (we call the set of such edges $(ab)^*$). This is coherent with the duality relation since the dual boundary conditions of the Dobrushin ones are induced by the configuration ξ^* equal to 1 on $(ab)^*$, and 0 elsewhere. Keep in mind that from this point of view, primal and dual models play symmetric roles with respect to Dobrushin boundary conditions.

We now explain how to construct the loop configuration, which is defined on another graph, called the *medial graph*. This graph is defined as follows. Let $(\mathbb{Z}^2)^\diamond$ be the *medial* lattice defined as follows. The set of vertices is given by the midpoints of edges of $\mathbb{Z}^2$. The edges are pairs of nearest vertices (i.e. vertices at a distance $\sqrt{2}/2$ of each other). It is a rotated and rescaled version of $\mathbb{Z}^2$, see Fig. 13. For future reference, note that the edges of the medial lattice can be oriented in a counterclockwise way around faces that are centered on a vertex of $\mathbb{Z}^2$ (the dark faces on Fig. 13). Let $\Omega^\diamond$ be the subgraph of $(\mathbb{Z}^2)^\diamond$ made of vertices corresponding to an edge of Ω or Ω^*. Let e_a and e_b be the two medial edges entering and exiting $\Omega^\diamond$ between the arc (ba) and $(ab)^*$ (see Fig. 14).

Draw self-avoiding loops on $\Omega^\diamond$ as follows: a loop arriving at a vertex of the medial lattice always takes a $\pm\pi/2$ turn at vertices so as not to cross the edges of ω or ω^*, see Fig. 15. The loop configuration is defined in an unequivocal way since:

- there is either an edge of ω or an edge of ω^* crossing non-boundary vertices in $\Omega^\diamond$, and therefore there is exactly one coherent way for the loop to turn at non-boundary vertices.
- the edges of ω in (ba) and the edges of ω^* in $(ab)^*$ are such that the loops at boundary vertices turn in order to remain in $\Omega^\diamond$.

From now on, the loop configuration associated with ω is denoted by $\overline{\omega}$. Beware that the denomination is slightly misleading: $\overline{\omega}$ is made of loops together with a self-avoiding path going from e_a to e_b, see Figs. 15. This curve is called the *exploration path* and is denoted by $\gamma = \gamma(\omega)$.

We allow ourselves a slight abuse of notation: below, $\phi^{a,b}_{\Omega,p,q}$ denotes the measure on percolation configurations as well as its push forward by the map $\omega \mapsto \overline{\omega}$. Therefore, the measure $\phi^{a,b}_{\Omega,p,q}$ will sometimes refer to a measure on loop configurations.

Proposition 12 *Let Ω be a connected finite subgraph of $\mathbb{Z}^2$ connected complement in $\mathbb{Z}^2$. Let $p \in [0, 1]$ and $q > 0$. For any configuration ω,*

$$\phi^{a,b}_{\Omega,p,q}[\overline{\omega}] = \frac{x^{o(\omega)}\sqrt{q}^{\ell(\overline{\omega})}}{\overline{Z}_{\Omega,p,q}},$$

where $x := \frac{p}{\sqrt{q}(1-p)}$, $\ell(\overline{\omega})$ is the number of loops[16] *in $\overline{\omega}$ and $\overline{Z}_{\Omega,p,q}$ is a normalizing constant.*

In particular, $x = 1$ when $p = p_c(q)$ and the probability of a loop configuration is expressed in terms of the number of loops only.

Proof Let v be the number of vertices of the graph Ω where (ba) has been contracted to a point. Induction on the number of open edges shows that

$$\ell(\overline{\omega}) = 2k(\omega) + o(\omega) - v. \tag{5.21}$$

Indeed, if there is no open edge, then $\ell(\overline{\omega}) = k(\omega) = v$ since there is a loop around each one of the vertices of $\Omega\setminus(ba)$, and one exploration path. Now, adding an edge can either:

- join two clusters of ω, thus decreasing both the numbers of loops and clusters by 1,
- close a cycle in ω, thus increasing the number of loops by 1 and not changing the number of clusters.

Equation (5.21) implies that

$$\begin{aligned} p^{o(\omega)}(1-p)^{c(\omega)}q^{k(\omega)} &= p^{o(\omega)}(1-p)^{|E|-o(\omega)}q^{k(\omega)} \\ &= (1-p)^{|E|}\sqrt{q}^{v}\Big(\frac{p}{(1-p)\sqrt{q}}\Big)^{o(\omega)}\sqrt{q}^{2k(\omega)+o(\omega)-v} \\ &= (1-p)^{|E|}\sqrt{q}^{v}x^{o(\omega)}\sqrt{q}^{\ell(\overline{\omega})}. \end{aligned}$$

The proof follows readily. □

We are now ready to define the parafermionic observable. Recall that $\gamma = \gamma(\omega)$ is the exploration path in the loop configuration $\overline{\omega}$. The *winding* $\mathrm{W}_\gamma(e, e')$ of the exploration path γ between two medial-edges e and e' of the medial graph is equal to $\pi/2$ times the number of left turns minus the number of right turns done by the curve between e and e'. When e or e' are not on γ, we set the winding to be equal to 0.

Definition 2 The *parafermionic observable* $F = F(\Omega, p, q, a, b)$ in a Dobrushin domain (Ω, a, b) is defined for any (medial) edge e of $\Omega^\diamond$ by

$$F(e) := \phi^{a,b}_{\Omega,p,q}[\mathrm{e}^{\mathrm{i}\sigma \mathrm{W}_\gamma(e,e_b)}\mathbb{1}_{e\in\gamma}],$$

where σ is a solution of the equation

$$\sin(\sigma\pi/2) = \sqrt{q}/2. \tag{5.22}$$

[16]The exploration path γ is considered as a loop and counts as 1 in $\ell(\overline{\omega})$.

Note that σ belongs to $\mathbb{R}$ for $q \leq 4$ and to $1 + i\mathbb{R}$ for $q > 4$. This suggests that the critical behavior of random-cluster model is different for $q > 4$ and $q \leq 4$. For $q \in [0, 4]$, σ has the physical interpretation of a spin, which is fractional in general, hence the name parafermionic.[17] For $q > 4$, σ is not real anymore and does not have any physical interpretation.

These observables first appeared in the context of the Ising model (there they are called order-disorder operators) and dimer models. They were later on extended to the random-cluster model and the loop $O(n)$-model by Smirnov [120] (see [58] for more detail). Since then, these observables have been at the heart of the study of these models. They also appeared in a slightly different form in several physics papers going back to the early eighties [18, 66]. They have been the focus of much attention in recent years: physicists exhibited such observables in a large class of models of two-dimensional statistical physics [32, 85, 86, 112, 114].

Contour integrals of the parafermionic observable The parafermionic observable satisfies a very special property at criticality.

Theorem 18 (Vanishing contour integrals) *Fix $q > 0$ and $p = p_c$, For any Dobrushin domain (Ω, a, b) and any vertex of $\Omega^\diamond$ with four incident edges in $\Omega^\diamond$,*

$$F(e_1) - F(e_3) = \mathrm{i}F(e_2) - \mathrm{i}F(e_4), \tag{5.23}$$

where e_1, e_2, e_3 and e_4 are the four edges incident to this vertex, indexed in clockwise order.

As in the case of the self-avoiding walk, interpret (5.23) as follows: the integral of F along a small square around a face is equal to 0. One may also sum this relation on every vertex to obtain that discrete contour integrals vanish.

Proof We follow a strategy close to the Proof of Lemma 13 and pair configurations in such a way that sums of contributions cancel.

Let e be an edge of $\Omega^\diamond$ and let

$$\mathsf{X}_e(\omega) := \mathrm{e}^{\mathrm{i}\sigma \mathrm{W}_{\gamma(\omega)}(e,e_b)} \mathbb{1}_{e\in\gamma(\omega)} \phi^{a,b}_{\Omega,p_c,q}[\omega]$$

be the contribution of the configuration ω to $F(e)$. Let ω' be the configuration obtained from ω by switching the state open or closed of the edge in ω passing through v. Since $\omega \mapsto \omega'$ is an involution, the following relation holds:

$$F(e) = \sum_\omega \mathsf{X}_e(\omega) = \tfrac{1}{2}\sum_\omega \big[\mathsf{X}_e(\omega) + \mathsf{X}_e(\omega')\big].$$

To prove (5.23), it is thus sufficient to show that for any configuration ω,

[17] Fermions have half-integer spins while bosons have integer spins, there are no particles with fractional spin, but the use of such fractional spins at a theoretical level has been very fruitful in physics.

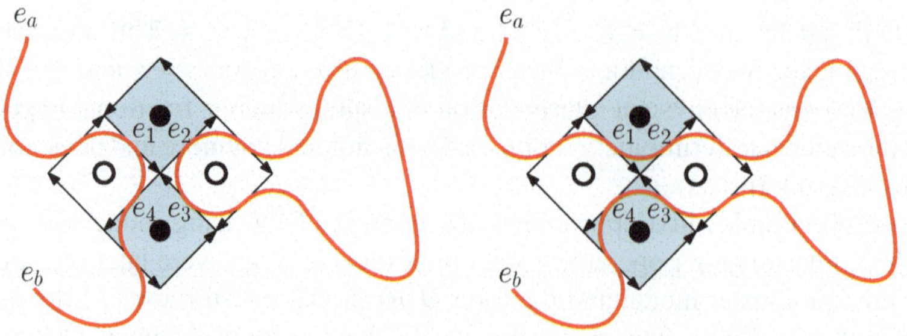

Fig. 16 **Left**. The neighborhood of v for two associated configurations ω and ω'

$$\mathsf{X}_{e_1}(\omega) + \mathsf{X}_{e_1}(\omega') - \mathsf{X}_{e_3}(\omega) - \mathsf{X}_{e_3}(\omega') \;=\; \mathrm{i}[\mathsf{X}_{e_2}(\omega) + \mathsf{X}_{e_2}(\omega') - \mathsf{X}_{e_4}(\omega) - \mathsf{X}_{e_4}(\omega')]. \tag{5.24}$$

There are three possible cases:

Case 1. No edge incident to v belongs to $\gamma(\omega)$. Then, none of these edges is incident to $\gamma(\omega')$ either. For any e incident to v, the contribution to (5.24) is equal to 0 so that (5.24) trivially holds.

Case 2. Two edges incident to v belong to $\gamma(\omega)$, see Fig. 16. Since $\gamma(\omega)$ and the medial lattice possess a natural orientation, $\gamma(\omega)$ enters through either e_1 or e_3 and leaves through e_2 or e_4. Assume that $\gamma(\omega)$ enters through the edge e_1 and leaves through the edge e_4. It is then possible to compute the contributions for ω and ω' of all the edges incident to v in terms of $\mathsf{X} = \mathsf{X}_{e_1}(\omega)$. Indeed, since ω' has one less loop, we find

$$\phi^{a,b}_{\Omega,p_c,q}[\omega'] = \frac{1}{\sqrt{q}}\phi^{a,b}_{\Omega,p_c,q}[\omega].$$

Furthermore, windings of $\gamma(\omega)$ and $\gamma(\omega')$ at e_2, e_3 and e_4 can be expressed using the winding at e_1 (for instance, $W_{\gamma(\omega)}(e_2, e_b) = W_{\gamma(\omega)}(e_1, e_b) - \pi/2$—the other cases are treated similarly). The contributions are given in the following table.

configuration	e_1	e_2	e_3	e_4
ω	X	0	0	$\mathrm{e}^{\mathrm{i}\sigma\pi/2}\mathsf{X}$
ω'	$\frac{\mathsf{X}}{\sqrt{q}}$	$\mathrm{e}^{\mathrm{i}\sigma\pi}\frac{\mathsf{X}}{\sqrt{q}}$	$\mathrm{e}^{-\mathrm{i}\sigma\pi/2}\frac{\mathsf{X}}{\sqrt{q}}$	$\mathrm{e}^{\mathrm{i}\sigma\pi/2}\frac{\mathsf{X}}{\sqrt{q}}$

Using the identity $\mathrm{e}^{\mathrm{i}\sigma\pi/2} - \mathrm{e}^{-\mathrm{i}\sigma\pi/2} = \mathrm{i}\sqrt{q}$, we deduce (5.24) by summing (with the right weight) the contributions of all the edges incident to v.

Case 3. The four edges incident to v belong to $\gamma(\omega)$. Then only two of these edges belong to $\gamma(\omega')$ and the computation is similar to Case 2 by exchanging the weights of ω' and ω.

In conclusion, (5.24) is always satisfied and the claim is proved. □

Continuous phase transition for random-cluster models with $q \in [1, 4]$ This section is devoted to the proof of the following result.

Theorem 19 (Duminil-Copin [41]) *For $q \in [1, 4]$, the property* **P4a** *is satisfied.*

As a consequence, we deduce from Theorem 16 that the properties **P1–P5** also are.[18] This gives "one half" of Theorem 14.

We first focus on the case $q \leq 2$. The proof follows an argument similar to the computation for self-avoiding walks: we will use that the discrete contour integral along the boundary of a domain vanishes together with the fact that windings are deterministic on the boundary.

Proof (*Theorem* 19 *in the case $q \in [1, 2]$*) In this proof, the first and second coordinates of a vertex $x \in \mathbb{Z}^2$ are denoted by x_1 and x_2. Also, define $\widetilde{\Lambda}_n := \{x \in \mathbb{Z}^2 : |x_1| + |x_2| \leq n\}$.

Fix n odd. Consider a degenerated case of Dobrushin domain in which

$$\Omega := \{x \in \widetilde{\Lambda}_n \text{ such that } x_1 + x_2 \leq 0\}$$

and $(ba) = \{0\}$ as well as $(ab) = \partial\Omega$. In this case, the parafermionic observable F still makes sense: e_a and e_b are the edges of $\Omega^\diamond$ north-west and south-east of 0, and $\gamma(\omega)$ is the loop going around 0 (and therefore through e_a and e_b). Note that, by definition, the Dobrushin boundary conditions are coinciding with the free boundary conditions in this context since the arc (ba) is restricted to a point.

Summing (5.23) on every vertex $v \in \Omega^\diamond$, we obtain that

$$\sum_{e\in\alpha} F(e) = \sum_{e\in\beta} F(e) + \mathrm{i}\sum_{e\in\varepsilon} F(e) - \mathrm{i}\sum_{e\in\bar{\varepsilon}} F(e),$$

where α, ε, β and $\bar{\varepsilon}$ are respectively the sets of medial edges intersecting the north-east, north-west, south-west and south-east boundaries of $\Omega^\diamond$. This immediately leads to

$$\Big|\sum_{e\in\alpha} F(e)\Big| \leq \sum_{e\notin\alpha} |F(e)|, \tag{5.25}$$

where the sum on the right is on edges of $\Omega^\diamond$ intersecting the boundary only. Any such edge e is bordering a vertex $x \in \partial\Omega$. Also, $\gamma(\omega)$ goes through e if and only if x and 0 are connected by a path of edges in ω. We deduce that

$$|F(e)| = \phi^0_{\Omega,p_c,q}[0 \longleftrightarrow x] \overset{\text{(CBC)}}{\leq} \phi^0_{p_c,q}[0 \longleftrightarrow x]. \tag{5.26}$$

[18] We did not prove that **P4b** implies **P5**, but since **P4a** implies **P4b** and **P5**, this follows readily.

Since there are exactly two medial edges bordering a prescribed vertex, and that each such vertex x is in $\partial\widetilde{\Lambda}_n$, (5.25) becomes

$$\Big|\sum_{e\in\alpha} F(e)\Big| \le 2\sum_{x\in\partial\widetilde{\Lambda}_n} \phi^0_{p_c,q}[0 \longleftrightarrow x]. \tag{5.27}$$

Let us now focus on the term on the left. First, note that since $\gamma(\omega)$ deterministically goes through e_a and e_b, we get

$$F(e_a) + F(e_b) = 1 + \mathrm{e}^{\mathrm{i}\pi\sigma} = 2\cos(\tfrac{\pi}{2}\sigma)\mathrm{e}^{\mathrm{i}\sigma\pi/2}. \tag{5.28}$$

Second, pick an edge $e \in \alpha\backslash\{e_a, e_b\}$. Since the winding of the loop is deterministic, we may improve the equality in (5.26) into

$$F(e) = \mathrm{e}^{\mathrm{i}\sigma W(e)}\phi^0_{\Omega,p_c,q}[0 \longleftrightarrow x], \tag{5.29}$$

where x is the vertex of $\partial\Omega$ bordered by e, and $W(e) \in \{-\pi, 0, \pi, 2\pi\}$ depending on which side of 0 the edge e is, and whether it is pointing inside or outside of $\Omega^\diamond$.

Define

$$S := \{x \in \partial\Omega\backslash\partial\widetilde{\Lambda}_n : x_1 > 0\}.$$

By gathering the contributions of edges bordering a vertex $x \in S$ and its symmetric $-x$, and using the symmetry of Ω with respect to the line $x_1 = x_2$, we deduce from (5.28) and the previous displayed equation that

$$\begin{aligned}\sum_{e\in\alpha\backslash\{e_a,e_b\}} F(e) &= \sum_{x\in S}(\mathrm{e}^{2\mathrm{i}\pi\sigma} + \mathrm{e}^{\mathrm{i}\pi\sigma} + 1 + \mathrm{e}^{-\mathrm{i}\pi\sigma})\phi^0_{\Omega,p_c,q}[0 \longleftrightarrow x]\\ &= \frac{\sin(\sigma 2\pi)}{\sin(\sigma\pi/2)}\mathrm{e}^{\mathrm{i}\sigma\pi/2}\sum_{x\in S}\phi^0_{\Omega,p_c,q}[0 \longleftrightarrow x].\end{aligned}$$

For $q \in [1, 2]$, $\cos(\sigma\pi/2) > 0$ and $\frac{\sin(2\pi\sigma)}{\sin(\frac{\pi}{2}\sigma)} \ge 0$. We deduce that

$$\Big|\sum_{e\in\alpha} F(e)\Big| \ge 2\cos(\tfrac{\pi}{2}\sigma) > 0.$$

Plugging this lower bound in (5.27) and then summing over odd n gives

$$\sum_{x\in\mathbb{Z}^2}\phi^0_{p_c,q}[0 \longleftrightarrow x] = \infty,$$

which is **P3**. Since **P3** implies **P4a**, the proof follows. □

Observe that for $q > 2$, the value of σ is such that $\sin(2\pi\sigma)$ becomes negative so that we may not conclude directly anymore. One may wonder whether this is just a technical problem, or whether something deeper is hidden behind this. It is natural to predict that the following quantity decays like a power law:

$$\phi^0_{\Omega,p_c,q}[0 \longleftrightarrow \partial\Lambda_{n/2}] = n^{-\alpha(q,\pi)+o(1)},$$

where $\alpha(q, \pi)$ is a constant depending on q only (π refers to the "angle of the opening" of Ω at 0), and $o(1)$ denotes a quantity tending to 0 as n tends to infinity. Moreover, one may argue using **P5** (which we believe is true) that the event that $x \longleftrightarrow 0$ in Ω has a probability close to the probability that 0 and x are connected to distance $n/2$ in Ω (see also Exercise 45). For x not too close to the corners, the boundary of Ω looks like a straight line and it is therefore natural to predict that

$$\phi^0_{\Omega,p_c,q}[0 \longleftrightarrow x] = n^{-2\alpha(q,\pi)+o(1)}.$$

Summing over all x (the vertices near the corner do not contribute substantially) we should find

$$\sum_{\|x\|_1=n} \phi^0_{\Omega,p_c,q}[0 \longleftrightarrow x] = n^{1-2\alpha(q,\pi)+o(1)}. \tag{5.30}$$

Now, it is conjectured in physics that

$$\alpha(q, \pi) = 1 - 2\frac{\arccos(\sqrt{q}/2)}{\pi}.$$

Therefore, for $q \in (2, 4]$, the quantity on the left-hand side of (5.30) converges to 0 as $n \to \infty$ and the strategy consisting in proving that it remains bounded away from 0 is hopeless for $q > 2$.

Nevertheless, we did not have to consider a flat boundary near 0 in the first place. For instance, one may consider Ω' obtained by taking the set of $x = (x_1, x_2)$ with $\|x\|_1 \leq n$ and $(x_1, x_2) \neq (n, 0)$ with $n \geq 0$. Then, one expects that

$$\phi^0_{\Omega',p_c,q}[0 \longleftrightarrow \partial\Lambda_{n/2}] = n^{-\alpha(q,2\pi)+o(1)},$$

where $\alpha(q, 2\pi)$ is a value which is a priori smaller than $\alpha(q, \pi)$ since $\mathbb{S}$ is larger (2π refers this time to the "opening angle" of Ω' at 0). Therefore, if one applies the same reasoning as above, we may prove that

$$\sum_{\|x\|_1=n} \phi^0_{\Omega',p_c,q}[0 \longleftrightarrow x] = n^{1-\alpha(q,\pi)-\alpha(q,2\pi)+o(1)}.$$

In fact, we know how to predict $\alpha(q, 2\pi)$: the map $z \mapsto z^2$ maps $\mathbb{R}_+^* \times \mathbb{R}$ to $\mathbb{R}^2 \setminus -\mathbb{R}_+$, conformal invariance (see Sect. 6 for more detail) predicts that $\alpha(q, 2\pi) = \alpha(q, \pi)/2$. As a consequence,

$$\sum_{\|x\|_1=n} \phi^0_{\Omega',p_c,q}[0 \longleftrightarrow x] = n^{1-\frac{3}{2}\alpha(q,\pi)+o(1)},$$

so that this quantity can indeed be larger or equal to 1 provided that $q \leq 3$.

The previous discussion remained at the level of predictions. It relies on conformal invariance, which is extremely hard to get, and definitely much more advanced than what we are seeking. However, it is very good news that the strategy of the previous proof can indeed be applied to Ω' instead of Ω to give that for $q \leq 3$, there exists $c = c(q) > 0$ such that for any $n \geq 1$,

$$\sum_{\|x\|_1=n} \phi^0_{\Omega', p_c, q}[0 \longleftrightarrow x] \geq c.$$

Since Ω' is a subset of $\mathbb{Z}^2$, the comparison between boundary conditions implies that for any $q \leq 3$.

$$\sum_{x \in \mathbb{Z}^2} \phi^0_{p_c, q}[0 \longleftrightarrow x] = \infty,$$

thus extending the result to every $q \leq 3$. We leave the details to Exercise 53.

Exercise 53 Fill up the details of the $q \leq 3$ case by considering Ω' instead of Ω.

This reasoning does not directly extend to $q > 3$ since $\frac{3}{2}\alpha(q, \pi) > 1$ in this case. Nevertheless, one could consider a graph generalizing Ω and Ω' with a "larger opening than 2π" at 0. In fact, one may even consider a graph with "infinite opening" at 0 by considering subgraphs of the universal cover $\mathbb{U}$ of the plane minus a face of $\mathbb{Z}^2$, see Fig. 17. This is what was done in [41]. The drawback of taking this set $\mathbb{U}$ is that it is not a subset of $\mathbb{Z}^2$ anymore. Thus, one has to translate the information obtained for the random-cluster model on $\mathbb{U}$ into information for the random-cluster model on $\mathbb{Z}^2$, which is a priori difficult since there is no easy comparison between the two graphs (for instance the comparison between boundary conditions is not sufficient). This is the reason why in general one uses **P4a** instead of **P3**.

Discontinuous phase transition for the random-cluster model with $q > 4$ The goal of this section is to briefly discuss the following theorem. This completes the results of the previous sections and determines the continuous/discontinuous nature of the phase transition for every $q \geq 1$. Below, we keep the notation $\widetilde{\Lambda}_n$ for the box of size n for the graph distance.

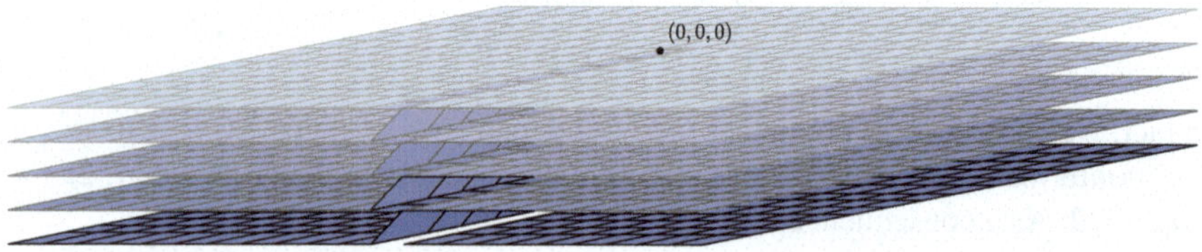

Fig. 17 The graph $\mathbb{U}$

Theorem 20 (Duminil-Copin et al. [47]) *For $q > 4$, the properties* **P1–5** *are not satisfied. In particular*

$$\lim_{n\to\infty} -\tfrac{1}{n} \log \phi^0_{p_c,q}[0 \longleftrightarrow \partial\widetilde{\Lambda}_n] = \lambda + 2\sum_{k=1}^{\infty} \tfrac{(-1)^k}{k} \tanh(k\lambda) > 0, \tag{5.31}$$

where $\lambda > 0$ satisfies $\cosh(\lambda) = \frac{\sqrt{q}}{2}$.

Note that in particular, one may get the asymptotics in (5.31) as $q \searrow 4$: it behaves asymptotically as $8\exp\left(-\pi^2/\sqrt{q-4}\right)$. Physically, that means that the correlation length of the models explodes very quickly (much faster than any polynomial) as q approaches 4.

Before sketching the ideas involved in the proof of this statement, let us make a small detour and prove that **P1–5** cannot be satisfied for $q \gg 1$ (see [44] for details).

Proof (discontinuity for $q > 256$) Consider a loop L of the medial lattice $(\mathbb{Z}^2)^\diamond$ surrounding the origin. We assume that L is oriented counterclockwise. Let n be the number of edges of $(\mathbb{Z}^2)^\diamond$ on L and consider a graph Ω containing the full loop. Let $\mathcal{E}_L$ be the event that the loop L is a loop of the configuration $\overline{\omega}$ (Figs. 18, 19 and 20).

Our goal is to bound $\phi^0_{\Omega,p_c,q}[\mathcal{E}_L]$. In order to do so, we construct a one-to-one "repair map" f_L from $\mathcal{E}_L$ to the set of loop configurations on Ω such that the image $f_L(\overline{\omega})$ has much larger probability than the probability of $\overline{\omega}$. This will imply a bound on the probability of $\mathcal{E}_L$ (see below).

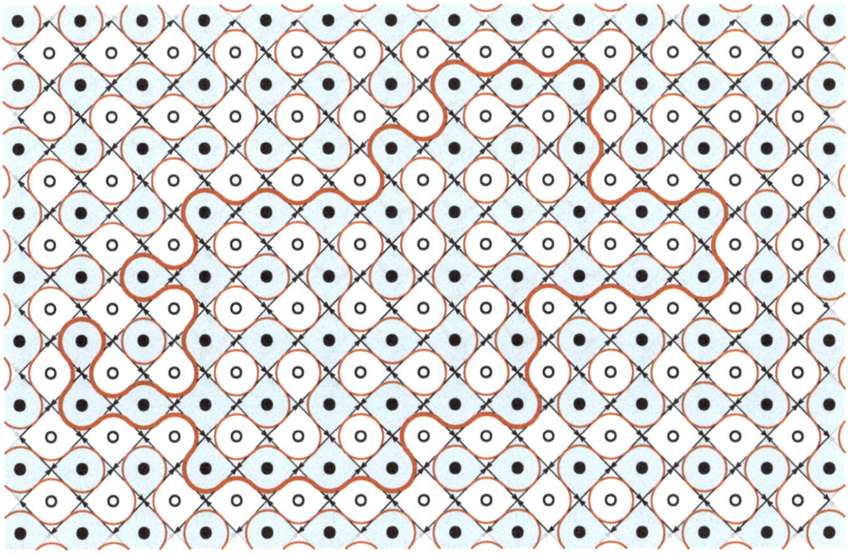

Fig. 18 Consider a loop configuration $\overline{\omega}$ containing the loop L (in bold)

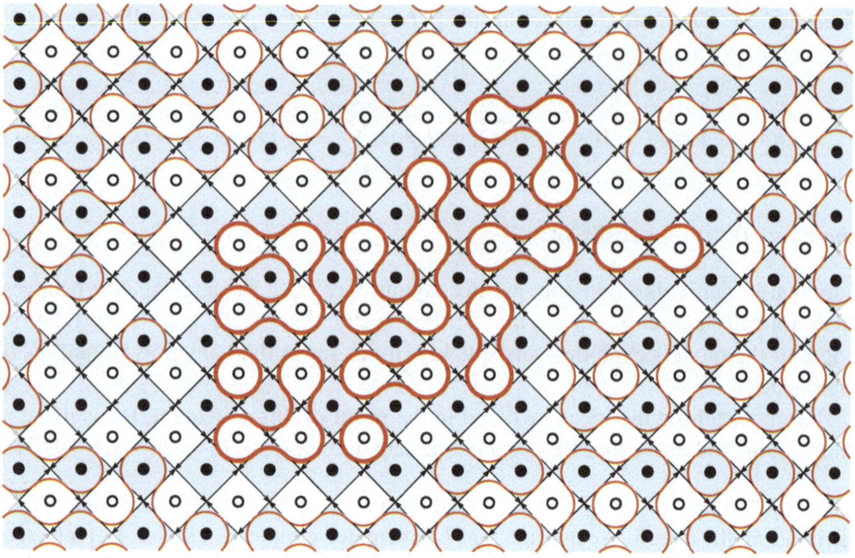

Fig. 19 (Step 1) Remove the loop L from $\overline{\omega}$. The loops inside L are depicted in bold

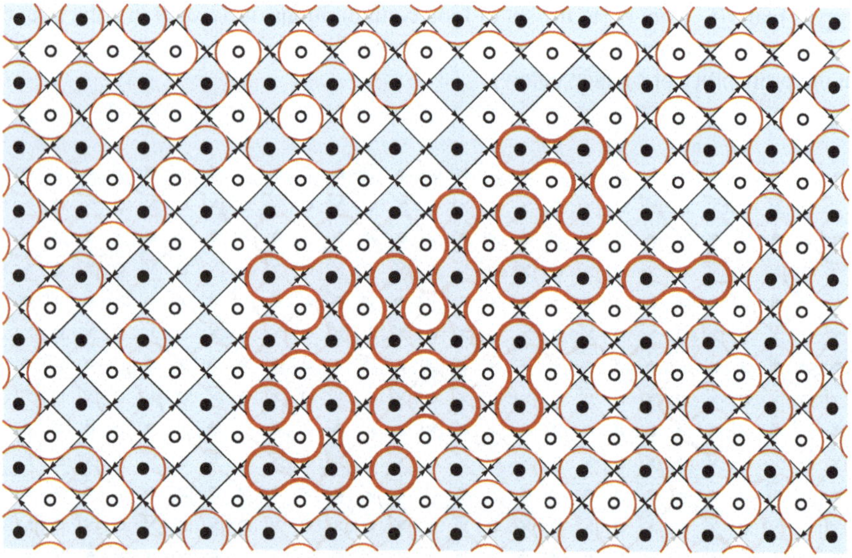

Fig. 20 (Step 2) Translate the loops inside L in the south-east direction

Let $\overline{\omega}$ be a loop configuration in $\mathcal{E}_L$. A loop of $\overline{\omega}$ is said to be *inside* (resp. *outside*) L if it is included in the bounded connected component of 0 in $\mathbb{R}^2\setminus L$. Perform the following three successive modifications on $\overline{\omega}$ (See Fig. 21 for an illustration.) to obtain a configuration $f_L(\overline{\omega})$:

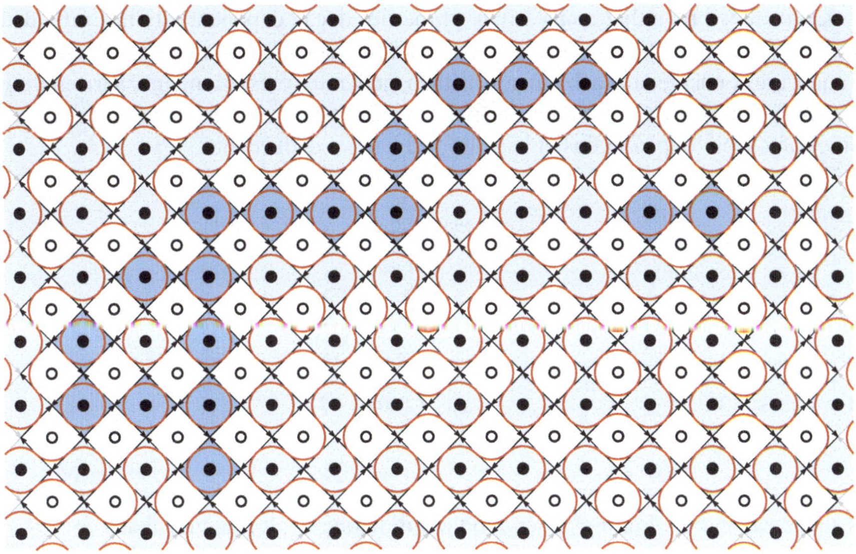

Fig. 21 (Step 3) Fill the "holes" (depicted in darker gray) with loops of length four

Step 1. Remove the loop L from $\overline{\omega}$.
Step 2. Translate the loops of $\overline{\omega}$ which are inside L by the vector $\frac{1-i}{2}$.
Step 3. Complete the configuration thus obtained by putting loops of length four around black faces of $\Omega^\diamond$ bordered by an edge which is not covered by any loop after Step 2.
The configuration $f_L(\overline{\omega})$ is a loop configuration on $\Omega^\diamond$ (Exercise 54). Furthermore, Step 1 of the construction removes a loop from $\overline{\omega}$, but Step 3 adds one loop per edge of L pointing south-west. Since the number of edges added in the last step is four times this number, and that the final configuration has as many edges as the first one, we deduce that this number is equal to $n/4$. Thus, we have

$$\phi^0_{\Omega,p_c,q}[\overline{\omega}] = \sqrt{q}^{\ell(\overline{\omega})-\ell(f_L(\overline{\omega}))}\phi^0_{\Omega,p_c,q}[f_L(\overline{\omega})] = \sqrt{q}^{1-n/4}\phi^0_{\Omega,p_c,q}[f_L(\overline{\omega})].$$

Using the previous equality in the second line and the fact that f_L is one-to-one in the third (this uses the fact that L is fixed at the beginning of the proof), we deduce that

$$\begin{aligned}\phi^0_{\Omega,p_c,q}[\mathcal{E}_L] &= \sum_{\overline{\omega}\in\mathcal{E}_L}\phi^0_{\Omega,p_c,q}[\overline{\omega}]\\ &= q^{1/2-n/8}\sum_{\overline{\omega}\in\mathcal{E}_L}\phi^0_{\Omega,p_c,q}[f_L(\overline{\omega})]\\ &= q^{1/2-n/8}\phi^0_{\Omega,p_c,q}[f_L(\mathcal{E}_L)] \le q^{1/2-n/8}.\end{aligned}$$

Let us now prove that connectivity properties decay exponentially fast provided that $q > 256$. Consider two vertices 0 and x and a graph Ω containing both 0 and x. If 0 and x are connected to each other in ω, then there must exist a loop in $\overline{\omega}$ surrounding 0 and x which is oriented counterclockwise (simply take the exterior-most such loop). Since any such loop contains at least $\|x\|$ edges, we deduce that

$$\begin{aligned}\phi^0_{\Omega,p_c,q}[0 \longleftrightarrow x] &\le \sum_{\substack{L \text{ surrounding} \\ 0 \text{ and } x}} \phi^0_{\Omega,p_c,q}[\mathcal{E}_L] \\ &\le \sum_{n \ge \|x\|} \sum_{\substack{L \text{ of length } n \\ \text{surrounding } 0}} q^{1/2-n/8} \\ &\le \sum_{n \ge \|x\|} n2^n \cdot q^{1/2-n/8}.\end{aligned}$$

In the last line we used that the number of loops surrounding 0 with n edges on $\Omega^\diamond$ is smaller than $n2^n$. Letting Ω tend to the full lattice $\mathbb{Z}^2$, we deduce that

$$\phi^0_{p_c,q}[0 \longleftrightarrow x] \le \sum_{n \ge \|x\|} n2^n \cdot q^{1/2-n/8} \le \exp(-c\|x\|).$$

The existence of $c > 0$ follows from the assumption $2q^{-1/8} < 1$. □

Exercise 54 Prove that the repair map f_L actually yields a loop configuration.

Mapping to the six-vertex model and sketch of the proof for $q > 4$ We do not discuss the exact computation of the correlation length. The proof is based on a relation between the random-cluster model on a graph Ω and the six-vertex model on its medial graph $\Omega^\diamond$.

The six-vertex model was initially proposed by Pauling in 1931 for the study of the thermodynamic properties of ice. While we are mainly interested in it for its connection to the random-cluster model, the six-vertex model is a major object of study on its own right. We do not attempt to give here an overview of the model and we rather refer to [113] and Chap. 8 of [8] (and references therein) for a bibliography on the subject.

The mapping between the random-cluster model and the six-vertex model being very sensitive to boundary conditions, we will work on a torus. As in the previous section, the first and second coordinates of $x \in \mathbb{Z}^2$ are denoted by x_1 and x_2. For M and N, consider the subgraph $\mathbb{T} = \mathbb{T}(M, N)$ of the square lattice induced by the set of vertices

$$\{x \in \mathbb{Z}^2 : 0 \le x_1 + x_2 \le M \text{ and } |x_1 - x_2| \le N\}.$$

Introduce the periodic boundary conditions per in which x and y on $\partial\mathbb{T}$ are identified together iff $x_1 + x_2 = y_1 + y_2$ or $x_1 - x_2 = y_1 - y_2$. Together with these boundary conditions, $\mathbb{T}$ may be seen as a torus.

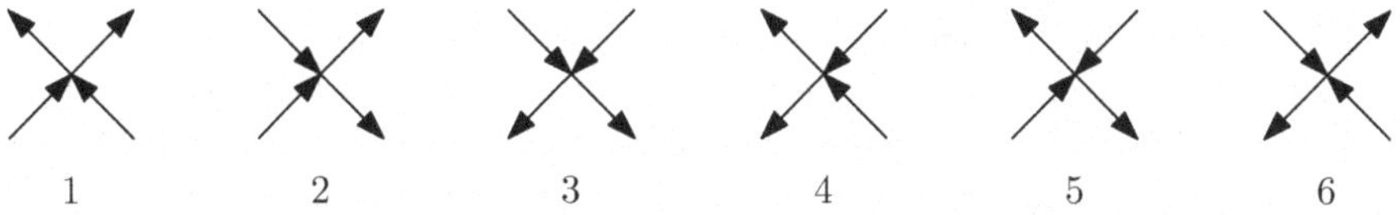

Fig. 22 The 6 possibilities for vertices in the six-vertex model. Each possibility comes with a weight a, b or c

An *arrow configuration* $\vec{\omega}$ on $\mathbb{T}^\diamond$ (the medial graph is defined in an obvious fashion here) is a map attributing to each edge $xy \in E$ one of the two oriented edges (x, y) and (y, x). We say that an arrow configuration satisfies the *ice rule* if each vertex of $\mathbb{T}^\diamond$ is incident to two edges pointing toward it (and therefore to two edges pointing outwards from it). The ice rule leaves six possible configurations at each vertex, depicted in Fig. 22, whence the name of the model. Each arrow configuration $\vec{\omega}$ receives a weight

$$w_{6V}(\vec{\omega}) := \begin{cases} a^{n_1+n_2} \cdot b^{n_3+n_4} \cdot c^{n_5+n_6} & \text{if } \vec{\omega} \text{ satisfies the ice rule,} \\ 0 & \text{otherwise,} \end{cases} \tag{5.32}$$

where a, b, c are three positive numbers, and n_i denotes the number of vertices with configuration $i \in \{1, \ldots, 6\}$ in $\vec{\omega}$. In what follows, we focus on the case $a = b = 1$ and $c > 2$, and will therefore only consider such weights from now on.

In our context, the interest of the six-vertex model stems from its solvability using the transfer-matrix formalism. More precisely, the partition function of a toroidal six-vertex model may be expressed as the trace of the M-th power of a matrix V called the *transfer matrix*, whose leading eigenvalues can be computed using the so-called *Bethe-Ansatz*. This part does not invoke probability at all, and relies heavily on exact computations. For more detail on the subject, we refer the curious reader to [46, 47]. Here, we will only use the following consequence of the study.

For a six-vertex configuration $\vec{\omega}$ on $\mathbb{T}^\diamond$, write $|\vec{\omega}|$ for the number of north-east arrows intersecting the line $x_1 + x_2 = 0$ (this number is the same for all lines $x_1 + x_2 = k$ with $-M \leq k \leq M$). The total number of arrows in each line is $2N$. It can be shown that typical configurations have N such arrows. In fact, one may prove a more refined statement. Set

$$Z_{6V}(N, M) = \sum_{\vec{\omega}} w_{6V}(\vec{\omega}) \qquad \text{and} \qquad \widetilde{Z}_{6V}(N, M) = \sum_{\vec{\omega}:\, |\vec{\omega}| = N-1} w_{6V}(\vec{\omega}).$$

Theorem 21 *For $c > 2$ and $r > 0$ integer, fix $\lambda > 0$ satisfying $e^\lambda + e^{-\lambda} = c^2$. Then,*

$$\lim_{N\to\infty} \lim_{M\to\infty} -\tfrac{1}{M} \log\left(\frac{\widetilde{Z}_{6V}(N, M)}{Z_{6V}(N, M)}\right) = \lambda + 2\sum_{k=1}^{\infty} \tfrac{(-1)^k}{k} \tanh(k\lambda) > 0. \tag{5.33}$$

Our goal now is to explain how one deduces discontinuity of the phase transition for random-cluster models from this theorem. In order to do so, we relate the random-cluster model to the six-vertex model. We denote the random-cluster measure on $\mathbb{T}$

by $\phi^{\text{per}}_{\mathbb{T},p_c,q}$ (there are no boundary conditions since $\mathbb{T}$ has no boundary). Let $k_{\text{nc}}(\omega)$ be the number of non-retractable clusters of ω, and $\mathcal{A}$ the event that both ω and ω^* contain exactly one cluster winding around the torus in the south-west north-east direction. Also, let $s(\omega)$ be the indicator function of the event that all clusters of ω^* are retractable.

Proposition 13 *Let $q > 4$ and set $c = \sqrt{2+\sqrt{q}}$. For N, M even,*

$$\phi^{\text{per}}_{\mathbb{T},p_c,q}[\mathcal{A}] = q\,\frac{\widetilde{Z}_{6V}(N,M)}{Z_{6V}(N,M)}\,\phi^{\text{per}}_{\mathbb{T},p_c,q}\Big[\big(\tfrac{4}{q}\big)^{k_{\text{nc}}(\omega)}4^{-s(\omega)}\Big].$$

Proof Define $w_{\text{RC}}(\omega) = p^{o(\omega)}(1-p)^{c(\omega)}q^{k(\omega)}$. As in Proposition 12, we may prove by induction that

$$\sqrt{q}^{\,\ell(\overline{\omega})+2s(\omega)} = c_0\, w_{\text{RC}}(\omega), \tag{5.34}$$

where $c_0 > 0$ is independent of the configuration.

Write $\omega^{\circlearrowleft}$ for oriented loop configurations, i.e. configurations of loops to which we associated an orientation. Let $\ell_-(\omega^{\circlearrowleft})$ and $\ell_+(\omega^{\circlearrowleft})$ for the number of retractable loops of $\omega^{\circlearrowleft}$ which are oriented clockwise and counterclockwise, respectively. Introduce $e^{\mu} + e^{-\mu} = \sqrt{q}$ and write, for an oriented loop configuration $\omega^{\circlearrowleft}$,

$$w_{\circlearrowleft}(\omega^{\circlearrowleft}) = e^{\mu\ell_+(\omega^{\circlearrowleft})}\,e^{-\mu\ell_-(\omega^{\circlearrowleft})}.$$

Fix ω a random-cluster configuration and consider its associated loop configuration $\overline{\omega}$. In summing the $2^{\ell(\omega)}$ oriented loop configurations $\omega^{\circlearrowleft}$ obtained from $\overline{\omega}$ by orienting loops, we find

$$\sum_{\omega^{\circlearrowleft}} w_{\circlearrowleft}(\omega^{\circlearrowleft}) = \big(1+1\big)^{\ell_0(\overline{\omega})}\big(e^{\mu}+e^{-\mu}\big)^{\ell(\overline{\omega})-\ell_0(\overline{\omega})} = c_0\big(\tfrac{4}{q}\big)^{k_{\text{nc}}(\omega)}4^{-s(\omega)}w_{\text{RC}}(\omega), \tag{5.35}$$

where $\ell_0(\overline{\omega})$ is the number of non-retractable loops of $\overline{\omega}$. In the last equality, we used (5.34) and the fact when $s(\omega) = 0$, any non-retractable cluster corresponds to two non-retractable loops. We also used that when $s(\omega) = 1$, there is no non-retractable loop and exactly one non-retractable cluster (Fig. 23).

Notice now that an oriented loop configuration gives rise to 8 different configurations at each vertex. These are depicted in Fig. 24. For an oriented loop configuration $\omega^{\circlearrowleft}$, write $n_i(\omega^{\circlearrowleft})$ for the number of vertices of type i in $\omega^{\circlearrowleft}$, with $i = 1, 2, 3, 4, 5A, 5B, 6A, 6B$.

The retractable loops of $\omega^{\circlearrowleft}$ which are oriented clockwise have total winding -2π, while those oriented counterclockwise have winding 2π. Loops which are not retractable have total winding 0. Write $W(\ell)$ for the winding of a loop $\ell \in \omega^{\circlearrowleft}$. Then

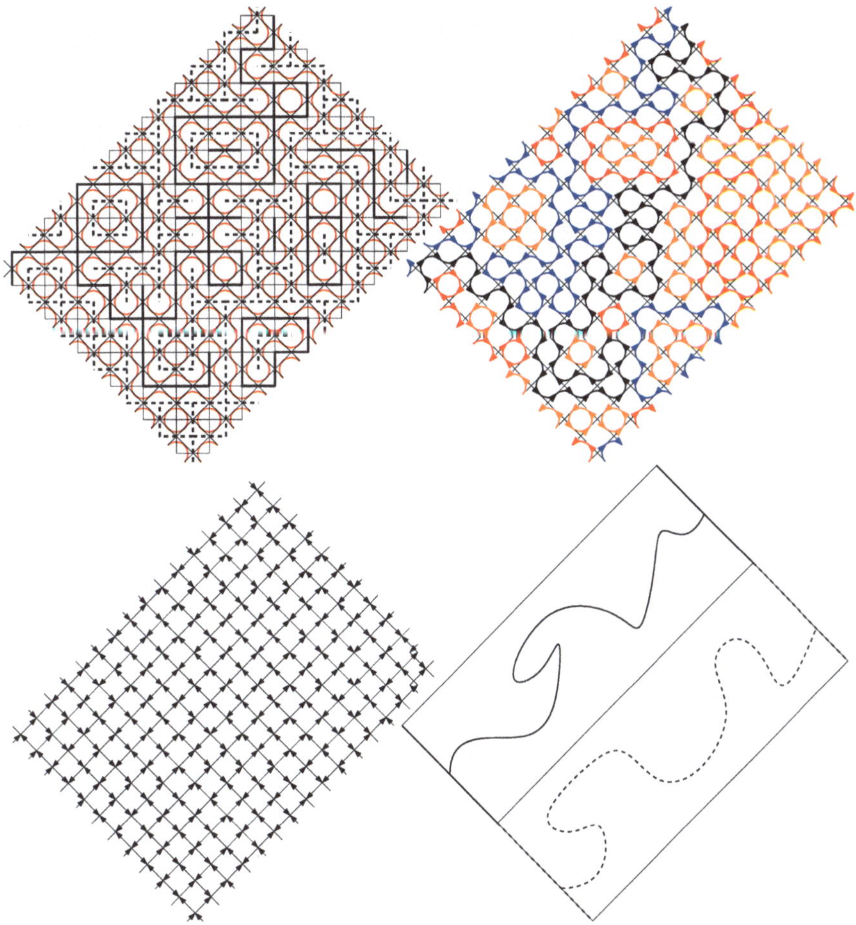

Fig. 23 The different steps in the correspondence between the random-cluster model and the six-vertex model on a torus. **Top-left**. A random-cluster configuration and its dual, as well as the corresponding loop configuration. **Top-right**. An orientation of the loop configuration (retractable loops oriented counterclockwise in red, clockwise in orange, in blue and black, the two non-retractable loops). **Bottom-left**. The resulting six-vertex configuration. Note that in the first picture, there exist both a primal and dual component winding vertically around the torus; this leads to two loops that wind vertically (see second picture); if these loops are oriented in the same direction (as in the third picture) then the number of up arrows on every row of the six-vertex configuration is equal to $N \pm 1$. **Bottom-right**. The intersection of the events $\mathcal{E}$, $\mathcal{E}'$, $\mathcal{F}$ and $\mathcal{F}'$ implies the event $\mathcal{A}$

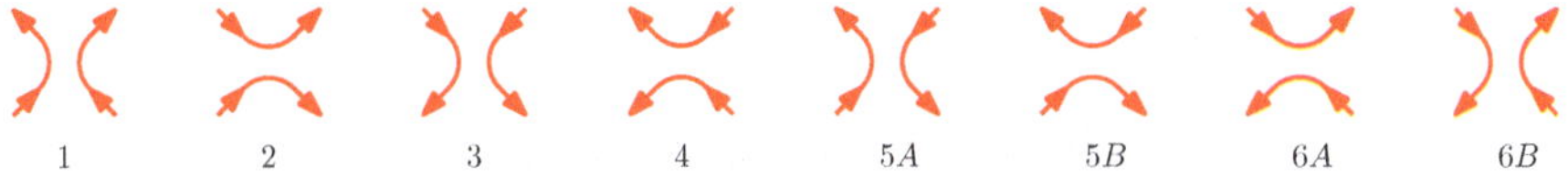

Fig. 24 The 8 different types of vertices encountered in an oriented loop configuration

$$w_{\frown}(\omega^{\frown}) = \exp\Big(\frac{\mu}{2\pi}\sum_{\ell\in\omega^{\frown}} W(\ell)\Big),$$

where the sum is over all loops ℓ of $\omega^{\frown}$. The winding of each loop may be computed by summing up the windings of turns along the loop. The compounded winding of the two pieces of paths appearing in the different configurations in Fig. 24 are

- vertices of type $1, \ldots, 4$: total winding 0;
- vertices of type $5A$ and $6A$: total winding π;
- vertices of type $5B$ and $6B$: total winding $-\pi$.

The total winding of all loops may therefore be expressed as

$$\sum_{\ell\in\omega^{\frown}} W(\ell) = \pi\left[n_{5A}(\omega^{\frown}) + n_{6A}(\omega^{\frown}) - n_{5B}(\omega^{\frown}) - n_{6B}(\omega^{\frown})\right].$$

We therefore deduce that for any oriented loop configuration $\omega^{\frown}$,

$$w_{\frown}(\omega^{\frown}) = e^{\frac{\mu}{2}[n_{5A}(\omega^{\frown})+n_{6A}(\omega^{\frown})]}\, e^{-\frac{\mu}{2}[n_{5B}(\omega^{\frown})+n_{6B}(\omega^{\frown})]}. \tag{5.36}$$

For the final step of the correspondence, notice that each diagram in Fig. 24 corresponds to a six-vertex local configuration (as those depicted in Fig. 22). Indeed, configurations $5A$ and $5B$ correspond to configuration 5 in Fig. 22 and configurations $6A$ and $6B$ correspond to configuration 6 in Fig. 22. The first four configurations of Fig. 24 correspond to the first four in Fig. 22, respectively.

Thus, to each oriented loop configuration $\omega^{\frown}$ is associated a six vertex configuration $!$. Note that the map associating $!$ to $\omega^{\frown}$ is not injective since there are $2^{n_5(!)+n_6(!)}$ oriented loop configurations corresponding to each $!$. In fact, for a six-vertex configuration $!$, if $N_{5,6}(!)$ is the set of vertices of type 5 and 6 in $!$, then the choice of $c = \sqrt{2+\sqrt{q}} = e^{\frac{\mu}{2}} + e^{-\frac{\mu}{2}}$ gives that

$$w_{6V}(!) = \prod_{u\in N_{5,6}(!)} \left(e^{\frac{\mu}{2}} + e^{-\frac{\mu}{2}}\right) = \sum_{\varepsilon\in\{\pm1\}^{N_{5,6}(!)}} \prod_{u\in N_{5,6}(!)} e^{\frac{\mu}{2}\varepsilon(u)} \overset{(5.36)}{=} \sum_{\omega^{\frown}} w_{\frown}(\omega^{\frown}). \tag{5.37}$$

We are now in a position to prove the statement of the proposition. First,

$$c_0 \sum_{\omega} \left(\tfrac{4}{q}\right)^{k_{\rm nc}(\omega)} 4^{-s(\omega)} w_{\rm RC}(\omega) \overset{(5.35)}{=} \sum_{\omega^{\frown}} w_{\frown}(\omega^{\frown}) \overset{(5.37)}{=} \sum_{!} w_{6V}(!) = Z_{6V}(N, M).$$

Second, using that $s(\omega) = 0$ and $k_{\rm nc}(\omega) = 1$ on the event $\mathcal{A}$, we find

$$c_0 \sum_{\omega\in\mathcal{A}} w_{\rm RC}(\omega) = c_0 \frac{q}{4} \sum_{\omega\in\mathcal{A}} w_{\rm RC}(\omega) \left(\tfrac{4}{q}\right)^{k_{\rm nc}(\omega)} 4^{-s(\omega)}$$

$$= q \sum_{|!|=N-1} w_{6V}(!) = q\, \widetilde{Z}_{6V}(N, M).$$

In the second step, we used that there are four ways of orienting the two loops bordering the unique non-retractable cluster of $\omega \in \mathcal{A}$, and that one of them leads to $|\omega^{\circ}| = N - 1$. Dividing by the partition function of the random-cluster model and then taking the ratio of the two last displayed equations leads to the result. □

Theorem 20 now follows pretty easily. Indeed, one has obviously that

$$\psi^{\text{per}}_{\mathbb{T},p_c,q}\Big[\Big(\tfrac{4}{q}\Big)^{k_{\text{nc}}(\omega)} 4^{-s(\omega)}\Big] \le 1.$$

Thus, Proposition 13 and Theorem 21 give the existence of $c_0 > 0$ such that for all fixed N large enough and $M \ge M_0(N)$,

$$\phi^{\text{per}}_{\mathbb{T},p_c,q}[\mathcal{A}] \le \exp(-c_0 M). \tag{5.38}$$

Now, consider the "rotated rectangles" $R = \{x \in \mathbb{T} : x_1 \le x_2\}$ and $R' = \{x \in \mathbb{T} : x_1 > x_2\}$. Assume that **P5** is satisfied, one obtains easily by combining crossings that

$$\phi^0_{R,p_c,q}[\mathcal{F}] \ge c_0^{M/N} \quad \text{and} \quad \phi^1_{R',p_c,q}[\mathcal{F}'] \ge c_0^{M/N}, \tag{5.39}$$

where $\mathcal{F}$ is the event that there exists a path in $\omega \cap R$ going from the line $x_1 + x_2 = 0$ to the line $x_1 + x_2 = M$, and $\mathcal{F}'$ is the event that there exists a path in $\omega^* \cap (R')^*$ from the line $x_1 + x_2 = 0$ to the line $x_1 + x_2 = M$. Now, let $\mathcal{E}$ be the event that all the edges in R with one endpoint in $x_1 + x_2 = 0$ are open, and $\mathcal{E}'$ be the event that all the edges in R' with one endpoint in $x_1 + x_2 = 0$ are closed. Note that on $\mathcal{E} \cap \mathcal{F} \cap \mathcal{E}' \cap \mathcal{F}'$, there exists exactly one cluster in ω and one cluster in ω^* winding around the torus; see Fig. 23.

The comparison between boundary conditions implies that conditionally on $\mathcal{E} \cap \mathcal{V}(R)$, the boundary conditions in R' are dominated by wired boundary conditions. We obtain

$$\begin{aligned}
\phi^{\text{per}}_{\mathbb{T},p_c,q}[\mathcal{A}] &\ge \phi^{\text{per}}_{\mathbb{T},p_c,q}[\mathcal{E} \cap \mathcal{F} \cap \mathcal{E}' \cap \mathcal{F}'] \\
&\ge \phi^0_{R,p_c,q}[\mathcal{E} \cap \mathcal{F}]\phi^1_{R',p_c,q}[\mathcal{E}' \cap \mathcal{F}'] \\
&\overset{\text{(FKG)}}{\ge} \phi^0_{R,p_c,q}[\mathcal{E}]\phi^0_{R,p_c,q}[\mathcal{F}]\phi^1_{R',p_c,q}[\mathcal{E}']\phi^1_{R',p_c,q}[\mathcal{F}'] \\
&\ge c_{\text{FE}}^N\, c_0^{2M/N},
\end{aligned}$$

where in the last line, we used (FE) and (5.39). By picking N large enough and then letting M go to infinity, we obtain a contradiction with (5.38), so that **P5** cannot be satisfied and the phase transition is discontinuous.

Remark 5 In fact, one may even prove directly that **P4b** does not hold (this is of value for these lectures since we did not formally prove that **P4b** was equivalent to **P5**). We refer to Exercise 56 for details.

Exercise 55 We wish to prove that for all $\delta > 0$, for N and M large enough,

$$\phi^{\rm per}_{\mathbb{T},p_c,q}\Big[\big(\tfrac{4}{q}\big)^{k_{\rm nc}(\omega)}\Big] \geq \exp(-\delta M).$$

1. Show that there exists $c_0 > 0$ depending on q only such that for all M and N, if $n = \delta M - N$, then

$$\phi_{\mathbb{T},p_c,q}(k_{\rm nc}(\omega) \geq \delta M) \leq c_0^{M+N}\phi^0_{\mathbb{T},p_c,q}[\exists n \text{ disjoint clusters crossing } \mathbb{T} \text{ from north-west to south-east}]. \tag{5.40}$$

2. Consider the event $\mathcal{E}(x_1, \dots, x_n)$ that the points $x_1, \dots, x_n$ on the north-west side of $\mathbb{T}$ are connected to the bottom-east side by open paths, and $x_1, \dots, x_n$ are all in different clusters. Conditioning inductively on clusters crossing $\mathbb{T}$ from north-west to south-east, show that

$$\phi^0_{\mathbb{T},p_c,q}[\mathcal{E}(x_1, \dots, x_n)] \leq \phi^0_{p_c,q}[0 \longleftrightarrow \partial\Lambda_N]^n.$$

3. Conclude.

Exercise 56 We wish to prove that **P4b** cannot hold if (6.5) is true.
1. Show that if **P4b** does not hold, then for every $\delta > 0$ there exists an infinite number of n such that

$$\phi^0_{\mathbb{T},p_c,q}[(0,0) \longleftrightarrow (n,n)] \geq \exp(-\delta n).$$

Hint. One may follow the same strategy as in Exercise 48.
2. Deduce that for N large enough, $\phi^0_{\mathbb{T},p_c,q}[\mathcal{F}] \geq c\exp(-\delta M)$ for some constant $c > 0$ depending on N only.
3. Conclude as in the proof that **P5** does not hold.

6 Conformal Invariance of the Ising Model on $\mathbb{Z}^2$

We will also adopt an important convention in this section. We now focus on the random-cluster model with cluster-weight $q = 2$. Also, we define $\mathbb{L}$ to be the rotation by $\pi/4$ of the graph $\sqrt{2}\mathbb{Z}^2$. Generically, (Ω, a, b) will be a Dobrushin subdomain of $\mathbb{L}$ with the additional assumption that $e_b \in \mathbb{R}_+$ (where e_b is seen as a complex number). Note that in this case e_b is simply equal to 1.

For a discrete Dobrushin domain (Ω, a, b), denote $e \ni v$ if v is one of the endpoints of e, and set $\partial\Omega^\diamond$ for the set of vertices of $\Omega^\diamond$ incident to exactly two edges of $\Omega^\diamond$. Define the *vertex fermionic observable* on vertices of $\Omega^\diamond$ by the formula

$$f(v) := \begin{cases} \frac{1}{2}\sum_{e \ni v} F(e) & \text{if } v \in \Omega^\diamond \setminus \partial\Omega^\diamond, \\ \frac{2}{2+\sqrt{2}}\sum_{e \ni v} F(e) & \text{if } v \in \partial\Omega^\diamond, \end{cases}$$

where F is the (edge) fermionic observable on (Ω, a, b) defined in Definition 2.

We are interested in the geometry at large scale of the critical Ising model on $\mathbb{L}$ (in particular the asymptotics of the vertex fermionic observable). A Dobrushin domain $(\Omega_\delta, a_\delta, b_\delta)$ will be a Dobrushin domain defined as a subgraph of the $\delta\mathbb{L}$, still with the convention that seen as a complex number, $e_b \in \mathbb{R}_+$. In particular, the length of the edges of $\Omega^\diamond$ is δ. We extend the notions of Dobrushin domain, edge and vertex fermionic observables to this context.

We will focus on discrete Dobrushin domains $(\Omega_\delta, a_\delta, b_\delta)$ approximating in a better and better way a simply connected domain $\boldsymbol{\Omega} \subseteq \mathbb{C}$ with two points $\mathbf{a}$ and $\mathbf{b}$ on the boundary. We choose the notion of Carathéodory convergence for these approximations, i.e. that $\psi_\delta \longrightarrow \psi$ on any compact subset $K \subseteq \mathbb{R} \times (0, \infty)$, where ψ is the unique conformal map from the upper half-plane $\mathbb{R} \times (0, \infty)$ to $\boldsymbol{\Omega}$ sending 0 to $\mathbf{a}$, ∞ to $\mathbf{b}$, and with derivative at infinity equal to 1, and ψ_δ is the unique conformal map from $\mathbb{H}$ to $\Omega_\delta^\diamond$ sending 0 to $a_\delta^\diamond$, ∞ to $b_\delta^\diamond$ and with derivative at infinity equal to 1. Here, we consider $\Omega_\delta^\diamond$ as a simply connected domain of $\mathbb{C}$ by taking the union of its faces.[19]

The first result of this section deals with the limit of the parafermionic observable (which we call *fermionic* observable in this case).

Theorem 22 (Smirnov [121]) *Fix $q = 2$ and $p = p_c$. Let $(\Omega_\delta, a_\delta, b_\delta)$ be Dobrushin domains approximating a simply connected domain $\boldsymbol{\Omega}$ with two marked points $\mathbf{a}$ and $\mathbf{b}$ on its boundary. If f_δ denotes the vertex fermionic observable on $(\Omega_\delta, a_\delta, b_\delta)$, then*

$$\lim_{\delta\to 0} \frac{1}{\sqrt{2\delta}} f_\delta = \sqrt{\phi'},$$

where ϕ is a conformal map from $\boldsymbol{\Omega}$ to the strip $\mathbb{R} \times (0, 1)$ mapping $\mathbf{a}$ to $-\infty$ and $\mathbf{b}$ to ∞.

Above, the convergence of functions is the uniform convergence on every compact subset of $\boldsymbol{\Omega}$. Since functions f_δ are defined on the graph $\Omega_\delta^\diamond$ only, we perform an implicit extension of the function to the whole graph, for instance by setting $f_\delta(y) = f_\delta(x)$ for the whole face above $x \in \Omega^\diamond$. Note that the constraint that $e_b = \delta$ is not really relevant. We could relax this constraint by simply renormalizing f_δ by $1/\sqrt{2e_b}$ where e_b is seen as a complex number. One word of caution here, δ is not the mesh size of the original lattice on which the random-cluster model is defined, but the mesh size of the medial lattice. Also notice that the map ϕ is not unique a priori since one could add any real constant to ϕ, but this modification does not change its derivative.

The second result we will prove deals with the limit of the exploration path (we postpone the discussion to Sect. 6.2).

Theorem 23 (Chelkak et al. [35]]) *Fix $q = 2$ and $p = p_c$. Let $(\Omega_\delta, a_\delta, b_\delta)$ be Dobrushin domains approximating a simply connected domain $\boldsymbol{\Omega}$ with two marked*

---

[19] If it has "pinched" points, we add a tiny ball of size $\varepsilon \ll \delta$. The very precise definition is not relevant here since the definition is a complicated way of phrasing an intuitive notion of convergence.

points $\mathbf{a}$ *and* $\mathbf{b}$ *on its boundary. The exploration path* $\gamma_{(\Omega_\delta,a_\delta,b_\delta)}$ *in* $(\Omega_\delta, a_\delta, b_\delta)$ *converges weakly to the Schramm-Loewner Evolution with parameter* $\kappa = 16/3$ *as* δ *tends to* 0.

Above, the topology of the weak convergence is given by the metric d on the set X of continuous parametrized curves defined for $\gamma_1 : I \to \mathbb{C}$ and $\gamma_2 : J \to \mathbb{C}$ by

$$d(\gamma_1, \gamma_2) = \min_{\substack{\varphi_1:[0,1]\to I\\ \varphi_2:[0,1]\to J}} \sup_{t\in[0,1]} |\gamma_1(\varphi_1(t)) - \gamma_2(\varphi_2(t))|,$$

where the minimization is over *increasing bijective functions* φ_1 and φ_2.

A fermionic observable for the Ising model itself (and not of its random-cluster representation) was proved to be conformally invariant in [38]. Since then, many other quantities of the model were proved to be conformally invariant.[20] Let us focus on one important case, namely the spin-spin correlations.

Theorem 24 (Chelkak et al. [36]) *Let* Ω_δ *be domains approximating a simply connected domain* $\mathbf{\Omega}$. *Consider also* $a_\delta^1, \dots, a_\delta^k$ *in* Ω_δ *converging to points* $\mathbf{a^1}, \dots, \mathbf{a^k}$ *in* $\mathbf{\Omega}$. *Then,*

$$\lim_{\delta\to 0} \delta^{-n/8} \mu^{\mathrm{f}}_{\Omega_\delta,\beta_c}\big[\sigma_{a_\delta^1} \cdots \sigma_{a_\delta^k}\big] = \langle \sigma_{\mathbf{a^1}} \cdots \sigma_{\mathbf{a^k}} \rangle_{\mathbf{\Omega}},$$

where $\langle \sigma_{\mathbf{a^1}} \cdots \sigma_{\mathbf{a^k}} \rangle_{\mathbf{\Omega}}$ *satisfies*

$$\langle \sigma_{\mathbf{a^1}} \cdots \sigma_{\mathbf{a^k}} \rangle_{\mathbf{\Omega}} = |\phi'(\mathbf{a^1})|^{1/8} \cdots |\phi'(\mathbf{a^k})|^{1/8} \langle \sigma_{\phi(\mathbf{a^1})} \cdots \sigma_{\phi(\mathbf{a^k})} \rangle_{\phi(\mathbf{\Omega})}$$

for any conformal map ϕ *on* $\mathbf{\Omega}$.

Note that this theorem shows that the critical exponent of the spin-spin correlations is 1/8, i.e. that

$$\mu_{\beta_c}[\sigma_0\sigma_x] = \|x\|^{-1/4+o(1)}. \tag{6.1}$$

In fact, this result is simpler to obtain and goes back to the middle of the 20th century (see [103] and references therein).

The general form of $\langle -- \rangle_{\mathbf{\Omega}}$ was predicted by means of Conformal Field Theory in [30]. The method of [36] gives another formula (which is slightly less explicit). The proof relies on similar ideas as the Proof of Theorem 22 (namely s-holomorphicity), but is substantially harder. We do not include it here and refer to [36] for details.

In the next two sections, we prove Theorems 22 and 23.

[20]Let us mention crossing probabilities [15, 91], interfaces with different boundary conditions [35, 82], full family of interfaces [16, 94], the energy fields [81, 83]. The observable has also been used off criticality, see [12, 48].

6.1 Conformal Invariance of the Fermionic Observable

In this section, we prove Theorem 22. We do so in two steps. We first prove that the vertex fermionic observable satisfies a certain boundary value problem on $\Omega^\diamond$. Then, we show that this boundary value problem has a unique solution converging to $\sqrt{\phi'}$ when taking Dobrushin domains $(\Omega_\delta, a_\delta, b_\delta)$ converging in the Carathéodory sense to $(\mathbf{\Omega}, \mathbf{a}, \mathbf{b})$.

s-holomorphic functions and connection to a boundary value problem We will use a very specific property of $q = 2$, which is that $\sigma = \frac{1}{2}$ in this case. This special value of σ enables us to prove the following:

Lemma 15 *Fix a Dobrushin domain (Ω, a, b). For any edge e of $\Omega^\diamond$, the edge fermionic observable $F(e)$ belongs to $\sqrt{\overline{e}}\,\mathbb{R}$.*

Note that the definition of the square root is irrelevant since we are only interested in its value up to a ± 1 multiplicative factor.

Proof The winding $W_{\gamma(\omega)}(e, e_b)$ at an edge e can only take its value in the set $W + 2\pi\mathbb{Z}$ where W is the winding at e of an arbitrary oriented path going from e to e_b. Therefore, the winding weight involved in the definition of $F(e)$ is always equal to $\mathrm{e}^{\mathrm{i}W/2}$ or $-\mathrm{e}^{\mathrm{i}W/2}$, ergo $F(e) \in \mathrm{e}^{\mathrm{i}W/2}\mathbb{R}$, which is the claim by the definition of the square root and the fact that $e_b = 1$. □

Together with the relations (5.23), the previous lemma has an important implication: while there were half the number of relations necessary to determine F in the general $q > 0$ case, we now know sufficiently many additional relations to hope to be able to compute F. We will harvest this new fact by introducing the notion of *s-holomorphic* functions, which was developed in [37, 38, 121]. For any edge e (recall that e is oriented and can therefore be seen as a complex number), define

$$\mathrm{P}_e[x] = \tfrac{1}{2}(x + \overline{e}\,\overline{x}),$$

which is nothing but the projection of x on the line $\sqrt{\overline{e}}\,\mathbb{R}$.

Definition 3 *(Smirnov).* A function $f : \Omega^\diamond \to \mathbb{C}$ is *s-holomorphic* if for any edge $e = uv$ of $\Omega^\diamond$, we have

$$\mathrm{P}_e[f(u)] = \mathrm{P}_e[f(v)].$$

The notion of s-holomorphicity is related to the classical notion of discrete holomorphic functions. On $\Omega^\diamond$, f is discrete holomorphic if if satisfies the *discrete Cauchy-Riemann equations*

$$f(v_1) - \mathrm{i}f(v_2) - f(v_3) + \mathrm{i}f(v_4) = 0 \tag{6.2}$$

for every $x \in \Omega \cup \Omega^*$, where the v_i are the four vertices around x indexed in counterclockwise order. Discrete holomorphic functions f distinctively appeared for the first time in the papers [88, 89] of Isaacs. Note that a s-holomorphic function is discrete holomorphic, since the definition of s-holomorphicity gives that for every $e = uv$,

$$e[f(u) - f(v)] = \overline{f(v)} - \overline{f(u)}, \tag{6.3}$$

and that summing this relation for the four edges around x gives (6.2).

The reason why s-holomorphic functions are easier to handle than discrete holomorphic function will become clear in the next section. In this section, we stick to the proof that the vertex fermionic observable is s-holomorphic, and that it satisfies some specific boundary conditions.

For a Dobrushin domain (Ω, a, b), let $b^\diamond$ be the vertex of $\Omega^\diamond$ at the beginning of the oriented edge e_b. Also, let $\nu_v = e + e'$ with e and e' the two edges of $\Omega^\diamond$ incident to v. The vector ν_v can be interpreted as a discrete version of the tangent vector along the boundary, when going from a to b.

Theorem 25 *Let (Ω, a, b) be a Dobrushin domain. The vertex fermionic observable f is s-holomorphic and satisfies $\mathrm{P}_{e_b}[f(b^\diamond)] = 1$ and $\nu_v f(v)^2 \in \mathbb{R}_+$ for any $v \in \partial\Omega^\diamond$.*

Proof The key to the proof is the following claim: for any $e \ni v$,

$$\mathrm{P}_e[f(v)] = F(e). \tag{6.4}$$

To prove this claim, consider v with four medial edges e_1, e_2, e_3 and e_4 incident to it (we index them in counterclockwise order). Note that (5.23) reads

$$e_1 F(e_1) + e_3 F(e_3) = e_2 F(e_2) + e_4 F(e_4).$$

Furthermore, Lemma 15 gives that

$$\overline{F(e)} = eF(e). \tag{6.5}$$

Plugging this in the previous equality and using the conjugation, we find

$$F(e_1) + F(e_2) = F(e_3) + F(e_4) \left(= \tfrac{1}{2} \sum_{e \ni v} F(e)\right).$$

The term under parentheses is nothing else but $f(v)$. Using Lemma 15 again, we see that $F(e_1)$ and $F(e_3)$ are two orthogonal vectors belonging to $\sqrt{\overline{e_1}}\mathbb{R}$ and $\sqrt{\overline{e_3}}\mathbb{R}$ respectively whose sum is $f(v)$, so that the claim follows readily for e_1 and e_3. One proves the claim for e_2 and e_4 in a similar way.

Let us now treat the case of $v \in \partial\Omega^\diamond$ (the normalization $2/(2 + \sqrt{2})$ will play a role here). Let e and e' be the two edges of $\Omega^\diamond$ incident to v. Recalling that the winding on the boundary is deterministic, and that $e \in \gamma$ if and only if $e' \in \gamma$, gives

$$\sqrt{e'} F(e') = \phi^{a,b}_{\Omega, p_c, 2}[e' \in \gamma] = \phi^{a,b}_{\Omega, p_c, 2}[e \in \gamma] = \sqrt{e} F(e). \tag{6.6}$$

(Here, we choose the square root so that $\sqrt{e'} = \pm e^{i\pi/4}\sqrt{e}$.) This gives

$$\tfrac{2+\sqrt{2}}{2} f(v) = F(e) + F(e') = (\sqrt{\bar{e}} + \sqrt{\bar{e}'})\, \phi^{a,b}_{\Omega,p_c,2}[e \in \gamma]. \tag{6.7}$$

We deduce that $f(v) \in \sqrt{\bar{e} + \bar{e}'}\mathbb{R}$. Since $e = \pm ie'$, a quick study of the complex arguments of $f(v)$, $F(e)$ and $F(e')$ immediately gives that $\mathrm{P}_e[f(v)] = F(e)$ and $\mathrm{P}_{e'}[f(v)] = F(e')$.

Now that (6.4) is proved, we can conclude. First, observe that the s-holomorphicity is trivial, since for any edge $e = uv$, the claim shows that $\mathrm{P}_e[f(u)] = F(e) = \mathrm{P}_e[f(v)]$. Second, $\mathrm{P}_{e_b}[f(b)] = F(e_b) = 1$. The last property follows from $f(v) \in \sqrt{\bar{e} + \bar{e}'}\mathbb{R}$. □

Theorem 22 therefore follows from the following result, which is a general statement on s-holomorphic functions.

Theorem 26 *For a family of Dobrushin domains $(\Omega_\delta, a_\delta, b_\delta)$ approximating a simply connected domain $\mathbf{\Omega}$ with two points* **a** *and* **b** *on its boundary, let f_δ be a s-holomorphic function satisfying $\mathrm{P}_{e_b}[f_\delta(b)] = 1$ and $\nu_v f_\delta(v)^2 \in \mathbb{R}_+$ for any $v \in \partial\Omega_\delta^\diamond$. Then,*

$$\lim_{\delta\to 0} \frac{1}{\sqrt{2\delta}} f_\delta = \sqrt{\phi'},$$

where ϕ is a conformal map from $\mathbf{\Omega}$ to the strip $\mathbb{R} \times (0, 1)$ mapping **a** *to $-\infty$ and* **b** *to ∞.*

We now turn to the proof of this statement, which will not involve the random-cluster anymore.

Remark 6 Let us discuss the general $q \neq 2$ case. Equation (6.2) looks similar to (5.23). Therefore, one may think of the (edge) parafermionic observable as a function defined on vertices of the medial graph $\Omega^{\diamond\diamond}$ of $\Omega^\diamond$ satisfying half of the discrete Cauchy-Riemann equations—namely those around faces of $\Omega^{\diamond\diamond}$ corresponding to vertices of $(\Omega^\diamond)^*$ (for the other faces, we do not know how to get the corresponding relations, which probably are not even true at the discrete level for $q \neq 2$). Such an interpretation is nonetheless slightly misleading, since the edge parafermionic observable does not really converge to a function in the scaling limit. Indeed, in the case of the fermionic observable ($q = 2$), the edge fermionic observable is the projection of the vertex fermionic observable, and therefore converges to different limits depending on the orientation of the edge of $\Omega^\diamond$ associated with the corresponding vertex of $\Omega^{\diamond\diamond}$.

Proof of Theorem 26 The idea of the Proof of Theorem 26 will be to prove that solutions of this discrete Boundary value problem (with Riemann-Hilbert type boundary conditions on the boundary, i.e. conditions on the function being parallel to a certain power of the tangent vector) must converge to the solution of their analog in the continuum. Unfortunately, treating this discrete boundary value problem directly is a mess, and we prefer to transport our problem as follows. The function $\mathrm{Im}(\phi)$ is the unique harmonic function in $\mathbf{\Omega}$ equal to 1 on the arc (**ab**), and 0 on the arc (**ba**). Therefore, one may try to prove that a discrete version H_δ of the imaginary part of the primitive of $\frac{1}{2\delta} f_\delta^2$ satisfies some approximate Dirichlet boundary value problem in the discrete, and that therefore this function must converge to $\mathrm{Im}(\phi)$ as δ tends to 0. This has a much greater chance to work, since Dirichlet boundary value problems are easier to handle.

For now, let us start by studying s-holomorphic functions on a domain $\Omega^\diamond$ with $e_b = 1$. For any such s-holomorphic function f, we associate the function $F = F_f$ defined on edges $e = uv$ of $\Omega^\diamond$ by

$$F(e) := \mathrm{P}_e[f(v)] = \mathrm{P}_e[f(u)]. \tag{6.8}$$

We also introduce the (unique) function $H = H_f : \Omega \cup \Omega^* \to \mathbb{C}$ such that $H(b) = 1$ and

$$H(x) - H(y) = |F(e)|^2 \tag{6.9}$$

for every $x \in \Omega$ and $y \in \Omega^*$, where e is the medial edge bordering both x and y. To justify the existence of such a function, construct $H(x)$ by summing increments along an arbitrary path from b to x. The fact that this function satisfies (6.9) for all neighboring x and y comes from the fact that the definition does not depend on the choice of the path.

This last fact can be justified as follows: the domain is the union of all the faces of the medial lattice within it. As a consequence, the property that the definition does not depend on the choice of the path is equivalent to the property that for any vertex $v \in \Omega^\diamond \setminus \partial\Omega^\diamond$, if $e_1, \ldots, e_4$ denote the four medial edges with endpoint v indexed in counterclockwise order, then the paths going through e_1 and e_2, and the one going through e_4 and e_3 contribute the same (see Fig. 25), i.e.

$$|F(e_1)|^2 - |F(e_2)|^2 = |F(e_4)|^2 - |F(e_3)|^2,$$

Since $F(e_1)$ and $F(e_3)$ are orthogonal (idem for $F(e_2)$ and $F(e_4)$), the previous equality follows from

$$|F(e_1)|^2 + |F(e_3)|^2 = |f(v)|^2 = |F(e_2)|^2 + |F(e_4)|^2. \tag{6.10}$$

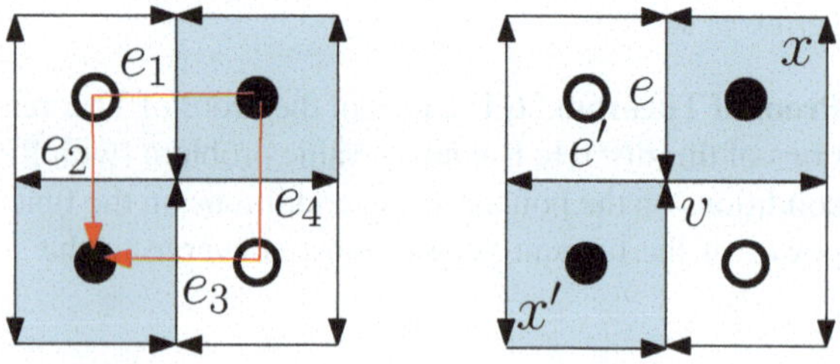

Fig. 25 On the left the two paths going through e_1 and e_2, and e_4 and e_3. On the right, the notation for the proof of (6.11)

The existence of H is the main reason why it is more convenient to work with s-holomorphic functions rather than the less constraining notion of discrete holomorphicity. Also, we hope that the brief discussion on boundary value problems above provides sufficient motivation for the introduction of H: as shown in the following theorem, the function H should be interpreted as the discrete analogue of $\mathrm{Im}\left(\int^z \frac{1}{2}f^2\right)$, which satisfies some nice property of sub and super harmonicity.

(NotBelow, the discrete Laplacian of H is defined by the formula

$$\Delta H(x) := \sum_y [H(y) - H(x)],$$

where the sum is over neighbors of x in Ω (or Ω^* if $x \in \Omega^*$).

Theorem 27 *If $x, x' \in \Omega \cup \Omega^*$ correspond to two opposite faces of $\Omega^\diamond$ bordered by $v \in \Omega^\diamond$,*

$$H(x) - H(x') \;=\; \tfrac{1}{2}\mathrm{Im}\left[f(v)^2 \cdot (x - x')\right]. \tag{6.11}$$

Furthermore, $\Delta H(x) \geq 0$ for every $x \in \Omega \backslash \partial\Omega$ and $\Delta H(y) \leq 0$ for every $y \in \Omega^* \backslash \partial\Omega^*$.

Proof ((6.11)) Assume that x and x' belong to Ω (the case of x and x' belonging to Ω^* is the same). Let e and e' two edges of $\Omega^\diamond$ incident to v bordering the same white face. We further assume that e and e' are respectively bordering the faces of x and x'; see Fig. 25. The s-holomorphicity implies that

$$|F(e)|^2 = \tfrac{1}{4}[ef(v)^2 + \overline{ef(v)}^2 + 2|f(v)|^2].$$

Using a similar relation for $|F(e')|^2$, we obtain

$$\begin{aligned} H(x) - H(x') &= |F(e)|^2 - |F(e')|^2 \\ &= \tfrac{1}{4}[(e - e')f(v)^2 + \overline{(e - e')f(v)}^2] = \tfrac{1}{2}\mathrm{Re}[f(v)^2(e - e')]. \end{aligned}$$

The proof follows by observing that $e - e' = \mathrm{i}(x - x')$.

Proof of sub-harmonicity. Fix $x \in \Omega$. Let A, B, C and D be the values of f on the vertices of $\Omega^\diamond$ north-east, north-west, south-west and south-east of x. Recall that

(a) $A - B = \overline{A} - \overline{B}$ by s-holomorphicity at the medial edge north of x (equal to i),
(b) $C - D = \overline{D} - \overline{C}$ by s-holomorphicity at the medial edge south of x (equal to 1),
(c) $A - C = \mathrm{i}(D - B)$ by discrete holomorphicity (6.2) around x.

Then,

$$A^2 + \mathrm{i}B^2 - C^2 - \mathrm{i}D^2 = (A - C)(A + C) + \mathrm{i}(B - D)(B + D)$$

$$\stackrel{(c)}{=} (A - C)(A + C - B - D)$$
$$\stackrel{(a,b)}{=} (A - C)(\overline{A} - \overline{B} + \overline{D} - \overline{C})$$
$$\stackrel{(c)}{=} (1 + \mathrm{i})|A - C|^2. \tag{6.12}$$

Taking the imaginary part of the quantity obtained by multiplying the previous expression by $\frac{1+\mathrm{i}}{2}$ (which is equal to $\frac{1}{2}(x' - x)$, where x' is the vertex of Ω north-east of x), (6.11) gives

$$\Delta H(x) = |A - C|^2 \geq 0.$$

Similarly, one may check that $\Delta H(x) = -|A - C|^2 \leq 0$ for $x \in \Omega^*$. □

Until now, we treated general s-holomorphic functions, but from this point we focus on the implications of boundary conditions. Let us start with the following easy lemma.

Lemma 16 *Consider a s-holomorphic function f satisfying $F(e_b) = 1$ and $\nu_v f(v)^2 \in \mathbb{R}_+$ for all $v \in \partial\Omega^\diamond$. Then, the function H is equal to 1 on (ba) and 0 on $(ab)^*$.*

Proof Equation (6.11) and the condition $(x - x')f(v)^2 = \pm\nu_v f(v)^2 \in \mathbb{R}$ give that H is constant on (ba) and $(ab)^*$ respectively. The fact that $H = 1$ on (ba) thus follows from the definition $H(b) = 1$. The claim that $H = 0$ on $(ab)^*$ follows from the fact that for $w \in (ab)^*$ neighboring b,

$$H(w) \stackrel{(6.9)}{=} H(b) - |F(e_b)|^2 = 1 - 1 = 0.$$ □

On the other part (ab) of the boundary of Ω, we would like to say that H is roughly 0. This is true but not so simple to prove. In order to circumvent this difficulty, we choose another path: we add a "layer" or additional vertices, and fix the value of H to be 0 on these new vertices (for simplicity, we consider all these vertices as one single ghost vertex $\mathfrak{g}$). With this definition, H is not quite super-harmonic on $\Omega \cup \{\mathfrak{g}\}$ but it almost is: one can define a modified Laplacian on the boundary for which H is super-harmonic. This procedure is explained formally below (we do a similar construction for Ω^*).

Introduce two additional ghost vertices $\mathfrak{g}$ and $\mathfrak{g}^*$ to Ω and Ω^* respectively. Define the continuous-time random walk X^x starting at x and jumping with rate 1 on edges of Ω and rate $\frac{2}{1+\sqrt{2}}N_x$ to $\mathfrak{g}$, where N_x is the number of vertices of $\partial\Omega^\diamond$ bordering x. Note that X^x jumps to $\mathfrak{g}$ with positive rate only when it is on the boundary of Ω. Also, from now on the Laplacian $\widehat{\Delta}$ on Ω denotes the generator of the random walk, which is defined by

$$\widehat{\Delta}H(x) := \Delta H(x) + \tfrac{2}{1+\sqrt{2}}N_x[H(\mathfrak{g}) - H(x)].$$

Similarly, we denote by X^y the continuous-time random walk starting at y and jumping with rate 1 on edges of Ω^* and with rate $\frac{2}{1+\sqrt{2}}N_y$ to $\mathfrak{g}^*$. We extend H to $\mathfrak{g}$ and $\mathfrak{g}^*$ by setting $H(\mathfrak{g}) = 0$ and $H(\mathfrak{g}^*) = 1$.

Lemma 17 *Consider a s-holomorphic function f satisfying $F(e_b) = 1$ and $\nu_v f(v)^2 \in \mathbb{R}_+$ for all $v \in \partial\Omega^\diamond$. Then, $\widehat{\Delta}H \geq 0$ on $\Omega\backslash(ba)$ and $\widehat{\Delta}H \leq 0$ on $\Omega^*\backslash(ab)^*$.*

Proof Let us prove that $\widehat{\Delta}H(x) \geq 0$ for $x \in \Omega\backslash(ba)$ (the proof for $x \in \Omega^*\backslash(ab)^*$ follows the same lines). If $x \notin \partial\Omega$, one has $\widehat{\Delta} = \Delta$ and the result follows from Theorem 27. We therefore focus our attention on $x \in (ab)$. We use the same computation as in (6.12), except that for $v \in \partial\Omega^\diamond$, we replace the expression

$$\mathrm{Im}[f(v)^2 \cdot (v - x)] = \tfrac{1}{2}\mathrm{Im}[f(v)^2 \cdot (x' - x)] = H(x') - H(x)$$

given by (6.11) by the expression

$$\mathrm{Im}[f(v)^2(v - x)] = \tfrac{2}{1+\sqrt{2}}[H(\mathfrak{g}) - H(x)]. \tag{6.13}$$

In order to prove (6.13), use that $v - x = -\frac{i}{2}\nu_v$ (since $x \in (ab)$) and $\nu_v f(v)^2 \in \mathbb{R}_+$ to get

$$\mathrm{Im}[f(v)^2(v - x)] = -\tfrac{\sqrt{2}}{2}|f(v)|^2.$$

Using the same reasoning as for (6.7) and the fact that ν_v has length $\sqrt{2}$, we find that

$$|f(v)|^2 = \tfrac{4|1+e^{i\pi/4}|^2}{(2+\sqrt{2})^2}|F(e)|^2 = \tfrac{2\sqrt{2}}{1+\sqrt{2}}|F(e)|^2.$$

Therefore, (6.13) follows from the two previous equalities together with $H(\mathfrak{g}) = 0$ and $H(x) = |F(e)|^2$ (which is true since there is $y \in (ab)^*$ neighboring x, which satisfies $H(y) = 0$). □

We are now in a position to prove Theorem 26.

Proof (*Theorem* 26) For f_δ, let H_δ constructed via the relation (6.9) and the condition $H_\delta(b_\delta) = 1$. Note that all the previous properties of H extend to H_δ (with trivial modification of the definition of Δ and $\widetilde{\Delta}$), except (6.11), which becomes

$$H_\delta(x') - H_\delta(x) = \tfrac{1}{2\delta}\mathrm{Im}[f_\delta(v)^2(x' - x)] \tag{6.14}$$

since the edge $x' - x$ does not have length $\sqrt{2}$ anymore but $\sqrt{2}\delta$ instead.

We start by proving that (H_δ) converges.[21] We set $H_\delta^\bullet$ and H_δ° for the restrictions of H_δ to Ω_δ and Ω_δ^*. Define

$$\mathbf{Hm}_\delta^\bullet(x) := \mathbb{P}[X^x \text{ hits } (b_\delta a_\delta) \text{ before } \mathfrak{g}] \text{ and}$$

[21] Recall that here and below, we consider the convergence on every compact subset of $\mathbf{\Omega}$.

$$\mathbf{Hm}^\circ_\delta(y) := \mathbb{P}[X^y \text{ hits } (a_\delta b_\delta)^* \text{ before } \mathfrak{g}^*].$$

The function $\mathbf{Hm}^\bullet_\delta$ is the harmonic solution on Ω_δ of the discrete Dirichlet problem with boundary conditions 1 on $(b_\delta a_\delta)$ and 0 on $\mathfrak{g}$. Since the random walk jumps on $\mathfrak{g}$ only when it is on $(a_\delta b_\delta)$, one may show that it converges to the harmonic solution of the Dirichlet problem with boundary conditions 1 on $(\mathbf{ba})$ and 0 on $(\mathbf{ab})$—i.e. to $\text{Im}(\phi)$—as δ tends to 0 (see Exercise 57 for details). Since $H^\bullet_\delta$ is sub-harmonic by Lemmata 16 and 17, one has $H^\bullet_\delta \le \mathbf{Hm}^\bullet_\delta$ and therefore

$$\limsup_{\delta\to 0} H^\bullet_\delta \le \text{Im}(\phi).$$

Similarly, $\mathbf{Hm}^\circ_\delta$ tends to $\text{Im}(\phi)$. Since H°_δ is super-harmonic, $H^\circ_\delta \ge \mathbf{Hm}^\circ_\delta$ and

$$\liminf_{\delta\to 0} H^\circ_\delta \ge \text{Im}(\phi).$$

Since $H^\bullet(x) \ge H^\circ(y)$ for y neighboring x, we deduce that H_δ converges to $\text{Im}(\phi)$.

Let us now prove that (f_δ) converges. Consider a holomorphic sub-sequential limit f (if it exists) of $f_\delta/\sqrt{2\delta}$. Also set F to be a primitive of f^2. By (6.14), H_δ is equal to the imaginary part of the primitive of $\frac{1}{2\delta}f_\delta^2$, so that by passing to the limit and using the first part of the proof, $\text{Im}(F) = \text{Im}(\phi) + C$. Since f is holomorphic, we know that F also is, so that it must be equal to ϕ up to an additive (real valued) constant. By differentiating and taking the square root, we deduce that $f = \sqrt{\phi'}$. To conclude, it only remains to prove that (f_δ) is pre-compact and that any sub-sequential limit is holomorphic, which is done in the next lemma. □

Lemma 18 *The family of functions* $(\frac{1}{\sqrt{2\delta}}f_\delta)$ *is pre-compact for the uniform convergence on every compact. Furthermore, any sub-sequential limit is holomorphic on* $\mathbf{\Omega}$.

In the next proof, we postpone three facts to exercises. We want to highlight the fact that we do not sweep any difficulty under the carpet: these statements are very simple and educational to prove and we therefore prefer to leave them to the reader.

Proof Since the functions f_δ is discrete holomorphic, the statement follows (see Exercise 60 for details) from the fact that $(\frac{1}{\sqrt{2\delta}}f_\delta)$ is square integrable, i.e. that for any compact subset $\mathbf{K}$ of $\mathbf{\Omega}$, there exists a constant $C = C(\mathbf{K}) > 0$ such that for all δ,

$$\delta \sum_{x\in\delta\mathbb{L}\cap\mathbf{K}} |f_\delta(x)|^2 \le C. \tag{6.15}$$

In particular, (6.11) implies that

$$\begin{aligned}\tfrac{\sqrt{2}}{2}|f_\delta(v)|^2 &= \tfrac{1}{2}\text{Im}[f_\delta(v)^2(x'-x)] + \tfrac{1}{2}\text{Re}[f_\delta(v)^2(x'-x)]\\ &= H^\bullet(x') - H^\bullet(x) + H^\circ(y') - H^\circ(y),\end{aligned} \tag{6.16}$$

where $x, x' \in \Omega_\delta$ and $y, y' \in \Omega_\delta^*$ are the four faces bordering v indexed so that $x' - x = \mathrm{i}(y' - y)$. Since $H_\delta^\bullet$ is bounded and sub-harmonic, Exercise 61 implies that

$$\delta \sum_{x \in \delta\mathbb{L} \cap \mathbf{K}} |H_\delta^\bullet(x) - H_\delta^\bullet(x')| \le C, \tag{6.17}$$

where the sum is over edges x' with xx' an edge of $\delta\mathbb{L}$. Similarly, one obtains the same bound for H_δ°. This, together with (6.16), implies (6.15). □

Exercise 57 (*Dirichlet problem*) 1. Prove that there exists $\alpha > 0$ such that for any $0 < r < \frac{1}{2}$ and any curve γ inside $\mathbb{D} := \{z : |z| < 1\}$ from $\{z : |z| = 1\}$ to $\{z : |z| = r\}$, the probability that a random walk on $\mathbb{D} \cap \delta\mathbb{L}$ starting at 0 exits $\mathbb{D} \cap \delta\mathbb{L}$ without crossing γ is smaller than r^α uniformly in $\delta > 0$.
2. Deduce that $\mathbf{Hm}_\delta^\bullet$ tends to 0 on (ab).
3. Using the convergence of the simple random walk to Brownian motion, prove the convergence of $\mathbf{Hm}_\delta^\bullet$ to the solution of the Dirichlet problem with 0 boundary conditions on (ab), and 1 on (ba).

Exercise 58 (*Regularity of discrete harmonic functions*) 1. Consider $\Lambda := [-1, 1]^2$. Show that there exists $C > 0$ such that, for each $\delta > 0$, one may couple two lazy random walks X and Y starting from 0 and its neighbor x in $\Lambda \cap \delta\mathbb{L}$ in such a way that $\mathbb{P}[X_\tau \neq Y_\tau] \le C\delta$, where τ is the hitting time of the box of the boundary of Λ.
2. Deduce that a bounded harmonic function h on Λ satisfies $|h(x) - h(y)| \le C\delta$.
3. Let $H_\Lambda(x, y)$ be the probability that the random walk starting from x exits Λ by y. Show that $H_\Lambda(x, y) \le C'\delta$.

Exercise 59 (*Limit of discrete holomorphic functions*) Prove that a discrete holomorphic function f on $\delta\mathbb{Z}^2$ is discrete harmonic for the leapfrog Laplacian, i.e. that $\Delta f_\delta(x) = 0$, where

$$\Delta f_\delta(x) = \sum_{\varepsilon, \varepsilon' \in \{\pm\delta\}} (f_\delta(x + (\varepsilon, \varepsilon')) - f(x)).$$

Prove that a convergent family of discrete holomorphic functions f_δ on $\delta\mathbb{Z}^2$ converges to a holomorphic function f. *Hint.* Observe that all the discrete versions of the partial derivatives with respect to x and y converge using Exercise 58.

Exercise 60 (*Precompactness criteria for discrete harmonic functions*) Below, $\|f\|_\infty = \sup\{|f(x) : x \in \mathbf{\Omega} \cap \delta\mathbb{Z}^2\}$ and $\|f\|_2 = \delta^2 \sum_{x \in \mathbf{\Omega} \cap \delta\mathbb{Z}^2} f(x)$.
1. Show that a family of $\|\cdot\|_\infty$-bounded harmonic functions (f_δ) on $\mathbf{\Omega}$ is precompact for the uniform convergence on compact subsets. *Hint.* Use the second question of Exercise 58.
2. Show that a family of $\|\cdot\|_2$-bounded harmonic functions (f_δ) on $\mathbf{\Omega}$ is precompact for the uniform convergence on compact subsets. *Hint.* Use the third question of Exercise 58 and the Cauchy-Schwarz inequality.

Exercise 61 (*Regularity of sub-harmonic functions*) Let H be a sub-harmonic function on $\Omega_\delta := \mathbf{\Omega} \cap \delta\mathbb{L}$, with 0 boundary conditions on $\partial\Omega_\delta$.
1. Show that $H(x) = \sum_{y \in \Omega_\delta} G_{\Omega_\delta}(x, y)\Delta H(y)$, where $G_{\Omega_\delta}(x, y)$ is the expected time a random walk starting at x spends at y before exiting Ω_δ.
2. Prove that G_{Ω_δ} is harmonic in $x \neq y$. Deduce that for two neighbors x and x' on Ω_δ,

$$|G_{\Omega_\delta}(x, y) - G_{\Omega_\delta}(x', y)| \le \frac{C\delta}{|x - y| \wedge d(x, \partial\mathbf{\Omega})}.$$

3. Deduce that for any compact subset $\mathbf{K}$ of $\mathbf{\Omega}$, there exists $C(\mathbf{K}) > 0$ such that for any δ,

$$\delta \sum_{x \in \mathbf{K} \cap \delta\mathbb{L}} |H(x) - H(x')| \le C,$$

where x' is an arbitrary choice of a neighbor of x.
4. What can we say for bounded boundary conditions?
5. Deduce (6.17) for $H_\delta^\bullet$.

6.2 Conformal Invariance of the Exploration Path

Conformal field theory leads to the prediction that the exploration path $\gamma_{(\Omega_\delta,a_\delta,b_\delta)}$ in the Dobrushin domains $(\Omega_\delta, a_\delta, b_\delta)$ mentioned before converges as $\delta \to 0$ to a random, continuous, non-self-crossing curve $\gamma_{(\mathbf{\Omega},\mathbf{a},\mathbf{b})}$ from $\mathbf{a}$ to $\mathbf{b}$ staying in $\overline{\mathbf{\Omega}}$, and which is expected to be conformally invariant in the following sense.

Definition 4 A family of random non-self-crossing continuous curves $\gamma_{(\mathbf{\Omega},\mathbf{a},\mathbf{b})}$, going from $\mathbf{a}$ to $\mathbf{b}$ and contained in $\overline{\mathbf{\Omega}}$, indexed by simply connected domains $\mathbf{\Omega}$ with two marked points $\mathbf{a}$ and $\mathbf{b}$ on the boundary is *conformally invariant* if for any $(\mathbf{\Omega}, \mathbf{a}, \mathbf{b})$ and any conformal map $\psi : \mathbf{\Omega} \to \mathbb{C}$,

$$\psi(\gamma_{(\mathbf{\Omega},\mathbf{a},\mathbf{b})}) \text{ has the same law as } \gamma_{(\psi(\mathbf{\Omega}),\psi(\mathbf{a}),\psi(\mathbf{b}))}.$$

In 1999, Schramm proposed a natural candidate for the possible conformally invariant families of continuous non-self-crossing curves. He noticed that interfaces of discrete models further satisfy the *domain Markov property* which, together with the assumption of conformal invariance, determines a one-parameter family of possible random curves. In [117], he introduced the Stochastic Loewner evolution (SLE for short) which is now known as the Schramm–Loewner evolution. Our goal is not to present in detail this well-studied model, and we rather refer the reader to the following expositions and references therein [99]. Here, we wish to prove Theorem 23 and therefore briefly remind the definition of SLEs.

Set $\mathbb{H}$ to be the upper half-plane $\mathbb{R} \times (0, \infty)$. Fix a simply connected subdomain H of $\mathbb{H}$ such that $\mathbb{H} \setminus H$ is compact. Riemann's mapping theorem guarantees[22] the existence of a *unique* conformal map g_H from H onto $\mathbb{H}$ such that

$$g_H(z) := z + \tfrac{C}{z} + O\left(\tfrac{1}{z^2}\right).$$

The constant C is called the *h-capacity* of H.

There is a natural way to parametrize certain continuous non-self-crossing curves $\Gamma : \mathbb{R}_+ \to \overline{\mathbb{H}}$ with $\Gamma(0) = 0$ and with $\Gamma(s)$ going to ∞ when $s \to \infty$. For every s, let H_s be the connected component of $\mathbb{H} \setminus \Gamma[0, s]$ containing ∞, and denote its h-capacity by C_s. The continuity of the curve guarantees that C_s grows continuously, so that it is possible to parametrize the curve via a time change $s(t)$ in such a way that $C_{s(t)} = 2t$. This parameterization is called the *h-capacity parameterization*. Below, we will assume that the parameterization is the h-capacity, and reflect this by using the letter t for the time parameter.

Let $(W_t)_{t>0}$ be a continuous real-valued function.[23] Fix $z \in \mathbb{H}$ and consider the map $t \mapsto g_t(z)$ satisfying the following differential equation up to its explosion time:

[22] The proof of the existence of this map is not completely obvious and requires Schwarz's reflection principle.

[23] Again, one usually requires a few things about this function, but let us omit these technical conditions here.

$$\partial_t g_t(z) = \frac{2}{g_t(z) - W_t}. \tag{6.18}$$

For every fix t, let H_t be the set of z for which the explosion time of the differential equation above is strictly larger than t. One may verify that H_t is a simply connected open set and that $\overline{\mathbb{H}}\setminus H_t$ is compact. Furthermore, the map $z \mapsto g_t(z)$ is a conformal map from H_t to $\mathbb{H}$. If there exists a parametrized curve $(\Gamma_t)_{t>0}$ such that for any $t > 0$, H_t is the connected component of $\mathbb{H}\setminus\Gamma[0, t]$ containing ∞, the curve $(\Gamma_t)_{t>0}$ is called *(the curve generating) the Loewner chain with driving process* $(W_t)_{t>0}$.

The Loewner chain in $(\mathbf{\Omega}, \mathbf{a}, \mathbf{b})$ with driving function $(W_t)_{t>0}$ is simply the image of the Loewner chain in $(\mathbb{H}, 0, \infty)$ by a conformal from $(\mathbb{H}, 0, \infty)$ to $(\mathbf{\Omega}, \mathbf{a}, \mathbf{b})$.

Definition 5 For $\kappa > 0$ and $(\mathbf{\Omega}, \mathbf{a}, \mathbf{b})$, $\mathsf{SLE}(\kappa)$ is the random Loewner evolution in $(\mathbf{\Omega}, \mathbf{a}, \mathbf{b})$ with driving process $\sqrt{\kappa}B_t$, where (B_t) is a standard Brownian motion.

The strategy of the Proof of Theorem 23 is the following. The first step consists in proving that the family $(\gamma_{(\Omega_\delta,a_\delta,b_\delta)})$ is tight for the weak convergence and that any sub-sequential limit γ is a curve generating a Loewner chain for a continuous driving process (W_t) satisfying some integrability conditions. The proof of this fact is technical and can be found in [35, 58, 95]. It is based on a Aizenman-Burchard type argument based on crossing estimates obtained in Property **P5** of Theorem 16 (see also [34, 49] for a stronger statement in the case of the Ising model).

The second step of the proof is based on the fermionic observable, which can be seen as a martingale for the exploration process. This fact implies that its limit is a martingale for γ. This martingale property, together with Itô's formula, allows to prove that W_t and $W_t^2 - \kappa t$ are martingales (where κ equals 16/3). Lévy's theorem thus implies that $W_t = \sqrt{\kappa}B_t$. This identifies $\mathsf{SLE}(\kappa)$ as being the only possible sub-sequential limit, which proves that $(\gamma_{(\Omega_\delta,a_\delta,b_\delta)})$ converges to $\mathsf{SLE}(\kappa)$. We now provide more detail for this second step.

Below, $\Omega\setminus\gamma[0, n]$ is the slit domain obtained from Ω by removing all the edges crossed by the exploration path up to time n. Also, $\gamma(n)$ denotes the vertex of Ω bordered by the last edge of $\gamma[0, n]$.

Lemma 19 *Let $\delta > 0$. The random variable $M_n(z) := f_{\Omega\setminus\gamma[0,n],\gamma(n),b}(z)$ is a martingale with respect to $(\mathcal{F}_n)$ where $\mathcal{F}_n$ is the σ-algebra generated by $\gamma[0, n]$.*

Proof The random variable $M_n(z)$ is a linear combination of the random variables $M_n(e) := F_{\Omega\setminus\gamma[0,n],\gamma(n),b}(e)$ for $e \ni z$ so that we only need to treat the latter random variables. The fact that conditionally on $\gamma[0, n]$, the law in $\Omega\setminus\gamma[0, n]$ is a random-cluster model with Dobrushin boundary conditions implies that $M_n(e)$ is equal to $\mathrm{e}^{\frac{1}{2}\mathrm{i}W_\gamma(e,e_b)}\mathbb{1}_{e\in\gamma}$ conditionally on $\mathcal{F}_n$, therefore it is automatically a closed martingale. □

Proof (*Theorem* 23) We treat the case of the upper half-plane $\mathbf{\Omega} = \mathbb{H}$ with $\mathbf{a} = 0$ and $\mathbf{b} = \infty$. The general case follows by first applying a conformal map from $(\mathbf{\Omega}, \mathbf{a}, \mathbf{b})$ to $(\mathbb{H}, 0, \infty)$. Consider γ a sub-sequential limit of $\gamma_{(\Omega_\delta,a_\delta,b_\delta)}$ and assume that its driving process is equal to (W_t). Define g_t as above. For $z \in \mathbb{H}$ and $\delta > 0$, define $M_n^\delta(z)$ for $\gamma_{(\Omega_\delta,a_\delta,b_\delta)}$ as above too.

The stopping time theorem implies that $M_{\tau_t}^\delta(z)$ is a martingale with respect to $\mathcal{F}_{\tau_t}$, where τ_t is the first time at which $\gamma_{(\Omega_\delta,a_\delta,b_\delta)}$ has a h-capacity larger than t. Now, if $M_{\tau_t}^\delta(z)$ converges uniformly as δ tends to 0, then, the limit $M_t(z)$ is a martingale with respect to the σ-algebra $\mathcal{G}_t$ generated by the curve γ up to the first time its h-capacity exceeds t. By definition of the parameterization, this time is t, and $\mathcal{G}_t$ is the σ-algebra generated by $\gamma[0,t]$.

Since the conformal map from $\mathbb{H}\backslash\gamma[0,t]$ to $\mathbb{R}\times(0,1)$, normalized to send γ_t to $-\infty$ and ∞ to ∞ is $\frac{1}{\pi}\ln(g_t - W_t)$, Theorem 22 gives that $M_t^\delta(z)$ converges to

$$\sqrt{\pi}M_t(z) = \sqrt{[\ln(g_t(z) - W_t)]'} = \Big(\frac{g_t'(z)}{g_t(z) - W_t}\Big)^{1/2}, \tag{6.19}$$

which is therefore a martingale for the filtration $(\mathcal{G}_t)$. Formally, in order to apply Theorem 22, one needs z and $\gamma[0,\tau_t]$ to be well apart. For this reason, we only obtain that $M_{t\wedge\sigma}^z$ is a martingale for $\mathcal{G}_{t\wedge\sigma}$, where σ is the hitting time of the boundary of the ball of size $R < |z|$ by the curve γ.

Recall that $g_t(z) = z + \frac{2t}{z} + O\left(\frac{1}{z^2}\right)$ and $g_t'(z) = 1 - \frac{2t}{z^2} + O\left(\frac{1}{z^3}\right)$ so that for t,

$$\begin{aligned}\sqrt{\pi z}\,M_t(z) &= \Big(\frac{1 - \frac{2t}{z^2} + O(\frac{1}{z^3})}{1 - \frac{W_t}{z} + \frac{2t}{z^2} + O(\frac{1}{z^3})}\Big)^{1/2}\\ &= 1 + \tfrac{1}{2z}W_t + \tfrac{1}{8z^2}(3W_t^2 - 16t) + O\left(\tfrac{1}{z^3}\right).\end{aligned}$$

Taking the conditional expectation against $\mathcal{G}_{s\wedge\sigma}$ (with $s \le t$) gives

$$\begin{aligned}\sqrt{\pi z}\,\mathbb{E}[M_{t\wedge\sigma}(z)|\mathcal{G}_{s\wedge\sigma}] = 1 &+ \tfrac{1}{2z}\mathbb{E}[W_{t\wedge\sigma}|\mathcal{G}_{s\wedge\sigma}]\\ &+ \tfrac{1}{8z^2}\mathbb{E}[3W_{t\wedge\sigma}^2 - 16(t\wedge\sigma)|\mathcal{G}_{s\wedge\sigma}] + O\left(\tfrac{1}{z^3}\right).\end{aligned}$$

Since $M_{t\wedge\sigma}(z)$ is a martingale, $\mathbb{E}[M_{t\wedge\sigma}(z)|\mathcal{G}_{s\wedge\sigma}] = M_{s\wedge\sigma}(z)$. Therefore, the terms of the previous asymptotic developments (in $1/z$) can be matched together by letting z tend to infinity so that

$$\mathbb{E}[W_{t\wedge\sigma}|\mathcal{G}_{s\wedge\sigma}] = W_{s\wedge\sigma} \quad \text{and} \quad \mathbb{E}[W_{t\wedge\sigma}^2 - \tfrac{16}{3}(t\wedge\sigma)|\mathcal{G}_{s\wedge\sigma}] = W_{s\wedge\sigma}^2 - \tfrac{16}{3}(s\wedge\sigma).$$

One can now let R and thus σ go to infinity to obtain

$$\mathbb{E}[W_t|\mathcal{G}_s] = W_s \quad \text{and} \quad \mathbb{E}[W_t^2 - \tfrac{16}{3}t|\mathcal{G}_s] = W_s^2 - \tfrac{16}{3}s.$$

(Note that some integrability condition on W_t is necessary to justify passing to the limit here.) The driving process W_t being continuous, Lévy's theorem implies that $W_t = \sqrt{16/3}B_t$ where B_t is a standard Brownian motion. Since we considered an arbitrary sub-sequential limit, this directly proves that $(\gamma_{(\Omega_\delta,a_\delta,b_\delta)})$ converges weakly to $\mathsf{SLE}(16/3)$. □

Note that despite the fact that the fermionic observable may not seem like a very natural choice at first sight, it is in fact corresponding to a discretization of a very natural martingale of $\mathsf{SLE}(16/3)$.

7 Where Are We Standing? And More Conjectures..

It is time to conclude these lectures. To summarize, we proved that the Potts model and its random-cluster representation undergo phase transitions between ordered and disordered phases. We also showed that the long-range order and the spontaneous magnetization phases of the Potts model coincide. Then, we proceeded to prove that the phase transition was sharp, meaning that correlations decay exponentially fast below the critical inverse-temperature.

After this study of the phases $\beta < \beta_c$ and $\beta > \beta_c$, we moved to the study of the $\beta = \beta_c$ phase. We determined that the phase transition of the Potts model is continuous in any dimension if $q = 2$ (i.e. for the Ising model), and that it is continuous if $q \le 4$ and discontinuous for $q > 4$ in two dimensions. This gives us the opportunity of mentioning the first major question left open by this manuscript:

Conjecture 2 *Prove that the phase transition of the nearest-neighbor Potts model on $\mathbb{Z}^d$ (with $d \ge 3$) is discontinuous for any $q \ge 3$.*

Let us mention that this conjecture is proved in special cases, namely

- if d is fixed and $q \ge q_c(d) \gg 1$ [98],
- if $q \ge 3$ is fixed and $d \ge d_c(q) \gg 1$ [19],
- if $q \ge 3$ and $d \ge 2$, but the range of the interactions is sufficiently spread-out [21, 73].

When the phase transition is continuous, there should be some conformally invariant scaling limit. In two dimensions, this concerns any $q \le 4$, and not only the $q = 2$ case mentioned previously in these lectures. One may formulate the conformal invariance conjecture for random-cluster models with $q \le 4$ in the following way.

Conjecture 3 (Schramm) *Fix $0 < q \le 4$ and $p = p_c$. Let $(\Omega_\delta, a_\delta, b_\delta)$ be Dobrushin domains approximating a simply connected domain $\mathbf{\Omega}$ with two marked points* **a** *and* **b** *on its boundary. The exploration path $\gamma_{(\Omega_\delta, a_\delta, b_\delta)}$ in $(\Omega_\delta, a_\delta, b_\delta)$ converges weakly to $\mathsf{SLE}(\kappa)$ as δ tends to 0, where*

$$\kappa = \frac{8}{\sigma+1} = \frac{4\pi}{\pi - \arccos(\sqrt{q}/2)}.$$

The values of κ range from $\kappa = 4$ for $q = 4$ to $\kappa = 8$ for $q = 0$. Also note that $\kappa = 6$ corresponds to $q = 1$, as expected. Following the same strategy as in the previous section, the previous conjecture would follow from the convergence of vertex parafermionic observables (they are defined for general q as the vertex fermionic observable).

Conjecture 4 (Smirnov) *Fix* $0 < q \le 4$ *and* $p = p_c$. *Let* $(\Omega_\delta, a_\delta, b_\delta)$ *be Dobrushin domains approximating a simply connected domain* $\mathbf{\Omega}$ *with two marked points* **a** *and* **b** *on its boundary. If* f_δ *denotes the vertex parafermionic observable on* $(\Omega_\delta, a_\delta, b_\delta)$ *defined as the average of the edge fermionic observable on neighboring edges, then*

$$\lim_{\delta\to 0}(2\delta)^{-\sigma} f_\delta = (\phi')^\sigma,$$

where ϕ *is a conformal map from* $\mathbf{\Omega}$ *to the strip* $\mathbb{R} \times (0, 1)$ *mapping* **a** *to* $-\infty$ *and* **b** *to* ∞.

For the Ising model in higher dimensions (the other Potts models are predicted to have a discontinuous phase transition by Conjecture 2), the model still undergoes a continuous phase transition and it therefore makes sense to study the critical phase in more detail.

It is believed that the critical exponent of the spin-spin correlations of the Ising model in dimension four and higher is the mean-field one, i.e. that

$$\mu_{\beta_c}[\sigma_x\sigma_y] \approx \frac{1}{\|x-y\|^{d-2+\delta}}$$

with $\delta = 0$. Note that in two dimensions this is not the case since by (6.1), $\delta = 1/4$. In three dimensions, the best known result is Theorem 13, which gets rephrased as $\delta \in [0, 1]$. The following improvement would be of great value.

Conjecture 5 *Consider the three dimensional Ising model. There exists* $\varepsilon > 0$ *and* $c_0, c_1 \in (0, \infty)$ *such that for all* $x, y \in \mathbb{Z}^3$,

$$\frac{c_0}{\|x-y\|^{2-\varepsilon}} \le \mu_{\beta_c}[\sigma_x\sigma_y] \le \frac{c_1}{\|x-y\|^{1+\varepsilon}}.$$

Another question of interest is the question of triviality/non-triviality of the scaling limit of the spin-field. In other words, the question is to measure whether the spin-spin correlations factorize like Gaussian field (i.e. whether they satisfy the Wick's rule or not). One usually defines the *renormalized coupling constant*[24]

$$g(\beta) := \sum_{x_2,x_3,x_4\in\mathbb{Z}^d} \frac{U_4(0, x_2, x_3, x_4)}{\chi(\beta)^2\xi(\beta)^d}, \tag{7.1}$$

where $U_4(x_1, x_2, x_3, x_4)$ was defined in Exercise 35 and (e_1 is a unit vector in $\mathbb{Z}^d$)

$$\chi(\beta) := \sum_{x\in\mathbb{Z}^d} \mu^{\mathrm{f}}_\beta[\sigma_0\sigma_x] \quad \text{and} \quad \xi(\beta) := \big(\lim_{n\to\infty} -\tfrac{1}{n}\log \mu^{\mathrm{f}}_\beta[\sigma_0\sigma_{ne_1}]\big)^{-1}. \tag{7.2}$$

[24]Since Wick's rule is equivalent to the fact that $U_4(x_1, x_2, x_3, x_4)$ vanishes (see the definition in Exercise 35), this quantity is a measure of how non-Gaussian the field $(\sigma_x : x \in \mathbb{Z}^d)$ is.

If $g(\beta)$ tends to 0 as $\beta \nearrow \beta_c$, the field is said to be *trivial*. Otherwise, it is said to be *non-trivial*. Aizenman [1] and Fröhlich [68] proved that the Ising model is trivial for $d \geq 5$. In two dimensions, one can use Theorem 24 to prove that the Ising model is non-trivial (in fact one can prove this result in a simpler way, but let us avoid discussing this here). Interestingly enough, Aizenman's proof of triviality is one of the first uses of the random current at its full power and it is therefore fair to say that proving this result was one of the motivations for the use of such currents. This leaves the following conjecture open.

Conjecture 6 *Prove that the three-dimensional Ising model is non-trivial.*

Physics predictions go much further. One expects conformal invariance in any dimension (in fact as soon as the phase transition is continuous). Conformal symmetry brings less information on the model in dimensions greater than 2, but recent developments in conformal bootstrap illustrate that still much can be said using these techniques, see [116]. It therefore motivates the question of proving conformal invariance in dimension three, which looks like a tremendously difficult problem.

Exercise 62 *(Triviality of Ising in dimension $d \geq 5$)*
Consider a graph G and denote by P_G^A the measure on currents (here we mean one current, not two) on G with a set of sources equal to A. Set σ_i for the spin at x_i.
1. Show that

$$U_4(x_1,\dots,x_4) = -2\mu^{\mathrm{f}}_{G,\beta}[\sigma_1\sigma_2]\mu^{\mathrm{f}}_{G,\beta}[\sigma_3\sigma_4]\cdot \mathrm{P}_G^{\{x_1,x_2\}}\otimes \mathrm{P}_G^{\{x_3,x_4\}}[x_1,x_2,x_3,x_4 \text{ all connected}].$$

2. Prove that for any $y \in \mathbb{Z}^d$, $\mathrm{P}_G^{\{x_1,x_2\}}\otimes \mathrm{P}_G^{\emptyset}[x_1 \overset{\widehat{\mathbf{n}_1+\mathbf{n}_2}}{\longleftrightarrow} y] = \dfrac{\mu^{\mathrm{f}}_{G,\beta}[\sigma_1\sigma_y]\mu^{\mathrm{f}}_{G,\beta}[\sigma_y\sigma_2]}{\mu^{\mathrm{f}}_{G,\beta}[\sigma_1\sigma_2]}$.
3. Show, using a reasoning similar to the switching lemma that

$$\mathrm{P}_G^{\{x_1,x_2\}}\otimes \mathrm{P}_G^{\{x_3,x_4\}}[x \overset{\widehat{\mathbf{n}_1+\mathbf{n}_2}}{\longleftrightarrow} y, z \overset{\widehat{\mathbf{n}_1+\mathbf{n}_2}}{\longleftrightarrow} y] \leq \mathrm{P}_G^{\{x_1,x_2\}}\otimes \mathrm{P}_G^{\{x_3,x_4\}}\otimes \mathrm{P}_G^{\emptyset}[x \overset{\widehat{\mathbf{n}_1+\mathbf{n}_3}}{\longleftrightarrow} y, z \overset{\widehat{\mathbf{n}_2}}{\longleftrightarrow} y].$$

4. Use another sourceless current $\mathbf{n}_4$ and the union bound to prove that

$$0 \leq -U_4(x_1,\dots,x_4) \leq 2\sum_{y\in\mathbb{Z}^d} \mu^{\mathrm{f}}_{G,\beta}[\sigma_y\sigma_1]\mu^{\mathrm{f}}_{G,\beta}[\sigma_y\sigma_2]\mu^{\mathrm{f}}_{G,\beta}[\sigma_y\sigma_3]\mu^{\mathrm{f}}_{G,\beta}[\sigma_y\sigma_4].$$

5. Deduce that $0 \leq -g(\beta) \leq \frac{\chi(\beta)^2}{\xi(\beta)^d}$.
6. Show that for every $x \in \mathbb{Z}^d$, $\mu^{\mathrm{f}}_{G,\beta}[\sigma_0\sigma_x] \leq \exp(-\|x\|_\infty/\xi(\beta))$.
7. Using (IR), show that $\chi(\beta) \leq C\xi(\beta)^2 \log \xi(\beta)^2$ and conclude that $g(\beta)$ tends to 0 when $d \geq 5$.

Acknowledgements This research was funded by an IDEX Chair from Paris Saclay and by the NCCR SwissMap from the Swiss NSF. These lecture notes describe the content of a class given at the PIMS-CRM probability summer school on the behavior of lattice spin models near their critical point. The author would like to thank the organizers warmly for offering him the opportunity to give this course. Also, special thanks to people who sent comments to me, especially Timo Hirscher and Franco Severo.

References

1. M. Aizenman. Geometric analysis of φ^4 fields and Ising models. I, II. *Comm. Math. Phys.*, 86(1):1–48, 1982.
2. M. Aizenman and D. J. Barsky. Sharpness of the phase transition in percolation models. *Comm. Math. Phys.*, 108(3):489–526, 1987.
3. M. Aizenman, J. T. Chayes, L. Chayes, and C. M. Newman. Discontinuity of the magnetization in one-dimensional $1/|x-y|^2$ Ising and Potts models. *J. Statist. Phys.*, 50(1–2):1–40, 1988.
4. M. Aizenman, H. Duminil-Copin, and V. Sidoravicius. Random Currents and Continuity of Ising Model's Spontaneous Magnetization. *Communications in Mathematical Physics*, 334:719–742, 2015.
5. M. Aizenman and Roberto Fernández. On the critical behavior of the magnetization in high-dimensional Ising models. *J. Statist. Phys.*, 44(3–4):393–454, 1986.
6. M. Aizenman, H. Kesten, and C. M. Newman. Uniqueness of the infinite cluster and continuity of connectivity functions for short and long range percolation. *Comm. Math. Phys.*, 111(4):505–531, 1987.
7. D. J. Barsky, G. R. Grimmett, and Charles M. Newman. Percolation in half-spaces: equality of critical densities and continuity of the percolation probability. *Probab. Theory Related Fields*, 90(1):111–148, 1991.
8. Rodney J. Baxter. *Exactly solved models in statistical mechanics*. Academic Press Inc. [Harcourt Brace Jovanovich Publishers], London, 1989. Reprint of the 1982 original.
9. N. R. Beaton, M. Bousquet-Mélou, J. de Gier, H. Duminil-Copin, and A. J. Guttmann. The critical fugacity for surface adsorption of self-avoiding walks on the honeycomb lattice is $1+\sqrt{2}$. *Comm. Math. Phys.*, 326(3):727–754, 2014.
10. W. Beckner. Inequalities in fourier analysis. *Ann. of Math*, 102(1):159–182, 1975.
11. V. Beffara and H. Duminil-Copin. The self-dual point of the two-dimensional random-cluster model is critical for $q \geq 1$. *Probab. Theory Related Fields*, 153(3-4):511–542, 2012.
12. V. Beffara and H. Duminil-Copin. Smirnov's fermionic observable away from criticality. *Ann. Probab.*, 40(6):2667–2689, 2012.
13. V. Beffara and H. Duminil-Copin. Lectures on planar percolation with a glimpse of Schramm Loewner Evolution. *Probability Surveys*, 10:1–50, 2013.
14. I. Benjamini, Russell Lyons, Y. Peres, and Oded Schramm. Critical percolation on any nonamenable group has no infinite clusters. *Ann. Probab.*, 27(3):1347–1356, 1999.
15. S. Benoist, H. Duminil-Copin, and C. Hongler. Conformal invariance of crossing probabilities for the Ising model with free boundary conditions. *Annales de l'Institut Henri Poincaré*, 52(4):1784–1798, 2016.
16. S. Benoist and C. Hongler. The scaling limit of critical Ising interfaces is CLE(3). arXiv:1604.06975.
17. V.L. Berezinskii. Destruction of long-range order in one-dimensional and two-dimensional systems possessing a continuous symmetry group. ii. quantum systems. *Soviet Journal of Experimental and Theoretical Physics*, 34:610, 1972.
18. Denis Bernard and André LeClair. Quantum group symmetries and nonlocal currents in 2D QFT. *Comm. Math. Phys.*, 142(1):99–138, 1991.
19. M. Biskup and L. Chayes. Rigorous analysis of discontinuous phase transitions via mean-field bounds. *Comm. Math. Phys*, (1):53–93, 2003.
20. Marek Biskup. Reflection positivity and phase transitions in lattice spin models. In *Methods of contemporary mathematical statistical physics*, volume 1970 of *Lecture Notes in Math.*, pages 1–86. Springer, Berlin, 2009.
21. Marek Biskup, Lincoln Chayes, and Nicholas Crawford. Mean-field driven first-order phase transitions in systems with long-range interactions. *J. Stat. Phys.*, 122(6):1139–1193, 2006.
22. Béla Bollobás. *Random Graphs* (2nd ed.). Cambridge University Press, 2001.
23. Béla Bollobás and Oliver Riordan. The critical probability for random Voronoi percolation in the plane is 1/2. *Probab. Theory Related Fields*, 136(3):417–468, 2006.

24. Béla Bollobás and Oliver Riordan. A short proof of the Harris-Kesten theorem. *Bull. London Math. Soc.*, 38(3):470–484, 2006.
25. Béla Bollobás and Oliver Riordan. Percolation on self-dual polygon configurations. In *An irregular mind*, volume 21 of *Bolyai Soc. Math. Stud.*, pages 131–217. János Bolyai Math. Soc., Budapest, 2010.
26. A. Bonami. Etude des coefficients de Fourier des fonctions de $L^p(G)$. *Ann. Inst. Fourier*, 20(2):335–402, 1970.
27. Jean Bourgain, Jeff Kahn, Gil Kalai, Yitzhak Katznelson, and Nathan Linial. The influence of variables in product spaces. *Israel J. Math.*, 77(1-2):55–64, 1992.
28. S. R. Broadbent and J. M. Hammersley. Percolation processes. I. Crystals and mazes. *Proc. Cambridge Philos. Soc.*, 53:629–641, 1957.
29. D. Brydges, and T. Spencer. Self-avoiding walk in 5 or more dimensions. *Comm. Math. Phys.*, 97(1–2):125–148, 1985.
30. Theodore W. Burkhardt and Ihnsouk Guim. Bulk, surface, and interface properties of the Ising model and conformal invariance. *Phys. Rev. B (3)*, 36(4):2080–2083, 1987.
31. R. M. Burton and M. Keane. Density and uniqueness in percolation. *Comm. Math. Phys.*, 121(3):501–505, 1989.
32. J. Cardy. Discrete Holomorphicity at Two-Dimensional Critical Points. *Journal of Statistical Physics*, 137:814–824, 2009.
33. D. Chelkak, D. Cimasoni, and A. Kassel. Revisiting the combinatorics of the 2D Ising model. *Ann. Inst. Henri Poincaré D*, 4(3):309–385, 2017.
34. D. Chelkak, H. Duminil-Copin, and C. Hongler. Crossing probabilities in topological rectangles for the critical planar FK-Ising model. *Electron. J. Probab*, 5:28pp, 2016.
35. D. Chelkak, H. Duminil-Copin, C. Hongler, A. Kemppainen, and S. Smirnov. Convergence of Ising interfaces to Schramm's SLE curves. *C. R. Acad. Sci. Paris Math.*, 352(2):157–161, 2014.
36. Dmitry Chelkak, Clément Hongler, and Konstantin Izyurov. Conformal invariance of spin correlations in the planar Ising model. *Ann. of Math. (2)*, 181(3):1087–1138, 2015.
37. Dmitry Chelkak and Stanislav Smirnov. Discrete complex analysis on isoradial graphs. *Adv. Math.*, 228(3):1590–1630, 2011.
38. Dmitry Chelkak and Stanislav Smirnov. Universality in the 2D Ising model and conformal invariance of fermionic observables. *Invent. Math.*, 189(3):515–580, 2012.
39. D. Cimasoni and H. Duminil-Copin. The critical temperature for the Ising model on planar doubly periodic graphs. *Electron. J. Probab*, 18(44):1–18, 2013.
40. H. Duminil-Copin. Phase transition in random-cluster and O(n)-models. *archive-ouverte.unige.ch/unige:18929*, page 360 p, 2011.
41. H. Duminil-Copin. Divergence of the correlation length for critical planar FK percolation with $1 \leq q \leq 4$ via parafermionic observables. *Journal of Physics A: Mathematical and Theoretical*, 45(49):494013, 2012.
42. H. Duminil-Copin. *Parafermionic observables and their applications to planar statistical physics models*, volume 25 of *Ensaios Matematicos*. Brazilian Mathematical Society, 2013.
43. H. Duminil-Copin. Geometric representations of lattice spin models. book, Edition Spartacus, 2015.
44. H. Duminil-Copin. A proof of first order phase transition for the planar random-cluster and Potts models with $q \gg 1$. *Proceedings of Stochastic Analysis on Large Scale Interacting Systems in RIMS kokyuroku Besssatu*, 2016.
45. H. Duminil-Copin. Random currents expansion of the Ising model. arXiv:1607:06933, 2016.
46. H. Duminil-Copin, M. Gagnebin, M. Harel, I. Manolescu, and V. Tassion. The Bethe ansatz for the six-vertex and XXZ models: an exposition. *arXiv preprint* arXiv:1611.09909, 2016.
47. H. Duminil-Copin, M. Gagnebin, M. Harel, I. Manolescu, and V. Tassion. Discontinuity of the phase transition for the planar random-cluster and Potts models with $q > 4$. *arXiv preprint* arXiv:1611.09877, 2016.
48. H. Duminil-Copin, C. Garban, and G. Pete. The near-critical planar FK-Ising model. *Comm. Math. Phys.*, 326(1):1–35, 2014.

49. H. Duminil-Copin, C. Hongler, and P. Nolin. Connection probabilities and RSW-type bounds for the two-dimensional FK Ising model. *Comm. Pure Appl. Math.*, 64(9):1165–1198, 2011.
50. H. Duminil-Copin and I. Manolescu. The phase transitions of the planar random-cluster and Potts models with $q \geq 1$ are sharp. *Probability Theory and Related Fields*, 164(3):865–892, 2016.
51. H. Duminil-Copin, R. Peled, W. Samotij, and Y. Spinka. Exponential decay of loop lengths in the loop $O(n)$ model with large n. *Communications in Mathematical Physics*, 349(3):777–817, 12 2017.
52. H. Duminil-Copin, A. Raoufi, and V. Tassion. A new computation of the critical point for the planar random-cluster model with $q \geq 1$. arXiv:1604.03702, 2016.
53. H. Duminil-Copin, A. Raoufi, and V. Tassion. Exponential decay of connection probabilities for subcritical Voronoi percolation in $\mathbb{R}^d$. arXiv:1705.07978, 2017.
54. H. Duminil-Copin, A. Raoufi, and V. Tassion. Sharp phase transition for the random-cluster and Potts models via decision trees. arXiv:1705.03104, 2017.
55. H. Duminil-Copin, A. Raoufi, and V. Tassion. Subcritical phase of d-dimensional Poisson-boolean percolation and its vacant set. in preparation, 2017.
56. H. Duminil-Copin, V. Sidoravicius, and V. Tassion. Absence of infinite cluster for critical Bernoulli percolation on slabs. *Communications in Pure and Applied Mathematics*, 69(7):1397–1411, 2016.
57. H. Duminil-Copin, V. Sidoravicius, and V. Tassion. Continuity of the phase transition for planar random-cluster and Potts models with $1 \leq q \leq 4$. *Communications in Mathematical Physics*, 349(1):47–107, 2017.
58. H. Duminil-Copin and S. Smirnov. Conformal invariance of lattice models. In *Probability and statistical physics in two and more dimensions*, volume 15 of *Clay Math. Proc.*, pages 213–276. Amer. Math. Soc., Providence, RI, 2012.
59. H. Duminil-Copin and S. Smirnov. The connective constant of the honeycomb lattice equals $\sqrt{2+\sqrt{2}}$. *Ann. of Math. (2)*, 175(3):1653–1665, 2012.
60. H. Duminil-Copin and V. Tassion. RSW and Box-Crossing Property for Planar Percolation. IAMP proceedings, 2015.
61. H. Duminil-Copin and V. Tassion. A new proof of the sharpness of the phase transition for Bernoulli percolation and the Ising model. *Communications in Mathematical Physics*, 343(2):725–745, 2016.
62. H. Duminil-Copin and V. Tassion. A new proof of the sharpness of the phase transition for Bernoulli percolation on $\mathbb{Z}^d$. *Enseignement Mathématique*, 62(1-2):199–206, 2016.
63. P. Erdős and A. Rényi. On random graphs i. *Publicationes Mathematicae*, 6:290–297, 1959.
64. R. Fitzner and R. van der Hofstad. Mean-field behavior for nearest-neighbor percolation in $d > 10$. *Electron. J. Probab.*, 22(43) 65 pp, 2017.
65. C. M. Fortuin and P. W. Kasteleyn. On the random-cluster model. I. Introduction and relation to other models. *Physica*, 57:536–564, 1972.
66. Eduardo Fradkin and Leo P Kadanoff. Disorder variables and para-fermions in two-dimensional statistical mechanics. *Nuclear Physics B*, 170(1):1–15, 1980.
67. S. Friedli and Y. Velenik. *Statistical Mechanics of Lattice Systems: a Concrete Mathematical Introduction*. Cambridge University Press, 2017.
68. J. Fröhlich. On the triviality of $\lambda\phi^4$ theories and the approach to the critical point in $d \geq 4$ dimensions. *Nuclear Physics B*, 200(2):281–296, 1982.
69. J. Fröhlich, B. Simon, and Thomas Spencer. Infrared bounds, phase transitions and continuous symmetry breaking. *Comm. Math. Phys.*, 50(1):79–95, 1976.
70. Jürg Fröhlich and Thomas Spencer. The Kosterlitz-Thouless transition in two-dimensional abelian spin systems and the Coulomb gas. *Comm. Math. Phys.*, 81(4):527–602, 1981.
71. Hans-Otto Georgii. *Gibbs measures and phase transitions*, volume 9 of *de Gruyter Studies in Mathematics*. Walter de Gruyter and Co., Berlin, second edition, 2011.
72. A. Glazman. Connective constant for a weighted self-avoiding walk on $\mathbb{Z}^2$. *Electron. Commun. Probab.*, 20(86):1–13, 2015.

73. T. Gobron and I. Merola. First-order phase transition in Potts models with finite-range interactions. *Journal of Statistical Physics*, 126(3):507–583, 2007.
74. B. T. Graham and G. R. Grimmett. Influence and sharp-threshold theorems for monotonic measures. *Ann. Probab.*, 34(5):1726–1745, 2006.
75. Robert B. Griffiths, C. A. Hurst, and S. Sherman. Concavity of magnetization of an Ising ferromagnet in a positive external field. *J. Mathematical Phys.*, 11:790–795, 1970.
76. G. Grimmett. *Percolation*, volume 321 of *Grundlehren der Mathematischen Wissenschaften [Fundamental Principles of Mathematical Sciences]*. Springer-Verlag, Berlin, second edition, 1999.
77. G. Grimmett. *The random-cluster model*, volume 333 of *Grundlehren der Mathematischen Wissenschaften [Fundamental Principles of Mathematical Sciences]*. Springer-Verlag, Berlin, 2006.
78. G. R. Grimmett. Inequalities and entanglements for percolation and random-cluster models. In *Perplexing problems in probability*, volume 44 of *Progr. Probab.*, pages 91–105. Birkhäuser Boston, Boston, MA, 1999.
79. J. M. Hammersley and D. J. A. Welsh. Further results on the rate of convergence to the connective constant of the hypercubical lattice. *Quart. J. Math. Oxford Ser. (2)*, 13:108–110, 1962.
80. Takashi Hara and Gordon Slade. Mean-field critical behaviour for percolation in high dimensions. *Comm. Math. Phys.*, 128(2):333–391, 1990.
81. C. Hongler. *Conformal invariance of Ising model correlations*. PhD thesis, université de Genève, 2010.
82. Clément Hongler and Kalle Kytölä. Ising interfaces and free boundary conditions. *J. Amer. Math. Soc.*, 26(4):1107–1189, 2013.
83. Clément Hongler and Stanislav Smirnov. The energy density in the planar Ising model. *Acta Math.*, 211(2):191–225, 2013.
84. Tom Hutchcroft. Critical percolation on any quasi-transitive graph of exponential growth has no infinite clusters. *Comptes Rendus Mathematique*, 354(9):944–947, 2016.
85. Y. Ikhlef and J.L. Cardy. Discretely holomorphic parafermions and integrable loop models. *J. Phys. A*, 42(10):102001, 11, 2009.
86. Y. Ikhlef, R. Weston, M. Wheeler, and P. Zinn-Justin. Discrete holomorphicity and quantized affine algebras. arxiv:1302.4649, 2013.
87. D. Ioffe, S. Shlosman, and Y. Velenik. 2D models of statistical physics with continuous symmetry: the case of singular interactions. *Comm. Math. Phys.*, 226(2):433–454, 2002.
88. R.P. Isaacs. Monodiffric functions. Construction and applications of conformal maps. In *Proceedings of a symposium*, National Bureau of Standards, Appl. Math. Ser., No. 18, pages 257–266, Washington, D. C., 1952. U. S. Government Printing Office.
89. Rufus Philip Isaacs. A finite difference function theory. *Univ. Nac. Tucumán. Revista A.*, 2:177–201, 1941.
90. E. Ising. Beitrag zur Theorie des Ferromagnetismus. *Z. Phys.*, 31:253–258, 1925.
91. K. Izyurov. Smirnov's observable for free boundary conditions, interfaces and crossing probabilities. *Communications in Mathematical Physics*, 337(1):225–252, 2015.
92. Joel L. Lebowitz and Anders Martin Löf. On the uniqueness of the equilibrium state for Ising spin systems. *Comm. Math. Phys.*, 25:276–282, 1972.
93. J. Kahn, G. Kalai, and N. Linial. The influence of variables on boolean functions. In *29th Annual Symposium on Foundations of Computer Science*, pages 68–80, 1988.
94. A Kemppainen and S. Smirnov. Conformal invariance in random cluster models. ii. full scaling limit as a branching sle. arXiv:1609.08527.
95. Antti Kemppainen and Stanislav Smirnov. Random curves, scaling limits and loewner evolutions. arXiv:1212.6215, 2012.
96. H. Kesten. The critical probability of bond percolation on the square lattice equals $\frac{1}{2}$. *Comm. Math. Phys.*, 74(1):41–59, 1980.
97. JM Kosterlitz and DJ Thouless. Ordering, metastability and phase transitions in two-dimensional systems. *Journal of Physics C: Solid State Physics*, 6(7):1181–1203, 1973.

98. R. Kotecký and S. B. Shlosman. First-order phase transitions in large entropy lattice models. *Comm. Math. Phys.*, 83(4):493–515, 1982.
99. Gregory F. Lawler. *Conformally invariant processes in the plane*, volume 114 of *Mathematical Surveys and Monographs*. American Mathematical Society, Providence, RI, 2005.
100. W. Lenz. Beitrag zum Verständnis der magnetischen Eigenschaften in festen Körpern. *Phys. Zeitschr.*, 21:613–615, 1920.
101. M. Lis. The fermionic observable in the Ising model and the inverse Kac-Ward operator. *Annales Henri Poincaré*, 15(10):1945–1965, 2013.
102. M. Lis. A short proof of the Kac-Ward formula. *Ann. Inst. Henri Poincaré Comb. Phys. Interact.*, 3:45–53, 2016.
103. B. M. McCoy and T. T. Wu. Ising model correlation functions: difference equations and applications to gauge theory. In *Nonlinear integrable systems—classical theory and quantum theory (Kyoto, 1981)*, pages 121–134. World Sci. Publishing, Singapore, 1983.
104. M. V. Menshikov. Coincidence of critical points in percolation problems. *Dokl. Akad. Nauk SSSR*, 288(6):1308–1311, 1986.
105. N.D. Mermin and H. Wagner. Absence of ferromagnetism or antiferromagnetism in one- or two-dimensional isotropic heisenberg models. *Phys. Rev. Lett.*, 17:1133–1136, 1966.
106. A. Messager and S. Miracle-Sole. Correlation functions and boundary conditions in the ising ferromagnet. *Journal of Statistical Physics*, 17(4):245–262, 1977.
107. B. Nienhuis. Coulomb gas description of 2D critical behaviour. *J. Statist. Phys.*, 34:731–761, 1984.
108. Bernard Nienhuis. Exact Critical Point and Critical Exponents of O(n) Models in Two Dimensions. *Physical Review Letters*, 49(15):1062–1065, 1982.
109. R. O'Donnell, M. Saks, O. Schramm, and R. Servedio. Every decision tree has an influential variable. *FOCS*, 2005.
110. A Polyakov. Interaction of goldstone particles in two dimensions. Applications to ferromagnets and massive Yang-Mills fields. *Physics Letters B*, 59(1):79–81, 1975.
111. Renfrey Burnard Potts. Some generalized order-disorder transformations. In *Proceedings of the Cambridge Philosophical Society*, volume 48(2), pages 106–109. Cambridge Univ Press, 1952.
112. M. A. Rajabpour and J.L. Cardy. Discretely holomorphic parafermions in lattice Z_N models. *J. Phys. A*, 40(49):14703–14713, 2007.
113. N. Reshetikhin. Lectures on the integrability of the 6-vertex model. *arXiv1010.5031*, 2010.
114. V. Riva and J. Cardy. Holomorphic parafermions in the Potts model and stochastic Loewner evolution. *J. Stat. Mech. Theory Exp.*, (12):P12001, 19 pp. (electronic), 2006.
115. L. Russo. A note on percolation. *Z. Wahrscheinlichkeitstheorie und Verw. Gebiete*, 43(1):39–48, 1978.
116. El-Showk S., Paulos M. F., Poland D., Rychkov S., Simmons-Duffin D., and Vichi A. Solving the 3d Ising model with the conformal bootstrap. *Physical Review D*, 86(2), 2012.
117. Oded Schramm. Scaling limits of loop-erased random walks and uniform spanning trees. *Israel J. Math.*, 118:221–288, 2000.
118. P. D. Seymour and D. J. A. Welsh. Percolation probabilities on the square lattice. *Ann. Discrete Math.*, 3:227–245, 1978. Advances in graph theory (Cambridge Combinatorial Conf., Trinity College, Cambridge, 1977).
119. Stanislav Smirnov. Critical percolation in the plane: conformal invariance, Cardy's formula, scaling limits. *C. R. Acad. Sci. Paris Sér. I Math.*, 333(3):239–244, 2001.
120. Stanislav Smirnov. Towards conformal invariance of 2D lattice models. In *International Congress of Mathematicians. Vol. II*, pages 1421–1451. Eur. Math. Soc., Zürich, 2006.
121. Stanislav Smirnov. Conformal invariance in random cluster models. I. Holomorphic fermions in the Ising model. *Ann. of Math. (2)*, 172(2):1435–1467, 2010.
122. H.E. Stanley. Dependence of critical properties on dimensionality of spins. *Physical Review Letters*, 20(12):589–592, 1968.
123. M. Talagrand. On Russo's approximate zero-one law. *Ann. Probab.*, 22(3):1576–1587, 1994.
124. Vincent Tassion. *Planarité et localité en percolation*. PhD thesis, ENS Lyon, 2014.

125. Vincent Tassion. Crossing probabilities for Voronoi percolation. *Annals of Probability*, 44(5):3385–3398, 2016. arXiv:1410.6773.
126. B. L. van der Waerden. Die lange Reichweite der regelmassigen Atomanordnung in Mischkristallen. *Z. Physik*, 118:473–488, 1941.
127. Wendelin Werner. Lectures on two-dimensional critical percolation. In *Statistical mechanics*, volume 16 of *IAS/Park City Math. Ser.*, pages 297–360. Amer. Math. Soc., Providence, RI, 2009.
128. A. C. Yao. Probabilistic computations: Toward a unified measure of complexity. In *Foundations of Computer Science, 1977., 18th Annual Symposium on*, pages 222–227. IEEE, 1977.

Extrema of the Two-Dimensional Discrete Gaussian Free Field

Marek Biskup

Abstract These lecture notes offer a gentle introduction to the two-dimensional Discrete Gaussian Free Field with particular attention paid to the scaling limits of the level sets at heights proportional to the absolute maximum. The bulk of the text is based on recent joint papers with O. Louidor and with J. Ding and S. Goswami. Still, new proofs of the tightness and distributional convergence of the centered DGFF maximum are presented that bypass the use of the modified Branching Random Walk. The text contains a wealth of instructive exercises and a list of open questions and conjectures for future research.

Keywords Extreme value theory · Gaussian processes · Gaussian multiplicative chaos · Liouville quantum gravity · Ballot problem · Entropic repulsion · Random walk in random environment · Random resistor network

Contents

M. Biskup (✉)
Department of Mathematics, UCLA, Los Angeles, CA 90095-1555, USA
e-mail: biskup@math.ucla.edu.edu
URL: http://www.math.ucla.edu/~biskup/index.html

M. T. Barlow and G. Slade (eds.), *Random Graphs, Phase Transitions, and the Gaussian Free Field*, Springer Proceedings in Mathematics & Statistics 304,
https://doi.org/10.1007/978-3-030-32011-9_3

Photo: Miguel Eichelberger (PIMS)

What This Course Is About (and What It Is Not)

This is a set of lecture notes for a course on *Discrete Gaussian Free Field* (DGFF) delivered at the 2017 PIMS-CRM Summer School in Probability at the University of British Columbia in June 2017. The course has been quite intense with the total of sixteen 90-min lectures spanning over 4 weeks. Still, the subject of the DGFF has become so so developed that we could mainly focus only on one specific aspect: *extremal values*. The text below stays close to the actual lectures although later additions have been made to make the exposition self-contained. Each lecture contains a number of exercises that address reasonably accessible aspects of the presentation.

In Lecture 1 we give an introduction to the DGFF in general spatial dimension and discuss possible limit objects. The aim here is to show that the scaling properties single out the *two-dimensional* DGFF as a case of special interest to which the rest of the text is then exclusively devoted.

Lecture 2 opens up by some earlier developments that captured the leading-order behavior of the maximum of the DGFF in growing lattice domains. This sets the scale for the study (in Lectures 2–4) of what we call *intermediate level sets*—namely, those where the DGFF takes values above a constant multiple of the absolute maximum. The technique of proof is of interest here as it will be reused later: we first establish tightness, then extract a subsequential limit and, finally, identify the limit object uniquely, thus proving the existence of the limit. The limit is expressed using the concept of *Liouville Quantum Gravity* (LQG) which we introduce via Kahane's theory of Gaussian Multiplicative Chaos.

Our next item of interest is the behavior of the DGFF maximum. For this we recall (in Lectures 5 and 6) two basic, albeit technically advanced, techniques from the theory of Gaussian processes: correlation inequalities (Kahane, Borell-TIS, Sudakov-Fernique, FKG) and Fernique's majorization bound for the expected maximum.

Once the generalities have been dispensed with, we return (in Lectures 7 and 8) to the DGFF and relate the tightness of its centered maximum to that of a *Branching Random Walk* (BRW). A novel proof of tightness of the DGFF maximum is presented that avoids the so-called modified BRW; instead we rely on the Sudakov-Fernique inequality and the Dekking–Host argument applied to the DGFF coupled with a BRW. This handles the upper tail tightness; for the lower tail we develop the concept of a *concentric decomposition* of the DGFF that will be useful later as well.

In Lectures 9–11 we move to the *extremal level sets*—namely, those where the DGFF is within order-unity of the absolute maximum. Similarly to the intermediate levels, we encode these via a three-coordinate point measure that records the scaled spatial position, the centered value and the "shape" of the configuration at the local maxima. Using the above proof strategy we then show that, in the scaling limit, this measure tends in law to a Poisson Point Process with a *random* intensity measure whose spatial part can be identified with the *critical* LQG.

A key technical tool in this part is Liggett's theory of *non-interacting* particle systems which we discuss in full detail (in Lecture 9); the uniqueness of the limit measure is linked to the convergence in law of the centered DGFF maximum (Lecture 10). Another key tool is the concentric decomposition which permits us to control (in Lecture 11) the local structure of the extremal points and to give (in Lecture 12) independent proofs of spatial tightness of the extremal level sets and convergence in law of the centered DGFF maximum. Interesting corollaries (stated at the end of Lecture 11) give existence of *supercritical* LQG measures and the limit law for the Gibbs measure naturally associated with the DGFF.

The final segment of the course (specifically, Lectures 13–15) is devoted to a *random walk driven by the DGFF*. After the statement of main theorems we proceed to develop (in Lecture 13) the main technique of the proofs: an electric network associated with the DGFF. In Lecture 14 we give variational characterizations of the effective resistance/conductance using only geometric objects such as paths and cuts. Various duality relations (conductance/resistance reciprocity, path/cut planar duality) along with concentration of measure are invoked to control the effective resistivity across rectangles. These are the key ingredients for controlling (in Lecture 15) the relevant aspects of the random walk.

A standalone final lecture (Lecture 16) discusses some open research-level problems that stem directly from the topics discussed in these notes.

To keep the course at a reasonable level of mathematical depth, considerable sacrifices in terms of scope had to be made. The course thus omits a large body of literature on Gaussian Multiplicative Chaos and Liouville Quantum Gravity. We do not address the level sets at heights of order unity and their connections to the Schramm–Loewner evolution and the Conformal Loop Ensemble. We ignore recent developments in Liouville First Passage percolation nor do we discuss the close cousin of our random walk, the Liouville Brownian Motion. We pass over the connections to random-walk local time. In many of these cases, better resources exist to obtain the relevant information elsewhere.

Thanks to our focused approach we are able to present many difficult proofs in nearly complete detail. A patient reader will thus have a chance to learn a number of useful arguments specific to the DGFF as well as several general techniques relevant for probability at large. The hope is that interlacing specifics with generalities will make these notes interesting for newcomers as well as advanced players in this field. The exposition should be sufficiently self-contained to serve as the basis for a one-semester graduate course.

Acknowledgments

These notes are largely based on research (supported, in part, by NSF grant DMS-1407558 and GAČR project P201/16-15238S) that was done jointly with

Oren Louidor

and, in the last part, with

Subhajit Goswami and **Jian Ding**

I would like to thank these gentlemen for all what I have learned from them, much of which is now reproduced here.

Next I wish to thank the organizers of the PIMS-CRM summer school, Omer Angel, Mathav Murugan, Ed Perkins and Gordon Slade, for the opportunity to be one of the main lecturers. These notes would not have been written without the motivation (and pressure) arising from that task. Special thanks go to Ed Perkins for finding a cozy place to stay for my whole family over the duration of the school; the bulk of this text was typed over the long summer nights there.

I have had the opportunity to lecture on various parts of this material at other schools and events; namely, in Bonn, Hejnice, Lyon, Haifa, Atlanta and Salt-Lake City. I am grateful to the organizers of these events, as well as participants thereof, for helping me find a way to parse some of the intricate details into (hopefully) a reasonably digestible form.

Finally, I wish to express my gratitude to Jian Ding, Jean-Christophe Mourrat and Xinyi Li for comments and corrections and to Saraí Hernández Torres and, particularly, Yoshi Abe for pointing out numerous errors in early versions of this manuscript. I take the blame for all the issues that went unnoticed.

Marek Biskup
Los Angeles
May 2019

Lecture 1: Discrete Gaussian Free Field and Scaling Limits

In this lecture we define the main object of interest in this course: the Discrete Gaussian Free Field (henceforth abbreviated as DGFF). By studying its limit properties we are naturally guided towards the two-dimensional case where we describe, in great level of detail, its scaling limit. The limit object, the continuum Gaussian Free Field (CGFF), will underlie, albeit often in disguise, most of the results discussed in the course.

1.1 Definitions

For $d \geq 1$ integer, let $\mathbb{Z}^d$ denote the d-dimensional hypercubic lattice. This is an unoriented graph with vertices at the points of $\mathbb{R}^d$ with integer coordinates and an edge between any pair of vertices at unit Euclidean distance. Denoting the set of all edges (with both orientations identified) by $\mathsf{E}(\mathbb{Z}^d)$, we put forward:

Definition 1.1 *(DGFF, explicit formula) Let $V \subseteq \mathbb{Z}^d$ be finite. The DGFF in V is a process $\{h_x^V : x \in \mathbb{Z}^d\}$ indexed by the vertices of $\mathbb{Z}^d$ with the law given (for any measurable $A \subseteq \mathbb{R}^{\mathbb{Z}^d}$) by*

$$P(h^V \in A) := \frac{1}{\text{norm.}} \int_A \mathrm{e}^{-\frac{1}{4d}\sum_{(x,y)\in\mathsf{E}(\mathbb{Z}^d)}(h_x-h_y)^2} \prod_{x\in V} \mathrm{d}h_x \prod_{x\notin V} \delta_0(\mathrm{d}h_x)\,. \tag{1.1}$$

Here δ_0 is the Dirac point-mass at 0 *and* "norm." *is a normalization constant.*

Notice that the definition forces the values of h outside V to zero—we thus talk about *zero boundary condition*. To see that this definition is good, we pose:

Exercise 1.2 *Prove that the integral in* (1.1) *is finite for* $A := \mathbb{R}^{\mathbb{Z}^d}$ *and so the measure can be normalized to be a probability.*

The appearance of the $4d$ factor in the exponent is a convention used in probability; physicists would write $\frac{1}{2}$ instead of $\frac{1}{4d}$. Without this factor, the definition extends readily from $\mathbb{Z}^d$ to any locally-finite graph but since $\mathbb{Z}^d$ (in fact $\mathbb{Z}^2$) will be our main focus, we keep the normalization as above.

Definition 1.1 gives the law of the DGFF in the form of a *Gibbs measure*; namely, a measure of the form

$$\frac{1}{\text{norm.}}\,\mathrm{e}^{-\beta H(h)}\,\nu(\mathrm{d}h) \tag{1.2}$$

where "norm." is again a normalization constant, H is the Hamiltonian, β is the inverse temperature and ν is an a priori (typically a product) measure. Many models of statistical mechanics are cast this way; a key feature of the DGFF is that the Hamiltonian is a positive-definite quadratic form and ν a product Lebesgue measure which makes the law of h a multivariate Gaussian. This offers the possibility to define the law directly by prescribing its mean and covariance.

Let $\{X_n : n \geq 0\}$ denote the path of a simple symmetric random walk on $\mathbb{Z}^d$. For $V \subseteq \mathbb{Z}^d$, we recall the definition of the *Green function* in V:

$$G^V(x,y) := E^x\left(\sum_{n=0}^{\tau_{V^c}-1} 1_{\{X_n=y\}}\right), \tag{1.3}$$

where E^x is the expectation with respect to the law of X with $X_0 = x$ a.s. and $\tau_{V^c} := \inf\{n \geq 0 : X_n \notin V\}$ is the first exit time of the walk from V. We note:

Exercise 1.3 *Prove that, for any* $x, y \in \mathbb{Z}^d$,

$$V \mapsto G^V(x,y) \text{ is non-decreasing with respect to set inclusion.} \tag{1.4}$$

In particular, for any $V \subsetneq \mathbb{Z}^d$ *(in any* $d \geq 1$*), we have* $G^V(x,y) < \infty$ *for all* $x, y \in \mathbb{Z}^d$.

As is also easy to check, $G^V(x,y) = 0$ unless $x, y \in V$ (in fact, unless x and y can be connected by a path on $\mathbb{Z}^d$ that lies entirely in V). The following additional properties of G^V will be important in the sequel:

Exercise 1.4 *Let* Δ *denote the discrete Laplacian on* $\mathbb{Z}^d$ *acting on finitely-supported functions* $f : \mathbb{Z}^d \to \mathbb{R}$ *as*

$$\Delta f(x) := \sum_{y:\,(x,y)\in \mathsf{E}(\mathbb{Z}^d)} \bigl(f(y) - f(x)\bigr). \tag{1.5}$$

Show that for any $V \subsetneq \mathbb{Z}^d$ *and any* $x \in \mathbb{Z}^d$, $y \mapsto G^V(x,y)$ *is the solution to*

$$\begin{cases} \Delta G^V(\cdot, x) = -2d\delta_x(\cdot), & \text{on } V, \\ G^V(\cdot, x) = 0, & \text{on } V^c, \end{cases} \tag{1.6}$$

where δ_x is the Kronecker delta at x.

Another way to phrase this exercise is by saying that the Green function is a $(2d)$-multiple of the inverse of the negative Laplacian on $\ell^2(V)$—with Dirichlet boundary conditions on V^c. This functional-analytic representation of the Green function allows us to solve:

Exercise 1.5 *Prove that for any $V \subsetneq \mathbb{Z}^d$ we have:*

(1) for any $x, y \in \mathbb{Z}^d$,

$$G^V(x, y) = G^V(y, x), \tag{1.7}$$

(2) for any $f : \mathbb{Z}^d \to \mathbb{R}$ with finite support,

$$\sum_{x,y \in \mathbb{Z}^d} G^V(x, y) f(x) f(y) \geq 0. \tag{1.8}$$

We remark that purely probabilistic ways to solve this (i.e., using solely considerations of random walks) exist as well. What matters for us is that properties (1–2) make G^V a covariance of a Gaussian process. This gives rise to:

Definition 1.6 (DGFF, via the Green function) Let $V \subsetneq \mathbb{Z}^d$ be given. The DGFF in V is a multivariate Gaussian process $\{h^V_x : x \in \mathbb{Z}^d\}$ with law determined by

$$E(h^V_x) = 0 \quad \text{and} \quad E(h^V_x h^V_y) = G^V(x, y), \qquad x, y \in \mathbb{Z}^d, \tag{1.9}$$

or, written concisely,

$$h^V := \mathcal{N}(0, G^V). \tag{1.10}$$

Here and henceforth, $\mathcal{N}(\mu, C)$ denotes the (multivariate) Gaussian with mean μ and covariance C. The dimensions of these objects will usually be clear from the context.

In order to avoid accusations of logical inconsistency, we pose:

Exercise 1.7 *Prove that for $V \subseteq \mathbb{Z}^d$ finite, Definitions 1.1 and 1.6 coincide.*

The advantage of Definition 1.6 over Definition 1.1 is that it works for infinite V as well. The functional-analytic connection can be pushed further as follows. Given a finite set $V \subseteq \mathbb{Z}^d$, consider the Hilbert space $\mathcal{H}^V := \{f : \mathbb{Z}^d \to \mathbb{R},\ \text{supp}(f) \subseteq V\}$ endowed with the *Dirichlet inner product*

$$\langle f, g \rangle_\nabla := \frac{1}{2d} \sum_{x \in \mathbb{Z}^d} \nabla f(x) \cdot \nabla g(x), \tag{1.11}$$

where $\nabla f(x)$, the discrete gradient of f at x, is the vector in $\mathbb{R}^d$ whose ith component is $f(x+e_i)-f(x)$, for e_i the ith unit coordinate vector in $\mathbb{R}^d$. (Since the supports of f and g are finite, the sum is effectively finite. The normalization is for consistence with Definition 1.1.) We then have:

Lemma 1.8 (DGFF as Hilbert-space Gaussian) *For the setting as above with V finite, let $\{\varphi_n : n = 1,\dots,|V|\}$ be an orthonormal basis in $\mathcal{H}^V$ and let $Z_1,\dots,Z_{|V|}$ be i.i.d. standard normals. Then $\{\widetilde{h}_x : x \in \mathbb{Z}^d\}$, where*

$$\widetilde{h}_x := \sum_{n=1}^{|V|} \varphi_n(x) Z_n, \quad x \in \mathbb{Z}^d, \tag{1.12}$$

has the law of the DGFF in V.

Exercise 1.9 *Prove the previous lemma.*

Lemma 1.8 thus gives yet another way to define DGFF. (The restriction to finite V was imposed only for simplicity.) As we will see at the end of this lecture, this is the definition that generalizes seamlessly to a continuum underlying space. Writing $\{\psi_n : n = 1,\dots,|V|\}$ for the orthonormal set in $\ell^2(V)$ of eigenfunctions of the negative Laplacian with the corresponding eigenvalue written as $2d\lambda_n$, i.e., $-\Delta\psi_n = 2d\lambda_n\psi_n$ on V, the choice

$$\varphi_n(x) := \frac{1}{\sqrt{\lambda_n}}\psi_n(x), \tag{1.13}$$

produces an orthonormal basis in $\mathcal{H}^V$. This is useful when one wishes to generate samples of the DGFF efficiently on a computer.

The fact that $\mathbb{Z}^d$ is an additive group, and the random walk is translation invariant, implies that the Green function and thus the law of h^V are *translation invariant* in the sense that, for all $z \in \mathbb{Z}^2$ and for $z + V := \{z + x : x \in V\}$,

$$G^{z+V}(x+z, y+z) = G^V(x,y), \quad x,y \in \mathbb{Z}^d, \tag{1.14}$$

and

$$\{h^{z+V}_{x+z} : x \in \mathbb{Z}^2\} \overset{\text{law}}{=} \{h^V_x : x \in \mathbb{Z}^2\}. \tag{1.15}$$

A similar rule applies to rotations by multiples of $\frac{\pi}{2}$ around any vertex of $\mathbb{Z}^2$.

We finish this section by a short remark on notation: Throughout these lectures we will write h^V to denote the whole configuration of the DGFF in V and write h^V_x for the value of h^V at x. We may at times write $h^V(x)$ instead of h^V_x when the resulting expression is easier to parse.

1.2 Why $d = 2$ Only?

As noted above, this course will focus on the DGFF in $d = 2$. Let us therefore explain what makes the two-dimensional DGFF special. We begin by noting:

Lemma 1.10 (Green function growth rate) *Let $V_N := (0, N)^d \cap \mathbb{Z}^d$ and, for any $\epsilon \in (0, 1/2)$, denote $V_N^\epsilon := (\epsilon N, (1-\epsilon)N)^d \cap \mathbb{Z}^d$. Then for any $x \in V_N^\epsilon$,*

$$G^{V_N}(x,x) \underset{N\to\infty}{\sim} \begin{cases} N, & d = 1, \\ \log N, & d = 2, \\ 1, & d \geq 3, \end{cases} \tag{1.16}$$

where "$\sim$" means that the ratio of the left and right-hand side tends to a positive and finite number as $N \to \infty$ (which may depend on where x is asymptotically located in V_N).

Proof (*sketch*). We will only prove this in $d = 1$ and $d \geq 3$ as the $d = 2$ case will be treated later in far more explicit terms. First note that a routine application of the Strong Markov Property for the simple symmetric random walk yields

$$G^V(x,x) = \frac{1}{P^x(\hat\tau_x > \tau_{V^c})}, \tag{1.17}$$

where $\hat\tau_x := \inf\{n \geq 1 : X_n = x\}$ is the first *return* time to x. Assuming $x \in V$, in $d = 1$ we then have

$$P^x(\hat\tau_x > \tau_{V^c}) = \frac{1}{2}\Big[P^{x+1}(\tau_x > \tau_{V^c}) + P^{x-1}(\tau_x > \tau_{V^c})\Big], \tag{1.18}$$

where $\tau_x := \inf\{n \geq 0 : X_n = x\}$ is the first *hitting* time of x. For $V := V_N$, the Markov property of the simple random walk shows that $y \mapsto P^y(\tau_x > \tau_{V_N^c})$ is discrete harmonic (and thus piecewise linear) on $\{1, \dots, x-1\} \cup \{x+1, \dots, N-1\}$ with value zero at $y = x$ and one at $y = 0$ and $y = N$. Hence

$$P^{x+1}(\tau_x > \tau_{V_N^c}) = \frac{1}{N-x} \quad \text{and} \quad P^{x-1}(\tau_x > \tau_{V_N^c}) = \frac{1}{x}. \tag{1.19}$$

As $x \in V_N^\epsilon$, both of these probabilities are order $1/N$ and scale nicely when x grows proportionally to N. This proves the claim in $d = 1$.

In $d \geq 3$ we note that the transience and translation invariance of the simple symmetric random walk imply

$$G^{V_N}(x,x) \underset{N\to\infty}{\longrightarrow} \frac{1}{P^0(\hat\tau_0 = \infty)} \tag{1.20}$$

uniformly in $x \in V_N^\epsilon$. (Transience is equivalent to $P^0(\hat\tau_0 = \infty) > 0$.) □

Let us now proceed to examine the law of the whole DGFF configuration in the limit as $N \to \infty$. Focusing for the moment on $d = 1$, the fact that the variance blows up suggests that we normalize the DGFF by the square-root of the variance, i.e., $\sqrt{N}$, and attempt to extract a limit. This does work and yields:

Theorem 1.11 (Scaling to Brownian Bridge in $d = 1$) *Suppose $d = 1$ and let h^{V_N} be the DGFF in $V_N := (0, N) \cap \mathbb{Z}$. Then*

$$\left\{\frac{1}{\sqrt{N}} h^{V_N}_{\lfloor tN \rfloor} : t \in [0, 1]\right\} \xrightarrow[N\to\infty]{\text{law}} \left\{\sqrt{2}\, W_t : t \in [0, 1]\right\}, \tag{1.21}$$

where W is the standard Brownian Bridge on $[0, 1]$.

We leave it to the reader to solve:

Exercise 1.12 *Prove Theorem 1.11 with the convergence in the sense of finite-dimensional distributions or, if you like the challenge, in Skorokhod topology. Hint: Note that $\{h^{V_N}_{x+1} - h^{V_N}_x : x = 0, \dots, N-1\}$ are i.i.d. $\mathcal{N}(0, 2)$ conditioned on their total sum being zero.*

We remark that the limit taken in Theorem 1.11 is an example of a *scaling (or continuum) limit*—the lattice spacing is taken to zero while keeping the overall (continuum) domain fixed. In renormalization group theory, taking the scaling limit corresponds to the removal of an ultraviolet cutoff.

Moving to $d \geq 3$, in the proof of Lemma 1.10 we observed enough to conclude:

Theorem 1.13 (Full-space limit in $d \geq 3$) *Suppose $d \geq 3$ and let $\widetilde{V}_N := (-N/2, N/2)^d \cap \mathbb{Z}^d$. Then for any $x, y \in \mathbb{Z}^d$,*

$$G^{\widetilde{V}_N}(x, y) \underset{N\to\infty}{\longrightarrow} G^{\mathbb{Z}^d}(x, y) := \int \frac{\mathrm{d}k}{(2\pi)^d} \frac{\cos(k \cdot (x - y))}{\frac{2}{d}\sum_{j=1}^d \sin(k_j/2)^2}. \tag{1.22}$$

In particular, $h^{\widetilde{V}_N} \to \mathcal{N}(0, G^{\mathbb{Z}^d})$ = full space DGFF.

This means that the DGFF in large enough (finite) domains is well approximated by the full-space DGFF as long as we are far away from the boundary. This is an example of a *thermodynamic limit*—the lattice stays fixed and the domain boundaries slide off to infinity. In renormalization group theory, taking the thermodynamic limit corresponds to the removal of an infrared cutoff.

From Lemma 1.10 it is clear that the thermodynamic limit is meaningless for the two-dimensional DGFF (indeed, variances blow up and, since we are talking about Gaussian random variables, there is no tightness). Let us attempt to take the scaling limit just as in $d = 1$: normalize the field by the square-root of the variance (i.e., $\sqrt{\log N}$) and extract a distributional limit (for which it suffices to prove the limit of the covariances). Here we note that, for all $\epsilon > 0$,

$$\sup_{N\geq 1}\sup_{\substack{x,y\in V_N\\ |x-y|\geq\epsilon N}} G^{V_N}(x,y)<\infty, \tag{1.23}$$

a fact that we will prove later. For any $s,t\in(0,1)^2$ we thus get

$$\mathrm{Cov}\left(\frac{h^{V_N}_{\lfloor tN\rfloor}}{\sqrt{\log N}},\frac{h^{V_N}_{\lfloor sN\rfloor}}{\sqrt{\log N}}\right)\underset{N\to\infty}{\longrightarrow}\begin{cases}c(t)>0, & \text{if } s=t,\\ 0, & \text{else,}\end{cases} \tag{1.24}$$

where, here and henceforth,

$$\lfloor tN\rfloor := \text{the unique } z\in\mathbb{Z}^2 \text{ such that } tN\in z+[0,1)^2\,. \tag{1.25}$$

The only way to read (1.24) is that the limit distribution is a collection of independent normals indexed by $t\in(0,1)^2$—an object too irregular and generic to retain useful information from before the limit was taken.

As we will see, the right thing to do is to take the limit of the DGFF *without* any normalization. That will lead to a singular limit as well but one that captures better the behavior of the DGFF. Moreover, the limit object exhibits additional symmetries (e.g., conformal invariance) not present at the discrete level.

1.3 Green Function Asymptotic

Let us now analyze the situation in $d=2$ in more detail. Our aim is to consider the DGFF in *sequences* of lattice domains $\{D_N : N\geq 1\}$ that somehow correspond to the scaled-up versions of a given continuum domain $D\subseteq\mathbb{C}$. The assumptions on the continuum domains are the content of:

Definition 1.14 *(Admissible domains) Let $\mathfrak{D}$ be the class of sets $D\subseteq\mathbb{C}$ that are bounded, open and such that their topological boundary ∂D is the union of a finite number of connected components each of which has a positive (Euclidean) diameter.*

For the sake of future use we note:

Exercise 1.15 *All bounded, open and simply connected $D\subseteq\mathbb{C}$ belong to $\mathfrak{D}$.*

As to what sequences of discrete approximations of D we will permit, a natural choice would be to work with plain discretizations $\{x\in\mathbb{Z}^2 : x/N\in D\}$. However, this is too crude because parts of ∂D could be missed out completely; see Fig. 1. We thus have to qualify admissible discretizations more precisely:

Definition 1.16 *(Admissible lattice approximations) A family of subsets $\{D_N : N\geq 1\}$ of $\mathbb{Z}^2$ is a sequence of admissible lattice approximations of a domain $D\in\mathfrak{D}$ if*

$$D_N\subseteq\left\{x\in\mathbb{Z}^2 : \operatorname{dist}_\infty(x/N,D^{\mathrm{c}})>{}^1\!/\!_N\right\}, \tag{1.26}$$

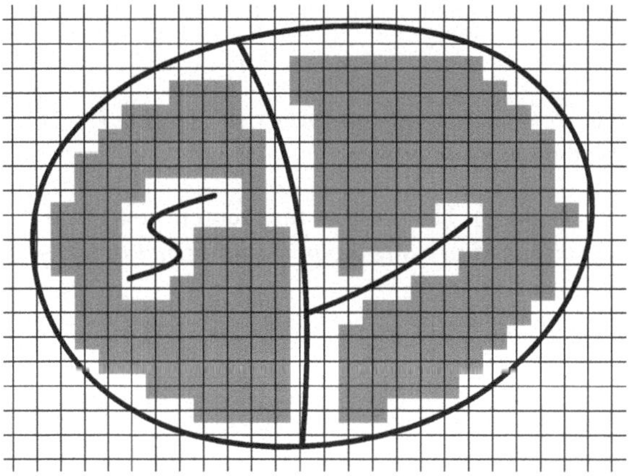

Fig. 1 An example of an admissible discretization of a continuum domain. The continuum domain is the region in the plane bounded by the solid curves. The discrete domain is the set of lattice vertices in the shaded areas

where dist_∞ *denotes the* ℓ^∞*-distance on* $\mathbb{Z}^2$*, and if, for each* $\delta > 0$,

$$D_N \supseteq \{x \in \mathbb{Z}^2 \colon \text{dist}_\infty(x/N, D^c) > \delta\} \tag{1.27}$$

holds once N *is sufficiently large (depending possibly on* δ*).*

Note that this still ensures that $x \in D_N$ implies $x/N \in D$. For the statement of the next theorem, consider the (standard) Brownian motion $\{B_t \colon t \geq 0\}$ on $\mathbb{R}^2$ and let $\tau_{D^c} := \inf\{t \geq 0 \colon B_t \notin D\}$ be the first exit time from D. Denote by $\Pi^D(x, \cdot)$ the law of the exit point from D of the Brownian motion started from x; i.e.,

$$\Pi^D(x, A) := P^x(B_{\tau_{D^c}} \in A), \tag{1.28}$$

for all Borel $A \subseteq \mathbb{R}^2$. This measure is supported on ∂D and is also known as the *harmonic measure from* x *in* D. The following then holds:

Theorem 1.17 (Green function asymptotic in $d = 2$) *Suppose* $d = 2$ *and let* $g := 2/\pi$*. There is* $c_0 \in \mathbb{R}$ *such that for all domains* $D \in \mathfrak{D}$*, all sequences* $\{D_N \colon N \geq 1\}$ *of admissible lattice approximations of* D*, and all* $x \in D$*,*

$$G^{D_N}(\lfloor xN \rfloor, \lfloor xN \rfloor) = g \log N + c_0 + g \int_{\partial D} \Pi^D(x, \mathrm{d}z) \log|x - z| + o(1), \tag{1.29}$$

where $o(1) \to 0$ *as* $N \to \infty$ *locally uniformly in* $x \in D$*. Moreover, for all* $x, y \in D$ *with* $x \neq y$*, we also have*

$$G^{D_N}\big(\lfloor xN\rfloor,\lfloor yN\rfloor\big)=-g\log|x-y|+g\int_{\partial D}\Pi^D(x,\mathrm{d}z)\log|y-z|+o(1),\tag{1.30}$$

where $o(1)\to 0$ as $N\to\infty$ locally uniformly in $(x,y)\in D\times D$ with $x\neq y$ and $|\cdot|$ denotes the Euclidean norm on $\mathbb{R}^2$.

Proof (modulo two lemmas). The proof of the theorem starts by a convenient representation of the Green function using the *potential kernel* $\mathfrak{a}\colon\mathbb{Z}^2\to[0,\infty)$ defined, e.g., by the explicit formula

$$\mathfrak{a}(x):=\int_{(-\pi,\pi)^2}\frac{\mathrm{d}k}{(2\pi)^2}\,\frac{1-\cos(k\cdot x)}{\sin(k_1/2)^2+\sin(k_2/2)^2}\,.\tag{1.31}$$

A characteristic (although not completely characterizing) property of the potential kernel is the content of:

Exercise 1.18 *Show that $\mathfrak{a}$ solves the Poisson problem*

$$\begin{cases}\Delta\mathfrak{a}(\cdot)=4\delta_0(\cdot), & \text{on } \mathbb{Z}^2,\\ \mathfrak{a}(0)=0,\end{cases}\tag{1.32}$$

where, as before, δ_0 is the Kronecker delta at 0.

Writing ∂V for external vertex boundary of V, i.e., the set of vertices in V^{c} that have an edge to a vertex in V, we now get:

Lemma 1.19 (Green function from potential kernel) *For any finite set $V\subseteq\mathbb{Z}^2$ and any vertices $x,y\in V$,*

$$G^V(x,y)=-\mathfrak{a}(x-y)+\sum_{z\in\partial V}H^V(x,z)\mathfrak{a}(z-y)\,,\tag{1.33}$$

where $H^V(x,z):=P^x(X_{\tau_{V^{\mathrm{c}}}}=z)$ is the probability that the simple symmetric random walk X started at x exists V at $z\in\partial V$.

Proof. Fix $x\in V$ and let X denote a path of the simple symmetric random walk. In light of (1.6), (1.32) and the translation invariance of the lattice Laplacian,

$$\phi(y):=G^V(x,y)+\mathfrak{a}(x-y)\tag{1.34}$$

is discrete harmonic in V. It follows that $M_n:=\phi(X_{\tau_{V^{\mathrm{c}}}\wedge n})$ is a martingale for the usual filtration $\sigma(X_0,\dots,X_n)$. The finiteness of V ensures that M is bounded and $\tau_{V^{\mathrm{c}}}<\infty$ a.s. The Optional Stopping Theorem then gives

$$\phi(y)=E^y\big(\phi(X_{\tau_{V^{\mathrm{c}}}})\big)=\sum_{z\in\partial V}H^V(y,z)\phi(z).\tag{1.35}$$

Since $G^V(x,\cdot)$ vanishes on ∂V, this along with the symmetry of $x, y \mapsto G^V(x,y)$ and $x, y \mapsto \mathfrak{a}(x-y)$ readily imply the claim. □

We note that the restriction to finite V was not a mere convenience. Indeed:

Exercise 1.20 *Show that* $G^{\mathbb{Z}^2\smallsetminus\{0\}}(x,y) = \mathfrak{a}(x) + \mathfrak{a}(y) - \mathfrak{a}(x-y)$. *Use this to conclude that, in particular,* (1.33) *is generally false for infinite* V.

As the next step of the proof, we invoke:

Lemma 1.21 (Potential kernel asymptotic) *There is* $c_0 \in \mathbb{R}$ *such that*

$$\mathfrak{a}(x) = g\log|x| + c_0 + O\bigl(|x|^{-2}\bigr), \tag{1.36}$$

where, we recall, $g = 2/\pi$ *and* $|x|$ *is the Euclidean norm of* x.

This asymptotic form was apparently first proved by Stöhr [119] in 1950. In a 2004 paper, Kozma and Schreiber [82] analyzed the behavior of the potential kernel on other lattices and identified the constants g and c_0 in terms of specific geometric attributes of the underlying lattice. In our case, we can certainly attempt to compute the asymptotic explicitly:

Exercise 1.22 *Prove Lemma 1.21 by asymptotic analysis of* (1.31).

Using (1.33) and (1.36) together, for $x, y \in D$ with $x \neq y$ we now get

$$G^{D_N}\bigl(\lfloor xN\rfloor, \lfloor yN\rfloor\bigr) = g\log|x-y| + g\sum_{z\in\partial D_N} H^{D_N}\bigl(\lfloor xN\rfloor, z\bigr)\log|y - z/N| + O(1/N)\,, \tag{1.37}$$

where the $O(1/N)$ term arises from the approximation of (the various occurrences of) $\lfloor xN\rfloor$ by xN and also from the error in (1.36). To get (1.30), we thus need to convert the sum into the integral. Here we will need:

Lemma 1.23 (Weak convergence of discrete harmonic measures) *For any domain* $D \in \mathfrak{D}$ *and any sequence* $\{D_N : N \geq 1\}$ *of admissible lattice approximations of* D,

$$\sum_{z\in\partial D_N} H^{D_N}\bigl(\lfloor xN\rfloor, z\bigr)\,\delta_{z/N}(\cdot) \;\underset{N\to\infty}{\overset{\text{weakly}}{\longrightarrow}}\; \Pi^D(x,\cdot). \tag{1.38}$$

We will omit the proof as it would take us too far on a tangent; the reader is instead referred to (the Appendix of) Biskup and Louidor [24]. The idea is to use Donsker's Invariance Principle to extract a coupling of the simple random walk and the Brownian motion that ensures that once the random walk exits D_N, the Brownian motion will exit D at a "nearby" point, and vice versa. This is where we find it useful that the boundary components have a positive diameter.

Since $u \mapsto \log|y-u|$ is bounded and continuous in a neighborhood of ∂D (whenever $y \in D$), the weak convergence in Lemma 1.23 implies

$$\sum_{z\in\partial D_N} H^{D_N}\bigl(\lfloor xN\rfloor, z\bigr)\log|y-z/N| = \int_{\partial D} \Pi^D(x,\mathrm{d}z)\log|y-z| + o(1)\,, \quad (1.39)$$

with $o(1)\to 0$ as $N\to\infty$. This proves (1.30). The proof of (1.29) only amounts to the substitution of $-g\log|x-y|$ in (1.37) by $g\log N + c_0$; the convergence of the sum to the corresponding integral is then handled as before. □

We remark that the convergence of the discrete harmonic measure to the continuous one in Lemma 1.23 is where the above assumptions on the continuum domain and its lattice approximations crucially enter. (In fact, we could perhaps even *define* admissible lattice approximations by requiring (1.38) to hold.) The reader familiar with conformal mapping theory may wish to think of (1.38) as a version of Carathéodory convergence for discrete planar domains.

The objects appearing on the right-hand side of (1.29)–(1.30) are actually well known:

Definition 1.24 *(Continuum Green function and conformal radius) For bounded, open $D\subseteq\mathbb{C}$, we define the continuum Green function in D from x to y by*

$$\widehat{G}^D(x,y) := -g\log|x-y| + g\int_{\partial D}\Pi^D(x,\mathrm{d}z)\log|y-z|\,. \quad (1.40)$$

Similarly, for $x\in D$ we define

$$r_D(x) := \exp\left\{\int_{\partial D}\Pi^D(x,\mathrm{d}z)\log|x-z|\right\} \quad (1.41)$$

to be the conformal radius of D from x.

The continuum Green function is usually defined as the fundamental solution to the Poisson equation; i.e., a continuum version of (1.6). We will not need this characterization in what follows so we will content ourselves with the explicit form above. Similarly, for open simply connected $D\subseteq\mathbb{C}$, the conformal radius of D from x is defined as the value $|f'(x)|^{-1}$ for f any conformal bijection of D onto $\{z\in\mathbb{C}\colon |z|<1\}$ such that $f(x)=0$. (The result does not depend on the choice of the bijection.) The reader will find it instructive to solve:

Exercise 1.25 *Check that this coincides with $r_D(x)$ above.*

For this as well as later derivations it may be useful to know:

Lemma 1.26 (Conformal invariance of harmonic measure) *For any conformal bijection $f\colon D\mapsto f(D)$ between $D, f(D)\in\mathfrak{D}$,*

$$\Pi^D(x,A) = \Pi^{f(D)}\bigl(f(x),f(A)\bigr) \quad (1.42)$$

for any measurable $A\subseteq\partial D$.

This can be proved using conformal invariance of the Brownian motion, although more direct approaches to prove this exist as well. The proof is fairly straightforward when both f and f^{-1} extend continuously to the boundaries; the general case is handled by stopping the Brownian motion before it hits the boundary and invoking approximation arguments.

1.4 Continuum Gaussian Free Field

Theorem 1.17 reveals two important facts: First, a pointwise limit of the unscaled DGFF is meaningless as, in light of (1.29), there is no tightness. Notwithstanding, by (1.30), the off-diagonal covariances of the DGFF do have a limit which is given by the continuum Green function. This function is singular on the diagonal, but the singularity is only logarithmic and thus relatively mild. This permits us to derive:

Theorem 1.27 (Scaling limit of DGFF) *Let $D \in \mathfrak{D}$ and consider a sequence $\{D_N : N \geq 1\}$ of admissible lattice approximations of D. For any bounded measurable function $f : D \to \mathbb{R}$, let*

$$h^{D_N}(f) := \int_D \mathrm{d}x\, f(x) h^{D_N}_{\lfloor xN \rfloor}. \tag{1.43}$$

Then

$$h^{D_N}(f) \underset{N\to\infty}{\overset{\text{law}}{\longrightarrow}} \mathcal{N}(0, \sigma_f^2)\,, \tag{1.44}$$

where

$$\sigma_f^2 := \int_{D\times D} \mathrm{d}x\mathrm{d}y\, f(x) f(y) \widehat{G}^D(x, y)\,. \tag{1.45}$$

Proof. The random variable $h^{D_N}(f)$ is Gaussian with mean zero and variance

$$E\big[h^{D_N}(f)^2\big] = \int_{D\times D} \mathrm{d}x\mathrm{d}y\, f(x) f(y) G^{D_N}\big(\lfloor xN \rfloor, \lfloor yN \rfloor\big). \tag{1.46}$$

The monotonicity from Exercise 1.3 and the reasoning underlying the proof of Theorem 1.17 show the existence of $c \in \mathbb{R}$ such that for all N large enough and all $x, y \in D_N$,

$$G^{D_N}(x, y) \leq g \log\biggl(\frac{N}{|x - y| \vee 1}\biggr) + c. \tag{1.47}$$

Since D is bounded, this gives

$$G^{D_N}\big(\lfloor xN \rfloor, \lfloor yN \rfloor\big) \leq -g \log|x - y| + \tilde{c} \tag{1.48}$$

for some $\tilde{c} \in \mathbb{R}$, uniformly in $x, y \in D$. Using this bound, we can estimate (and later neglect) the contributions to the integral in (1.46) from pairs (x, y) with $|x - y| < \epsilon$ for any given $\epsilon > 0$. The convergence of the remaining part of the integral is then treated using the pointwise convergence (1.30) (which is locally uniform on the set $\{(x, y) \in D \times D : x \neq y\}$) and the Bounded Convergence Theorem. □

Exercise 1.28 *Give a detailed proof of* (1.47).

Formula (1.43) can be viewed as a projection of the field configuration onto a test function. Theorem 1.27 then implies that these projections admit a joint distributional limit. This suggests we could regard the limit object as a *linear functional* on a suitable space of test functions, which leads to:

Definition 1.29 *(CGFF as a function space-indexed Gaussian) A continuum Gaussian Free Field (CGFF) on a bounded, open $D \subseteq \mathbb{C}$ is an assignment $f \mapsto \Phi(f)$ of a random variable to each bounded measurable $f : D \to \mathbb{R}$ such that*

(1) Φ is a.s. linear, i.e.,

$$\Phi(af + bg) = a\Phi(f) + b\Phi(g) \quad a.s. \tag{1.49}$$

for all bounded measurable f and g and each $a, b \in \mathbb{R}$, and

(2) for all bounded measurable f,

$$\Phi(f) \overset{\text{law}}{=} \mathcal{N}(0, \sigma_f^2), \tag{1.50}$$

where σ_f^2 is as in (1.45).

Theorem 1.27 shows that the DGFF scales to the CGFF in this sense—which, modulo a density argument, also entails that a CGFF as defined above exists! (Linearity is immediate from (1.43). An independent construction of a CGFF will be performed in Exercise 2.16.)

We remark that definitions given in the literature usually require that the CGFF is even a.s. *continuous* in a suitable topology over a suitable (and suitably completed) class of functions. We will not need such continuity considerations here so we do not pursue them in any detail. However, they are important when one tries to assign meaning to $\Phi(f)$ for singular f (e.g., those vanishing Lebesgue a.e.) or for f varying continuously with respect to some parameter. This is for example useful in the study of the *disc-average process* $t \mapsto \Phi(f_t)$, where

$$f_t(y) := \frac{1}{\pi e^{-2t}} 1_{B(x,e^{-t})}(y) \quad \text{for} \quad B(x, r) := \big\{y \in \mathbb{C} : |y - x| < r\big\}. \tag{1.51}$$

Here it is quite instructive to note:

Exercise 1.30 *(Disc average process) For CGFF as defined above and for any $x \in D$ at Euclidean distance $r > 0$ from D^c, show that for $t > \log(1/r)$, the process $t \mapsto \Phi(f_t)$ has independent increments with*

$$\mathrm{Var}\big(\Phi(f_t)\big) = gt + c_1 + g\log r_D(x) \tag{1.52}$$

for some $c_1 \in \mathbb{R}$ *independent of* x. *In particular,* $t \mapsto \Phi(f_t)$ *admits a continuous version whose law is (for* $t > \log(1/r)$*) that of a Brownian motion.*

The same conclusion is obtained for the *circle average process*, where Φ is projected, via a suitable regularization procedure, onto the indicator of a circle $\{y \in \mathbb{C}\colon |x-y| = r\}$. This is because the circle and disc average of the continuum Green function (in one variable with the other one fixed) coincide for small-enough radii, due to the fact that the Green function is harmonic away from the diagonal.

We refer to the textbook by Janson [79] for a thorough discussion of various aspects of generalized Gaussian processes. The Gaussian Free Field existed in physics for a long time where it played the role of a "trivial," which in the jargon of physics means "non-interacting," field theory. Through various scaling limits as well as in its own right, it has recently come to the focus of a larger mathematical community as well. A pedestrian introduction to the subject of the CGFF can be found in Sheffield [114] and also in Chap. 5 of the recent posting by Armstrong, Kuusi and Mourrat [14].

Lecture 2: Maximum and Intermediate Values

In this lecture we begin to discuss the main topic of interest in this course: extremal properties of the DGFF sample paths. After some introduction and pictures, we focus attention on the behavior of the absolute maximum and the level sets at heights proportional to the absolute maximum. We then state the main theorem on the scaling limit of such level sets and link the limit object to the concept of Liouville Quantum Gravity. The proof of the main theorem is relegated to the forthcoming lectures.

2.1 *Level Set Geometry*

The existence of the scaling limit established in Theorem 1.27 indicates that the law of the DGFF is asymptotically scale invariant. Scale invariance of a random object usually entails one of the following two possibilities:

- either the object is trivial and boring (e.g., degenerate, flat, non-random),
- or it is very interesting (e.g., chaotic, rough, fractal).

As attested by Fig. 2, the two-dimensional DGFF seems, quite definitely, to fall into the latter category.

Looking at Fig. 2 more closely, a natural first question is to understand the behavior of the (implicit) boundaries between warm and cold colors. As the field averages to zero, and should thus take both positive and negative values pretty much equally likely, this amounts to looking at the contour lines *between* the regions where the

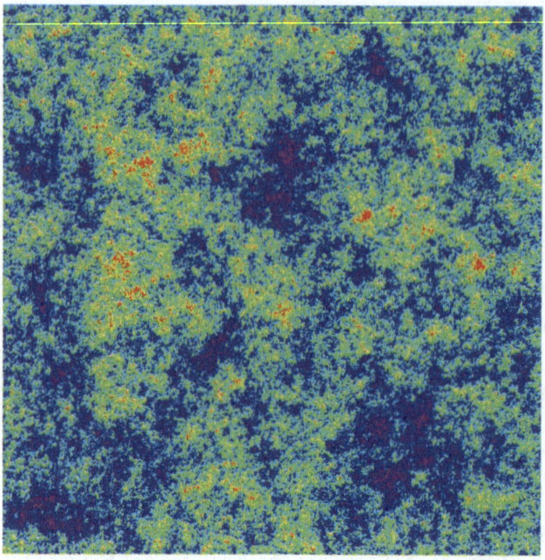

Fig. 2 A sample of the DGFF on 300×300 square in $\mathbb{Z}^2$. The cold colors (purple and blue) indicate low values, the warm colors (yellow and red) indicate large values. The fractal nature of the sample is quite apparent

field is positive and where it is negative. This has been done and constitutes the beautiful work started by Schramm and Sheffield [110] and continued in Sheffield and Werner [116], Sheffield [115] and Sheffield and Miller [95, 97–99]. We thus know that the contour lines admit a scaling limit to a process of nested collections of loops called the *Conformal Loop Ensemble* with the individual curves closely related to the *Schramm–Loewner process* SLE_4.

Our interest in these lectures is somewhat different as we wish to look at the level sets at heights that scale proportionally to the absolute maximum. We call these the *intermediate level sets* although the term *thick points* is quite common as well. Samples of such level sets are shown in Fig. 3.

The self-similar nature of the plots in Fig. 3 is quite apparent. This motivates the following questions:

- Is there a way to take a scaling limit of the samples in Fig. 3?
- And if so, is there a way to characterize the limit object directly?

Our motivation for these questions stems from Donsker's Invariance Principle for random walks. There one first answers the second question by constructing a limit process; namely, the Brownian Motion. Then one proves that, under the diffusive scaling of space and time, all random walks whose increments have zero mean and finite second moment scale to that Brownian motion.

The goal of this and the next couple of lectures is to answer the above questions for the intermediate level sets of the DGFF. We focus only on one underlying discrete process so this can hardly be sold as a full-fledged analogue of Donsker's Invariance

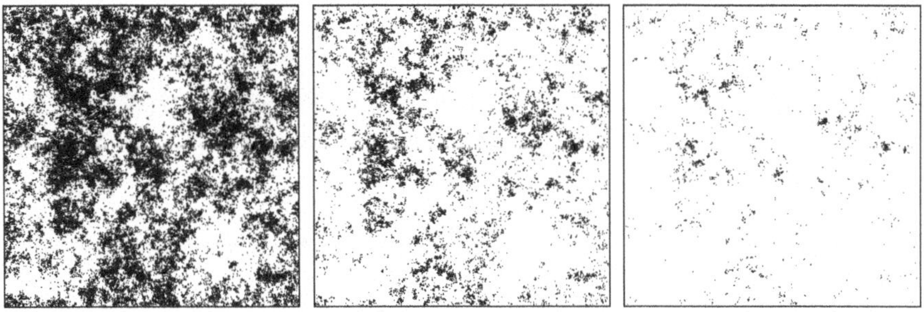

Fig. 3 Plots of the points in the sample of the DGFF in Fig. 2 at heights (labeled left to right) above 0.1, 0.3 and 0.5-multiples of the absolute maximum, respectively. Higher level sets are too sparse to produce a visible effect

Principle. Still, we will perform the analysis over a whole class of continuum domains thus achieving some level of *universality*. The spirit of the two results is thus quite similar.

2.2 Growth of Absolute Maximum

In order to set the scales for our future discussion, we first have to identify the growth rate of the absolute maximum. Here an early result of Bolthausen, Deuschel and Giacomin [30] established the leading-order asymptotic in square boxes. Their result reads:

Theorem 2.1 (Growth of absolute maximum) *For* $V_N := (0, N)^2 \cap \mathbb{Z}^2$,

$$\max_{x\in V_N} h_x^{V_N} = \bigl(2\sqrt{g} + o(1)\bigr)\log N, \tag{2.1}$$

where $o(1) \to 0$ *in probability as* $N \to \infty$.

Proof of upper bound in (2.1). We start by noting the well-known tail estimate for centered normal random variables:

Exercise 2.2 *(Standard Gaussian tail bound) Prove that, for any* $\sigma > 0$,

$$Z \overset{\text{law}}{=} \mathcal{N}(0,\sigma^2) \quad\Rightarrow\quad P(Z > a) \le \frac{\sigma}{\sigma + a}\,\mathrm{e}^{-\frac{a^2}{2\sigma^2}}, \qquad a > 0. \tag{2.2}$$

We want to use this for Z replaced by $h_x^{V_N}$ but for that we need to bound the variance of $h_x^{V_N}$ uniformly in $x \in V_N$. Here we observe that, by the monotonicity of $V \mapsto G^V(x,x)$ and translation invariance (1.14), denoting $\widetilde{V}_N := (-N/2, N/2)^2 \cap \mathbb{Z}^2$, there is a $c \ge 0$ such that

$$\max_{x\in V_N} \mathrm{Var}(h_x^{V_N}) \le \mathrm{Var}\big(h_0^{\widetilde{V}_{2N}}\big) \le g\log N + c, \tag{2.3}$$

where the last bound follows from the asymptotic in Theorem 1.17. (We have just given away the solution to Exercise 1.28.) Plugging this in (2.2), for any $\theta > 0$ we thus get

$$P\big(h_x^{V_N} > \theta\log N\big) \le \exp\left\{-\frac{1}{2}\frac{\theta^2(\log N)^2}{g\log N + c}\right\}. \tag{2.4}$$

Using that $(1+\lambda)^{-1} \ge 1-\lambda$ for $\lambda \in (0,1)$ we obtain

$$\frac{1}{g\log N + c} \ge \frac{1}{g\log N} - \frac{c}{(g\log N)^2} \tag{2.5}$$

as soon as N is sufficiently large. Then

$$\max_{x\in V_N} P\big(h_x^{V_N} > \theta\log N\big) \le c' N^{-\frac{\theta^2}{2g}} \tag{2.6}$$

for $c' := \mathrm{e}^{\theta^2 c/(2g^2)}$ as soon as N is large enough. The union bound and the fact that $|V_N| \le N^2$ then give

$$\begin{aligned} P\Big(\max_{x\in V_N} h_x^{V_N} > \theta\log N\Big) &\le \sum_{x\in V_N} P\big(h_x^{V_N} > \theta\log N\big) \\ &\le c'|V_N|N^{-\frac{\theta^2}{2g}} = c' N^{2-\frac{\theta^2}{2g}}. \end{aligned} \tag{2.7}$$

This tends to zero as $N \to \infty$ for any $\theta > 2\sqrt{g}$ thus proving "$\le$" in (2.1). □

The proof of the complementary lower bound is considerably harder. The idea is to use the second-moment method but that requires working with a scale decomposition of the DGFF and computing the second moment under a suitable truncation. We will not perform this calculation here as the result will follow as a corollary from Theorem 7.3. (Section 4.5 gives some hints how the truncation is performed.)

Building on [30], Daviaud [51] was able to extend the control to the level sets of the form

$$\big\{x \in V_N : h_x^{V_N} \ge 2\sqrt{g}\,\lambda\log N\big\}, \tag{2.8}$$

where $\lambda \in (0,1)$. The relevant portion of his result reads:

Theorem 2.3 (Size of intermediate level sets) *For any* $\lambda \in (0,1)$,

$$\#\big\{x \in V_N : h_x^{V_N} \ge 2\sqrt{g}\,\lambda\log N\big\} = N^{2(1-\lambda^2)+o(1)}, \tag{2.9}$$

where $o(1) \to 0$ *in probability as* $N \to \infty$.

Proof of "$\le$" in (2.9). Let L_N denote the cardinality of the set in (2.8). Using the Markov inequality and the reasoning (2.6)–(2.7),

$$
\begin{aligned}
P\bigl(L_N \geq N^{2(1-\lambda^2)+\epsilon}\bigr) &\leq N^{-2(1-\lambda^2)-\epsilon} E(L_N) \\
&\leq c' N^{-2(1-\lambda^2)-\epsilon} N^{2-2\lambda^2} = c' N^{-\epsilon}.
\end{aligned}
\tag{2.10}
$$

This tends to zero as $N \to \infty$ for any $\epsilon > 0$ thus proving "$\leq$" in (2.9). □

We will not give a full proof of the lower bound for all $\lambda \in (0, 1)$ as that requires similar truncations as the corresponding bound for the maximum. However, these truncations are avoidable for λ sufficiently small, so we will content ourselves with: *Proof of "$\geq$" in* (2.9) *with positive probability for* $\lambda < 1/\sqrt{2}$. Define

$$
Y_N := \sum_{x \in V_N^\epsilon} e^{\beta h_x^{V_N}}, \tag{2.11}
$$

where $\beta > 0$ is a parameter to be adjusted later and $V_N^\epsilon := (\epsilon N, (1-\epsilon)N)^2 \cap \mathbb{Z}^2$ for some $\epsilon \in (0, 1/2)$ to be fixed for the rest of the calculation. The quantity Y_N may be thought of as the normalizing factor (a.k.a. the partition function) for the Gibbs measure on V_N where the "state" x gets "energy" $h_x^{V_N}$. (See Sect. 11.4 for more on this.) Our first observation is: □

Lemma 2.4 *For $\beta > 0$ such that $\beta^2 g < 2$ there is $c = c(\beta) > 0$ such that*

$$
P\bigl(Y_N \geq c N^{2+\frac{1}{2}\beta^2 g}\bigr) \geq c \tag{2.12}
$$

once N is sufficiently large.

Proof. We will prove this by invoking the *second moment method* whose driving force is the following inequality: □

Exercise 2.5 *(Second moment estimate) Let $Y \in L^2$ be a non-negative random variable with $EY > 0$. Prove that*

$$
P\bigl(Y \geq qEY\bigr) \geq (1-q)^2 \frac{[E(Y)]^2}{E(Y^2)}, \quad q \in (0, 1). \tag{2.13}
$$

In order to make use of (2.13) we have to prove that the second moments of Y_N is of the same order as the first moment squared. We begin by the first moment of Y_N. The fact that $Ee^X = e^{EX + \frac{1}{2}\mathrm{Var}(X)}$ for any X normal yields

$$
EY_N = \sum_{x \in V_N^\epsilon} e^{\frac{1}{2}\beta^2 \mathrm{Var}(h_x^{V_N})}. \tag{2.14}
$$

Writing $\widetilde{V}_N := (-N/2, N/2)^2 \cap \mathbb{Z}^2$, the monotonicity of $V \mapsto G^V(x, x)$ with respect to set inclusion gives $\mathrm{Var}(h_0^{\widetilde{V}_{\epsilon N}}) \leq \mathrm{Var}(h_x^{V_N}) \leq \mathrm{Var}(h_0^{\widetilde{V}_{2N}})$. Theorem 1.17 then implies

$$
\sup_{N \geq 1} \max_{x \in V_N^\epsilon} \bigl|\mathrm{Var}(h_x^{V_N}) - g \log N\bigr| < \infty \tag{2.15}
$$

As $|V_N^\epsilon|$ is of order N^2, using this in (2.14) we conclude that

$$cN^{2+\frac{1}{2}\beta^2 g} \le EY_N \le c^{-1}N^{2+\frac{1}{2}\beta^2 g} \tag{2.16}$$

holds for some constant $c \in (0, 1)$ and all $N \ge 1$.

Next we will estimate the second moment of Y_N. Recalling the notation G^{V_N} for the Green function in V_N, we have

$$E(Y_N^2) = \sum_{x,y\in V_N^\epsilon} \mathrm{e}^{\frac{1}{2}\beta^2[G^{V_N}(x,x)+G^{V_N}(y,y)+2G^{V_N}(x,y)]}\,. \tag{2.17}$$

Invoking (2.15) and (1.47) we thus get

$$E(Y_N^2) \le c' N^{\beta^2 g} \sum_{x,y\in V_N^\epsilon} \left(\frac{N}{|x-y|\vee 1}\right)^{\beta^2 g}\,. \tag{2.18}$$

For $\beta^2 g < 2$ the sum is dominated by pairs x and y with $|x-y|$ of order N. The sum is thus of order N^4 and so we conclude

$$E(Y_N^2) \le c'' N^{\beta^2 g+4} \tag{2.19}$$

for some constant $c'' > 0$. By (2.16), this bound is proportional to $[EY_N]^2$ so using this in (2.13) (with, e.g., $q := 1/2$) readily yields (2.12). □

Next we will need to observe that the main contribution to Y_N comes from the set of points where the field roughly equals $\beta g \log N$:

Lemma 2.6 *For any* $\delta > 0$,

$$P\left(\sum_{x\in V_N^\epsilon} 1_{\{|h_x^{V_N}-\beta g\log N|>(\log N)^{1/2+\delta}\}}\,\mathrm{e}^{\beta h_x^{V_N}} \ge \delta N^{2+\frac{1}{2}\beta^2 g}\right) \underset{N\to\infty}{\longrightarrow} 0. \tag{2.20}$$

Proof. By (2.15), we may instead prove this for $\beta g\log N$ replaced by $\beta\mathrm{Var}(h_x^{V_N})$. Using the Markov inequality, the probability is then bounded by

$$\frac{1}{\delta N^{2+\frac{1}{2}\beta^2 g}} \sum_{x\in V_N^\epsilon} E\left(1_{\{|h_x^{V_N}-\beta\mathrm{Var}(h_x^{V_N})|>(\log N)^{1/2+\delta}\}}\,\mathrm{e}^{\beta h_x^{V_N}}\right). \tag{2.21}$$

Changing variables inside the (single-variable) Gaussian integral gives

$$E\Big(1_{\{|h_x^{V_N}-\beta\mathrm{Var}(h_x^{V_N})|>(\log N)^{1/2+\delta}\}}\,\mathrm{e}^{\beta h_x^{V_N}}\Big)$$
$$=\mathrm{e}^{\frac{1}{2}\beta^2\mathrm{Var}(h_x^{V_N})}P\big(|h_x^{V_N}|>(\log N)^{1/2+\delta}\big)$$
$$\le cN^{\frac{1}{2}\beta^2 g}\mathrm{e}^{-c''(\log N)^{2\delta}} \tag{2.22}$$

for some $c, c'' > 0$, where we used again (2.15). The probability in the statement is thus at most a constant times $\delta^{-1}\mathrm{e}^{-c''(\log N)^{2\delta}}$, which vanishes as $N \to \infty$. □

Continuing the proof of "$>$" in (2.9), the above lemmas yield

$$P\Bigg(\sum_{x\in V_N^\epsilon} 1_{\{h_x^{V_N}\ge\beta g\log N-(\log N)^{1/2+\delta}\}} \ge \frac{c}{2}N^{2+\frac{1}{2}\beta^2 g-\beta^2 g}\mathrm{e}^{-\beta(\log N)^{1/2+\delta}}\Bigg)\ge\frac{c}{2} \tag{2.23}$$

as soon as N is sufficiently large. Choosing β so that

$$\beta g\log N-(\log N)^{1/2+\delta}=2\sqrt{g}\lambda\log N \tag{2.24}$$

gives $2-\frac{1}{2}\beta^2 g = 2(1-\lambda^2)+O((\log N)^{-1/2+\delta})$ and so, since $V_N^\epsilon \subseteq V_N$, the cardinality L_N of the level set (2.8) obeys

$$P\big(L_N \ge N^{2(1-\lambda^2)-c'(\log N)^{1/2+\delta}}\big)\ge\frac{c}{2} \tag{2.25}$$

for some constant $c' \in \mathbb{R}$ once N is large enough. This implies "$\ge$" in (2.9) with $o(1) \to 0$ with a uniformly positive probability. The proof used that $\beta^2 g < 2$, which means that it applies only to $\lambda < 1/\sqrt{2}$. □

Having the lower bound with a uniformly positive probability is actually sufficient to complete the proof of (2.9) as stated (when $\lambda < 1/\sqrt{2}$). The key additional tool needed for this is the Gibbs-Markov decomposition of the DGFF which will be discussed in the next lecture. (See Exercise 3.5.)

It is actually quite remarkable that the first-moment calculation alone is able to nail the correct leading order of the maximum as well as the asymptotic size of the level set (2.8). As that calculation did not involve correlations between the DGFF at distinct vertices, the same estimate would apply to i.i.d. Gaussians with the same growth rate of the variances. This (and many subsequent derivations) may lead one to think that the extreme values behave somehow like those of i.i.d. Gaussians. However, although some connection does exist, this is very far from the truth, as seen in Fig. 4.

The factor $1-\lambda^2$ in the exponent is ubiquitous in this subject area. Indeed, it appears in the celebrated analysis of the Brownian *fast points* (Orey and Taylor [103]) and, as was just noted, for i.i.d. Gaussians with variance $g\log N$. A paper by Chatterjee, Dembo and Ding [41] gives (generous) conditions under which such a factor should be expected. For suitably formulated analogue of Daviaud's level sets, called the *thick points*, of the two-dimensional CGFF, this has been shown by Hu, Miller and Peres [78].

2.3 Intermediate Level Sets

The main objective in this part of the course is to show that the intermediate level sets (2.8) admit a non-trivial *scaling limit* whose law can be explicitly characterized. We start by pondering about the formulation that makes taking a scaling limit meaningful. Not all natural ideas may work; for instance, scaling the box down to a unit size, the set (2.8) becomes increasingly dense everywhere so taking its limit using, e.g., the topology of Hausdorff convergence does not seem useful. A better idea here is to encode the set into the point measure on $[0, 1]^2 \times \mathbb{R}$ of the form

$$\sum_{x\in V_N} \delta_{x/N} \otimes \delta_{h_x^{V_N} - a_N}, \tag{2.26}$$

where, to allow for some generalizations, a_N is a scale sequence such that

$$\frac{a_N}{\log N} \underset{N\to\infty}{\longrightarrow} 2\sqrt{g}\,\lambda \tag{2.27}$$

for some $\lambda \in (0, 1)$. A sample of the measure (2.26) is shown on the left of Fig. 4.

By Theorem 2.3, the measures in (2.26) assign unbounded mass to bounded intervals in the second variable, and so a normalization is required. We will show that this can be done (somewhat surprisingly) by a *deterministic* sequence of the form

$$K_N := \frac{N^2}{\sqrt{\log N}}\, \mathrm{e}^{-\frac{a_N^2}{2g\log N}}\,. \tag{2.28}$$

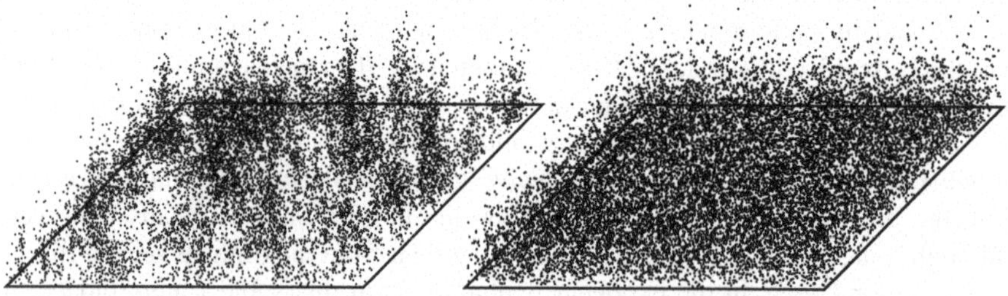

Fig. 4 Left: a sample of the level set (2.8), or rather the point measure (2.26), on a square of side $N := 300$ with $\lambda := 0.2$. Right: a corresponding sample for i.i.d. normals with mean zero and variance $g \log N$. Although both samples live on the same "vertical scale", their local structure is very different

Note that (2.27) implies $K_N = N^{2(1-\lambda^2)+o(1)}$ so the normalization is consistent with Theorem 2.3. Our main result, proved in Biskup and Louidor [26], is then:

Theorem 2.7 (Scaling limit of intermediate level sets) *For each* $\lambda \in (0,1)$ *and each* $D \in \mathfrak{D}$ *there is an a.s.-finite random Borel measure* Z_λ^D *on* D *such that for any* a_N *satisfying* (2.27) *and any admissible sequence* $\{D_N : N \geq 1\}$ *of lattice approximations of* D*, the normalized point measure*

$$\eta_N^D := \frac{1}{K_N}\sum_{x\in D_N}\delta_{x/N}\otimes\delta_{h_x^{D_N}-a_N} \tag{2.29}$$

obeys

$$\eta_N^D \ \underset{N\to\infty}{\overset{\text{law}}{\longrightarrow}}\ Z_\lambda^D(\mathrm{d}x)\otimes \mathrm{e}^{-\alpha\lambda h}\mathrm{d}h, \tag{2.30}$$

where $\alpha := 2/\sqrt{g}$. *Moreover,* $Z_\lambda^D(A) > 0$ *a.s. for every non-empty open* $A \subseteq D$.

A remark is perhaps in order on what it means for random measures to converge in law. The space of Radon measures on $D\times\mathbb{R}$ (of which η_N^D is an example) is naturally endowed with the topology of vague convergence. This topology makes the space of Radon measures a Polish space which permits discussion of distributional convergence. We pose:

Exercise 2.8 *Let* $\mathscr{X}$ *be a Polish space and* $\mathcal{M}(\mathscr{X})$ *the space of Radon measures on* $\mathscr{X}$ *endowed with the vague topology. Prove that a sequence of random measures* $\eta_N \in \mathcal{M}(\mathscr{X})$ *converges in law to* η *if and only if for each* $f \in C_c(\mathcal{X})$,

$$\langle \eta_N, f\rangle \ \underset{N\to\infty}{\overset{\text{law}}{\longrightarrow}}\ \langle \eta, f\rangle, \tag{2.31}$$

where $\langle \eta, f\rangle$ *denotes the integral of* f *with respect to* η *and the convergence in law is in the sense of ordinary* $\mathbb{R}$*-valued random variables.*

A subtlety of the above theorem is that the convergence actually happens over a larger "base" space, namely, $\overline{D}\times(\mathbb{R}\cup\{+\infty\})$ implying, in particular, $Z_\lambda^D(\partial D) = 0$ a.s. This means that we get weak convergence of the said integral even for (continuous) functions that take non-zero values on ∂D in the x-variable and/or at $+\infty$ in the h-variable. This implies:

Corollary 2.9 *For the setting of Theorem 2.7,*

$$\frac{1}{K_N}\#\bigl\{x\in D_N : h_x^{D_N} \geq a_N\bigr\} \ \underset{N\to\infty}{\overset{\text{law}}{\longrightarrow}}\ (\alpha\lambda)^{-1}\,Z_\lambda^D(D). \tag{2.32}$$

Proof (*idea*). Use Theorem 2.7 for η_N^D integrated against $f(x,y) := 1_{[0,\infty)}(h)$. $\square$

Exercise 2.10 *Apply suitable monotone limits to check that the convergence in* (2.30)*—which involves a priori only integrals of these measures with respect to compactly-supported continuous functions—can be applied to functions of the form*

$$f(x,h) := \tilde{f}(x)1_{[a,b]}(h) \tag{2.33}$$

for $\tilde{f}\colon \overline{D} \to \mathbb{R}$ *continuous and* $a < b$ *(including* $b := \infty$*).*

We remark that Corollary 2.9 extends, quite considerably, Theorem 2.3 originally proved by Daviaud [51]. In order to get some feeling for what Theorem 2.7 says about the positions of the points in the level set, we also state:

Corollary 2.11 *For* a_N *as in* (2.27)*, given a sample of* h^{D_N}*, let* X_N *be a point chosen uniformly from* $\{x \in D_N\colon h^{D_N}_x \ge a_N\}$*. Then*

$$\frac{1}{N}X_N \underset{N\to\infty}{\overset{\text{law}}{\longrightarrow}} \widehat{X} \;\; \text{with} \;\; law(\widehat{X}) = E\left[\frac{Z^D_\lambda(\cdot)}{Z^D_\lambda(D)}\right]. \tag{2.34}$$

In fact, the joint law of X_N *and* η^D_N *obeys*

$$(N^{-1}X_N, \eta^D_N) \underset{N\to\infty}{\overset{\text{law}}{\longrightarrow}} (\widehat{X}, \eta^D), \tag{2.35}$$

where the marginal law of η^D *is that of* $Z^D_\lambda(\mathrm{d}x) \otimes \mathrm{e}^{-\alpha\lambda h}\mathrm{d}h$ *and the law of* $\widehat{X}$ *conditional on* η^D *is given by* $Z^D_\lambda(\cdot)/Z^D_\lambda(D)$*.*

Proof. We easily check that, for any $f\colon \overline{D} \to \mathbb{R}$ continuous (and thus bounded),

$$E\big[\,f(X_N/N)\big] = E\left[\frac{\langle \eta^D_N, f \otimes 1_{[0,\infty)}\rangle}{\langle \eta^D_N, 1_{[0,\infty)}\rangle}\right], \tag{2.36}$$

where $(f \otimes 1_{[0,\infty)})(x,h) := f(x)1_{[0,\infty)}(h)$ and the brackets denote the integral of the function with respect to the measure. Applying Theorem 2.7, we get

$$\frac{\langle \eta^D_N, f \otimes 1_{[0,\infty)}\rangle}{\langle \eta^D_N, 1 \otimes 1_{[0,\infty)}\rangle} \underset{N\to\infty}{\overset{\text{law}}{\longrightarrow}} \frac{\int_D Z^D_\lambda(\mathrm{d}x)\, f(x)}{Z^D_\lambda(D)}. \tag{2.37}$$

This yields (2.34). The more general clause (2.35) is then proved by considering test functions in both variables in (2.36) and proceeding as above. □

Exercise 2.12 *The statement* (2.37) *harbors a technical caveat: we are taking the distributional limit of a ratio of two random variables, each of which converges (separately) in law. Fill the details needed to justify the conclusion.*

We conclude that the spatial part of the right-hand side of (2.30) thus tells us about the local "intensity" of the sets in the samples in Fig. 4. Concerning the values of the field, we may be tempted to say that these are Gumbel "distributed" with decay exponent $\alpha\lambda$. This is not justified by the statement per se as the measure on the right of (2.30) is not a probability (it is not even finite). Still, one can perhaps relate this to the corresponding problem for i.i.d. Gaussians with variance $g \log N$. Indeed, as an exercise in extreme-value statistics, we pose:

Exercise 2.13 *Consider the measure η_N^D for h^{D_N} replaced by i.i.d. Gaussians with mean zero and variance $g \log N$. Prove the same limit as in* (2.30), *with the same K_N and but with Z_λ^D replaced by a multiple of the Lebesgue measure on D.*

We rush to add that (as we will explain later) Z_λ^D is a.s. singular with respect to the Lebesgue measure. Compare, one more time, the two samples in Fig. 4.

2.4 Link to Liouville Quantum Gravity

Not too surprisingly, the random measures $\{Z_\lambda^D : D \in \mathfrak{D}\}$ (or, rather, their laws) are very closely related. We will later give a list of properties that characterize these laws uniquely. From these properties one can derive the following transformation rule for conformal maps between admissible domains:

Theorem 2.14 (Conformal covariance) *Let $\lambda \in (0, 1)$. Under any conformal bijection $f : D \to f(D)$ between the admissible domains $D, f(D) \in \mathfrak{D}$, the laws of the above measures transform as*

$$Z_\lambda^{f(D)} \circ f(\mathrm{d}x) \overset{\text{law}}{=} |f'(x)|^{2+2\lambda^2} Z_\lambda^D(\mathrm{d}x). \tag{2.38}$$

Recall that $r_D(x)$, defined in (1.41), denotes the conformal radius of D from x. The following is now a simple consequence of the above theorem:

Exercise 2.15 *Show that in the class of admissible D, the law of*

$$\frac{1}{r_D(x)^{2+2\lambda^2}} Z_\lambda^D(\mathrm{d}x) \tag{2.39}$$

is invariant under conformal maps.

In light of Exercise 1.15, the law of Z_λ^D for any bounded, open, simply connected D can thus be reconstructed from the law on, say, the open unit disc. As we will indicate later, such a link exists for all admissible D. However, that still leaves us with the task of determining the law of Z^D for at least one admissible domain. We will instead give an independent construction of the law of Z_λ^D that works for general D. This will also give us the opportunity to review some of the ideas of Kahane's theory of multiplicative chaos.

Let $\mathrm{H}_0^1(D)$ denote the closure of the set of smooth functions with compact support in D in the topology of the Dirichlet inner product

$$\langle f, g\rangle_\nabla := \frac{1}{4}\int_D \nabla f(x)\cdot\nabla g(x)\,\mathrm{d}x\,, \tag{2.40}$$

where ∇f is now the ordinary (continuum) gradient and $\cdot$ denotes the Euclidean scalar product in $\mathbb{R}^2$. For $\{X_n : n \ge 1\}$ i.i.d. standard normals and $\{f_n : n \ge 1\}$ an orthonormal basis in $\mathsf{H}_0^1(D)$, let

$$\varphi_n(x) := \sum_{k=1}^n X_k f_k(x). \tag{2.41}$$

These are to be thought of as regularizations of the CGFF:

Exercise 2.16 *(Hilbert space definition of CGFF) For any smooth function $f \in \mathsf{H}_0^1(D)$, let $\Phi_n(f) := \int_D f(x)\varphi_n(x)\mathrm{d}x$. Show that $\Phi_n(f)$ converges, as $n \to \infty$, in L^2 to a CGFF in the sense of Definition 1.29.*

Using the above random fields, for each $\beta \in [0,\infty)$, we then define the random measure

$$\mu_n^{D,\beta}(\mathrm{d}x) := 1_D(x)\mathrm{e}^{\beta\varphi_n(x) - \frac{\beta^2}{2}E[\varphi_n(x)^2]}\,\mathrm{d}x. \tag{2.42}$$

The following observation goes back to Kahane [81] in 1985:

Lemma 2.17 (Gaussian Multiplicative Chaos) *There exists a random, a.s.-finite (albeit possibly trivial) Borel measure $\mu_\infty^{D,\beta}$ on D such that for each Borel $A \subseteq D$,*

$$\mu_n^{D,\beta}(A) \underset{n\to\infty}{\longrightarrow} \mu_\infty^{D,\beta}(A), \quad a.s. \tag{2.43}$$

Proof. Pick $A \subseteq D$ Borel measurable. First we will show that $\mu_n^{D,\beta}(A)$ converge a.s. To this end, for each $n \in \mathbb{N}$, define

$$M_n := \mu_n^{D,\beta}(A) \quad\text{and}\quad \mathcal{F}_n := \sigma(X_1,\dots,X_n). \tag{2.44}$$

We claim that $\{M_n : n \ge 1\}$ is a martingale with respect to $\{\mathcal{F}_n : n \ge 1\}$. Indeed, using the regularity of the underlying measure space (to apply Fubini–Tonelli)

$$E(M_{n+1}|\mathcal{F}_n) = E\bigl(\mu_{n+1}^{D,\beta}(A)\,\big|\,\mathcal{F}_n\bigr) = \int_A \mathrm{d}x\, E\bigl(\mathrm{e}^{\beta\varphi_{n+1}(x) - \frac{\beta^2}{2}E[\varphi_{n+1}(x)^2]}\,\big|\,\mathcal{F}_n\bigr). \tag{2.45}$$

The additive structure of φ_n now gives

$$\begin{aligned} &E\bigl(\mathrm{e}^{\beta\varphi_{n+1}(x) - \frac{\beta^2}{2}E[\varphi_{n+1}(x)^2]}\,\big|\,\mathcal{F}_n\bigr)\\ &\quad= \mathrm{e}^{\beta\varphi_n(x) - \frac{\beta^2}{2}E[\varphi_n(x)^2]}E\bigl(\mathrm{e}^{\beta f_{n+1}(x)X_{n+1} - \frac{1}{2}\beta^2 f_{n+1}(x)^2E[X_{n+1}^2]}\bigr)\\ &\qquad\qquad= \mathrm{e}^{\beta\varphi_n(x) - \frac{\beta^2}{2}E[\varphi_n(x)^2]}. \end{aligned} \tag{2.46}$$

Using this in (2.45), the right-hand side then wraps back into $\mu_n^{D,\beta}(A) = M_n$ and so $\{M_n : n \geq 1\}$ is a martingale as claimed.

The martingale $\{M_n : n \geq 1\}$ is non-negative and so the Martingale Convergence Theorem yields $M_n \to M_\infty$ a.s. (with the implicit null event depending on A). In order to identify the limit in terms of a random measure, we have to rerun the above argument as follows: For any bounded measurable $f : D \to \mathbb{R}$ define

$$\phi_n(f) := \int f \, \mathrm{d}\mu_n^{D,\beta}. \tag{2.47}$$

Then the same argument as above shows that $\phi_n(f)$ is a bounded martingale and so $\phi_n(f) \to \phi_\infty(f)$ a.s. Specializing to continuous f, the bound

$$\bigl|\phi_n(f)\bigr| \leq \mu_n^{D,\beta}(D) \, \|f\|_{C(\overline{D})} \tag{2.48}$$

along with the above a.s. convergence $M_n \to M_\infty$ (for $A := D$) yields

$$\bigl|\phi_\infty(f)\bigr| \leq M_\infty \|f\|_{C(\overline{D})}. \tag{2.49}$$

Restricting to a countable dense subclass of $f \in C(\overline{D})$ to manage the proliferation of null sets, $f \mapsto \phi_\infty(f)$ extends to a continuous linear functional ϕ'_∞ on $C(\overline{D})$ a.s. such that $\phi_\infty(f) = \phi'_\infty(f)$ a.s. for each $f \in C(\overline{D})$ (the null set possibly depending on f). On the event that ϕ'_∞ is well-defined and continuous, the Riesz Representation Theorem yields existence of a (random) Borel measure $\mu_\infty^{D,\beta}$ such that

$$\phi_\infty(f) \overset{\text{a.s.}}{=} \phi'_\infty(f) = \int f \, \mathrm{d}\mu_\infty^{D,\beta} \tag{2.50}$$

for each $f \in C(\overline{D})$.

To identify the limit in (2.43) (which we already showed to exist a.s.) with $\mu_\infty^{D,\beta}(A)$ we proceed as follows. First, given a $\mathcal{G}_\delta$-set $A \subseteq D$, we can find functions $f_k \in C(\overline{D})$ such that $f_k \downarrow 1_A$ as $k \to \infty$. Since $\mu_n^{D,\beta}(A) \leq \phi_n(f_k) \to \phi'_\infty(f_k)$ a.s. as $n \to \infty$ and $\phi'_\infty(f_k) \downarrow \mu_\infty^{D,\beta}(A)$ as $k \to \infty$ by the Bounded Convergence Theorem, we get

$$\lim_{n\to\infty} \mu_n^{D,\beta}(A) \leq \mu_\infty^{D,\beta}(A) \quad \text{a.s.} \tag{2.51}$$

Next, writing $L^1(D)$ for the space of Lebesgue absolutely integrable $f : D \to \mathbb{R}$, Fatou's lemma shows

$$E\bigl|\phi'_\infty(f)\bigr| \leq \|f\|_{L^1(D)}, \quad f \in C(\overline{D}). \tag{2.52}$$

The above approximation argument, along with the fact the Lebesgue measure is outer regular now shows that $\mathrm{Leb}(A) = 0$ implies $\mu_\infty^{D,\beta}(A) = 0$ a.s. This gives (2.51) for all Borel $A \subseteq D$. The string of equalities

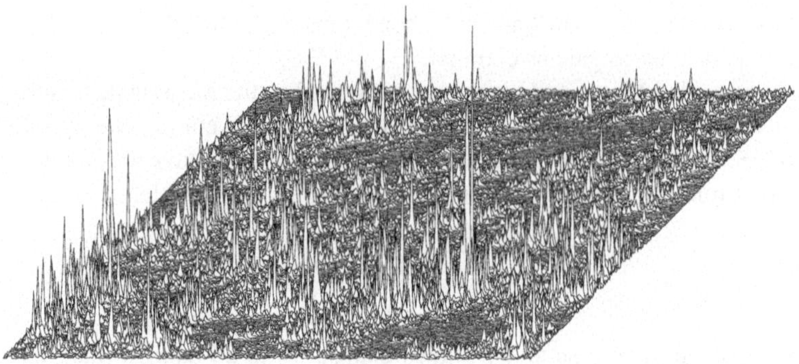

Fig. 5 A sample of the LQG measure $r_D(x)^{2\lambda^2}\mu_\infty^{D,\lambda\alpha}(\mathrm{d}x)$ for D a unit square and $\lambda := 0.3$. The high points indicate places of high local intensity

$$\mu_n^{D,\beta}(A) + \mu_n^{D,\beta}(A^{\mathrm{c}}) = \phi_n(1) \underset{n\to\infty}{\overset{\text{a.s.}}{\longrightarrow}} \phi'_\infty(1) = \mu_\infty^{D,\beta}(A) + \mu_\infty^{D,\beta}(A^{\mathrm{c}}) \tag{2.53}$$

now shows that equality must hold in (2.51). □

An interesting question is how the limit object $\mu_\infty^{D,\beta}$ depends on β and, in particular, for what β it is non-trivial. We pose:

Exercise 2.18 *Use Kolmogorov's zero-one law to show that, for each $A \subseteq D$ measurable,*

$$\big\{\mu_\infty^{D,\beta}(A) = 0\big\} \text{ is a zero-one event.} \tag{2.54}$$

Exercise 2.19 *Prove that there is $\beta_{\mathrm{c}} \in [0, \infty]$ such that*

$$\mu_\infty^{D,\beta}(D) \begin{cases} >0, & \text{if } 0 \le \beta < \beta_{\mathrm{c}}, \\ =0, & \text{if } \beta > \beta_{\mathrm{c}}. \end{cases} \tag{2.55}$$

Hint: Use Gaussian interpolation and conditional Fatou's lemma to show that the Laplace transform of $\mu_n^{D,\beta}(A)$ is non-decreasing in β.

We will show later that, in our setting (and with α from Theorem 2.7),

$$\beta_{\mathrm{c}} := \alpha = 2/\sqrt{g}. \tag{2.56}$$

As it turns out, the law of the measure $\mu_\infty^{D,\beta}$ is independent of the choice of the underlying basis in $\mathrm{H}_0^1(D)$. This has been proved gradually starting with somewhat restrictive Kahane's theory [81] (which we will review later) and culminating in a recent paper by Shamov [113]. (See Theorems 4.14 and 5.5.)

The measure $\mu_\infty^{D,\beta}$ is called the *Gaussian multiplicative chaos* associated with the continuum Gaussian Free Field. We now claim:

Theorem 2.20 (Z^D_λ-measure as LQG measure) *Assume the setting of Theorem 2.7 with $\lambda \in (0,1)$. Then there is $\hat{c} \in (0,\infty)$ such that for all $D \in \mathfrak{D}$,*

$$Z^D_\lambda(\mathrm{d}x) \overset{\text{law}}{=} \hat{c}\, r_D(x)^{2\lambda^2}\, \mu^{D,\lambda\alpha}_\infty(\mathrm{d}x). \tag{2.57}$$

where, we recall, $r_D(x)$ denotes the conformal radius of D from x.

The measure on the right of (2.57) (without the constant $\hat{c}$) is called the *Liouville Quantum Gravity* (LQG) measure in D for parameter $\beta := \lambda\alpha$. This object is currently heavily studied in connection with random conformally-invariant geometry (see, e.g., Miller and Sheffield [96, 100]). See Fig. 5.

An alternative construction of the LQG measure was given by Duplantier and Sheffield [65] using disc/circle averages (cf Exercise 1.30). This construction is technically more demanding (as it is not based directly on martingale convergence theory) but, as a benefit, one gets some regularity of the limit.

We note (and will prove this for $\lambda < 1/\sqrt{2}$ in Corollary 3.17) that,

$$E\mu^{D,\lambda\alpha}_\infty(A) = \mathrm{Leb}(A), \quad \lambda \in [0,1). \tag{2.58}$$

For each $\lambda < 1$, the total mass $\mu^{D,\lambda\alpha}_\infty(D)$ has moments up to $1+\epsilon(\lambda)$, for $\epsilon(\lambda) > 0$ that tends to zero as $\lambda \uparrow 1$.

We will not discuss Gaussian Multiplicative Chaos and/or the LQG measures as a stand-alone topic much in these lectures (although these objects will keep popping up in our various theorems) and instead refer the reader to Berestycki [20] and the review by Rhodes and Vargas [105]. The proofs of the Theorems 2.7, 2.14 and 2.20 will be given in the forthcoming lectures.

Lecture 3: Intermediate Level Sets: Factorization

The aim of this and the following lecture is to give a fairly detailed account of the proofs of the above theorems on the scaling limit of the intermediate level sets. We will actually do this only in the regime where the second-moment calculations work without the need for truncations; this requires restricting to $\lambda < 1/\sqrt{2}$. We comment on the changes that need to be made for the complementary set of λ's at the end of the next lecture.

3.1 Gibbs-Markov Property of DGFF

A number of forthcoming proofs will use a special property of the DGFF that addresses the behavior of the field restricted, via conditioning, to a subdomain. This property is the spatial analogue of the Markov property in one-parameter stochastic processes and is a consequence of the Gaussian decomposition into orthogonal subspaces along with the fact that the law of the DGFF is a Gibbs measure for a nearest-neighbor Hamiltonian (cf Definition 1.1). For this reason, we will attach the adjective *Gibbs-Markov* to this property, although the literature uses the term *domain-Markov* as well. Here is the precise statement:

Lemma 3.1 (Gibbs-Markov property) *For $U \subsetneq V \subsetneq \mathbb{Z}^2$, denote*

$$\varphi_x^{V,U} := E\big(h_x^V \,\big|\, \sigma(h_z^V : z \in V \smallsetminus U)\big). \tag{3.1}$$

where h^V is the DGFF in V. Then we have:

(1) A.e. sample of $x \mapsto \varphi_x^{V,U}$ is discrete harmonic on U with "boundary values" determined by $\varphi_x^{V,U} = h_x^V$ for each $x \in V \smallsetminus U$.
(2) The field $h^V - \varphi^{V,U}$ is independent of $\varphi^{V,U}$ and

$$h^V - \varphi^{V,U} \overset{\text{law}}{=} h^U. \tag{3.2}$$

Proof. Assume that V is finite for simplicity. Conditioning a multivariate Gaussian on part of the values preserves the multivariate Gaussian nature of the law. Hence $\varphi^{V,U}$ and $h^V - \varphi^{V,U}$ are multivariate Gaussians that are, by the properties of the conditional expectation, uncorrelated. It follows that $\varphi^{V,U} \perp\!\!\!\perp h^V - \varphi^{V,U}$.

Next let us prove that $\varphi^{V,U}$ has discrete-harmonic sample paths in U. To this end pick any $x \in U$ and note that the "smaller-always-wins" principle for nested conditional expectations yields

$$\varphi_x^{V,U} = E\Big(E\big(h_x^V \,\big|\, \sigma(h_{z'}^V : z' \neq x)\big) \,\Big|\, \sigma(h_z^V : z \in V \smallsetminus U)\Big). \tag{3.3}$$

In light of Definition 1.1, the inner conditional expectation admits the explicit form

$$E\big(h_x^V \,\big|\, \sigma(h_{z'}^V : z' \neq x)\big) = \frac{\displaystyle\int_{\mathbb{R}} h_x\, \mathrm{e}^{-\frac{1}{8}\sum_{y:\, y\sim x}(h_y^V - h_x)^2} \mathrm{d}h_x}{\displaystyle\int_{\mathbb{R}} \mathrm{e}^{-\frac{1}{8}\sum_{y:\, y\sim x}(h_y^V - h_x)^2} \mathrm{d}h_x}, \tag{3.4}$$

where $y \sim x$ abbreviates $(x, y) \in E(\mathbb{Z}^2)$. Now

$$\begin{aligned}\frac{1}{4}\sum_{y:\, y\sim x}(h_y - h_x)^2 &= h_x^2 - 2\frac{1}{4}\sum_{y:\, y\sim x} h_y + \frac{1}{4}\sum_{y:\, y\sim x} h_y^2 \\ &= \Big(h_x - \frac{1}{4}\sum_{y:\, y\sim x} h_y\Big)^2 + \frac{1}{4}\sum_{y:\, y\sim x} h_y^2 - \Big(\frac{1}{4}\sum_{y:\, y\sim x} h_y\Big)^2.\end{aligned} \tag{3.5}$$

The last two terms factor from both the numerator and denominator on the right of (3.4). Shifting h_x by the average of the neighbors then gives

$$E\big(h^V_x\,\big|\,\sigma(h^V_{z'}\colon z'\neq x)\big)=\frac{1}{4}\sum_{y\colon y\sim x}h^V_y\,. \tag{3.6}$$

Using this in (3.3) shows that $\varphi^{V,U}$ has the mean-value property, and is thus discrete harmonic, on U.

Finally, we need to show that $\widetilde{h}^U := h^V - \varphi^{V,U}$ has the law of h^U. The mean of $\widetilde{h}^U$ is zero so we just need to verify that the covariances match. Here we note that, using H^U to denote the discrete harmonic measure on U, the mean-value property of $\varphi^{V,U}$ yields

$$\widetilde{h}^U_x = h^V_x - \sum_{z\in\partial U} H^U(x,z)h^V_z\,,\quad x\in U. \tag{3.7}$$

For any $x,y\in U$, this implies

$$\begin{aligned}\mathrm{Cov}(\widetilde{h}^U_x,\widetilde{h}^U_y)=G^V(x,y)&-\sum_{z\in\partial U}H^U(x,z)G^V(z,y)\\&-\sum_{z\in\partial U}H^U(y,z)G^V(z,x)+\sum_{z,\tilde z\in\partial U}H^U(x,z)H^U(y,\tilde z)G^V(z,\tilde z)\,.\end{aligned} \tag{3.8}$$

Now recall the representation (1.33) which casts $G^V(x,y)$ as $-\mathfrak{a}(x-y)+\phi(y)$ with ϕ harmonic on V. Plugging this in (3.8), the fact that

$$\sum_{z\in\partial U}H^U(x,z)\phi(z)=\phi(x),\quad x\in U, \tag{3.9}$$

shows that all occurrences of ϕ in (3.8) cancel out. As $x\mapsto\mathfrak{a}(z-x)$ is discrete harmonic on U for any $z\in\partial U$, replacing $G^V(\cdot,\cdot)$ by $-\mathfrak{a}(\cdot-\cdot)$ in the last two sums on the right of (3.8) makes these sums cancel each other as well.

We are thus left with the first two terms on the right of (3.8) in which $G^V(\cdot,\cdot)$ is now replaced by $-\mathfrak{a}(\cdot-\cdot)$. The representation (1.33) then tells us that

$$\mathrm{Cov}(\widetilde{h}^U_x,\widetilde{h}^U_y)=G^U(x,y),\qquad x,y\in U. \tag{3.10}$$

Since both $\widetilde{h}^U$ and h^U vanish outside U, we have $\widetilde{h}^U \overset{\text{law}}{=} h^U$ as desired. □

Exercise 3.2 *Supply the missing (e.g., limiting) arguments to prove the Gibbs-Markov decomposition applies even to the situation when U and V are allowed to be infinite.*

A short way to write the Gibbs-Markov decomposition is as

$$h^V \overset{\text{law}}{=} h^U+\varphi^{V,U}\quad\text{where}\quad h^U \perp\!\!\!\perp \varphi^{V,U}. \tag{3.11}$$

with the law of h^U and $\varphi^{V,U}$ (often implicitly) as above.

We have seen that the monotonicity $V \mapsto G^V(x,y)$ allows for control of the variance of the DGFF in general domains by that in more regular ones. One of the important consequences of the Gibbs-Markov property is to give similar comparisons for various probabilities involving a finite number of vertices. The following examples will turn out to be useful:

Exercise 3.3 *Suppose $\emptyset \neq U \subseteq V \subsetneq \mathbb{Z}^2$. Prove that for every $a \in \mathbb{R}$,*

$$P\bigl(h_x^U \geq a\bigr) \leq 2P\bigl(h_x^V \geq a\bigr), \qquad x \in U. \tag{3.12}$$

Similarly, for any binary relation $\mathcal{R} \subseteq \mathbb{Z}^d \times \mathbb{Z}^d$, show that also

$$P\bigl(\exists x,y \in U\colon (x,y) \in \mathcal{R},\ h_x^U, h_y^U \geq a\bigr) \leq 4P\bigl(\exists x,y \in V\colon (x,y) \in \mathcal{R},\ h_x^V, h_y^V \geq a\bigr). \tag{3.13}$$

Similar ideas lead to:

Exercise 3.4 *Prove that for any $\emptyset \neq U \subseteq V \subsetneq \mathbb{Z}^2$ and any $a \in \mathbb{R}$,*

$$P\bigl(\max_{x\in U} h_x^U > a\bigr) \leq 2P\bigl(\max_{x\in V} h_x^V > a\bigr). \tag{3.14}$$

For finite V we get

$$E\bigl(\max_{x\in U} h_x^U\bigr) \leq E\bigl(\max_{x\in V} h_x^V\bigr), \tag{3.15}$$

and so $U \mapsto E(\max_{x\in U} h_x^U)$ is non-decreasing with respect to the set inclusion.

These estimates squeeze the maximum in one domain between that in smaller and larger domains. If crude bounds are sufficient, this permits reduction to simple domains, such as squares.

A typical setting for the application of the Gibbs-Markov property is depicted in Fig. 6. There each of the small boxes (the translates of V_K) has its "private" independent copy of the DGFF. By (3.11), to get h^{V_N} these copies are "bound together" by an independent Gaussian field φ^{V_N,V_N° that, as far as its law is concerned, is just the harmonic extension of the values of h^{V_N} on the dividing lines that separate the small boxes from each other. For this reason we sometimes refer to φ^{V_N,V_N° as the *binding field*. Note that φ^{V_N,V_N° has discrete-harmonic sample paths on V_N° yet it becomes quite singular on $V_N \smallsetminus V_N^\circ$; cf Fig. 7.

Iterations of the partitioning sketched in Fig. 6 lead to a hierarchical description of the DGFF on a square of side $N := 2^n$ as the sum (along root-to-leaf paths of length n) of a family of tree-indexed binding fields. If these binding fields could be regarded as constant on each of the "small" square, this would cast the DGFF as a Branching Random Walk. Unfortunately, the binding fields are not constant on relevant squares so this representation is only approximate. Still, it is extremely useful; see Lecture 7.

The Gibbs-Markov property can be used to bootstrap control from positive probability (in small boxes) to probability close to one (in a larger box, perhaps for a slightly modified event). An example of such a statement is:

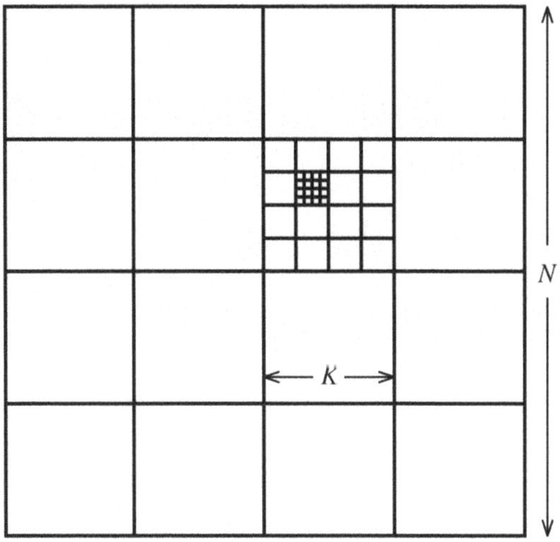

Fig. 6 A typical setting for the application of the Gibbs-Markov property. The box $V_N = (0, N)^2 \cap \mathbb{Z}^2$ is partitioned into $(N/K)^2$ translates of boxes $V_K := (0, K)^2 \cap \mathbb{Z}^2$ of side K (assuming K divides N). This leaves a "line of sites" between any two adjacent translates of V_K. The DGFF on V_N is then partitioned as $h^{V_N} = h^{V_N^\circ} + \varphi^{V_N, V_N^\circ}$ with $h^{V_N^\circ} \perp\!\!\!\perp \varphi^{V_N, V_N^\circ}$, where V_N° is the union of the shown translates of V_K and φ^{V_N, V_N° has the law of the harmonic extension to V_N° of the values of h^{V_N} on $V_N \smallsetminus V_N^\circ$. The translates of V_K can be further subdivided to produce a hierarchical description of the DGFF

Exercise 3.5 *Prove that "$\geq$" holds in* (2.9) *with probability tending to one as* $N \to \infty$ *(still assuming that* $\lambda < 1/\sqrt{2}$*).*

The term *sprinkling technique* is sometimes used for such bootstrap arguments in the literature although we prefer to leave it reserved for parameter manipulations of independent Bernoulli or Poisson random variables.

3.2 First Moment of Level-Set Size

Equipped with the Gibbs-Markov property, we are now ready to begin the proof of the scaling limit of the measures in Theorem 2.7. The key point is to estimate, as well as compute the asymptotic of, the first two moments of the size of the level set

$$\Gamma_N^D(b) := \{x \in D_N : h_x^{D_N} \geq a_N + b\}. \tag{3.16}$$

We begin with the first-moment calculation. Assume that $\lambda \in (0, 1)$, an admissible domain $D \in \mathfrak{D}$ and an admissible sequence $\{D_N : N \geq 1\}$ of domains approximating D are fixed. Our first lemma is then:

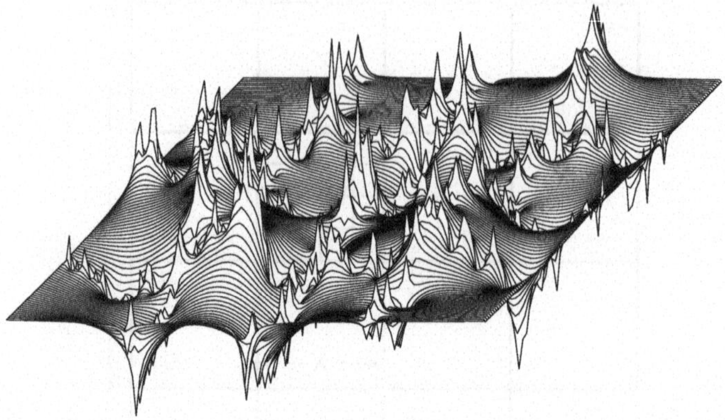

Fig. 7 A sample of the binding field φ^{V_N, V_N° for the (first-level) partition depicted in Fig. 6 with $N := 4K$. Here V_N° is the union of 16 disjoint translates of V_K. Note that while samples of the field are discrete harmonic inside the individual squares, they become quite rough on the dividing lines of sites

Lemma 3.6 (First moment upper bound) *For each $\delta \in (0, 1)$ there is $c \in (0, \infty)$ such that for all $N \geq 1$, all $b \in \mathbb{R}$ with $|b| \leq \log N$ and all a_N with $\delta \log N \leq a_N \leq \delta^{-1} \log N$, and all $A \subseteq D_N$,*

$$E\big|\Gamma_N^D(b) \cap A\big| \leq cK_N \frac{|A|}{N^2} \mathrm{e}^{-\frac{a_N}{g \log N} b}. \tag{3.17}$$

Proof. Similarly as in the proof of the upper bound in Theorem 2.1, the claim will follow by summing over $x \in A$ the inequality

$$P\big(h_x^{D_N} \geq a_N + b\big) \leq c \frac{1}{\sqrt{\log N}} \mathrm{e}^{-\frac{a_N^2}{2g \log N}} \mathrm{e}^{-\frac{a_N}{g \log N} b}, \tag{3.18}$$

which (we claim) holds with some $c > 0$ uniformly in $x \in D_N$ and b with $|b| \leq \log N$. By (3.12) and translation invariance of the DGFF, it suffices to prove (3.18) for $x := 0$ and D_N replaced by the box $\widetilde{D}_N$ of side length $4\mathrm{diam}_\infty(D_N)$ centered at the origin. (We will still write x for the vertex in question though.)

For this setting, Theorem 1.17 ensures that the variance of $h_x^{\widetilde{D}_N}$ is within a constant $\tilde{c} > 0$ of $g \log N$. Hence we get

$$P\big(h_x^{\widetilde{D}_N} \geq a_N + b\big) \leq \frac{1}{\sqrt{2\pi}} \frac{1}{\sqrt{g \log N - \tilde{c}}} \int_b^\infty \mathrm{e}^{-\frac{1}{2} \frac{(a_N + s)^2}{g \log N + \tilde{c}}} \mathrm{d}s. \tag{3.19}$$

Bounding $(a_N + s)^2 \geq a_N^2 + 2a_N s$ and noting that the assumptions on a_N (and the inequality $(1 + r)^{-1} \geq 1 - r$ for $0 < r < 1$) imply

$$\frac{a_N^2}{g\log N + c} \geq \frac{a_N^2}{g\log N} - \frac{c}{g^2\delta}\,, \tag{3.20}$$

we get

$$\int_b^\infty \mathrm{e}^{-\frac{1}{2}\frac{(a_N+s)^2}{g\log N+\tilde c}}\,\mathrm{d}s \leq c'\mathrm{e}^{-\frac{a_N^2}{2g\log N}}\,\mathrm{e}^{-\frac{a_N}{g\log N+\tilde c}b} \tag{3.21}$$

for some constant $c' > 0$. As $a_N \leq \delta^{-1}\log N$ and $|b| \leq \log N$, the constant $\tilde c$ in the exponent can be dropped at the cost of another multiplicative (constant) term popping up in the front. The claim follows. □

The fact that the estimate in Lemma 3.6 holds uniformly for all subsets of D_N will be quite useful for the following reason:

Exercise 3.7 *Show that for each $D \in \mathfrak{D}$, each $b \in \mathbb{R}$ and each $\epsilon > 0$ there is $\delta > 0$ such that for all N sufficiently large,*

$$E\big|\{x \in \Gamma_N^D(b)\colon \operatorname{dist}_\infty(x, D_N^{\mathrm{c}}) < \delta N\}\big| \leq \epsilon K_N\,. \tag{3.22}$$

We remark that no properties of ∂D other than those stated in Definition 1.14 should be assumed in the solution. Next let us note the following fact:

Exercise 3.8 *A sequence $\{\mu_n\colon n \geq 1\}$ of random Borel measures on a topological space $\mathscr{X}$ is tight with respect to the vague topology if and only if the sequence of random variables $\{\mu_N(K)\colon n \geq 1\}$ is tight for every compact $K \subseteq \mathscr{X}$.*

Lemma 3.6 then gives:

Corollary 3.9 (Tightness) *The family $\{\eta_N^D\colon N \geq 1\}$, regarded as measures on $\overline{D} \times (\mathbb{R} \cup \{+\infty\})$, is a tight sequence in the topology of vague convergence.*

Proof. Every compact set in $\overline{D} \times (\mathbb{R} \cup \{+\infty\})$ is contained in $K_b := \overline{D} \times [b, \infty]$ for some $b \in \mathbb{R}$. The definition of η_N^D shows

$$\eta_N^D(K_b) = \frac{1}{K_N}\big|\Gamma_N^D(b)\big|. \tag{3.23}$$

Lemma 3.6 shows these have uniformly bounded expectations and so are tight as ordinary random variables. The claim follows by Exercise 3.8. □

In probability, tightness is usually associated with "mass not escaping to infinity" or "the total mass being conserved." However, for convergence of random unnormalized measures in the vague topology, tightness does not prevent convergence to zero measure. In order to rule that out, we will need:

Lemma 3.10 (First moment asymptotic) *Assume that a_N obeys* (2.27) *and let c_0 be as in* (1.36)*. Then for all $b \in \mathbb{R}$ and all open $A \subseteq D$,*

$$E\big|\{x\in\Gamma_N^D(b)\colon x/N\in A\}\big| = \frac{\mathrm{e}^{2c_0\lambda^2/g}}{\lambda\sqrt{8\pi}}\mathrm{e}^{-\alpha\lambda b}\left[o(1)+\int_A \mathrm{d}x\, r_D(x)^{2\lambda^2}\right]K_N\,, \tag{3.24}$$

where $o(1)\to 0$ as $N\to\infty$ uniformly on compact sets of b.

Proof. Thanks to Exercise 3.7 we may remove a small neighborhood of ∂D from A and thus assume that $\mathrm{dist}_\infty(A, D^{\mathrm{c}}) > 0$. We will proceed by extracting an asymptotic expression for $P(h_x^{D_N} \ge a_N + b)$ with x such that $x/N \in A$. For such x, Theorem 1.17 gives

$$G^{D_N}(x,x) = g\log N + \theta_N(x), \tag{3.25}$$

where

$$\theta_N(x) = c_0 + g\log r_D(x/N) + o(1)\,, \tag{3.26}$$

with $o(1)\to 0$ as $N\to\infty$ uniformly in $x\in D_N$ with $x/N\in A$. Using this in the formula for the probability density of $h_x^{D_N}$ yields

$$P\big(h_x^{D_N} \ge a_N + b\big) = \frac{1}{\sqrt{2\pi}}\frac{1}{\sqrt{g\log N+\theta_N(x)}}\int_b^\infty \mathrm{e}^{-\frac12\frac{(a_N+s)^2}{g\log N+\theta_N(x)}}\,\mathrm{d}s\,. \tag{3.27}$$

The first occurrence of $\theta_N(x)$ does not affect the overall asymptotic as this quantity is bounded uniformly for all x under consideration. Expanding the square $(a_N+s)^2 = a_N^2 + 2a_N s + s^2$ and noting that (by decomposing the integration domain into $s \ll \log N$ and its complement) the s^2 term has negligible effect on the overall asymptotic of the integral, we find out

$$\int_b^\infty \mathrm{e}^{-\frac12\frac{(a_N+s)^2}{g\log N+\theta_N(x)}}\,\mathrm{d}s = \big(1+o(1)\big)(\alpha\lambda)^{-1}\mathrm{e}^{-\frac12\frac{a_N^2}{g\log N+\theta_N(x)}}\,\mathrm{e}^{-\alpha\lambda b+o(1)}\,. \tag{3.28}$$

We now use Taylor's Theorem (and the asymptotic of a_N) to get

$$\frac{a_N^2}{g\log N+\theta_N(x)} = \frac{a_N^2}{g\log N} - \frac{4\lambda^2}{g}\theta_N(x) + o(1) \tag{3.29}$$

with $o(1)\to 0$ again uniformly in all x under consideration. This yields

$$P\big(h_x^{D_N} \ge a_N + b\big) = \big(1+o(1)\big)\frac{\mathrm{e}^{2c_0\lambda^2/g}}{\lambda\sqrt{8\pi}}\,\mathrm{e}^{-\alpha\lambda b}\, r_D(x/N)^{2\lambda^2}\,\frac{K_N}{N^2}. \tag{3.30}$$

The result follows by summing this probability over x with $x/N\in A$ and using the continuity of r_D to convert the resulting Riemann sum into an integral. □

3.3 Second Moment Estimate

Our next task is to perform a rather tedious estimate on the second moment of the size of $\Gamma_N^D(b)$. It is here where we need to limit the range of possible λ.

Lemma 3.11 (Second moment bound) *Suppose $\lambda \in (0, 1/\sqrt{2})$. For each $b_0 > 0$ and each $D \in \mathfrak{D}$ there is $c_1 \in (0, \infty)$ such that for each $b \in [-b_0, b_0]$ and each $N \geq 1$,*

$$E\big(|\Gamma_N^D(b)|^2\big) \leq c_1 K_N^2 . \tag{3.31}$$

Proof. Assume $b := 0$ for simplicity (or absorb b into a_N). Writing

$$E\big(|\Gamma_N^D(0)|^2\big) = \sum_{x,y \in D_N} P\big(h_x^{D_N} \geq a_N,\, h_y^{D_N} \geq a_N\big). \tag{3.32}$$

we need to derive a good estimate for the probability on the right-hand side. In order to ensure uniformity, let $\widetilde{D}_N$ be a neighborhood of D_N of diameter twice the diameter of D_N. Exercise 3.3 then shows

$$P\big(h_x^{D_N} \geq a_N,\, h_y^{D_N} \geq a_N\big) \leq 4P\big(h_x^{\widetilde{D}_N} \geq a_N,\, h_y^{\widetilde{D}_N} \geq a_N\big). \tag{3.33}$$

We will now estimate the probability on the right by conditioning on $h_x^{\widetilde{D}_N}$.

First note that the Gibbs-Markov property yields a pointwise decomposition

$$h_y^{\widetilde{D}_N} = \mathfrak{g}_x(y) h_x^{\widetilde{D}_N} + \hat{h}_y^{\widetilde{D}_N \smallsetminus \{x\}}, \tag{3.34}$$

where

(1) $h_x^{\widetilde{D}_N}$ and $\hat{h}^{\widetilde{D}_N \smallsetminus \{x\}}$ are independent,
(2) $\hat{h}^{\widetilde{D}_N \smallsetminus \{x\}}$ has the law of the DGFF in $\widetilde{D}_N \smallsetminus \{x\}$, and
(3) $\mathfrak{g}_x$ is harmonic in $\widetilde{D}_N \smallsetminus \{x\}$, vanishes outside $\widetilde{D}_N$ and obeys $\mathfrak{g}_x(x) = 1$.

Using (3.34), the above probability is recast as

$$\begin{aligned} &P\big(h_x^{\widetilde{D}_N} \geq a_N,\, h_y^{\widetilde{D}_N} \geq a_N\big) \\ &\quad = \int_0^\infty P\Big(\hat{h}_y^{\widetilde{D}_N \smallsetminus \{x\}} \geq a_N(1 - \mathfrak{g}_x(y)) - s\mathfrak{g}_x(y)\Big) P\big(h_x^{\widetilde{D}_N} - a_N \in \mathrm{d}s\big). \end{aligned} \tag{3.35}$$

Given $\delta > 0$ we can always bound the right-hand side by $P(h_x^{\widetilde{D}_N} \geq a_N)$ when $|x - y| \leq \delta\sqrt{K_N}$. This permits us to assume that $|x - y| > \delta\sqrt{K_N}$ from now on. The $s \geq a_N$ portion of the integral is similarly bounded by $P(h_x^{\widetilde{D}_N} \geq 2a_N)$, so we will henceforth focus on $s \in [0, a_N]$.

Since x, y lie "deep" inside $\widetilde{D}_N$ and $|x - y| > \delta\sqrt{K_N} = N^{1-\lambda^2+o(1)}$, we have

$$
\begin{aligned}
\mathfrak{g}_x(y) = \frac{G^{\widetilde{D}_N}(x,y)}{G^{\widetilde{D}_N}(x,x)} &\le \frac{\log\frac{N}{|x-y|}+c}{\log N - c} \\
&\le 1-(1-\lambda^2)+o(1) = \lambda^2+o(1),
\end{aligned}
\tag{3.36}
$$

where $o(1)\to 0$ uniformly in $x,y\in D_N$. For $s\in[0,a_N]$, $\lambda<1/\sqrt{2}$ implies the existence of an $\epsilon>0$ such that, for N is large enough,

$$
\epsilon a_N \le a_N\big(1-\mathfrak{g}_x(y)\big)-s\mathfrak{g}_x(y)\le a_N,\quad x,y\in D_N. \tag{3.37}
$$

The Gaussian bound $P(X\ge t)\le \sigma t^{-1}\mathrm{e}^{-\frac{t^2}{2\sigma^2}}$ for $X=\mathcal{N}(0,\sigma^2)$ and any $t>0$ along with $G^{\widetilde{D}_N\smallsetminus\{x\}}(y,y)\le g\log N+c$ uniformly in $y\in D_N$ show

$$
\begin{aligned}
&P\Big(\hat h_y^{\widetilde{D}_N\smallsetminus\{x\}}\ge a_N\big(1-\mathfrak{g}_x(y)\big)-s\mathfrak{g}_x(y)\Big)\\
&\qquad\le\frac{\sqrt{G(y,y)}}{\epsilon a_N}\,\mathrm{e}^{-\frac{[a_N(1-\mathfrak{g}_x(y))-s\mathfrak{g}_x(y)]^2}{2G(y,y)}}\le c\frac{K_N}{N^2}\,\mathrm{e}^{\mathfrak{g}_x(y)\frac{a_N^2}{g\log N}+\frac{a_N}{g\log N}\mathfrak{g}_x(y)s},
\end{aligned}
\tag{3.38}
$$

where $G(y,y)$ abbreviates $G^{\widetilde{D}_N\smallsetminus\{x\}}(y,y)$. The first inequality in (3.36) gives

$$
\mathrm{e}^{\mathfrak{g}_x(y)\frac{a_N^2}{g\log N}}\le c\left(\frac{N}{|x-y|}\right)^{4\lambda^2+o(1)} \tag{3.39}
$$

with $o(1)\to 0$ uniformly in $x,y\in D_N$ with $|x-y|>\delta\sqrt{K_N}$.

The uniform upper bound on $G^{\widetilde{D}_N}(x,x)$ allows us to dominate the law of $h_x^{\widetilde{D}_N}-a_N$ on $[0,\infty)$ by

$$
P\big(h_x^{\widetilde{D}_N}-a_N\in\mathrm{d}s\big)\le c\frac{K_N}{N^2}\,\mathrm{e}^{-\frac{a_N}{g\log N}s}\mathrm{d}s. \tag{3.40}
$$

As $1-\mathfrak{g}_x(y)=1-\lambda^2+o(1)$, the $s\in[0,a_N]$ part of the integral in (3.35) is then readily performed to yield

$$
P\big(h_x^{\widetilde{D}_N}\ge a_N,\,h_y^{\widetilde{D}_N}\ge a_N\big)\le P\big(h_x^{\widetilde{D}_N}\ge 2a_N\big)+c\left(\frac{K_N}{N^2}\right)^2\left(\frac{N}{|x-y|}\right)^{4\lambda^2+o(1)} \tag{3.41}
$$

uniformly in $x,y\in D_N$ with $|x-y|>\delta\sqrt{K_N}$.

In order to finish the proof, we now use (3.33) to write

$$
\begin{aligned}
E\big(|\Gamma_N^D(0)|^2\big) \le & \sum_{\substack{x,y\in D_N\\ |x-y|\le\delta\sqrt{K_N}}} P\big(h_x^{D_N}\ge a_N\big) \\
& + \sum_{\substack{x,y\in D_N\\ |x-y|>\delta\sqrt{K_N}}} 4P\big(h_x^{\widetilde{D}_N}\ge a_N,\, h_y^{\widetilde{D}_N}\ge a_N\big).
\end{aligned}
\tag{3.42}
$$

Summing over y and invoking Lemma 3.6 bounds the first term by a factor of order $(\delta K_N)^2$. The contribution of the first term on the right of (3.41) to the second sum is bounded via Lemma 3.6 as well:

$$
\begin{aligned}
P\big(h_x^{D_N}\ge 2a_N\big) &\le \frac{c}{\sqrt{\log N}}\mathrm{e}^{-2\frac{a_N^2}{g\log N}} \\
&= c\left(\frac{K_N}{N^2}\right)^2 \mathrm{e}^{-\frac{a_N^2}{g\log N}}\sqrt{\log N}\le c\delta\left(\frac{K_N}{N^2}\right)^2.
\end{aligned}
\tag{3.43}
$$

Plugging in also the second term on the right of (3.41), we thus get

$$
E\big(|\Gamma_N^D(0)|^2\big)\le 2c\delta(K_N)^2 + c\left(\frac{K_N}{N^2}\right)^2 \sum_{\substack{x,y\in D_N\\ |x-y|>\delta\sqrt{K_N}}}\left(\frac{N}{|x-y|}\right)^{4\lambda^2+o(1)}. \tag{3.44}
$$

Dominating the sum by $cN^4\int_{D\times D}|x-y|^{-4\lambda^2+o(1)}\mathrm{d}x\mathrm{d}y$, with the integral convergent due to $4\lambda^2<2$, we find that also the second term on the right is of order $(K_N)^2$. □

As a corollary we now get:

Corollary 3.12 (Subsequential limits are non-trivial) *Let* $\lambda\in(0,1/\sqrt{2})$. *Then every subsequential limit* η^D *of* $\{\eta_N^D\colon N\ge 1\}$ *obeys*

$$
P\big(\eta^D(A\times[b,b'])>0\big)>0 \tag{3.45}
$$

for any open and non-empty $A\subseteq D$ *and every* $b<b'$.

Proof. Abbreviate $X_N:=\eta_N^D(A\times[b,b'])$. Then Lemma 3.10 implies

$$
E(X_N)\underset{N\to\infty}{\longrightarrow}\hat{c}\left[\int_A \mathrm{d}x\, r_D(x)^{2\lambda^2}\right]\big(\mathrm{e}^{-\lambda\alpha b}-\mathrm{e}^{-\lambda\alpha b'}\big), \tag{3.46}
$$

where $\hat{c}:=\mathrm{e}^{2c_0\lambda^2/g}/(\lambda\sqrt{8\pi})$. This is positive and finite for any A and b,b' as above. On the other hand, Lemma 3.11 shows that $\sup_{N\ge1}E(X_N^2)<\infty$. The second-moment estimate (Exercise 2.5) then yields the claim. □

3.4 Second-Moment Asymptotic and Factorization

At this point we know that the subsequential limits exist and are non-trivial (with positive probability). The final goal of this lecture is to prove:

Proposition 3.13 (Factorization) *Suppose $\lambda \in (0, 1/\sqrt{2})$. Then every subsequential limit η^D of $\{\eta^D_N : N \geq 1\}$ takes the form*

$$\eta^D(\mathrm{d}x\,\mathrm{d}h) = Z^D_\lambda(\mathrm{d}x) \otimes \mathrm{e}^{-\alpha\lambda h}\mathrm{d}h, \tag{3.47}$$

where Z^D_λ is a random, a.s.-finite Borel measure on D with $P(Z^D_\lambda(D) > 0) > 0$.

The proof relies on yet another (and this time quite lengthy) second-moment calculation. The result of this calculation is the content of:

Lemma 3.14 *For any $\lambda \in (0, 1/\sqrt{2})$, any open $A \subseteq D$, any $b \in \mathbb{R}$, and*

$$A_N := \{x \in \mathbb{Z}^2 : x/N \in A\} \tag{3.48}$$

we have

$$\lim_{N\to\infty} \frac{1}{K_N} E\Big|\big|\Gamma^D_N(0) \cap A_N\big| - \mathrm{e}^{\alpha\lambda b}\big|\Gamma^D_N(b) \cap A_N\big|\Big| = 0. \tag{3.49}$$

Proof (*modulo a computation*). By Lemma 3.6 we may assume $\mathrm{dist}_\infty(A, D^c) > \epsilon$ for some $\epsilon > 0$. We will prove

$$\lim_{N\to\infty} \frac{1}{K_N^2} E\bigg(\Big|\big|\Gamma^D_N(0) \cap A_N\big| - \mathrm{e}^{\alpha\lambda b}\big|\Gamma^D_N(b) \cap A_N\big|\Big|^2\bigg) = 0 \tag{3.50}$$

which implies the claim via the Cauchy–Schwarz inequality. Plugging

$$\big|\Gamma^D_N(\cdot) \cap A_N\big| = \sum_{x\in A_N} 1_{\{h^{D_N}_x \geq a_N + \cdot\}} \tag{3.51}$$

into (3.50) we get a sum of pairs of (signed) products of the various combinations of these indicators. The argument in the proof of Lemma 3.11 allows us to estimate the pairs where $|x - y| \leq \delta N$ by a quantity that vanishes as $N \to \infty$ and $\delta \downarrow 0$. It will thus suffice to show

$$\begin{aligned}\max_{\substack{x,y\in A_N\\|x-y|>\delta N}} \Big(&P\big(h^{D_N}_x \geq a_N,\, h^{D_N}_y \geq a_N\big) \\ &- \mathrm{e}^{\alpha\lambda b} P\big(h^{D_N}_x \geq a_N + b,\, h^{D_N}_y \geq a_N\big) \\ &- \mathrm{e}^{\alpha\lambda b} P\big(h^{D_N}_x \geq a_N,\, h^{D_N}_y \geq a_N + b\big) \\ &+ \mathrm{e}^{2\alpha\lambda b} P\big(h^{D_N}_x \geq a_N + b,\, h^{D_N}_y \geq a_N + b\big)\Big) = o\Big(\frac{K_N^2}{N^4}\Big)\end{aligned} \tag{3.52}$$

as $N \to \infty$. A computation refining the argument in the proof of Lemma 3.11 to take into account the precise asymptotic of the Green function (this is where we get aided by the fact that $|x-y| > \delta N$ and $\text{dist}_\infty(A, D^c) > \epsilon$) now shows that, for any $b_1, b_2 \in \{0, b\}$,

$$\begin{aligned} P\bigl(h_x^{D_N} \geq a_N + b_1,\, h_y^{D_N} \geq a_N + b_2\bigr) \\ = \bigl(\mathrm{e}^{-\alpha\lambda(b_1+b_2)} + o(1)\bigr) P\bigl(h_x^{D_N} \geq a_N,\, h_y^{D_N} \geq a_N\bigr) \end{aligned} \tag{3.53}$$

with $o(1) \to 0$ as $N \to \infty$ uniformly in $x, y \in A_N$. This then implies (3.52) and thus the whole claim. □

Exercise 3.15 *Supply a detailed proof of* (3.53). *(Consult [24] if lost.)*

From Lemma 3.14 we get:

Corollary 3.16 *Suppose* $\lambda \in (0, 1/\sqrt{2})$. *Then any subsequential limit* η^D *of* $\{\eta_N^D\colon N \geq 1\}$ *obeys the following: For any open* $A \subseteq D$ *and any* $b \in \mathbb{R}$,

$$\eta^D\bigl(A \times [b, \infty)\bigr) = \mathrm{e}^{-\alpha\lambda b} \eta^D\bigl(A \times [0, \infty)\bigr), \quad a.s. \tag{3.54}$$

Proof. In the notation of Lemma 3.14,

$$\eta_N^D\bigl(A \times [b, \infty)\bigr) = \frac{1}{K_N} \bigl|\Gamma_N^D(b) \cap A_N\bigr|. \tag{3.55}$$

Taking a *joint* distributional limit of $\eta_N^D(A \times [b, \infty))$ and $\eta_N^D(A \times [0, \infty))$ along the given subsequence, Lemma 3.14 along with Fatou's lemma show

$$E\Bigl|\eta^D\bigl(A \times [0, \infty)\bigr) - \mathrm{e}^{\alpha\lambda b} \eta^D\bigl(A \times [b, \infty)\bigr)\Bigr| = 0. \tag{3.56}$$

(This requires a routine approximation of indicators of these events by continuous functions as in Exercise 2.10.) The claim follows. □

We now give:
Proof of Proposition 3.13. For each Borel $A \subseteq D$ define

$$Z_\lambda^D(A) := (\alpha\lambda)\eta^D\bigl(A \times [0, \infty)\bigr) \tag{3.57}$$

Then Z_λ^D is an a.s.-finite (random) Borel measure on D. Letting $\mathcal{A}$ be the set of all half-open dyadic boxes entirely contained in D, Corollary 3.16 and a simple limiting argument show that, for any $A \in \mathcal{A}$ and any $b \in \mathbb{Q}$,

$$\eta^D\bigl(A \times [b, \infty)\bigr) = (\alpha\lambda)^{-1} Z_\lambda^D(A) \mathrm{e}^{-\alpha\lambda b}, \quad \text{a.s.} \tag{3.58}$$

Since $\mathcal{A} \times \mathbb{Q}$ is countable, the null set in (3.58) can be chosen so that the equality in (3.58) holds for all $A \in \mathcal{A}$ and all $b \in \mathbb{Q}$ simultaneously, a.s. Note that the sets $\{A \times [b, \infty) \colon A \in \mathcal{A}, b \in \mathbb{Q}\}$ constitute a π-system that generates all Borel sets in $D \times \mathbb{R}$. In light of

$$(\alpha\lambda)^{-1} Z_\lambda^D(A) \mathrm{e}^{-\alpha\lambda b} = \int_{A \times [b,\infty)} Z_\lambda^D(\mathrm{d}x) \otimes \mathrm{e}^{-\alpha\lambda h} \mathrm{d}h \tag{3.59}$$

the claim follows from Dynkin's π-λ-theorem. □

We also record an important observation:

Corollary 3.17 *Assume $\lambda \in (0, 1/\sqrt{2})$ and denote*

$$\hat{c} := \frac{\mathrm{e}^{2c_0\lambda^2/g}}{\lambda\sqrt{8\pi}} \tag{3.60}$$

for c_0 as in (1.36)*. Then Z_λ^D from* (3.47) *obeys*

$$E\big[Z_\lambda^D(A)\big] = \hat{c} \int_A \mathrm{d}x\, r_D(x)^{2\lambda^2} \tag{3.61}$$

for each Borel $A \subseteq D$. Moreover, there is $c \in (0, \infty)$ such that for any open square $S \subseteq \mathbb{C}$,

$$E\big[Z_\lambda^S(S)^2\big] \le c\, \mathrm{diam}(S)^{4+4\lambda^2}. \tag{3.62}$$

Proof (sketch). Thanks to the uniform square integrability proved in Lemma 3.11, the convergence in probability is accompanied by convergence of the first moments. Then (3.61) follows from Lemma 3.10. To get also (3.62) we need a uniform version of the bound in Lemma 3.11. We will not perform the requisite calculation, just note that for a $c' \in (0, \infty)$ the following holds for all $D \in \mathfrak{D}$,

$$\limsup_{N\to\infty} \frac{1}{K_N^2} E\big(|\Gamma_N^D(0)|^2\big) \le c' \int_{D\times D} \mathrm{d}x \otimes \mathrm{d}y \left(\frac{[\mathrm{diam}\, D]^2}{|x-y|}\right)^{4\lambda^2}, \tag{3.63}$$

where $\mathrm{diam}\, D$ is the diameter of D in the Euclidean norm. We leave further details of the proof to the reader. □

This closes the first part of the proof of Theorem 2.7 which showed that every subsequential limit of the measures of interest factors into the desired product form. The proof continues in the next lecture.

Lecture 4: Intermediate Level Sets: Nailing the Limit

The goal of this lecture is to finish the proof of Theorem 2.7 and the results that follow thereafter. This amounts to proving a list of properties that the Z_λ^D-measures (still tied to a specific subsequence) satisfy and showing that these properties characterize the law of the Z^D_λ-measures uniquely. As part of the proof, we obtain a conformal transformation rule for Z^D_λ and a representation thereof as a Liouville Quantum Gravity measure. All proofs remain restricted to $\lambda < 1/\sqrt{2}$; the final section comments on necessary changes in the complementary regime of λ's.

4.1 Gibbs-Markov Property in the Scaling Limit

We have shown so far that every subsequential limit of the family of point measures $\{\eta^D_N : N \geq 1\}$ takes the form

$$Z^D_\lambda(\mathrm{d}x) \otimes \mathrm{e}^{-\alpha\lambda h}\mathrm{d}h \tag{4.1}$$

for *some* random Borel measure Z^D_λ on D. Our next goal is to identify properties of these measures that will ultimately nail their law uniquely. The most important of these is the behavior under restriction to a subdomain which arises from the Gibbs-Markov decomposition of the DGFF (which defines the concept of the binding field used freely below). However, as the Z^D_λ-measure appears only in the scaling limit, we have to first describe the scaling limit of the Gibbs-Markov decomposition itself.

The main observation to be made below is that, although the DGFF has no pointwise scaling limit, the binding field does. This is facilitated (and basically implied) by the fact that the binding field has discrete-harmonic sample paths. To define the relevant objects, let $\widetilde{D}, D \in \mathfrak{D}$ be two domains satisfying $\widetilde{D} \subseteq D$. For each $x, y \in \widetilde{D}$, set

$$C^{D,\widetilde{D}}(x,y) := g\int_{\partial D} \Pi^D(x,\mathrm{d}z)\log|y-z| - g\int_{\partial\widetilde{D}} \Pi^{\widetilde{D}}(x,\mathrm{d}z)\log|y-z|\,, \tag{4.2}$$

where Π^D is the harmonic measure from (1.28). Given any admissible approximations $\{D_N : N \geq 1\}$ and $\{\widetilde{D}_N : N \geq 1\}$ of domains D and $\widetilde{D}$, respectively, of which we also assume that $\widetilde{D}_N \subseteq D_N$ for each $N \geq 1$, we now observe:

Lemma 4.1 (Convergence of covariances) *Locally uniformly in $x, y \in \widetilde{D}$,*

$$G^{D_N}\big(\lfloor xN\rfloor, \lfloor yN\rfloor\big) - G^{\widetilde{D}_N}\big(\lfloor xN\rfloor, \lfloor yN\rfloor\big) \underset{N\to\infty}{\longrightarrow} C^{D,\widetilde{D}}(x,y). \tag{4.3}$$

We leave it to the reader to solve:

Exercise 4.2 *Prove Lemma 4.1 while noting that this includes uniform convergence on the diagonal $x = y$. Hint: Use the representation in Lemma 1.19.*

From here we get:

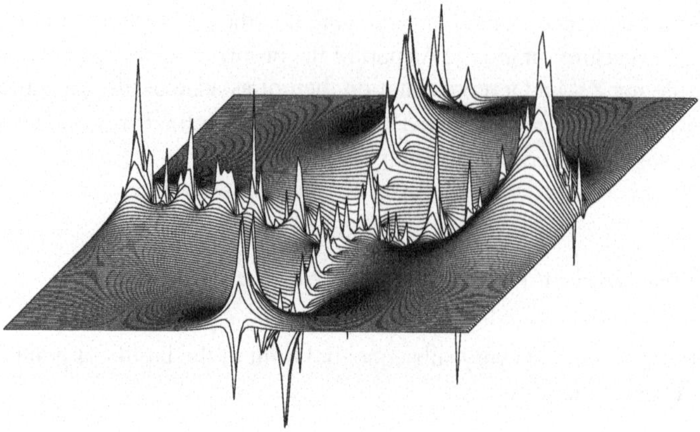

Fig. 8 A sample of $\varphi_N^{D,\widetilde{D}}$ where $D := (-1, 1)^2$ and $\widetilde{D}$ obtained from D by removing points on the coordinate axes

Lemma 4.3 (Limit binding field) *For any D and $\widetilde{D}$ as above, $x, y \mapsto C^{D,\widetilde{D}}(x, y)$ is a symmetric, positive semi-definite kernel on $\widetilde{D} \times \widetilde{D}$. In particular, there is a Gaussian process $x \mapsto \Phi^{D,\widetilde{D}}(x)$ on $\widetilde{D}$ with zero mean and covariance (Fig. 8)*

$$\mathrm{Cov}\big(\Phi^{D,\widetilde{D}}(x), \Phi^{D,\widetilde{D}}(y)\big) = C^{D,\widetilde{D}}(x, y), \qquad x, y \in \widetilde{D}. \tag{4.4}$$

Proof. Let $U \subseteq V$ be non-empty and finite. The Gibbs-Markov decomposition implies

$$\mathrm{Cov}(\varphi_x^{V,U}, \varphi_y^{V,U}) = G^V(x, y) - G^U(x, y). \tag{4.5}$$

Hence, $x, y \mapsto G^V(x, y) - G^U(x, y)$ is symmetric and positive semi-definite on $U \times U$. In light of (4.3), this extends to $C^{D,\widetilde{D}}$ on $\widetilde{D} \times \widetilde{D}$ by a limiting argument. Standard arguments then imply the existence of the Gaussian process $\Phi^{D,\widetilde{D}}$. □

In light of Exercise 1.3 and (4.5) we see that $C^{D,\widetilde{D}}(x, y) \geq 0$ for all $x, y \in \widetilde{D}$. When $\widetilde{D} \subsetneq D$, we will even have $C^{D,\widetilde{D}}(x, x) > 0$ for some $x \in \widetilde{D}$. We will call $\Phi^{D,\widetilde{D}}$ the *continuum binding field*. To justify this name, observe:

Lemma 4.4 (Coupling of binding fields) *$\Phi^{D,\widetilde{D}}$ has a version with continuous sample paths on $\widetilde{D}$. Moreover, for each $\delta > 0$ and each $N \geq 1$ there is a coupling of $\varphi^{D_N,\widetilde{D}_N}$ and $\Phi^{D,\widetilde{D}}$ such that*

$$\sup_{\substack{x\in\widetilde{D} \\ \mathrm{dist}(x,\partial\widetilde{D})>\delta}} \big|\Phi^{D,\widetilde{D}}(x) - \varphi_{\lfloor xN\rfloor}^{D_N,\widetilde{D}_N}\big| \underset{N\to\infty}{\longrightarrow} 0, \qquad \textit{in probability}. \tag{4.6}$$

Proof (assuming regularity of the fields). Abbreviate $\varphi_N^{D,\widetilde{D}}(x) := \varphi_{\lfloor xN\rfloor}^{D_N,\widetilde{D}_N}$. The convergence of the covariances from Lemma 4.1 implies $\varphi_N^{D,\widetilde{D}}(x) \to \Phi^{D,\widetilde{D}}(x)$ in law for each $x \in \widetilde{D}$, so the point is to extend this to the convergence of these fields as random functions. Fix $\delta > 0$ and denote

$$\widetilde{D}^\delta := \bigl\{x \in \mathbb{C}\colon \operatorname{dist}(x, \partial\widetilde{D}) > \delta\bigr\}. \tag{4.7}$$

Fix $r > 0$ small and let $x_1, \dots, x_k$ be an r-net in $\widetilde{D}^\delta$. As convergence of the covariances implies convergence in law, and convergence in law on $\mathbb{R}^k$ can be turned into convergence in probability under a suitable coupling measure, for each $N \geq 1$ there is a coupling of $\varphi^{D_N,\widetilde{D}_N}$ and $\Phi^{D,\widetilde{D}}$ with

$$P\Bigl(\max_{i=1,\dots,k} \bigl|\Phi^{D,\widetilde{D}}(x_i) - \varphi_N^{D,D}(x_i)\bigr| > \epsilon\Bigr) \underset{N\to\infty}{\longrightarrow} 0. \tag{4.8}$$

The claim will then follow if we can show that

$$\lim_{r\downarrow 0} P\Biggl(\sup_{\substack{x,y\in\widetilde{D}^\delta\\ |x-y|<r}} \bigl|\Phi^{D,\widetilde{D}}(x) - \Phi^{D,\widetilde{D}}(y)\bigr| > \epsilon\Biggr) = 0 \tag{4.9}$$

and similarly (with an additional *limes superior* as $N \to \infty$ preceding the limit $r \downarrow 0$) for $\Phi^{D,\widetilde{D}}(\cdot)$ replaced by $\varphi_N^{D,\widetilde{D}}$. This, along with the continuity of $\Phi^{D,\widetilde{D}}$, will follow from the regularity estimates on Gaussian processes to be proved in Lecture 6 (specifically, in Lemma 6.17). □

For future reference, we suggest that the reader solve the following exercises:

Exercise 4.5 *(Projection definition) Prove (by comparing covariances) that $\Phi^{D,\widetilde{D}}$ is the projection of the CGFF on D onto the subspace of functions in $\mathsf{H}_0^1(D)$ that are harmonic on (the connected components of) $\widetilde{D}$.*

We will also need to note that the binding field has a nesting property:

Exercise 4.6 *(Nesting property) Show that if $U \subseteq V \subseteq W$, then*

$$\varphi^{W,U} \overset{\text{law}}{=} \varphi^{W,V} + \varphi^{V,U} \quad \textit{with} \quad \varphi^{W,V} \perp\!\!\!\perp \varphi^{V,U}. \tag{4.10}$$

Similarly, if $D'' \subseteq D' \subseteq D$ are admissible, then (on D'')

$$\Phi^{D,D''} \overset{\text{law}}{=} \Phi^{D,D'} + \Phi^{D',D''} \quad \textit{with} \quad \Phi^{D,D'} \perp\!\!\!\perp \Phi^{D',D''}. \tag{4.11}$$

To demonstrate the benefits of taking the scaling limit we note that continuous binding field behaves nicely under conformal maps of the underlying domain. Indeed, we have:

Exercise 4.7 *Show that under a conformal bijection $f\colon D \to f(D)$, we have*

$$C^{f(D),f(\widetilde{D})}\bigl(f(x), f(y)\bigr) = C^{D,\widetilde{D}}(x,y), \qquad x, y \in \widetilde{D}, \tag{4.12}$$

for any $\widetilde{D} \subseteq D$. Prove that this implies

$$\Phi^{f(D),f(\widetilde{D})} \circ f \overset{\text{law}}{=} \Phi^{D,\widetilde{D}}. \tag{4.13}$$

Hint: Use Lemma 1.26.

4.2 Properties of Z^D_λ-Measures

We are now ready to move to the discussion of the properties of (the laws of) the Z^D_λ-measures. These will often relate the measures in different domains and so we need to ensure that the subsequential limit can be taken for all of these domains simultaneously. Applying Cantor's diagonal argument, we can definitely use the same subsequence for any *countable* collection $\mathfrak{D}_0 \subseteq \mathfrak{D}$ of admissible domains and thus assume that a subsequential limit η^D, and hence also the measure Z^D_λ, has been extracted for each $D \in \mathfrak{D}_0$. All of our subsequent statements will then be restricted to the domains in $\mathfrak{D}_0$. We will always assume that $\mathfrak{D}_0$ contains all finite unions of dyadic open and/or half-open squares plus any finite number of additional domains of present (or momentary) interest.

Some of the properties of the measures $\{Z^D_\lambda : D \in \mathfrak{D}_0\}$ are quite elementary consequences of the above derivations and so we relegate them to:

Exercise 4.8 *(Easy properties) Prove that:*

(1) for each $D \in \mathfrak{D}_0$, Z^D_λ is supported on D; i.e., $Z^D_\lambda(\partial D) = 0$ a.s.,
(2) if $A \subseteq D \in \mathfrak{D}_0$ is measurable with $\mathrm{Leb}(A) = 0$, then $Z^D_\lambda(A) = 0$ a.s.,
(3) if $D, \widetilde{D} \in \mathfrak{D}_0$ obey $D \cap \widetilde{D} = \emptyset$, then

$$Z^{D\cup\widetilde{D}}_\lambda(\mathrm{d}x) \overset{\text{law}}{=} Z^D_\lambda(\mathrm{d}x) + Z^{\widetilde{D}}_\lambda(\mathrm{d}x), \tag{4.14}$$

with the measures Z^D_λ and $Z^{\widetilde{D}}_\lambda$ on the right regarded as independent, and
(4) the law of Z^D_λ is translation invariant; i.e.,

$$Z^{a+D}_\lambda(a + \mathrm{d}x) \overset{\text{law}}{=} Z^D_\lambda(\mathrm{d}x) \tag{4.15}$$

for each $a \in \mathbb{C}$ such that $D, a + D \in \mathfrak{D}_0$.

As already mentioned, a key point for us is to prove the behavior under the restriction to a subdomain. We formulate this as follows:

Proposition 4.9 (Gibbs-Markov for Z^D_λ-measures) *For any $D, \widetilde{D} \in \mathfrak{D}_0$ satisfying $\widetilde{D} \subseteq D$ and* $\mathrm{Leb}(D \smallsetminus \widetilde{D}) = 0$,

$$Z^D_\lambda(\mathrm{d}x) \overset{\text{law}}{=} \mathrm{e}^{\alpha\lambda\Phi^{D,\widetilde{D}}(x)}\, Z^{\widetilde{D}}_\lambda(\mathrm{d}x), \tag{4.16}$$

where $\Phi^{D,\widetilde{D}} \perp\!\!\!\perp Z_\lambda^{\widetilde{D}}$ *with the laws as above.*

Note that, by Exercise 4.8(2), both sides of the expression assign zero mass to $D \smallsetminus \widetilde{D}$ a.s. It therefore does not matter that $\Phi^{D,\widetilde{D}}$ is not really defined there. (The decomposition in Fig. 6 shows that, even with $D, \widetilde{D} \in \mathfrak{D}_0$, non-trivial examples exist with $\widetilde{D} \subseteq D$ and $\text{Leb}(D \smallsetminus \widetilde{D}) = 0$.)

Proof. Suppose $D, \widetilde{D} \in \mathfrak{D}_0$ obey $\widetilde{D} \subseteq D$ and $\text{Leb}(D \smallsetminus \widetilde{D}) = 0$. Thanks to Exercise 4.8(2) and the Monotone Convergence Theorem, it suffices to prove equality in law for the integrals of the two measures with respect to any function bounded and continuous $f\colon D \times \mathbb{R} \to \mathbb{R}$ with support in $K \times [-b, b]$ for a compact $K \subseteq \widetilde{D}$ and some $b > 0$.

We start by invoking the Gibbs-Markov decomposition

$$h^{D_N} \overset{\text{law}}{=} h^{\widetilde{D}_N} + \varphi_N^{D,\widetilde{D}} \quad \text{where} \quad h^{\widetilde{D}_N} \perp\!\!\!\perp \varphi_N^{D,\widetilde{D}}. \tag{4.17}$$

A calculation then shows

$$\langle \eta_N^D, f\rangle \overset{\text{law}}{=} \langle \eta_N^{\widetilde{D}}, f_\varphi\rangle\,, \tag{4.18}$$

where

$$f_\varphi(x,h) := f\bigl(x, h + \varphi_{\lfloor xN\rfloor}^{D_N,\widetilde{D}_N}\bigr)\,. \tag{4.19}$$

Next consider the coupling of $\varphi^{D_N,\widetilde{D}_N}$ and $\Phi^{D,\widetilde{D}}$ from Lemma 4.4 where we may and will assume $\Phi^{D,\widetilde{D}} \perp\!\!\!\perp \eta_N^{\widetilde{D}}$. Our aim is to replace f_φ by

$$f_\Phi(x,h) := f\bigl(x, h + \Phi^{D,\widetilde{D}}(x)\bigr) \tag{4.20}$$

in (4.18). Given $\epsilon > 0$ let $\delta > 0$ be such that for all $x \in \widetilde{D}$ and all $h, h' \in [-b, b]$ with $|h - h'| < \delta$ we have $|f(x,h) - f(x,h')| < \epsilon$. Then, on the event

$$\Bigl\{\sup_{x\in K} |\Phi^{D,\widetilde{D}}(x)| \le M\Bigr\} \cap \Bigl\{\sup_{x\in K} |\varphi_{\lfloor xN\rfloor}^{D_N,\widetilde{D}_N} - \Phi^{D,\widetilde{D}}(x)| < \delta\Bigr\}, \tag{4.21}$$

we get

$$\bigl|\langle \eta_N^{\widetilde{D}}, f_\varphi\rangle - \langle \eta_N^{\widetilde{D}}, f_\Phi\rangle\bigr| \le \eta_N^{\widetilde{D}}\bigl(\widetilde{D} \times [-b - M - \delta, \infty)\bigr)\epsilon. \tag{4.22}$$

By Lemmas 3.6 and 4.4, the probability of the event in (4.21) tends to one while the right-hand side of (4.22) tends to zero in probability as $N \to \infty$ followed by $\epsilon \downarrow 0$ and $M \to \infty$. Any simultaneous subsequential limits η^D, resp., $\eta^{\widetilde{D}}$ of $\{\eta_N^D\colon N \ge 1\}$, resp., $\{\eta_N^{\widetilde{D}}\colon N \ge 1\}$ therefore obey

$$\langle \eta^D, f\rangle \overset{\text{law}}{=} \langle \eta^{\widetilde{D}}, f_\Phi\rangle\,, \tag{4.23}$$

where $\Phi^{D,\widetilde{D}}$ (implicitly contained in f_Φ) is independent of $\eta^{\widetilde{D}}$.

The representation from Proposition 3.13 now permits us to write

$$
\begin{aligned}
\int_{D\times\mathbb{R}} Z_\lambda^D(\mathrm{d}x)\otimes \mathrm{d}h\, \mathrm{e}^{-\alpha\lambda h}\, f(x,h) &= \langle \eta^D, f\rangle \overset{\text{law}}{=} \langle \eta^{\widetilde D}, f_\Phi\rangle \\
&= \int_{D\times\mathbb{R}} Z_\lambda^{\widetilde D}(\mathrm{d}x)\otimes \mathrm{d}h\, \mathrm{e}^{-\alpha\lambda h}\, f\bigl(x, h+\Phi^{D,\widetilde D}(x)\bigr) \qquad (4.24)\\
&= \int_{D\times\mathbb{R}} Z_\lambda^{\widetilde D}(\mathrm{d}x)\otimes \mathrm{d}h\, \mathrm{e}^{\alpha\lambda\Phi^{D,\widetilde D}(x)}\,\mathrm{e}^{-\alpha\lambda h}\, f(x,h).
\end{aligned}
$$

As noted at the beginning of the proof, this implies the claim. □

4.3 Representation via Gaussian Multiplicative Chaos

We will now show that the above properties determine the laws of $\{Z_\lambda^D : D \in \mathfrak{D}_0\}$ uniquely. To this end we restrict our attention to dyadic open squares, i.e., those of the form

$$2^{-n}z + (0, 2^{-n})^2 \quad \text{for} \quad z \in \mathbb{Z}^2 \text{ and } n \geq 0. \tag{4.25}$$

For a fixed $m \in \mathbb{Z}$, set $S := (0, 2^{-m})^2$ and let $\{S_{n,i} : i = 1, \dots, 4^n\}$ be an enumeration of the dyadic squares of side $2^{-(n+m)}$ that have a non-empty intersection with S. (See Fig. 6 for an illustration of this setting.) Recall that we assumed that $\mathfrak{D}_0$ contains all these squares which makes Z_λ^D defined on all of them.

Abbreviating

$$\widetilde S^n := \bigcup_{i=1}^{4^n} S_{n,i} \tag{4.26}$$

the Gibbs-Markov decomposition (4.16) gives

$$Z_\lambda^S(\mathrm{d}x) \overset{\text{law}}{=} \sum_{i=1}^{4^n} \mathrm{e}^{\alpha\lambda\Phi^{S,\widetilde S^n}(x)}\, Z_\lambda^{S_{n,i}}(\mathrm{d}x), \tag{4.27}$$

where the measures $\{Z_\lambda^{S_{n,i}} : i = 1, \dots, 4^n\}$ and the binding field $\Phi^{S,\widetilde S^n}$ on the right-hand side are regarded as independent. The expression (4.27) links Z_λ-measure in one set in terms of Z_λ-measures in a scaled version thereof. This suggests we think of (4.27) as a fixed point of a *smoothing transformation*; cf e.g., Durrett and Liggett [68]. Such fixed points are found by studying the variant of the fixed point equation where the object of interest on the right is replaced by its expectation. This leads to the consideration of the measure

$$Y_n^S(\mathrm{d}x) := \hat c \sum_{i=1}^{4^n} 1_{S_{n,i}}(x)\, \mathrm{e}^{\alpha\lambda\Phi^{S,\widetilde S^n}(x)}\, r_{\widetilde S^n}(x)^{2\lambda^2}\, \mathrm{d}x. \tag{4.28}$$

where $\hat c$ is as in (3.60). Indeed, in light of the fact that

$$r_{\tilde{S}^n}(x) = r_{S_{n,i}}(x) \quad \text{for} \quad x \in S_{n,i} \tag{4.29}$$

we have

$$Y_n^S(A) = E\big[Z_\lambda^S(A)\,\big|\,\sigma(\Phi^{S,\tilde{S}^n})\big] \tag{4.30}$$

for any Borel $A \subseteq \mathbb{C}$. The next point to observe is that these measures can be interpreted in terms of *Gaussian multiplicative chaos* (see Lemma 2.17):

Lemma 4.10 *There is a random measure Y_∞^S such that for all Borel $A \subseteq \mathbb{C}$,*

$$Y_n^S(A) \underset{n\to\infty}{\overset{\text{law}}{\longrightarrow}} Y_\infty^S(A). \tag{4.31}$$

Proof. Denoting $\tilde{S}^0 := S$, the nesting property of the binding field (Exercise 4.6) allows us to represent $\{\Phi^{S,\tilde{S}^n} : n \geq 1\}$ on the same probability space via

$$\Phi^{S,\tilde{S}^n} := \sum_{k=0}^{n-1} \Phi^{\tilde{S}^k,\tilde{S}^{k+1}}, \tag{4.32}$$

where the fields $\{\Phi^{\tilde{S}^k,\tilde{S}^{k+1}} : k \geq 0\}$ are independent with their corresponding laws. In this representation, the measures Y_n are defined all on the same probability space and so we can actually prove the stated convergence in almost-sure sense. Indeed, (1.41), (4.2) and $\alpha^2 g = 4$ imply

$$\widetilde{D} \subseteq D \quad \Rightarrow \quad r_{\widetilde{D}}(x)^{2\lambda^2} = r_D(x)^{2\lambda^2} \mathrm{e}^{\frac{1}{2}\alpha^2\lambda^2 \mathrm{Var}[\Phi^{D,\widetilde{D}}(x)]}, \quad x \in \widetilde{D}. \tag{4.33}$$

This permits us to rewrite (4.28) as

$$Y_n^S(\mathrm{d}x) := \hat{c}\, r_S(x)^{2\lambda^2} \sum_{i=1}^{4^n} 1_{S_{n,i}}(x)\, \mathrm{e}^{\alpha\lambda\Phi^{S,\tilde{S}^n}(x) - \frac{1}{2}\alpha^2\lambda^2 \mathrm{Var}[\Phi^{D,\widetilde{D}}(x)]}\, \mathrm{d}x \tag{4.34}$$

and thus cast Y_n^S in the form we encountered in the definition of the Gaussian Multiplicative Chaos. Adapting the proof of Lemma 2.17 (or using it directly with the help of Exercise 4.5), we get (4.31) any Borel $A \subseteq \mathbb{C}$. □

We now claim:

Proposition 4.11 (Characterization of Z_λ^D measure) *For any dyadic square $S \subseteq \mathbb{C}$ and any bounded and continuous function $f : S \to [0, \infty)$, we have*

$$E\big(\mathrm{e}^{-\langle Z_\lambda^S, f\rangle}\big) = E\big(\mathrm{e}^{-\langle Y_\infty^S, f\rangle}\big). \tag{4.35}$$

In particular,

$$Z_\lambda^S(\mathrm{d}x) \overset{\text{law}}{=} Y_\infty^S(\mathrm{d}x). \tag{4.36}$$

Proof of "≥" in (4.35). Writing Z_λ^S via (4.27) and invoking conditional expectation given $\Phi^{S,\tilde{S}^n}$ with the help of (4.30), the conditional Jensen inequality shows

$$
\begin{aligned}
E\big(\mathrm{e}^{-\langle Z_\lambda^S, f\rangle}\big) &= E\Big(E\big(\mathrm{e}^{-\langle Z_\lambda^S, f\rangle}\,\big|\,\sigma(\Phi^{S,\tilde{S}^n})\big)\Big)\\
&\ge E\big(\mathrm{e}^{-E[\langle Z_\lambda^S, f\rangle\,|\,\sigma(\Phi^{S,\tilde{S}^n})]}\big) = E\big(\mathrm{e}^{-\langle Y_n^S, f\rangle}\big)
\end{aligned}
\tag{4.37}
$$

for any continuous $f\colon S \to [0,\infty)$. The convergence in Lemma 4.10 implies

$$
E(\mathrm{e}^{-\langle Y_n^S, f\rangle}) \underset{n\to\infty}{\longrightarrow} E(\mathrm{e}^{-\langle Y_\infty^S, f\rangle})
\tag{4.38}
$$

and so we get "≥" in (4.35). □

For the proof of the opposite inequality in (4.35) we first note:

Lemma 4.12 (Reversed Jensen's inequality) *If $X_1, \dots, X_n$ are independent non-negative random variables, then for each $\epsilon > 0$,*

$$
E\left(\exp\left\{-\sum_{i=1}^n X_i\right\}\right) \le \exp\left\{-\mathrm{e}^{-\epsilon}\sum_{i=1}^n E(X_i\,;\,X_i \le \epsilon)\right\}.
\tag{4.39}
$$

Proof. In light of assumed independence, it suffices to prove this for $n = 1$. This is checked by bounding $E(\mathrm{e}^{-X}) \le E(\mathrm{e}^{-\widetilde{X}})$, where $\widetilde{X} := X1_{\{X\le\epsilon\}}$, writing

$$
-\log E(\mathrm{e}^{-\widetilde{X}}) = \int_0^1 \mathrm{d}s\,\frac{E(\widetilde{X}\mathrm{e}^{-s\widetilde{X}})}{E(\mathrm{e}^{-s\widetilde{X}})}
\tag{4.40}
$$

and invoking $E(\widetilde{X}\mathrm{e}^{-s\widetilde{X}}) \ge \mathrm{e}^{-\epsilon}E(\widetilde{X})$ and $E(\mathrm{e}^{-s\widetilde{X}}) \le 1$. □

We are now ready to give:

Proof of "≤" in (4.35). Pick n large and assume Z_λ^S is again represented via (4.27). We first invoke an additional truncation: Given $\delta > 0$, let $S_{n,i}^\delta$ be the translate of $(\delta 2^{-(n-m)}, (1-\delta)2^{-(n-m)})$ centered at the same point as $S_{n,i}$. Denote

$$
\tilde{S}_\delta^n := \bigcup_{i=1}^{4^n} S_{n,i}^\delta \quad\text{and}\quad f_{n,\delta}(x) := f(x)1_{\tilde{S}_\delta^n}(x).
\tag{4.41}
$$

Denoting also

$$
X_i := \int_{S_{n,i}} f_{n,\delta}(x)\,\mathrm{e}^{\alpha\lambda\Phi^{S,\tilde{S}^n}}\,Z_\lambda^{S_{n,i}}(\mathrm{d}x)
\tag{4.42}
$$

from $f \ge f_{n,\delta}$ we then have

$$E\left(\mathrm{e}^{-\langle Z_\lambda^S,f\rangle}\right)\le E\left(\mathrm{e}^{-\langle Z_\lambda^S,f_{n,\delta}\rangle}\right)=E\left(\exp\left\{-\sum_{i=1}^n X_i\right\}\right). \tag{4.43}$$

Conditioning on $\Phi^{S,\tilde S^n}$, the bound (4.39) yields

$$E\left(\mathrm{e}^{-\langle Z_\lambda^S,f\rangle}\right)\le E\left(\exp\left\{-\mathrm{e}^{-\epsilon}\sum_{i=1}^{4^n}E\big(X_i1_{\{X_i\le\epsilon\}}\,\big|\,\sigma(\Phi^{S,\tilde S^n})\big)\right\}\right). \tag{4.44}$$

Since (4.30) shows

$$\sum_{i=1}^{4^n}E\big(X_i\,\big|\,\sigma(\Phi^{S,\tilde S^n})\big)=\langle Y_n^S,\,f_{n,\delta}\rangle\,, \tag{4.45}$$

we will also need: □

Lemma 4.13 *Assume $\lambda\in(0,1/\sqrt{2})$. Then for each $\epsilon>0$,*

$$\lim_{n\to\infty}\sum_{i=1}^{4^n}E\big(X_i;\;X_i>\epsilon\big)=0\,. \tag{4.46}$$

Postponing the proof until after that of Proposition 4.11, from (4.44) and (4.46) we now get

$$E\left(\mathrm{e}^{-\langle Z_\lambda^S,f\rangle}\right)\le\limsup_{n\to\infty}E\big(\mathrm{e}^{-\mathrm{e}^{-\epsilon}\langle Y_n^S,f_{n,\delta}\rangle}\big)\,. \tag{4.47}$$

But

$$\langle Y_n^S,\,f_{n,\delta}\rangle\ge\langle Y_n^S,\,f\rangle-\|f\|\,Y_n^S(S\smallsetminus\tilde S_\delta^n) \tag{4.48}$$

and so

$$E\big(\mathrm{e}^{-\mathrm{e}^{-\epsilon}\langle Y_n^S,f_{n,\delta}\rangle}\big)\le\mathrm{e}^{\epsilon\|f\|}\,E\big(\mathrm{e}^{-\mathrm{e}^{-\epsilon}\langle Y_n^S,f\rangle}\big)+P\big(Y_n^S(S\smallsetminus\tilde S_\delta^n)>\epsilon\big)\,. \tag{4.49}$$

A calculation based on (4.34) shows

$$P\big(Y_n^S(S\smallsetminus\tilde S_\delta^n)>\epsilon\big)\le c\epsilon^{-1}\mathrm{Leb}(S\smallsetminus\tilde S_\delta^n)\le c'\epsilon^{-1}\delta. \tag{4.50}$$

Invoking also (4.31) we get

$$E\left(\mathrm{e}^{-\langle Z_\lambda^S,f\rangle}\right)\le\lim_{\epsilon\downarrow0}\,\limsup_{\delta\downarrow0}\,\limsup_{n\to\infty}E\big(\mathrm{e}^{-\mathrm{e}^{-\epsilon}\langle Y_n^S,f_{n,\delta}\rangle}\big)\le E\big(\mathrm{e}^{-\langle Y_\infty^S,f\rangle}\big). \tag{4.51}$$

This completes the proof of (4.35); (4.36) then directly follows. □

It remains to give:

Proof of Lemma 4.13. First we note

$$\sum_{i=1}^{4^n} E(X_i;\, X_i > \epsilon) \le \frac{1}{\epsilon}\sum_{i=1}^{4^n} E(X_i^2)$$
$$\le \frac{\|f\|^2}{\epsilon}\sum_{i=1}^{4^n} E\left(\int_{S_{n,i}^\delta\times S_{n,i}^\delta} E\bigl(\mathrm{e}^{\alpha\lambda[\Phi^{S,\tilde S^n}(x)+\Phi^{S,\tilde S^n}(y)]}\bigr) Z_\lambda^{S_{n,i}}(\mathrm{d}x) Z_\lambda^{S_{n,i}}(\mathrm{d}y)\right). \tag{4.52}$$

Denote $L := 2^n$. In light of the fact that, for some constant c independent of n,

$$\mathrm{Var}(\Phi^{S,\tilde S^n}(x)) = g\log\frac{r_S(x)}{r_{S_{n,i}}(x)} \le g\log(L) + c \tag{4.53}$$

holds uniformly in $x \in \tilde S^n_\delta$, (4.52) is bounded by $\|f\|^2/\epsilon^2$ times

$$c'\mathrm{e}^{4\frac{1}{2}\alpha^2\lambda^2 g\log(L)}\sum_{i=1}^{4^n} E\bigl[Z_\lambda^{S_{n,i}}(S_{n,i})^2\bigr] \le c'' L^{8\lambda^2+2-(4+4\lambda^2)} = c'' L^{-2(1-2\lambda^2)}, \tag{4.54}$$

where we also used $\alpha^2 g = 4$, invoked (3.62) and noted that there are L^2 terms in the sum. In light of $\lambda < 1/\sqrt{2}$, this tends to zero as $L \to \infty$. □

4.4 Finishing Touches

We are now ready to combine the above observations to get the main conclusion about the existence of the limit of processes $\{\eta_N^D : N \ge 1\}$:

Proof of Theorem 2.7 *for* $\lambda < 1/\sqrt{2}$. Pick $D \in \mathfrak{D}$ and assume, as discussed before, that $\mathfrak{D}_0$ used above contains D. For any subsequence of N's for which the limit of the measures in question exists for all domains in $\mathfrak{D}_0$ we then have the representation (3.47) as well as the properties stated in Exercise 4.8 and Proposition 4.9. We just have to show that the limit measure Z_λ^D is determined by these properties and the observation from Proposition 4.11. This will also prove the existence of the limit of $\{\eta_N^D : N \ge 1\}$.

By Exercise 4.8(1, 2) it suffices to show that the law of $\langle Z_\lambda^D, f\rangle$ is determined for any continuous $f : D \times \mathbb{R} \to \mathbb{R}$ with compact support. Let D^n be the union of all the open dyadic squares of side 2^{-n} entirely contained in D. Letting n be so large that $\mathrm{supp}(f) \subseteq D^n \times [-n, n]$ and denoting $\widetilde D^n := D \smallsetminus \partial D^n$, Proposition 4.9 and Exercise 4.8(3) then imply

$$\langle Z_\lambda^D, f\rangle \overset{\text{law}}{=} \bigl\langle Z_\lambda^{D^n}, \mathrm{e}^{\alpha\lambda\Phi^{D,\widetilde D^n}} f\bigr\rangle \quad\text{with}\quad Z_\lambda^{D^n} \perp\!\!\!\perp \Phi^{D,\widetilde D^n}. \tag{4.55}$$

It follows that the law of the left-hand side is determined once the law of $Z_\lambda^{D^n}$ is determined. But part (3) of Exercise 4.8 also shows (as in (4.27)) that the measure $Z_\lambda^{D^n}$ is the exponential of the continuum binding field times the sum of independent copies of Z_λ^S for S ranging over the dyadic squares constituting D^n. The laws of these Z_λ^S are determined uniquely by Proposition 4.11 and hence so are those of $Z_\lambda^{D^n}$ and Z_λ^D as well. □

Concerning the proof of conformal invariance and full characterization by the LQG measure, we will need the following result:

Theorem 4.14 (Uniqueness of the GMC/LQG measure) *The law of the Gaussian Multiplicative Chaos measure $\mu_\infty^{D,\beta}$ does not depend on the choice of the orthonormal basis in $\mathsf{H}_0^1(D)$ that was used to define it (see Lemma 2.17).*

We will not prove this theorem in these notes as that would take us on a tangent that we do not wish to follow. We remark that the result has a rather neat proof due to Shamov [113] which is made possible by his ingenious characterization of the GMC measures using Cameron-Martin shifts. An earlier work of Kahane [81] required uniform convergence of the covariances of the approximating fields; we state and prove this version in Theorem 5.5. This version also suffices to establish a conformal transformation rule for the limit (cf Exercise 5.7).

Equipped with the uniqueness claim in Theorem 4.14, let us now annotate the steps that identify Z_λ^D with the LQG-measure $\hat{c}r_D(x)^{2\lambda^2}\mu_\infty^{D,\lambda\alpha}(\mathrm{d}x)$. Let us start with a unit square S. For each $k \geq 0$, let $\{S_{k,i} : i = 1, \dots, n(k)\}$ be the collection of open dyadic squares of side 2^{-k} that are entirely contained in S. Denote

$$D^k := \bigcup_{j=1}^{n(k)} S_{k,j}, \quad k \geq 0, \tag{4.56}$$

with $D^{-1} := S$ and observe that $D^k \subseteq D^{k-1}$ for each $k \geq 0$. Observe that ∂D^k is a collection of horizontal and vertical lines. Note also that $\partial D^{k-1} \subseteq \partial D^k$.

Exercise 4.15 *For each $k \geq 0$, let H_k denote the subspace of functions in $\mathsf{H}_0^1(S)$ that are harmonic in D^k and vanish on ∂D^{k-1}. Prove that*

$$\mathsf{H}_0^1(S) = \bigoplus_{k=0}^{\infty} \mathsf{H}_k\,. \tag{4.57}$$

Next, for each $k \geq 1$, let $\{\widetilde{f}_{k,j} : j \geq 1\}$ be an orthonormal basis in H_k with respect to the Dirichlet inner product. Then show:

Exercise 4.16 *Prove that, for $\{X_{k,j} : k, j \geq 1\}$ i.i.d. standard normals,*

$$\Phi^{D^{k-1},D^k} \overset{\text{law}}{=} \sum_{j\geq 1} X_{k,j}\widetilde{f}_{k,j} \quad on\ D^k \tag{4.58}$$

holds for all $k \geq 1$ with the sums converging locally uniformly in D^k, a.s. Conclude that for all $m \geq 1$,

$$\Phi^{D,D^m} \overset{\text{law}}{=} \sum_{k=1}^{m} \sum_{j\ge1} X_{k,j} \tilde{f}_{k,j} \quad on\ D^m\,. \tag{4.59}$$

Hint: Define the $X_{k,j}$'s by suitable inner products and check the covariances.

From here and Theorem 4.14 we now infer the following:

Exercise 4.17 *Using suitable test functions, and a convenient enumeration of the above orthogonal basis in $\mathsf{H}_0^1(S)$, show that*

$$Z_\lambda^S(\mathrm{d}x) \overset{\text{law}}{=} Y_\infty^S(\mathrm{d}x) \overset{\text{law}}{=} \hat{c}\, r_S(x)^{2\lambda^2} \mu_\infty^{S,\lambda\alpha}(\mathrm{d}x). \tag{4.60}$$

Hint: Use either the argument based on Jensen's inequality from the previous section or invoke Kahane's convexity inequality from Proposition 5.6.

The point here is that although the right-hand side (4.58) casts the binding field from D to D^m in the form akin to (2.41), the sum over j is infinite. One thus has to see that a suitable truncation to a finite sum will do as well.

Exercise 4.17 identifies Z_λ^D with the LQG measure for D a dyadic square. To extend this to general domains, which can be reduced to dyadic squares using the Gibbs-Markov property, we also need to solve:

Exercise 4.18 *(Gibbs-Markov for LQG measure) Suppose that $\widetilde{D} \subseteq D$ obey* $\mathrm{Leb}(D \smallsetminus \widetilde{D}) = 0$. *Use* (4.33) *to prove that, for all $\beta \in (0, \alpha)$,*

$$r_D(x)^{2(\beta/\alpha)^2} \mu_\infty^{D,\beta}(\mathrm{d}x) \overset{\text{law}}{=} \mathrm{e}^{\beta\Phi^{D,\widetilde{D}}(x)}\, r_{\widetilde{D}}(x)^{2(\beta/\alpha)^2} \mu_\infty^{\widetilde{D},\beta}(\mathrm{d}x)\,, \tag{4.61}$$

where $\Phi^{D,\widetilde{D}}$ and $\mu_\infty^{\widetilde{D},\beta}$ are regarded as independent.

We remark that, in the regime $\lambda < 1/\sqrt{2}$, formula (4.61) would be enough to complete (4.60); indeed, our Proposition 4.11 only required the Gibbs-Markov property, the expectation formula (3.61) and the bounds on the second moment in (3.62)–(3.63), which are known to hold for the LQG measure as well. However, this does not apply in the complementary regime of λ's where we use a truncation on the underlying field that is hard to interpret in the language of the above random measures.

Once we identify the limit measure with the LQG-measure—and accept Theorem 4.14 without proof—the proof of the conformal transformation rule in Theorem 2.14 is quite easy. A key point is to solve:

Exercise 4.19 *(Conformal transform of GMC measure) Let $f\colon D \mapsto f(D)$ be a conformal map between bounded and open domains in $\mathbb{C}$. Show that if $\{g_n\colon n \ge 1\}$ is an orthonormal basis in $\mathsf{H}_0^1(D)$ with respect to the Dirichlet inner product, then $\{g_n \circ f^{-1}\colon n \ge 1\}$ is similarly an orthonormal basis in $\mathsf{H}_0^1(f(D))$. Prove that this implies*

$$\mu_\infty^{f(D),\beta} \circ f(\mathrm{d}x) \overset{\text{law}}{=} \left|f'(x)\right|^2 \mu_\infty^{D,\beta}(\mathrm{d}x). \tag{4.62}$$

To get the proof of Theorem 2.14, one then needs to observe:

Exercise 4.20 *For any conformal bijection* $f: D \to f(D)$,

$$r_{f(D)}\big(f(x)\big) = \big|f'(x)\big|\, r_D(x), \qquad x \in D. \tag{4.63}$$

We reiterate that a proof of conformal invariance avoiding the full strength of Theorem 4.14 will be posed as Exercise 5.7.

4.5 *Dealing with Truncations*

The above completes the proof of our results in the regime where second-moment calculations can be applied without truncations. To get some feeling for what happens in the complementary regime, $1/\sqrt{2} \le \lambda < 1$, let us at least introduce the basic definitions and annotate the relevant steps.

Denote $\Lambda_r(x) := \{y \in \mathbb{Z}^d : \operatorname{dist}_\infty(x, y) \le r\}$. Given a discretized version D_N of a continuum domain D, for each $x \in D_N$ define

$$\Delta^k(x) := \begin{cases} \emptyset & \text{for } k = 0\,, \\ \Lambda_{e^k}(x) & \text{for } k = 1, \dots, n(x) - 1\,, \\ D_N & \text{for } k = n(x)\,, \end{cases} \tag{4.64}$$

where $n(x) := \max\{n \ge 0 : \Lambda_{e^{n+1}}(x) \subseteq D_N\}$. See Fig. 9 for an illustration.

Using the definition of $\varphi^{D_N, \Delta^k}(x)$ as a conditional expectation of h^{D_N}, we now define the truncation event

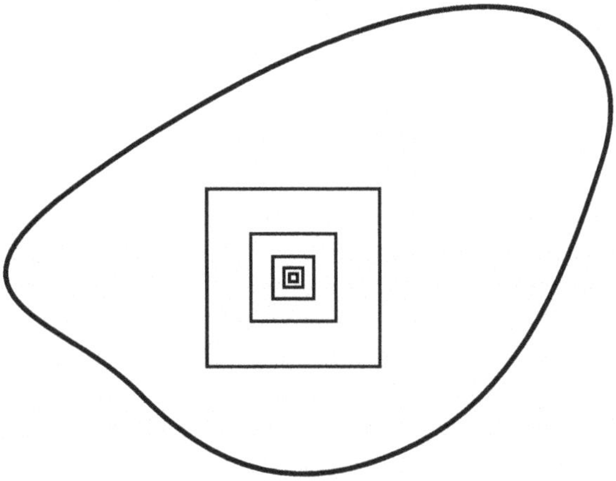

Fig. 9 An illustration of the collection of sets $\Delta^k(x)$ above. The domain D_N corresponds to the region marked by the outer curve

$$T_{N,M}(x) := \bigcap_{k=k_N}^{n(x)} \left\{ \left| \varphi_x^{D_N,\Delta^k} - a_N \frac{n(x)-k}{n(x)} \right| \le M\big[n(x)-k\big]^{3/4} \right\}, \tag{4.65}$$

where M is a parameter and $k_N := \frac{1}{8}\log(K_N) \approx \frac{1}{4}(1-\lambda^2)\log N$. Next we introduce the *truncated point measure*

$$\widehat{\eta}_N^{D,M} := \frac{1}{K_N} \sum_{x\in D_N} 1_{T_{N,M}(x)}\, \delta_{x/N} \otimes \delta_{h_x^{D_N}-a_N}. \tag{4.66}$$

The following elementary inequality will be quite useful:

$$\eta_N^D \ge \widehat{\eta}_N^{D,M}, \qquad M \in (0,\infty). \tag{4.67}$$

Indeed, the tightness of $\{\widehat{\eta}_N^D \colon N \ge 1\}$ is thus inherited from $\{\eta_N^D \colon N \ge 1\}$ and, as can be shown, the limit points of the former increase to those of the latter as $M \to \infty$. The requisite (now really ugly) second moment calculations are then performed which yield the following conclusions for all $M < \infty$ and all $\lambda \in (0,1)$:

(1) Defining

$$\widehat{\Gamma}_N^{D,M}(b) := \big\{x \in D_N \colon h_x^{D_N} \ge a_N + b,\ T_{N,M}(x) \text{ occurs}\big\}, \tag{4.68}$$

we have

$$\sup_{N\ge1} \frac{1}{K_N^2} E\big(|\widehat{\Gamma}_N^{D,M}(b)|^2\big) < \infty. \tag{4.69}$$

By a second-moment argument, the limits of $\{\widehat{\eta}_N^{D,M} \colon N \ge 1\}$ are non-trivial.

(2) The factorization property proved in Proposition 3.13 applies to limit points of $\{\widehat{\eta}_N^{D,M} \colon N \ge 1\}$ with Z_λ^D replaced by some $\widehat{Z}_\lambda^{D,M}$ instead.

The property (4.67) now implies that $M \mapsto \widehat{Z}_\lambda^{D,M}$ is pointwise increasing and so we may define

$$Z_\lambda^D(\cdot) := \lim_{M\to\infty} \widehat{Z}_\lambda^{D,M}(\cdot). \tag{4.70}$$

We then check that this measure has the properties in Exercise 4.8 as well as the Gibbs-Markov property from Proposition 4.9. However, although the limit in (4.70) exists in L^1 for all $\lambda \in (0,1)$, it does not exist in L^2 for $\lambda \ge 1/\sqrt{2}$—because, e.g., by Fatou's Lemma, the limit LQG measure would then be in L^2 as well which is known to be false—and so we have to keep using $\widehat{Z}_\lambda^{D,M}$ whenever estimates involving second moments are needed.

This comes up only in one proof: "$\le$" in (4.35). There we use the inequality $Z_\lambda^D(\cdot) \ge \widehat{Z}_\lambda^{D,M}(\cdot)$ and the fact that Z_λ^D satisfies the Gibbs-Markov property to dominate $Z_\lambda^D(\cdot)$ from below by the measure

$$\widetilde{Z}^S_\lambda(\mathrm{d}x) := \sum_{i=1}^{4^n} \mathrm{e}^{\alpha\lambda\Phi^{S,\widetilde{S}^n}} \widehat{Z}^{S_{n,i},M}_\lambda(\mathrm{d}x). \tag{4.71}$$

Then we perform the calculation after (4.27) with this measure instead of Z^S_λ modulo one change: In the proof of Lemma 4.13 we truncate to the event

$$\sup_{x\in\widetilde{S}^n_\delta} \Phi^{S,\widetilde{S}^n}(x) < 2\sqrt{g}\log(2^n) + c\sqrt{\log(2^n)}\,, \tag{4.72}$$

which has probability very close to one. On this event, writing again $L := 2^n$, the sum on the right-hand side of (4.52) is thus bounded by

$$c'\mathrm{e}^{2\sqrt{g}\alpha\lambda\log(L)+c\sqrt{\log(L)}}\,\mathrm{e}^{\frac{1}{2}\alpha^2\lambda^2 g\log(L)}L^2L^{-2(2+2\lambda^2)}. \tag{4.73}$$

Using the definition of α, this becomes $L^{-2(1-\lambda)^2+o(1)}$ which vanishes as $n\to\infty$ for all $\lambda\in(0,1)$. The rest of the proof is then more or less the same.

The full proof of Theorem 2.7 is carried out in [26] to which we refer the reader for further details.

Lecture 5: Gaussian Comparison Inequalities

In our discussion we have so far managed to get by using only elementary facts about Gaussian processes. The forthcoming derivations will require more sophisticated techniques and so it is time we addressed them properly. In this lecture we focus on Gaussian comparison inequalities, starting with Kahane's inequality and its corollaries called the Slepian Lemma and the Sudakov-Fernique inequality. We give an application of Kahane's inequality to uniqueness of the Gaussian Multiplicative Chaos, a subject touched upon before. In the last section, we will review the concepts of stochastic domination and the FKG inequality. The presentation draws on Adler [5], Adler and Taylor [6] and Liggett [88].

5.1 Kahane's Inequality

Let us first make a short note on terminology: We say that a (multivariate) Gaussian $X=(X_1,\dots,X_n)$ is *centered* if $E(X_i)=0$ for all $i=1,\dots,n$. A function $f\colon\mathbb{R}^n\to\mathbb{R}$ is said to have a *subgaussian growth* if for each $\epsilon>0$ there is $C>0$ such that $|f(x)|\le C\mathrm{e}^{\epsilon|x|^2}$ holds for all $x\in\mathbb{R}^n$.

In his development of the theory of Gaussian Multiplicative Chaos, Kahane made convenient use of inequalities that, generally, give comparison estimates of expectation of functions (usually convex in appropriate sense) of Gaussian random variables whose covariances can be compared in a pointwise sense. One version of this inequality is as follows:

Proposition 5.1 (Kahane inequality) *Let X, Y be centered Gaussian vectors on $\mathbb{R}^n$ and let $f \in C^2(\mathbb{R}^n)$ be a function whose second derivatives have a subgaussian growth. Assume*

$$\forall i,j=1,\dots,n: \quad \begin{cases} E(Y_iY_j) > E(X_iX_j) \ \Rightarrow \ \dfrac{\partial f}{\partial x_i \partial x_j}(x) \ge 0, \quad x \in \mathbb{R}^n, \\ E(Y_iY_j) < E(X_iX_j) \ \Rightarrow \ \dfrac{\partial f}{\partial x_i \partial x_j}(x) \le 0, \quad x \in \mathbb{R}^n. \end{cases} \tag{5.1}$$

Then

$$Ef(Y) \ge Ef(X). \tag{5.2}$$

Note also that for pairs i, j such that $E(Y_iY_j) = E(X_iX_j)$ the sign of $\frac{\partial f}{\partial x_i x_j}$ is not constrained. For the proof we will need the following standard fact:

Lemma 5.2 (Gaussian integration by parts) *Let X be a centered Gaussian vector on $\mathbb{R}^n$ and suppose $f \in C^1(\mathbb{R}^n)$ is such that $f, \nabla f$ have subgaussian growth. Then for each $i = 1, \dots, n$,*

$$\operatorname{Cov}\bigl(f(X), X_i\bigr) = \sum_{j=1}^n \operatorname{Cov}(X_i, X_j) E\left(\frac{\partial f}{\partial x_j}(X)\right). \tag{5.3}$$

Exercise 5.3 *Prove Lemma 5.2. Hint: The proof is an actual integration by parts for X one-dimensional. For the general case use the positive definiteness of the covariance to find an $n \times n$ matrix A such that $X = AZ$ for Z i.i.d. $\mathcal{N}(0, 1)$. Then apply the one-dimensional results to each coordinate of Z.*

Gaussian integration can also be proved by approximating f by polynomials and invoking the following identity:

Exercise 5.4 *(Wick pairing formula) Let $(X_1, \dots, X_{2n})$ be a centered multivariate Gaussian (with some variables possibly repeating). Show that*

$$E(X_1 \dots X_{2n}) = \sum_{\pi:\, pairing} \prod_{i=1}^n \operatorname{Cov}\bigl(X_{\pi_1(i)} X_{\pi_2(i)}\bigr), \tag{5.4}$$

where a "pairing" is a pair $\pi = (\pi_1, \pi_2)$ of functions $\pi_1, \pi_2 \colon \{1, \dots, n\} \to \{1, \dots, 2n\}$ such that

$$\pi_1(i) < \pi_2(i), \quad i = 1, \dots, n, \tag{5.5}$$

$$\pi_1(1) < \pi_1(2) \dots < \pi_1(n) \tag{5.6}$$

and

$$\{1, \ldots, 2n\} = \bigcup_{i=1}^{n} \{\pi_1(i), \pi_2(i)\}. \tag{5.7}$$

(Note that these force $\pi_1(1) = 1$.)

The pairing formula plays an important role in computations involving Gaussian fields; in fact, it is the basis of perturbative calculations of functionals of Gaussian processes and their organization in terms of Feynman diagrams.
Proof of Proposition 5.1. Suppose that X and Y are realized on the same probability space so that $X \perp\!\!\!\perp Y$. Consider the Gaussian interpolation $Z_t := \sqrt{1-t^2}\,X + tY$, $t \in [0, 1]$. Then $Z_0 = X$ and $Z_1 = Y$ and so

$$Ef(Y) - Ef(X) = \int_0^1 \frac{\mathrm{d}}{\mathrm{d}t} Ef(Z_t)\,\mathrm{d}t. \tag{5.8}$$

Using elementary calculus along with Lemma 5.2,

$$\begin{aligned} \frac{\mathrm{d}}{\mathrm{d}t} Ef(Z_t) &= \sum_{i=1}^{n} E\left(\left(\frac{-t}{\sqrt{1-t^2}} X_i + Y_i\right) \frac{\partial f}{\partial x_i}(Z_t)\right) \\ &= t \sum_{i,j=1}^{n} E\left(\bigl[E(Y_iY_j) - E(X_iX_j)\bigr] \frac{\partial^2 f}{\partial x_i \partial x_j}(Z_t)\right). \end{aligned} \tag{5.9}$$

Based on our assumptions, the expression under the expectation is non-negative for every realization of Z_t. Using this in (5.8) yields the claim. □

5.2 Kahane's Theory of Gaussian Multiplicative Chaos

We will find Proposition 5.1 useful later but Kahane's specific interest in Gaussian Multiplicative Chaos actually required a version that is not directly obtained from the one above. Let us recall the setting more closely.

Let $D \subseteq \mathbb{R}^d$ be a bounded open set and let ν be a finite Borel measure on D. Assume that $C\colon D \times D \to \mathbb{R} \cup \{\infty\}$ is a symmetric, positive semi-definite kernel in $L^2(\nu)$ which means that

$$\int_{D\times D} \nu(\mathrm{d}x) \otimes \nu(\mathrm{d}y)\, C(x, y) f(y) f(x) \ge 0 \tag{5.10}$$

holds for every bounded measurable $f\colon D \to \mathbb{R}$. If C is finite everywhere, then one can define a Gaussian process $\varphi = \mathcal{N}(0, C)$. Our interest is, however, in the situation when C is allowed to diverge on the diagonal $\{(x, x)\colon x \in D\} \subseteq D \times D$ which means that the Gaussian process exists only in a generalized sense—e.g., as a random distribution on a suitable space of test functions.

We will not try to specify the conditions on C that would make this setting fully meaningful; instead, we will just assume that C can be written as

$$C(x,y)=\sum_{k=1}^{\infty} C_k(x,y), \qquad x,y\in D, \tag{5.11}$$

where C_k is a continuous (and thus finite) covariance kernel for each k and the sum converges pointwise everywhere (including, possibly, to infinity when $x=y$). We then consider the Gaussian processes

$$\varphi_k=\mathcal{N}(0,C_k) \quad \text{with} \quad \{\varphi_k\colon k\ge 1\} \text{ independent}. \tag{5.12}$$

Letting

$$\varPhi_n(x):=\sum_{k=1}^{n}\varphi_k(x) \tag{5.13}$$

we define

$$\mu_n(\mathrm{d}x):=\mathrm{e}^{\varPhi_n(x)-\frac12\mathrm{Var}[\varPhi_n(x)]}\nu(\mathrm{d}x). \tag{5.14}$$

Lemma 2.17 (or rather its proof) gives the existence of a random Borel measure μ_∞ such that for each $A\subseteq D$ Borel,

$$\mu_n(A)\underset{n\to\infty}{\longrightarrow}\mu_\infty(A) \quad \text{a.s.} \tag{5.15}$$

As the covariances $\mathrm{Cov}(\varPhi_n(x),\varPhi_n(y))$ converge to $C(x,y)$, we take μ_∞ as our interpretation of the measure

$$\text{“ } \mathrm{e}^{\varPhi_\infty(x)-\frac12\mathrm{Var}[\varPhi_\infty(x)]}\nu(\mathrm{d}x) \text{ ”} \tag{5.16}$$

for $\varPhi_\infty$ being the centered generalized Gaussian field with covariance C. A key problem that Kahane had to deal with was the dependence of the limit measure on the above construction, and the uniqueness of the law of μ_∞ in general. This is, at least partially, resolved in:

Theorem 5.5 (Kahane's Uniqueness Theorem) *For $D\subseteq\mathbb{R}^d$ bounded and open, suppose there are covariance kernels $C_k,\widetilde{C}_k\colon D\times D\to\mathbb{R}$ such that*

(1) both C_k and $\widetilde{C}_k$ is continuous and non-negative everywhere on $D\times D$,
(2) for each $x,y\in D$,

$$\sum_{k=1}^{\infty}C_k(x,y)=\sum_{k=1}^{\infty}\widetilde{C}_k(x,y) \tag{5.17}$$

with both sums possibly simultaneously infinite, and

(3) *the fields* $\varphi_k = \mathcal{N}(0, C_k)$ *and* $\widetilde{\varphi}_k = \mathcal{N}(0, \widetilde{C}_k)$, *with* $\{\varphi_k, \widetilde{\varphi}_k \colon k \geq 1\}$ *all independent of one-another, have versions with continuous paths for each* $k \geq 1$.

Define, via (5.13)–(5.15), the random measures μ_∞ *and* $\widetilde{\mu}_\infty$ *associated with these fields. Then*

$$\mu_\infty(\mathrm{d}x) \overset{\text{law}}{=} \widetilde{\mu}_\infty(\mathrm{d}x). \tag{5.18}$$

In order to prove Theorem 5.5 we will need the following variation on Proposition 5.1 which lies at the heart of Kahane's theory:

Proposition 5.6 (Kahane's convexity inequality) *Let* $D \subseteq \mathbb{R}^n$ *be bounded and open and let* ν *be a finite Borel measure on* D. *Let* $C, \widetilde{C} \colon D \times D \to \mathbb{R}$ *be covariance kernels such that* $\varphi = \mathcal{N}(0, C)$ *and* $\widetilde{\varphi} = \mathcal{N}(0, \widetilde{C})$ *have continuous paths a.s. If*

$$\widetilde{C}(x, y) \geq C(x, y), \quad x, y \in D, \tag{5.19}$$

then for each convex $f \colon [0, \infty) \to \mathbb{R}$ *with at most polynomial growth at infinity,*

$$E\, f\left(\int_D \mathrm{e}^{\widetilde{\varphi}(x) - \frac{1}{2}\mathrm{Var}[\widetilde{\varphi}(x)]}\nu(\mathrm{d}x)\right) \geq E\, f\left(\int_D \mathrm{e}^{\varphi(x) - \frac{1}{2}\mathrm{Var}[\varphi(x)]}\nu(\mathrm{d}x)\right). \tag{5.20}$$

Proof. By approximation we may assume that $f \in C^2(\mathbb{R})$ (still convex). By the assumption of the continuity of the fields, it suffices to prove this for ν being the sum of a finite number of point masses, $\nu = \sum_{i=1}^n p_i \delta_{x_i}$ where $p_i > 0$. (The general case then follows by the weak limit of such measures to ν.)

Assume that the fields φ and $\widetilde{\varphi}$ are realized on the same probability space so that $\varphi \perp\!\!\!\perp \widetilde{\varphi}$. Consider the interpolated field

$$\varphi_t(x) := \sqrt{1 - t^2}\,\varphi(x) + t\widetilde{\varphi}(x), \qquad t \in [0, 1]. \tag{5.21}$$

Since $\varphi_0(x) = \varphi(x)$ and $\varphi_1(x) = \widetilde{\varphi}(x)$, it suffices to show

$$\frac{\mathrm{d}}{\mathrm{d}t} E\, f\left(\sum_{i=1}^n p_i \mathrm{e}^{\varphi_t(x_i) - \frac{1}{2}\mathrm{Var}[\varphi_t(x_i)]}\right) \geq 0\,. \tag{5.22}$$

For this we abbreviate $W_t(x) := \mathrm{e}^{\varphi_t(x) - \frac{1}{2}\mathrm{Var}[\varphi_t(x)]}$ and use elementary calculus to get

$$\begin{aligned} \frac{\mathrm{d}}{\mathrm{d}t} E\, f\Bigl(\sum_{i=1}^n p_i W_t(x_i)\Bigr) = \sum_{i=1}^n p_i E\biggl(\Bigl[-\frac{t}{\sqrt{1-t^2}}\varphi(x_i) + \widetilde{\varphi}(x_i) \\ + t\mathrm{Var}\bigl(\varphi(x_i)\bigr) - t\mathrm{Var}\bigl(\widetilde{\varphi}(x_i)\bigr)\Bigr] W_t(x_i) f'(\cdots)\biggr) \end{aligned} \tag{5.23}$$

Next we integrate by parts (cf Lemma 5.2) the terms involving $\varphi(x_i)$, which results in the $\varphi(x_j)$-derivative of $W_t(x_i)$ or $f(\cdots)$. A similar process is applied to the term $\widetilde{\varphi}(x_i)$. A key point is that the contribution from differentiating $W_t(x_i)$ exactly cancels that coming from the variances. Hence we get

$$\frac{\mathrm{d}}{\mathrm{d}t} E f\left(\sum_{i=1}^n p_i W_t(x_i)\right) = \sum_{i,j=1}^n p_i p_j \big[\widetilde{C}(x_i, x_j) - C(x_i, x_j)\big] E\Big(W_t(x_i) W_t(x_j) f''(\cdots)\Big). \tag{5.24}$$

As $f'' \geq 0$ by assumption and $W_t(x) \geq 0$ and $p_i, p_j \geq 0$ by inspection, (5.19) indeed implies (5.22). The claim follows by integration over t. □

This permits us to give:

Proof of Theorem 5.5. The claim (5.18) will follow once we show

$$\int g(x)\mu_\infty(\mathrm{d}x) \overset{\text{law}}{=} \int g(x)\widetilde{\mu}_\infty(\mathrm{d}x) \tag{5.25}$$

for any continuous function $g\colon D \to [0, \infty)$ supported in a compact set $A \subseteq D$. Let $\{C_k : k \geq 1\}$ and $\{\widetilde{C}_k : k \geq 1\}$ be the covariances in the statement. We now invoke a reasoning underlying the proof of Dini's Theorem: For each $\epsilon > 0$ and each $n \in \mathbb{N}$, there is $m \in \mathbb{N}$ such that

$$\sum_{k=1}^n C(x, y) < \epsilon + \sum_{k=1}^m \widetilde{C}(x, y), \qquad x, y \in A. \tag{5.26}$$

Indeed, fix $n \in \mathbb{N}$ and let F_m be the set of pairs $(x, y) \in A \times A$ where (5.26) fails. Then F_m is closed (and thus compact) by the continuity of the covariances; their non-negativity in turn shows that $m \mapsto F_m$ is decreasing with respect to set inclusion. The equality (5.17) translates into $\bigcap_{m \geq 1} F_m = \emptyset$ and so, by Heine–Borel, we must have that $F_m = \emptyset$ for m large enough thus giving us (5.26).

Interpreting the ϵ term on the right-hand side of (5.26) as the variance of the random variable $Z_\epsilon = \mathcal{N}(0, \epsilon)$ that is independent of $\widetilde{\varphi}$, Proposition 5.6 with the choice $f(x) := \mathrm{e}^{-\lambda x}$ for some $\lambda \geq 0$ gives us

$$E\left(\mathrm{e}^{-\lambda \mathrm{e}^{Z_\epsilon - \epsilon/2} \int g \,\mathrm{d}\widetilde{\mu}_m}\right) \geq E\left(\mathrm{e}^{-\lambda \int g \,\mathrm{d}\mu_n}\right). \tag{5.27}$$

Invoking the limit (5.15) and taking $\epsilon \downarrow 0$ afterwards yields

$$E\left(\mathrm{e}^{-\lambda \int g \,\mathrm{d}\widetilde{\mu}_\infty}\right) \geq E\left(\mathrm{e}^{-\lambda \int g \,\mathrm{d}\mu_\infty}\right). \tag{5.28}$$

Swapping C and $\widetilde{C}$ shows that equality holds in (5.28) and since this is true for every $\lambda \geq 0$, we get (5.25) as desired. □

As far as the GMC associated with the two-dimensional Gaussian Free Field is concerned, Theorem 5.5 shows that any decomposition of the continuum Green function $\widehat{G}^D(x, y)$ into the sum of positive covariance kernels will yield, through the construction in Lemma 2.17, the same limiting measure $\mu_\infty^{D,\beta}$. One example of such a decomposition is that induced by the Gibbs-Markov property upon reductions to a subdomain; this is the content of Exercises 4.15–4.17. Another example is the *white noise decomposition* that will be explained in Sect. 10.5.

We remark that (as noted above) uniqueness of the GMC measure has now been proved in a completely general setting by Shamov [113]. Still, Kahane's approach is sufficient to prove conformal transformation rule for the Z^D_λ-measures discussed earlier in these notes:

Exercise 5.7 *Prove that* $\{Z^D_\lambda\colon D \in \mathfrak{D}\}$ *obey the conformal transformation rule stated in Theorem 2.14. Unlike Exercise 4.19, do not assume the full uniqueness stated in Theorem 4.14.*

5.3 Comparisons for the Maximum

Our next task is to use Kahane's inequality from Proposition 5.1 to provide comparisons between the maxima of two Gaussian vectors with point-wise ordered covariances. We begin with a corollary to Proposition 5.1:

Corollary 5.8 *Suppose that X and Y are centered Gaussians on $\mathbb{R}^n$ such that*

$$E(X_i^2) = E(Y_i^2), \qquad i = 1, \dots, n \tag{5.29}$$

and

$$E(X_i X_j) \le E(Y_i Y_j), \qquad i, j = 1, \dots, n. \tag{5.30}$$

Then for any $t_1, \dots, t_n \in \mathbb{R}$,

$$P\bigl(X_i \le t_i : i = 1, \dots, n\bigr) \le P\bigl(Y_i \le t_i : i = 1, \dots, n\bigr). \tag{5.31}$$

Proof. Consider any collection $g_1, \dots, g_n \colon \mathbb{R} \to \mathbb{R}$ of non-negative bounded functions that are smooth and non-increasing. Define

$$f(x_1, \dots, x_n) := \prod_{i=1}^n g_i(x_i). \tag{5.32}$$

Then $\frac{\partial^2 f}{\partial x_i \partial x_j} \ge 0$ for each $i \ne j$. Hence, by Proposition 5.1, conditions (5.29)–(5.30) imply $Ef(Y) \ge Ef(X)$. The claim follows by letting g_i decrease to $1_{(-\infty,t_i]}$. □

From here we now immediately get:

Corollary 5.9 (Slepian's lemma) *Suppose X and Y are centered Gaussians on $\mathbb{R}^n$ with*

$$E(X_i^2) = E(Y_i^2), \qquad i = 1, \dots, n \tag{5.33}$$

and

$$E\big((X_i - X_j)^2\big) \le E\big((Y_i - Y_j)^2\big), \qquad i, j = 1, \ldots, n. \tag{5.34}$$

Then for each $t \in \mathbb{R}$,

$$P\Big(\max_{i=1,\ldots,n} X_i > t\Big) \le P\Big(\max_{i=1,\ldots,n} Y_i > t\Big). \tag{5.35}$$

Proof. Set $t_1 = \cdots = t_n := t$ in the previous corollary. □

Slepian's lemma (proved originally in [118]) has a nice verbal formulation using the following concept: Given a Gaussian process $\{X_t : t \in T\}$ on a set T,

$$\rho_X(t, s) := \sqrt{E\big((X_t - X_s)^2\big)} \tag{5.36}$$

defines a pseudometric on T. Indeed, we pose:

Exercise 5.10 *Verify that ρ_X is indeed a pseudo-metric on T.*

Disregarding the prefix "pseudo", we will call ρ_X the *canonical*, or *intrinsic*, metric associated with Gaussian processes. Slepian's lemma may then be verbalized as follows: *For two Gaussian processes with equal variances, the one with larger intrinsic distances has a stochastically larger maximum.*

The requirement of equal variances is often too much to ask for. One way to compensate for an inequality there is by adding suitable independent Gaussians to X and Y. However, it turns out that this inconvenience disappears altogether if we contend ourselves with the comparison of expectations only (which is, by way of integration, implied by (5.35)):

Proposition 5.11 (Sudakov-Fernique inequality) *Suppose that X and Y are centered Gaussians in $\mathbb{R}^n$ such that*

$$E\big((X_i - X_j)^2\big) \le E\big((Y_i - Y_j)^2\big), \qquad i, j = 1, \ldots, n. \tag{5.37}$$

Then

$$E\big(\max_{i=1,\ldots,n} X_i\big) \le E\big(\max_{i=1,\ldots,n} Y_i\big) \tag{5.38}$$

Proof. Consider the function

$$f_\beta(x_1, \ldots, x_n) := \frac{1}{\beta} \log\left(\sum_{i=1}^n \mathrm{e}^{\beta x_i}\right). \tag{5.39}$$

For readers familiar with statistical mechanics, f_β can be thought of as a free energy. Hölder's inequality implies that $x \mapsto f_\beta(x)$ is convex. In addition, we also get

$$\lim_{\beta\to\infty} f_\beta(x) = \max_{i=1,\dots,n} x_i. \tag{5.40}$$

Using Dominated Convergence, it therefore suffices to show that

$$Ef_\beta(X) \le Ef_\beta(Y), \qquad \beta \in (0,\infty). \tag{5.41}$$

The proof of this inequality will be based on a re-run of the proof of Kahane's inequality. Assuming again $X \perp\!\!\!\perp Y$ and letting $Z_t := \sqrt{1-t^2}\,X + tY$, differentiation yields

$$\frac{\mathrm{d}}{\mathrm{d}t}Ef_\beta(Z_t) = t\sum_{i,j=1}^n E\left(\big[E(Y_iY_j) - E(X_iX_j)\big]\frac{\partial^2 f_\beta}{\partial x_i \partial x_j}(Z_t)\right). \tag{5.42}$$

Now

$$\frac{\partial f_\beta}{\partial x_i} = \frac{\mathrm{e}^{\beta x_i}}{\sum_{j=1}^n \mathrm{e}^{\beta x_j}} =: p_i(x) \tag{5.43}$$

where $p_i \ge 0$ with $\sum_{i=1}^n p_i(x) = 1$. For the second derivatives we get

$$\frac{\partial^2 f_\beta}{\partial x_i \partial x_j} = \beta\big[p_i(x)\delta_{ij} - p_i(x)p_j(x)\big]. \tag{5.44}$$

Plugging this on the right of (5.42) (and omitting the argument Z_t of the second derivative as well as the p_i's) we then observe

$$\begin{aligned}
&\sum_{i,j=1}^n \big[E(Y_iY_j) - E(X_iX_j)\big]\frac{\partial^2 f_\beta}{\partial x_i \partial x_j} \\
&\qquad = \beta\sum_{i,j=1}^n \big[E(Y_iY_j) - E(X_iX_j)\big]\big[p_i\delta_{ij} - p_ip_j\big] \\
&\qquad = \beta\sum_{i,j=1}^n \big[E(Y_i^2) + E(X_i^2)\big]p_ip_j + \beta\sum_{i,j=1}^n \big[E(Y_iY_j) - E(X_iX_j)\big]p_ip_j \\
&\qquad = \frac{1}{2}\beta\sum_{i,j=1}^n \Big[E\big((Y_i - Y_j)^2\big) - E\big((X_i - X_j)^2\big)\Big]p_ip_j,
\end{aligned} \tag{5.45}$$

where we used that $\{p_i : i = 1,\dots,n\}$ are probabilities in the second line and then symmetrized the first sum under the exchange of i for j to wrap the various parts into the expression in the third line. Invoking (5.37), this is non-negative (pointwise) and so we get (5.41) by integration. The claim follows. □

The Sudakov-Fernique inequality may be verbalized as follows: *For two Gaussian processes, the one with larger intrinsic distances has a larger expected maximum.* Here are some other, rather elementary, facts related to the same setting:

Exercise 5.12 *Show that, for any centered Gaussians $X_1, \dots, X_n$,*

$$E\bigl(\max_{i=1,\dots,n} X_i\bigr) \geq 0. \tag{5.46}$$

Prove that equality occurs if and only if $X_i = X_1$ for all $i = 1, \dots, n$ a.s.

Exercise 5.13 *Suppose that X, resp., Y are centered Gaussian vectors on $\mathbb{R}^n$ with covariances C, resp., $\widetilde{C}$. Show that if $\widetilde{C} - C$ is positive semi-definite, then* (5.38) *holds.*

We will use the Sudakov-Fernique inequality a number of times in these notes. Unfortunately, the comparison between the expected maxima is often insufficient for the required level of precision; indeed, (5.38) only tells us that $\max_i X_i$ is not larger than a *large multiple* of $E \max_i Y_i$ with significant probability. In Lemma 8.2, which is a kind of cross-breed between the Slepian Lemma and the Sudakov-Fernique inequality, we will show how this can be boosted to control the upper tail of $\max_i X_i$ by that of $\max_i Y_i$, assuming that the latter maximum is concentrated.

5.4 Stochastic Domination and FKG Inequality

Yet another set of convenient inequalities that we will use on occasion is related to the (now) classical notions of stochastic domination. This topic has been covered in the concurrent class by Hugo Duminil-Copin [63], but we will still review its salient features for future (and independent) reference.

We have already invoked the following version of stochastic domination: Given real-valued random variables X and Y, we say that X *is stochastically larger than* Y or, equivalently, that X *stochastically dominates* Y if the cumulative distribution function of Y exceeds that of X at all points, i.e.,

$$P(X \leq t) \leq P(Y \leq t), \quad t \in \mathbb{R}. \tag{5.47}$$

Although this may seem just a statement about the laws of these random variables, there is a way to realize the "domination" pointwise:

Exercise 5.14 *Suppose that X stochastically dominates Y. Prove that there is a coupling of X and Y (i.e., a realization of both of them on the same probability space) with $P(Y \leq X) = 1$.*

The complete order of the real line plays a crucial role here. A natural question is what to do about random variables that take values in a set which admits only a (non-strict) *partial order*; i.e., a reflexive, antisymmetric and transitive binary relation $\preccurlyeq$.

An instance of this is the product space $\mathbb{R}^A$ where

$$x \preccurlyeq y \text{ means that } x_i \le y_i \text{ for every } i \in A. \tag{5.48}$$

In the more general context, we will rely on the following notion: A real-valued function f is called *increasing* if $x \preccurlyeq y$ implies $f(x) \le f(y)$. The same concept applies for Cartesian products of arbitrary ordered spaces (not just $\mathbb{R}$).

Definition 5.15 *(Stochastic domination) We say that X stochastically dominates Y, writing $Y \preccurlyeq X$, if $Ef(Y) \le Ef(X)$ holds for every bounded measurable increasing f.*

It is an easy exercise to check that, for X and Y real valued and $\preccurlyeq$ being the usual order of the reals, this coincides with the definition given above. Exercise 5.14 then turns into an instance of:

Theorem 5.16 (Strassen's lemma) *Suppose X and Y take values in a partially-ordered compact metric space $\mathscr{X}$ such that $\{(x, y) : x \preccurlyeq y\}$ is closed in $\mathscr{X} \times \mathscr{X}$. Then X stochastically dominates Y if and only if there is a coupling of these random variables such that $P(Y \preccurlyeq X) = 1$.*

For a proof we refer to Liggett [88, Theorem 2.4]. Stochastic domination is often interpreted as a property of probability measures; i.e., we write $\mu \preccurlyeq \nu$ if the random variables X and Y with respective laws ν and μ obey $Y \preccurlyeq X$.

Stochastic domination is closely related to the concept of *positive correlations*, also known as *positive association* or *(weak) FKG inequality*. Let us call an event A increasing if its indicator 1_A is an increasing function, and decreasing if 1_A is a decreasing function. The following concept inherits its acronym from the authors of Fortuin, Kasteleyn and Ginibre [72]:

Definition 5.17 *(Positive correlations a.k.a. FKG inequality) A probability measure μ on a partially-order space is said to have positive correlations, or satisfy the (weak) FKG inequality, if*

$$\mu(A \cap B) \ge \mu(A)\mu(B) \tag{5.49}$$

holds for any pair of increasing events A and B.

Positive correlations can be interpreted using the concept of stochastic domination as follows:

$$\forall B \text{ increasing with } \mu(B) > 0\colon \quad \mu \preccurlyeq \mu(\cdot|B). \tag{5.50}$$

In other words, a probability measure has positive correlations if (and only if) conditioning on an increasing event "increases" the measure.

As is readily checked, if (5.49) holds for pairs of increasing events then it also holds for pairs of decreasing events, and that the opposite inequality applies when one event is increasing and the other decreasing. A seemingly stronger, albeit equivalent, formulation of positive correlations is via increasing functions:

Exercise 5.18 *Show that μ has positive correlations if and only if*

$$E_\mu(fg) \ge E_\mu(f)E_\mu(g) \tag{5.51}$$

holds true for any pair of increasing (or decreasing) functions $f, g \in L^2(\mu)$. Hint: Write every such f as a limit of linear combinations of increasing indicators.

That positive correlations and stochastic domination are related is seen from:

Exercise 5.19 *Let μ and ν be probability measures on a partially-ordered space. If μ and ν have positive correlations and $\mu \preccurlyeq \nu$ (or $\nu \preccurlyeq \mu$), then their convex combination $t\mu + (1-t)\nu$ has positive correlations for all $t \in [0, 1]$.*

To see how the above concepts are related, let us recall the situation of independent random variables where these connections were observed first:

Lemma 5.20 (Harris' inequality) *For any set A, any product law on $\mathbb{R}^A$ (endowed with the product σ-algebra and the partial order* (5.48)*) has positive correlations.*

Proof. Let μ be a product law which (by the assumed product structure) we may think of as the distribution of independent random variables $\{X_i : i \in A\}$. We first prove (5.51) for $f, g \in L^2(\mu)$ that depend only on one of these random variables, say X_i. Let $\widetilde{X}_i$ be an independent copy of X_i. If f, g are increasing, then

$$\bigl[f(X_i) - f(\widetilde{X}_i)\bigr]\bigl[g(X_i) - g(\widetilde{X}_i)\bigr] \ge 0. \tag{5.52}$$

Taking expectation then yields (5.51).

Assuming (5.51) holds for $f, g \in L^2(\mu)$ that depend on random variables $X_1, \dots, X_k$, we will now show that it holds for any $f, g \in L^2(\mu)$ that depend on $X_1, \dots, X_{k+1}$. Denote

$$\mathcal{F}_k := \sigma(X_1, \dots, X_k), \quad f_k := E(f|\mathcal{F}_k) \ \text{ and } \ g_k := E(g|\mathcal{F}_k) \tag{5.53}$$

and write μ_k for the regular conditional probability $\mu(\cdot|\mathcal{F}_k)$. Then (5.51) for the one-dimensional case yields $E_{\mu_k}(fg) \ge f_k g_k$. Moreover, thanks to the product structure and Fubini–Tonelli, f_k, g_k are both increasing. They are also functions of $X_1, \dots, X_k$ only and so, by the induction assumption,

$$E_\mu(fg) = E_\mu\bigl(E_{\mu_k}(fg)\bigr) \ge E_\mu\bigl(f_k g_k\bigr) \ge E_\mu(f_k)E_\mu(g_k) = E_\mu(f)E_\mu(g). \tag{5.54}$$

We conclude that (5.51) applies to all $f, g \in L^2(\mu)$ depending only on a finite number of coordinates.

Now take a general $f \in L^2(\mu)$. By elementary measurability considerations, f is a function of at most a countable (sub)collection $\{X_1, X_2, \dots\}$ of the above random variables; Levy's Forward Theorem ensures $E_\mu(f|\mathcal{F}_k) \to f$ in $L^2(\mu)$ as $k \to \infty$. Since $E_\mu(f|\mathcal{F}_k)$ is also increasing, (5.51) for any $f, g \in L^2(\mu)$ follows from the finite-dimensional case by usual approximation arguments. □

The above proof actually showed more than (5.51); namely, that any product law has the following property:

Definition 5.21 *(strong-FKG property) A probability measure on $\mathbb{R}^A$ is said to be strong-FKG if the conditional law given the values for any finite number of coordinates has positive correlations.*

We remark that the expression "positive correlations" is sometimes used in the context when μ is a law of random variables $\{X_i : i \in I\}$ and "positivity of correlations" refers to $\operatorname{Cov}(X_i, X_j) \geq 0$—namely, a special case of (5.51) with $f(x) := x_i$ and $g(x) := x_j$. This is the reason why the term "positive association" is generally preferred to capture the full strength of (5.51). Notwithstanding, this is all the same for Gaussian random variables:

Proposition 5.22 (strong-FKG for Gaussians) *Suppose that μ is the law of a Gaussian vector X on $\mathbb{R}^n$. Then*

$$\mu \text{ is strong-FKG} \quad \Leftrightarrow \quad \operatorname{Cov}(X_i, X_j) \geq 0, \quad i, j = 1, \ldots, n. \tag{5.55}$$

Proof. To get $\Rightarrow$ we just use the definition of (weak) FKG along with the fact that $f(X) := X_i$ is increasing. Moving to the $\Leftarrow$ part, assume $\operatorname{Cov}(X_i, X_j) \geq 0$ for all $i, j = 1, \ldots, n$. Conditioning a multivariate Gaussian on part of the variables preserves the multivariate Gaussian structure as well as the covariances. It thus suffices to prove that μ satisfies the weak FKG inequality for which, by Exercise 5.18 and routine approximation arguments, it suffices to show $\operatorname{Cov}(f(X), g(X)) \geq 0$ for any non-decreasing smooth functions $f, g \colon \mathbb{R}^n \to \mathbb{R}$ with bounded gradients. This follows from an enhanced version of Gaussian integration by parts in Lemma 6.2 (to be proved soon) and the fact that the first partial derivatives of f and g are non-negative. □

We note that, since the DGFF covariance is given by the Green function which is non-negative everywhere, Proposition 5.22 shows that the DGFF is a strong-FKG process.

We close this lecture by noting that the above discussion of stochastic domination focused only on the topics that are needed for a full understanding of the arguments carried out in these notes. The reader is referred to, e.g., Liggett [88] or Grimmett [77] for a comprehensive treatment of this subject including its (interesting) history.

Lecture 6: Concentration Techniques

In this lecture we will establish bounds on the maximum of Gaussian random variables that are not be based on comparisons but rather on the metric properties of the covariance kernel. The first result to be proved here is the Borell-Tsirelson-Ibragimov-Sudakov inequality on the concentration of the maximum. Any use of this inequality will inevitably entail estimates on the expected maximum which we do via the Fernique majorization technique. Once these are stated and proved, we will infer some standard but useful consequences concerning boundedness and continuity of centered Gaussian processes. The presentation draws on that in Ledoux [86], Adler [5] and Adler and Taylor [6].

6.1 Inheritance of Gaussian Tails

Much of the present day probability hinges on the phenomenon of *concentration of measure*. For Gaussian random variables this is actually a very classical subject. The following inequality will come up frequently in the sequel:

Theorem 6.1 (Borell–TIS inequality) *Let X be a centered Gaussian on $\mathbb{R}^n$ and set*

$$\sigma_X^2 := \max_{i=1,\dots,n} E(X_i^2). \tag{6.1}$$

Then for each $t > 0$,

$$P\Bigl(\bigl|\max_{i=1,\dots,n} X_i - E(\max_{i=1,\dots,n} X_i)\bigr| > t\Bigr) \le 2\mathrm{e}^{-\frac{t^2}{2\sigma_X^2}}. \tag{6.2}$$

This result may be verbalized as: *The maximum of Gaussian random variables has a tail no heavier than the heaviest tail seen among these random variables.* Of course, the maximum is no longer centered (cf Exercise 5.12) and so any use of (6.2) requires information on its expectation as well.

The original proof of Theorem 6.1 was given by Borell [32] using a Gaussian isoperimetric inequality; the inequality was discovered independently in the Eastern block by Tsirelson, Ibragimov and Sudakov [124]. We will instead proceed using analytic techniques based on hypercontractivity. The following lemma offers a generalization of the formula on the Gaussian integration by parts:

Lemma 6.2 *Let X be a centered Gaussian vector on $\mathbb{R}^n$ and let $f, g \in C^1(\mathbb{R}^n)$ have subgaussian growth. Then*

$$\mathrm{Cov}\bigl(f(X), g(X)\bigr) = \int_0^1 \mathrm{d}t \sum_{i,j=1}^n \mathrm{Cov}(X_i, X_j) E\left(\frac{\partial f}{\partial x_i}(X)\frac{\partial g}{\partial x_j}\bigl(tX + \sqrt{1-t^2}\,Y\bigr)\right), \tag{6.3}$$

where $Y \overset{\text{law}}{=} X$ with $Y \perp\!\!\!\perp X$ on the right-hand side.

Proof. Since (6.3) is an equality between bilinear expressions for two functions of finitely-many variables, we may try to prove it by first checking it for a sufficiently large class of functions (e.g., the exponentials $x \mapsto \mathrm{e}^{k\cdot x}$) and then using extension arguments. We will instead rely on Gaussian integration by parts.

For X and Y as above and $t \in [0, 1]$, abbreviate

$$Z_t := tX + \sqrt{1-t^2}\,Y. \tag{6.4}$$

Approximation arguments permit us to assume $g \in C^2$ along with all the second partial derivatives having subgaussian growth. Then

$$\begin{aligned}\text{Cov}\big(f(X), g(X)\big) &= E\Big(f(X)\big[g(Z_1) - g(Z_0)\big]\Big) \\ &- \int_0^1 \mathrm{d}t\, E\left(f(X)\frac{\mathrm{d}}{\mathrm{d}t}g(Z_t)\right) \\ &= \int_0^1 \mathrm{d}t \sum_{i=1}^n E\left(\left[X_i - \frac{t}{\sqrt{1-t^2}}Y_i\right] f(X)\frac{\partial g}{\partial x_i}(Z_t)\right).\end{aligned} \tag{6.5}$$

The integration by parts (cf Lemma 5.2) will eliminate the square bracket and yield two contributions: one from the derivative hitting f and the other from the derivative hitting the partial derivative of g. The latter term equals the sum over j of $t\text{Cov}(X_i, X_j) - t\text{Cov}(Y_i, Y_j)$ times $E[f(X)\frac{\partial^2 g}{\partial x_i \partial x_j}(Z_t)]$. As $Y \overset{\text{law}}{=} X$, this term vanishes identically. The term where the derivative hits f produces the integrand in (6.3). □

As a side note, we notice that this implies:

Corollary 6.3 (Gaussian Poincaré inequality) *For $X_1, \dots, X_n$ i.i.d. copies of $\mathcal{N}(0, 1)$ and any $f \in C^1(\mathbb{R}^n)$ with $f, \nabla f \in L^2(\mathrm{e}^{-|x|^2/2}\mathrm{d}x)$,*

$$\text{Var}\big(f(X)\big) \le E\big(|\nabla f(X)|^2\big). \tag{6.6}$$

Proof. Apply Cauchy–Schwarz on the right-hand side of (6.3) while noting $tX + \sqrt{1-t^2}\,Y \overset{\text{law}}{=} X$. An alternative is to use Gaussian integration by parts formula instead of (6.3). □

Note that the bound (6.6) is of dimension-less nature—meaning: with no n dependence of the (implicit) constant on the right-hand side. This is quite in contrast to the Poincaré inequality on $\mathbb{R}^d$. (The generalization of (6.6) to non-i.i.d. Gaussian vectors is straightforward.) Moving along with the proof of the Borell–TIS inequality, next we will prove:

Lemma 6.4 (Concentration for Lipschitz functions) *Let $X_1, \dots, X_n$ be i.i.d. $\mathcal{N}(0, 1)$ and let $f : \mathbb{R}^n \to \mathbb{R}$ be Lipschitz in the sense that, for some $M \in (0, \infty)$,*

$$\big|f(x) - f(y)\big| \le M|x - y|, \quad x, y \in \mathbb{R}^n, \tag{6.7}$$

where $|\cdot|$ on the right-hand side is the Euclidean norm. Then for each $t > 0$,

$$P\big(f(X) - Ef(X) > t\big) \le \mathrm{e}^{-\frac{t^2}{2M^2}}. \tag{6.8}$$

Proof. By approximation we may assume that $f \in C^1$ with ∇f having the Euclidean norm at most M. By adding a suitable constant to f we may assume $Ef(X) = 0$. The exponential Chebyshev inequality then shows

$$P\big(f(X) - Ef(X) > t\big) \le \mathrm{e}^{-\lambda t} E\big(\mathrm{e}^{\lambda f(X)}\big) \tag{6.9}$$

for any $\lambda \ge 0$ and so just we need to bound the expectation on the right.

Here we note that Lemma 6.2 with $g(x) := \mathrm{e}^{\lambda f(x)}$ and (6.7) imply

$$E\big(f(X)\mathrm{e}^{\lambda f(X)}\big) = \int_0^1 \mathrm{d}t\, \lambda E\Big(\nabla f(X) \cdot \nabla f(Z_t)\mathrm{e}^{\lambda f(Z_t)}\Big) \overset{\lambda \ge 0}{\le} \lambda M^2 E\big(\mathrm{e}^{\lambda f(X)}\big). \tag{6.10}$$

The left-hand side is the derivative of the expectation on the right-hand side. It follows that the function

$$h(\lambda) := E\big(\mathrm{e}^{\lambda f(X)}\big) \tag{6.11}$$

obeys the differential inequality

$$h'(\lambda) \le \lambda M^2 h(\lambda), \qquad \lambda \ge 0. \tag{6.12}$$

As $h(0) = 1$, this is readily solved to give

$$E\big(\mathrm{e}^{\lambda f(X)}\big) \le \mathrm{e}^{\frac{1}{2}\lambda^2 M^2}. \tag{6.13}$$

Inserting this into (6.9) and optimizing over $\lambda \ge 0$ then yields the claim. □

In order to prove the Borell–TIS inequality, we will also need:

Exercise 6.5 *Denote* $f(x) := \max_{i=1,\dots,n} x_i$. *Prove that for any* $n \times n$*-matrix* A,

$$\big|f(Ax) - f(Ay)\big| \le \sqrt{\max_{i=1,\dots,n} (A^T A)_{ii}}\ |x - y|, \quad x, y \in \mathbb{R}^n, \tag{6.14}$$

with $|x - y|$ *denoting the Euclidean norm of* $x - y$ *on the right-hand side.*

Proof of Theorem 6.1. Let X be the centered Gaussian on $\mathbb{R}^n$ from the statement and let C denote its covariance matrix. In light of the symmetry and positive semi-definiteness of C, there is an $n \times n$-matrix A such that $C = A^{\mathrm{T}} A$. If $Z = (Z_1, \dots, Z_n)$ are i.i.d. copies of $\mathcal{N}(0, 1)$, then

$$X \overset{\text{law}}{=} AZ. \tag{6.15}$$

Denoting $f(x) := \max_{i=1,\dots,n} x_i$, Exercise 6.5 shows that $x \mapsto f(Ax)$ is Lipschitz with Lipschitz constant σ_X. The claim follows from (6.8) and a union bound. □

For a future reference, note that using (6.15), Theorem 6.1 generalizes to all functions that are Lipschitz with respect to the ℓ^∞-norm:

Corollary 6.6 (Gaussian concentration, a general case) *Let $f : \mathbb{R}^n \to \mathbb{R}$ be such that for some $M > 0$ and all $x, y \in \mathbb{R}^n$,*

$$\bigl|f(y) - f(x)\bigr| \le M \max_{i=1,\dots,n} |x_i - y_i|. \tag{6.16}$$

Then for any centered Gaussian X on $\mathbb{R}^n$ with σ_X as in (6.1) and any $t \ge 0$,

$$P\bigl(f(X) - Ef(X) > t\bigr) \le \mathrm{e}^{-\frac{t^2}{2M^2\sigma_X^2}}. \tag{6.17}$$

Proof. Let A be the $n \times n$ matrix such that (6.15) holds. From (6.16) and (6.14) we get

$$\bigl|f(Ay) - f(Ax)\bigr| \le M \max_{i=1,\dots,n} \bigl|(A(x-y))_i\bigr| \le M\sigma_X |x - y|. \tag{6.18}$$

Now apply Lemma 6.4. □

6.2 *Fernique Majorization*

As noted before, the Borell–TIS inequality is of little use unless we have a way to control the expected maximum of a large collection of Gaussian random variables. Our next task is to introduce a method for this purpose. We will actually do this for the supremum over a countable family of such variables as that requires no additional effort. A principal notion here is that of the canonical (pseudo) metric ρ_X associated via (5.36) with the Gaussian process $\{X_t : t \in T\}$ on any set T. Our principal result here is:

Theorem 6.7 (Fernique majorization) *There is $K \in (0, \infty)$ such that the following holds for any centered Gaussian process $\{X_t : t \in T\}$ over a countable set T for which (T, ρ_X) is totally bounded: For any probability measure μ on T, we have*

$$E\bigl(\sup_{t\in T} X_t\bigr) \le K \sup_{t\in T} \int_0^\infty \mathrm{d}r \sqrt{\log \frac{1}{\mu(B(t,r))}}, \tag{6.19}$$

where $B(t, r) := \{s \in T : \rho_X(t, s) < r\}$.

A measure μ for which the integral in (6.19) converges is called a *majorizing measure*. Note that the integral exists because the integrand is non-increasing and left-continuous. Also note that the domain of integration is effectively bounded because $\mu(B(t, r)) = 1$ whenever r exceeds the ρ_X-diameter of T, which is finite by the assumed total boundedness.

The above theorem takes its origin in Dudley's work [62] whose main result is the following theorem:

Theorem 6.8 (Dudley's inequality) *For the same setting as in the previous theorem, there is a universal constant $K \in (0, \infty)$ such that*

$$E\bigl(\sup_{t\in T} X_t\bigr) \le K \int_0^\infty \mathrm{d}r \sqrt{\log N_X(r)}\,, \tag{6.20}$$

where $N_X(r)$ is the minimal number of ρ_X-balls of radius r that are needed to cover T.

We will prove Dudley's inequality by modifying a few steps in the proof of Fernique's estimate. Dudley's inequality is advantageous as it sometimes easier to work with. To demonstrate its use we note that the setting of the above theorems is so general that they fairly seamlessly connect boundedness of Gaussian processes to sample-path continuity. Here is an exercise in this vain:

Exercise 6.9 *Apply Dudley's inequality to the process $X_{t,s} := X_t - X_s$ with $t, s \in T$ restricted by $\rho_X(t,s) \le R$ to prove, for K' a universal constant,*

$$E\Bigl(\sup_{\substack{t,s\in T\\ \rho_X(t,s)\le R}} |X_t - X_s|\Bigr) \le K' \int_0^R \mathrm{d}r \sqrt{\log N_X(r)}\,. \tag{6.21}$$

Conclude that if $r \mapsto \sqrt{\log N_X(r)}$ is integrable, then $t \mapsto X_t$ has a version with (uniformly) ρ_X-continuous sample paths a.s.

If this exercise seems too hard at first, we suggest that the reader first reads the proof of Theorem 6.14 and solves Exercise 6.15. To see (6.21) in action, it is instructive to solve:

Exercise 6.10 *Use Dudley's inequality* (6.21) *to prove the existence of a ρ_X-continuous version for the following Gaussian processes:*

(1) the standard Brownian motion, i.e., a Gaussian process $\{B_t : t \in [0,1]\}$ with zero mean and $E(B_t B_s) = t \wedge s$,
(2) the Brownian sheet, i.e., a centered Gaussian process $\{W_t : t \in [0,1]^d\}$ with

$$E(W_t W_s) = \prod_{i=1}^d (t_i \wedge s_i), \tag{6.22}$$

(3) any centered Gaussian process $\{X_t : t \in [0,1]\}$ such that

$$E\bigl([X_t - X_s]^2\bigr) \le c[\log(1/|t-s|)]^{-1-\delta} \tag{6.23}$$

for some $\delta > 0$ and $c > 0$ and $|t-s|$ sufficiently small.

The continuity of these processes can as well be proved via the Kolmogorov–Čenstov condition. Both techniques play small probability events against the entropy arising from their total count. (Roughly speaking, this is why the logarithm of $N_X(r)$ appears; the square root arises from Gaussian tails.) Both techniques offer an extension to the proof of uniform Hölder continuity.

Notwithstanding the above discussion, an advantage of Fernique's bound over Dudley's inequality is that it allows optimizing over the probability measure μ. This is in fact all one needs to get a *sharp* estimate. Indeed, a celebrated result of Talagrand [122] shows that a choice of μ exists such that the corresponding integral bounds the expectation from below modulo a universal multiplicative constant. This is known to fail for Dudley's inequality. The optimizing μ may be thought of as the distribution of the point t where the maximum of X_t is achieved, although this has not been made rigorous.

6.3 Proof of Fernique's Estimate

We will now give a proof of Fernique's bound but before we get embroiled in detail, let us outline the main idea. The basic strategy is simple: We identify an auxiliary centered Gaussian process $\{Y_t : t \in T\}$ whose intrinsic distance function ρ_Y dominates ρ_X. The Sudakov-Fernique inequality then bounds the expected supremum of X by that of Y.

For the reduction to be useful, the Y-process must be constructed with a lot of independence built in from the outset. This is achieved by a method called *chaining*. First we organize the elements of T in a kind of tree structure by defining, for each $n \in \mathbb{N}$, a map $\pi_n : T \to T$ whose image is a finite set such that the ρ_X-distance between $\pi_{n-1}(t)$ and $\pi_n(t)$ tends to zero *exponentially* fast with $n \to \infty$, uniformly in t. Assuming that $\pi_0(T)$ is a singleton, $\pi_0(T) = \{t_0\}$, the Borel–Cantelli estimate then allows us to write

$$X_t - X_{t_0} = \sum_{n=1}^{\infty} \bigl[X_{\pi_n(t)} - X_{\pi_{n-1}(t)}\bigr], \tag{6.24}$$

with the sum converging a.s. for each $t \in T$. We then define Y by replacing the increments $X_{\pi_n(t)} - X_{\pi_{n-1}(t)}$ by *independent* random variables with a similar variance. The intrinsic distances for Y can be computed quite explicitly and shown, thanks to careful choices in the definition of π_n, to dominate those for X. A computation then bounds the expected supremum of Y_t by the integral in (6.19). We will now begin with the actual proof:

Proof of Theorem 6.7. Assume the setting of the theorem and fix a probability measure μ on T. The proof (whose presentation draws on Adler [5]) comes in five steps.

STEP 1: Reduction to unit diameter. If $D := \operatorname{diam}(T)$ vanishes, T is effectively a singleton and the statement holds trivially. So we may assume $D > 0$. The process $\widetilde{X}_t := D^{-1/2}X_t$ has a unit diameter. In light of $\rho_{\widetilde{X}}(s,t) = D^{-1/2}\rho_X(s,t)$, the $\rho_{\widetilde{X}}$-ball of radius r centered at t coincides with $B(t, D^{-1/2}r)$. Passing from X to $\widetilde{X}$ in (6.19), both sides scale by factor $\sqrt{D}$.

STEP 2: Construction of the tree structure. Next we will define the aforementioned maps π_n subject to properties that will be needed later: □

Lemma 6.11 *For each $n \in \mathbb{N}$ there is $\pi_n : T \to T$ such that*

(1) $\pi_n(T)$ is finite,
(2) for each $t \in T$, we have $\rho(t, \pi_n(t)) < 2^{-n}$,
(3) for each $t \in T$,

$$\mu\big(B(\pi_n(t), 2^{-n-2})\big) \geq \mu\big(B(t, 2^{-n-3})\big), \tag{6.25}$$

(4) the sets $\{B(t, 2^{-n-2}) : t \in \pi_n(T)\}$ are (pairwise) disjoint.

Proof. Fix $n \in \mathbb{N}$ and, using the assumption of total boundedness, let $t_1, \dots, t_{r_n}$ be points such that

$$\bigcup_{i=1}^{r_n} B(t_i, 2^{-n-3}) = T. \tag{6.26}$$

Assume further that these points have been labeled so that

$$i \mapsto \mu\big(B(t_i, 2^{-n-2})\big) \quad \text{is non-increasing.} \tag{6.27}$$

We will now identify $\{C_1, C_2, \dots\} \subseteq \{\emptyset\} \cup \{B(t_i, 2^{-n-2}) : i = 1, \dots, r_n\}$ by progressively dropping balls that have a non-empty intersection with one of the lesser index. Formally, we set

$$C_1 := B(t_1, 2^{-n-2}) \tag{6.28}$$

and, assuming that $C_1, \dots, C_i$ have already been defined, let

$$C_{i+1} := \begin{cases} B(t_{i+1}, 2^{-n-2}), & \text{if } B(t_{i+1}, 2^{-n-2}) \cap \bigcup_{j=1}^{i} C_j = \emptyset, \\ \emptyset, & \text{else.} \end{cases} \tag{6.29}$$

Now we define π_n as the composition of two maps described informally as follows: Using the ordering induced by (6.27), first assign t to the point t_i of smallest index i such that $t \in B(t_i, 2^{-n-3})$. Then assign this t_i to the t_j with the largest $j \in \{1, \dots, i\}$ such that $B(t_i, 2^{-n-2}) \cap C_j \neq \emptyset$. In summary,

$$\begin{aligned} i = i(t) &:= \min\big\{i = 1, \dots, r_n : t \in B(t_i, 2^{-n-3})\big\} \\ j = j(t) &:= \max\big\{j = 1, \dots, i(t) : B(t_{i(t)}, 2^{-n-2}) \cap C_j \neq \emptyset\big\}, \end{aligned} \tag{6.30}$$

where we notice that, by the construction of $\{C_k\}$, the set in the second line is always non-empty. We then define

$$\pi_n(t) := t_j \quad \text{for} \quad j := j(t). \tag{6.31}$$

This implies $\pi_n(T) \subseteq \{t_1, \dots, t_{r_n}\}$ and so $\pi_n(T)$ is indeed finite, proving (1). For (2), using i and j for the given t as above, the construction gives

$$\begin{aligned}\rho_X(t, \pi_n(t)) &= \rho_X(t, t_j) \\ &\le \rho_X(t, t_i) + \rho_X(t_i, t_j) \\ &\le 2^{-n-3} + 2\,2^{-n-2} < 2^{-n}.\end{aligned} \tag{6.32}$$

For (3) we note that

$$B(t, 2^{-n-3}) \subseteq B(t_i, 2^{-n-2}) \tag{6.33}$$

and, by (6.27) and $j \le i$,

$$\mu\big(B(t_i, 2^{-n-2})\big) \le \mu\big(B(t_j, 2^{-n-2})\big). \tag{6.34}$$

Finally, $t_j \in \pi_n(T)$ only if $C_j \ne \emptyset$ and, when that happens, $C_j = B(t_j, 2^{-n-2})$. The construction ensures that the C_j's are disjoint, thus proving (4). □

STEP 3: Auxiliary process. We are now ready to define the aforementioned process $\{Y_t : t \in T\}$. For this, consider a collection $\{Z_n(t) \colon n \in \mathbb{N},\ t \in \pi_n(T)\}$ of i.i.d. standard normals and set

$$Y_t := \sum_{n \ge 1} 2^{-n} Z_n\big(\pi_n(t)\big). \tag{6.35}$$

The sum converges absolutely a.s. for each t due to the fact that the maximum of the first n terms in a sequence of i.i.d. standard normals grows at most like a constant times $\sqrt{\log n}$. We now state:

Lemma 6.12 *For any $t, s \in T$,*

$$E\big([X_t - X_s]^2\big) \le 6E\big([Y_t - Y_s]^2\big). \tag{6.36}$$

In particular,

$$E\big(\sup_{t \in T} X_t\big) \le \sqrt{6}\, E\big(\sup_{t \in T} Y_t\big). \tag{6.37}$$

Proof. We may assume $\rho_X(t, s) > 0$ as otherwise there is nothing to prove. Since $\mathrm{diam}(T) = 1$, there is an integer $N \ge 1$ such that $2^{-N} < \rho_X(t, s) \le 2^{-N+1}$. Lemma 6.11(2) and the triangle inequality then show $\pi_n(t) \ne \pi_n(s)$ for all $n \ge N + 1$. This is quite relevant because the independence built into Y_t yields

$$E\big([Y_t - Y_s]^2\big) = \sum_{n \ge 1} 2^{-2n} E\Big(\big[Z_n(\pi_n(t)) - Z_n(\pi_n(s))\big]^2\Big) \tag{6.38}$$

and the expectation on the right vanishes unless $\pi_n(t) \ne \pi_n(s)$. As that expectation is either zero or 2, we get

$$E\big([Y_t - Y_s]^2\big) \geq 2 \sum_{n\geq N+1} 2^{-2n} = 2\frac{4^{-(N+1)}}{3/4} = \frac{1}{6}4^{-N+1} \geq \frac{1}{6}E\big([X_t - X_s]^2\big), \tag{6.39}$$

where the last inequality follows from the definition of N. This is (6.36); the second conclusion then follows from the Sudakov-Fernique inequality. □

STEP 4: Majorizing $E(\sup_{t\in T} Y_t)$. For the following argument it will be convenient to have a random variable τ, taking values in T, that identifies the maximizer of $t \mapsto Y_t$. Such a random variable can certainly be defined when T is finite. For T infinite, one has to work with approximate maximizers only. To this end we pose:

Exercise 6.13 *Suppose there is $M \in (0, \infty)$ such that $E(Y_\tau) \leq M$ holds for any T-valued random variable τ that is measurable with respect to $\sigma(Y_t : t \in T)$. Prove that then also $E(\sup_{t\in T} Y_t) \leq M$.*

It thus suffices to estimate $E(Y_\tau)$ for any T-valued random variable τ. For this we first partition the expectation according to the values of $\pi_n(\tau)$ as

$$E(Y_\tau) = \sum_{n\geq 1} 2^{-n} \sum_{t\in\pi_n(T)} E\big(Z_n(t)1_{\{\pi_n(\tau)=t\}}\big). \tag{6.40}$$

We now estimate the expectation on the right as follows: Set $g(a) := \sqrt{2\log(1/a)}$ and note that, for $Z = \mathcal{N}(0, 1)$ and any $a > 0$,

$$E\big(Z1_{\{Z>g(a)\}}\big) = \frac{1}{\sqrt{2\pi}}\int_{g(a)}^{\infty} x\,\mathrm{e}^{-\frac{1}{2}x^2}\mathrm{d}x = \frac{1}{\sqrt{2\pi}}\mathrm{e}^{-\frac{1}{2}g(a)^2} = \frac{a}{\sqrt{2\pi}}. \tag{6.41}$$

Therefore,

$$\begin{aligned} E\big(Z_n(t)1_{\{\pi_n(\tau)=t\}}\big) &\leq E\big(Z_n(t)1_{\{Z_n(t)>g(a)\}}\big) + g(a)P\big(\pi_n(\tau) = t\big) \\ &= \frac{a}{\sqrt{2\pi}} + g(a)P\big(\pi_n(\tau) = t\big). \end{aligned} \tag{6.42}$$

Now set $a := \mu(B(t, 2^{-n-2}))$ and perform the sum over t and n. In the first term we use the disjointness claim from Lemma 6.11(4) to get

$$\frac{1}{\sqrt{2\pi}} \sum_{t\in\pi_n(T)} \mu(B(t, 2^{-n-2})) \leq \frac{1}{\sqrt{2\pi}} \tag{6.43}$$

while in the second term we invoke

$$g\big(\mu(B(\pi_n(t), 2^{-n-2}))\big) \leq g\big(\mu(B(\tau, 2^{-n-3}))\big), \tag{6.44}$$

as implied by Lemma 6.11(3), and the fact that g is non-increasing to get

$$\begin{aligned}\sum_{n\geq 1} 2^{-n} \sum_{t\in\pi_n(T)} g\big(\mu(B(t,2^{-n-2}))\big)P\big(\pi_n(\tau)=t\big)\\ &= E\left[\sum_{n\geq 1} 2^{-n} g\big(\mu(B(\pi_n(\tau),2^{-n-2}))\big)\right]\\ &\leq E\left[\sum_{n\geq 1} 2^{-n} g\big(\mu(B(\tau,2^{-n-3}))\big)\right] \leq \sup_{t\in T}\sum_{n\geq 1} 2^{-n} g\big(\mu(B(t,2^{-n-3}))\big)\,.\end{aligned} \tag{6.45}$$

Using the monotonicity of g,

$$2^{-n} g\big(\mu(B(t,2^{-n-3}))\big) \leq 16\int_{2^{-n-4}}^{2^{-n-3}} g\big(\mu(B(t,r))\big)\,\mathrm{d}r, \tag{6.46}$$

and so the last sum in (6.45) can now be dominated by 16-times the integral in the statement of the theorem. Putting the contribution of both terms on the right of (6.42) together, we thus conclude

$$E(Y_\tau) \leq \frac{1}{\sqrt{2\pi}} + 16\sup_{t\in T}\int_0^1 g\big(\mu(B(t,r))\big)\,\mathrm{d}r. \tag{6.47}$$

Exercise 6.13 now extends this to a bound on the expected supremum.

STEP 5: A final touch. In order to finish the proof, we need to show that the term $1/\sqrt{2\pi}$ is dominated by, and can thus be absorbed into, the integral. Here we use the fact that, since $\operatorname{diam}(T)=1$, there is $t\in T$ such that $\mu(B(t,1/2))\leq 1/2$. The supremum on the right of (6.47) is then at least $\frac{1}{2}\sqrt{2\log 2}$. The claim follows with $K:=[\frac{1}{\sqrt{2\pi}}\frac{2}{\sqrt{\log 2}}+16\sqrt{2}]\sqrt{6}$. □

The proof of Dudley's inequality requires only minor adaptations:

Proof of Theorem 6.8. We follow the previous proof *verbatim* (while disregarding all statements concerning μ) until (6.42) at which point we instead choose

$$a := \frac{1}{N_X(2^{-n-3})}. \tag{6.48}$$

As the number of balls in (6.26) could be assumed minimal for the given radius, we have $|\pi_n(T)|\leq N_X(2^{-n-3})$ and so the analogue of (6.43) applies. Hence,

$$E(Y_\tau) \leq \frac{1}{\sqrt{2\pi}} + \sum_{n\geq 1} 2^{-n} g\big(1/N_X(2^{-n-3})\big). \tag{6.49}$$

The sum is converted to the desired integral exactly as in (6.46). The additive prefactor is absorbed by noting that $N_X(1/2)\geq 2$ because $\operatorname{diam}(X)=1$. □

6.4 Consequences for Continuity

As already alluded to after the statement of Dudley's inequality, the generality of the setting in which Fernique's inequality was proved permits a rather easy extension to a criterion for continuity. The relevant statement is as follows:

Theorem 6.14 *There is a universal constant $K' \in (0, \infty)$ such that the following holds for every centered Gaussian process $\{X_t : t \in T\}$ on a countable set T such that (T, ρ_X) is totally bounded: For any probability measure μ on T and any $R > 0$,*

$$E\Bigl(\sup_{\substack{t,s\in T\\ \rho_X(t,s)\le R}} |X_t - X_s|\Bigr) \le K' \sup_{t\in T} \int_0^R \mathrm{d}r \sqrt{\log \frac{1}{\mu(B(t,r))}}\,. \tag{6.50}$$

Proof. We will reduce this to Theorem 6.7 but that requires preparations. Let

$$U \subseteq \bigl\{(t,s) \in T \times T : \rho_X(t,s) \le R\bigr\} \tag{6.51}$$

be a finite and symmetric set. Denote $Y_{s,t} := X_t - X_s$ and notice that

$$\rho_Y\bigl((s,t),(s',t')\bigr) := \sqrt{E\bigl([Y_{s,t} - Y_{s',t'}]^2\bigr)} \tag{6.52}$$

obeys

$$\rho_Y\bigl((s,t),(s',t')\bigr) \le \begin{cases} \rho(s,s') + \rho(t,t'), \\ \rho(s,t) + \rho(s',t'). \end{cases} \tag{6.53}$$

Writing B_Y for the balls (in $T \times T$) in the ρ_Y-metric and B_X for the balls (in T) in ρ_X-metric, the first line in (6.53) then implies

$$B_Y\bigl((s,t),r\bigr) \supseteq B_X(s,r/2) \times B_X(t,r/2) \tag{6.54}$$

while the second line shows $\mathrm{diam}_{\rho_Y}(U) \le 2R$. Now define $f : T \times T \to U$ by

$$f(y) := \begin{cases} y, & \text{if } y \in U, \\ \mathrm{argmin}_U \rho_Y(y,\cdot), & \text{else}, \end{cases} \tag{6.55}$$

where in the second line the minimizer exists because U is finite and, in case of ties, is chosen minimal in some a priori complete ordering of U. Then f is clearly measurable and so, given a probability measure μ on T

$$\nu(A) := \mu \otimes \mu\bigl(f^{-1}(A)\bigr)\,. \tag{6.56}$$

defines a probability measure on U. Theorem 6.7 then yields

$$E\Bigl(\sup_{(s,t)\in U} Y_{s,t}\Bigr) \le K \sup_{(s,t)\in U} \int_0^{2R} \sqrt{\log \frac{1}{\nu(B_Y((t,s),r))}}\, \mathrm{d}r\,. \tag{6.57}$$

Our next task is to bring the integral on the right to the form in the statement.

First observe that if $x \in U$ and $y \in B_Y(x,r)$, then

$$\rho_Y\bigl(x, f(y)\bigr) \le \rho_Y(x,y) + \rho_Y\bigl(y, f(y)\bigr) \overset{x\in U}{\le} 2\rho_Y(x,y)\,. \tag{6.58}$$

Hence we get

$$B_Y(x,r) \subseteq f^{-1}\bigl(B_Y(x,2r)\bigr), \qquad x \in U, \tag{6.59}$$

and so, in light of (6.54),

$$\begin{aligned} \nu\bigl(B_Y((s,t),2r)\bigr) &= \mu\otimes\mu\bigl(f^{-1}\bigl(B_Y((s,t),2r)\bigr)\bigr) \\ &\ge \mu\otimes\mu\bigl(B_Y((s,t),r)\bigr) \\ &\ge \mu\bigl(B_X(s,r/2)\bigr)\mu\bigl(B_X(t,r/2)\bigr)\,. \end{aligned} \tag{6.60}$$

Plugging this in (6.57) and invoking $\sqrt{a+b} \le \sqrt{a} + \sqrt{b}$, elementary calculus gives

$$E\Bigl(\sup_{(s,t)\in U} Y_{t,s}\Bigr) \le 4K \sup_{t\in T} \int_0^{R} \sqrt{\log \frac{1}{\mu(B_X(t,r))}}\, \mathrm{d}r\,. \tag{6.61}$$

Increasing U to $U_R := \{(s,t) \in T\times T\colon \rho_X(s,t) \le R\}$ and applying the Monotone Convergence Theorem, the bound holds for $U := U_R$ as well. To connect this to the expectation on the left of (6.50), the symmetry of $\rho_X(t,s)$ and antisymmetry of $|X_t - X_s|$ under the exchange of t and s shows

$$E\Bigl(\sup_{\substack{t,s\in T\\ \rho_X(t,s)\le R}} |X_t - X_s|\Bigr) = E\Bigl(\sup_{(s,t)\in U_R} Y_{s,t}\Bigr). \tag{6.62}$$

The claim follows with $K' := 4K$, where K is as in Theorem 6.7. □

Theorem 6.14 gives us means to prove continuity with respect to the intrinsic metric. However, more often than not, T has its own private metric structure and continuity is desired in the topology thereof. Here the following exercise—a version of which we already asked for Dudley's inequality (6.20)—helps:

Exercise 6.15 *Suppose (T,ρ) is a metric space, $\{X_t\colon t\in T\}$ a Gaussian process and ρ_X the intrinsic metric on T induced thereby. Assume*

(1) (T,ρ) is totally bounded, and
(2) $s,t \mapsto \rho_X(s,t)$ is uniformly ρ-continuous on $T\times T$.

Prove that, if there is a probability measure μ on T such that

$$\lim_{R\downarrow 0}\,\sup_{t\in T}\int_0^R \sqrt{\log\frac{1}{\mu(B(t,r))}}\,\mathrm{d}r = 0, \tag{6.63}$$

then X admits (uniformly) ρ-continuous sample paths on T, a.s.

We note that condition (2) is necessary for sample path continuity, but definitely not sufficient. To see this, solve:

Exercise 6.16 *Given a measure space $(\mathscr{X},\mathcal{F},\nu)$ with ν finite, consider the (centered) Gaussian white-noise process $\{W(A)\colon A\in\mathcal{F}\}$ defined by*

$$E\bigl(W(A)W(B)\bigr) = \nu(A\cap B). \tag{6.64}$$

This corresponds to the intrinsic metric $\rho_W(A,B)=\sqrt{\nu(A\triangle B)}$. Give a (simple) example of $(\mathscr{X},\mathcal{F},\nu)$ for which $A\mapsto W(A)$ does not admit ρ_W-continuous sample paths.

6.5 Binding Field Regularity

As our last item of concern in this lecture, we return to the problem of uniform continuity of the binding field for the DGFF and its continuum counterpart. (We used these in the proof of the coupling of the two processes in Lemma 4.4.) The relevant bounds are stated in:

Lemma 6.17 (Binding field regularity) *Let $\widetilde{D}, D\in\mathfrak{D}$ be domains such that $\widetilde{D}\subseteq D$ and* $\mathrm{Leb}(D\smallsetminus\widetilde{D})=0$. *For $\delta>0$, denote $\widetilde{D}^\delta := \{x\in D\colon \operatorname{dist}(x,D^{\mathrm{c}})>\delta\}$. Then for each $\epsilon,\delta>0$,*

$$\lim_{r\downarrow 0} P\Bigl(\sup_{\substack{x,y\in\widetilde{D}^\delta\\ |x-y|<r}} \bigl|\Phi^{D,\widetilde{D}}(x)-\Phi^{D,\widetilde{D}}(y)\bigr|>\epsilon\Bigr)=0. \tag{6.65}$$

Similarly, given an admissible sequence $\{D_N\colon N\ge 1\}$ of approximating lattice domains and denoting $\widetilde{D}_N^\delta := \{x\in\widetilde{D}_N\colon \operatorname{dist}(x,\widetilde{D}_N^{\mathrm{c}})>\delta N\}$, for each $\epsilon,\delta>0$,

$$\lim_{r\downarrow 0}\,\limsup_{N\to\infty} P\Bigl(\sup_{\substack{x,y\in\widetilde{D}_N^\delta\\ |x-y|<rN}} \bigl|\varphi_x^{D_N,\widetilde{D}_N}-\varphi_y^{D_N,\widetilde{D}_N}\bigr|>\epsilon\Bigr)=0. \tag{6.66}$$

Proof of (6.65). Consider the set $\widetilde{D}_1^\delta := \{x\in\mathbb{C}\colon \operatorname{dist}(x,\widetilde{D}^\delta)<\delta/2\}$. The intrinsic metric associated with $\{\Phi^{D,\widetilde{D}}(x)\colon x\in\widetilde{D}_1^\delta\}$ is given by

$$\rho_\Phi(x,y)=\sqrt{C^{D,\widetilde{D}}(x,x)+C^{D,\widetilde{D}}(y,y)-2C^{D,\widetilde{D}}(x,y)}. \tag{6.67}$$

Since $x \mapsto C^{D,\widetilde{D}}(x,y)$ is harmonic on $\widetilde{D}$, it is continuously differentiable and thus uniformly Lipschitz on $\widetilde{D}_1^\delta$. It follows that, for some constant $L = L(\delta) < \infty$,

$$\rho_\Phi(x,y) \le L\sqrt{|x-y|}, \qquad x, y \in \widetilde{D}_1^\delta. \tag{6.68}$$

Let $B(x,r) := \{y \in \mathbb{C} \colon |x-y| < r\}$, denote $B_\Phi(x,r) := \{y \in \widetilde{D}_1^\delta \colon \rho_\Phi(x,y) < r\}$ and let μ be the normalized Lebesgue measure on $\widetilde{D}_1^\delta$. Then

$$B_\Phi(x, L\sqrt{r}) \supseteq B(x,r), \qquad x \in \widetilde{D}_1^\delta, \tag{6.69}$$

while (by the choice of $\widetilde{D}_1^\delta$),

$$\mu\bigl(B(x,r)\bigr) \ge cr^2, \quad x \in \widetilde{D}_1^\delta, \tag{6.70}$$

for some $c = c(\delta) > 0$. Combining these observations, we get

$$\mu(B_\Phi(x,r)) \ge cL^{-2}r^4. \tag{6.71}$$

As $r \mapsto \log(1/r^4)$ is integrable at zero, (6.65) follows from Theorem 6.14, Exercise 6.15 and Markov's inequality. □

The above argument could be improved a bit by noting that ρ_Φ is itself Lipschitz although that does not change the main conclusion.

Exercise 6.18 *Using an analogous argument with the normalized counting measure replacing the Lebesgue measure, prove* (6.66).

Lecture 7: Connection to Branching Random Walk

In this lecture we return to the two-dimensional DGFF and study the behavior of its absolute maximum beyond the leading order discussed in Lecture 2. We begin by recounting the so-called Dekking–Host argument which yields, rather effortlessly, tightness of the maximum (away from its expectation) along a subsequence. Going beyond this will require development of a connection to Branching Random Walk and proving sharp concentration for the maximum thereof. These will serve as the main ingredients for our proof of the tightness of the DGFF maximum in Lecture 8.

7.1 Dekking–Host Argument for DGFF

In Lecture 2 we already noted that the maximum of the DGFF in a box of side-length N,

$$M_N := \max_{x \in V_N} h_x^{V_N}, \tag{7.1}$$

grows as $M_N \sim 2\sqrt{g}\log N$ in probability, with the same leading-order growth rate for EM_N. The natural follow-up questions are then:

(1) What is the growth rate of

$$EM_N - 2\sqrt{g}\log N \tag{7.2}$$

with N; i.e., what are the lower-order corrections?

(2) What is the size of the fluctuations, i.e., the growth rate of $M_N - EM_N$?

As observed by Bolthausen, Deuschel and Zeitouni [31] in 2011, an argument that goes back to Dekking and Host [52] from 1991 shows that, for the DGFF, these seemingly unrelated questions are tied closely together:

Lemma 7.1 (Dekking–Host argument) *For M_N as above and any $N \geq 2$,*

$$E\big|M_N - EM_N\big| \leq 2\big(EM_{2N} - EM_N\big). \tag{7.3}$$

Proof. We will use an idea underlying the solution of the second part of Exercise 3.4. Note that the box V_{2N} embeds four translates $V_N^{(1)}, \dots, V_N^{(4)}$ of V_N that are separated by two lines of sites in-between; see Fig. 10. Denoting by V_{2N}° the union of the four translates of V_N, the Gibbs-Markov property tells us that

$$h^{V_{2N}} := h^{V_{2N}^\circ} + \varphi^{V_{2N},V_{2N}^\circ}, \quad \text{with} \quad h^{V_{2N}^\circ} \perp\!\!\!\perp \varphi^{V_{2N},V_{2N}^\circ}, \tag{7.4}$$

has the law of the DGFF in V_{2N}. Writing M_N for the maximum of h^{V_N}, using X to denote the (a.s.-unique) vertex where $h^{V_{2N}^\circ}$ achieves its maximum on V_{2N}° and abbreviating

$$M_N^{(i)} := \max_{x \in V_N^{(i)}} h_x^{V_{2N}^\circ}, \quad i = 1, \dots, 4, \tag{7.5}$$

it follows that

$$\begin{aligned} EM_{2N} &= E\Big(\max_{x \in V_{2N}} (h_x^{V_{2N}^\circ} + \varphi_x^{V_{2N},V_{2N}^\circ})\Big) \\ &\geq E\Big(\max_{x \in V_{2N}^\circ} h_x^{V_{2N}^\circ} + \varphi_X^{V_{2N},V_{2N}^\circ}\Big) \\ &= E\Big(\max_{x \in V_{2N}^\circ} h_x^{V_{2N}^\circ}\Big) = E\big(\max_{i=1,\dots,4} M_N^{(i)}\big), \end{aligned} \tag{7.6}$$

where we used that $\varphi^{V_{2N},V_{2N}^\circ}$ is independent of $h^{V_{2N}^\circ}$ and thus also of X to infer that $E\varphi_X^{V_{2N},V_{2N}^\circ} = 0$. The portions of $h^{V_{2N}^\circ}$ restricted to $V_N^{(1)}, \dots, V_N^{(4)}$ are i.i.d. copies of h^{V_N} and so $\{M_N^{(i)} : i = 1, \dots, 4\}$ are i.i.d. copies of M_N. Dropping two out of the four terms from the last maximum in (7.6) then yields

$$M_N' \overset{\text{law}}{=} M_N, \; M_N' \perp\!\!\!\perp M_N \quad \Rightarrow \quad EM_{2N} \geq E\max\{M_N, M_N'\}. \tag{7.7}$$

Now use $2\max\{a,b\} = a + b + |a - b|$ to turn this into

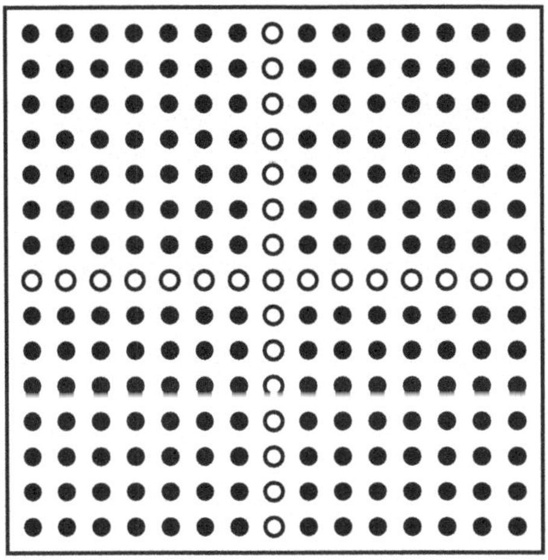

Fig. 10 The partition of box V_{2N} (both kinds of bullets) into four translates of V_N for $N := 8$ and two lines of sites (empty bullets) in the middle. The set V_{2N}° is the collection of all fat bullets

$$\begin{aligned} E\big|M_N - M_N'\big| &= 2E\max\{M_N, M_N'\} - E(M_N + M_N') \\ &\le 2EM_{2N} - 2EM_N \,. \end{aligned} \tag{7.8}$$

To get (7.3), we just apply Jensen's inequality to pass the expectation over M_N' inside the absolute value. □

From the growth rate of EM_N we then readily conclude:

Corollary 7.2 (Tightness along a subsequence) *There is a (deterministic) sequence* $\{N_k : k \ge 1\}$ *of integers with* $N_k \to \infty$ *such that* $\{M_{N_k} - EM_{N_k} : k \ge 1\}$ *is tight.*

Proof. Denote $a_n := EM_{2^n}$. From (7.3) we know that $\{a_n\}$ is non-decreasing. The fact that $EM_N \le c \log N$ (proved earlier by simple first-moment calculations) reads as $a_n \le c'n$ for $c' := c \log 2$. The increments of an increasing sequence with at most a linear growth cannot tend to infinity, so there must be $\{n_k : k \ge 1\}$ such that $n_k \to \infty$ and $a_{n_k+1} - a_{n_k} \le 2c'$. Setting $N_k := 2^{n_k}$, from Lemma 7.1 we get $E|M_{N_k} - EM_{N_k}| \le 4c'$. This gives tightness via Markov's inequality. □

Unfortunately, tightness along an (existential) subsequence seems to be all one is able to infer from the leading-order asymptotic of EM_N. If we hope to get any better along this particular line of reasoning, we need to control the asymptotic of EM_N up to terms of order unity. This was achieved by Bramson and Zeitouni [37] in 2012. Their main result reads:

Theorem 7.3 (Tightness of DGFF maximum) *Denote*

$$m_N := 2\sqrt{g}\,\log N - \frac{3}{4}\sqrt{g}\,\log\log(N \vee \mathrm{e}) \,. \tag{7.9}$$

Then

$$\sup_{N\ge 1} E\big|M_N - m_N\big| < \infty. \tag{7.10}$$

As a consequence, $\{M_N - m_N : N \ge 1\}$ *is tight.*

This and the next lecture will be spent on proving Theorem 7.3 using, however, a different (and, in the lecturer's view, easier) approach than that of [37].

7.2 Upper Bound by Branching Random Walk

The Gibbs-Markov decomposition underlying the proof of Lemma 7.1 can be iterated as follows: Consider a box $V_N := (0,N)^2 \cap \mathbb{Z}^2$ of side $N := 2^n$ for some large $n \in \mathbb{N}$. As illustrated in Fig. 11, the square V_{2^n} then contains four translates of $V_{2^{n-1}}$ separated by a "cross" of "lines of sites" in-between, and each of these squares contains four translates of $V_{2^{n-2}}$, etc. Letting $V_N^{(i)}$, for $i = 1,\dots,n-1$, denote the union of the resulting 4^i translates of $V_{2^{n-i}}$, and setting $V_N^{(0)} := V_N$ and $V_N^{(n)} := \emptyset$, we can then write

$$\begin{aligned} h^{V_N} &\overset{\text{law}}{=} h^{V_N^{(1)}} + \varphi^{V_N^{(0)},V_N^{(1)}} \\ &\overset{\text{law}}{=} h^{V_N^{(2)}} + \varphi^{V_N^{(0)},V_N^{(1)}} + \varphi^{V_N^{(1)},V_N^{(2)}} \\ &\;\;\vdots \qquad\quad \ddots \qquad\qquad \ddots \\ &\overset{\text{law}}{=} \varphi^{V_N^{(0)},V_N^{(1)}} + \dots + \varphi^{V_N^{(n-1)},V_N^{(n)}}, \end{aligned} \tag{7.11}$$

where $V_N^{(n)} := \emptyset$ gives $h^{V_N^{(n-1)}} = \varphi^{V_N^{(n-1)},V_N^{(n)}}$.

The fields in each line on the right-hand side of (7.11) are independent. Moreover, the binding field $\varphi^{V_N^{(i)},V_N^{(i+1)}}$ is a concatenation of 4^i independent copies of the binding field $\varphi^{U,V}$ for $U := V_{2^{n-i}}$ and $V := V_{2^{n-i-1}}$, with one for each translate of $V_{2^{n-i}}$ constituting $V_N^{(i)}$. A significant nuisance is that $\varphi^{V_N^{(i)},V_N^{(i+1)}}$ is not constant on these translates. If it were, we would get a representation of the DGFF by means of a Branching Random Walk that we will introduce next.

For an integer $b \ge 2$, consider a b-ary tree T^b which is a connected graph without cycles where each vertex except one (to be denoted by $\varnothing$) has exactly $b+1$ neighbors. The distinguished vertex $\varnothing$ is called the root; we require that the degree of the root is b. We will write L_n for the set of vertices at graph-theoretical distance n from the root—these are the *leaves* at depth n.

Every vertex $x \in L_n$ can be identified with a sequence

$$(x_1,\dots,x_n) \in \{1,\dots,b\}^n, \tag{7.12}$$

where x_i can be thought of as an instruction which "turn" to take at the ith step on the (unique) path from the root to x. The specific case of $b = 4$ can be linked with a binary

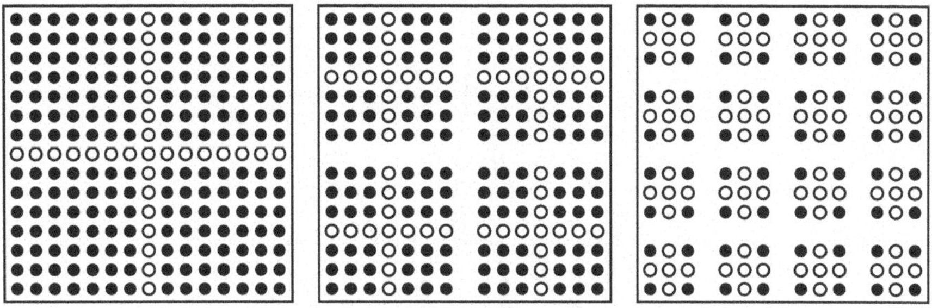

Fig. 11 The sets $V_N^{(0)} = V_N$, $V_N^{(1)}$ and $V_N^{(2)}$ underlying the hierarchical representation of the DGFF on V_N with $N := 16$. The empty bullets in $V_N^{(i)}$ mark the vertices that are being removed to define $V_N^{(i+1)}$. Boundary vertices (where the fields are set to zero by default) are not depicted otherwise. The binding field $\varphi^{V_N^{(1)},V_N^{(2)}}$ is independent on each of the four squares constituting $V_N^{(1)}$, but is neither constant nor independent on the squares constituting $V_N^{(2)}$

decomposition of $V_N := (0, N)^2 \cap \mathbb{Z}^2$ with $N := 2^n$ as follows: Every $x \in V_{2^n}$ has non-negative coordinates so it can be written in $\mathbb{R}^2$-vector notation as

$$x = \left(\sum_{i=0}^{n-1} \sigma_i 2^i, \sum_{i=0}^{n-1} \tilde{\sigma}_i 2^i \right), \tag{7.13}$$

for some unique $\sigma_i, \tilde{\sigma}_i \in \{0, 1\}$. Now set the ith instruction for the sequence $(x_1, \dots, x_n)$ as

$$x_i := 2\sigma_{n-i+1} + \tilde{\sigma}_{n-i+1} + 1 \tag{7.14}$$

to identify $x \in V_{2^n}$ with a point in L_n. Since V_N contains only $(N-1)^2$ vertices while, for $b := 4$, the cardinality of L_n is 4^n, we only get a subset of L_n. see Fig. 12.

We now come to:

Definition 7.4 *(Branching random walk) Given integers $b \geq 2$, $n \geq 1$ and a random variable Z, let $\{Z_x : x \in T^b\}$ be i.i.d. copies of Z indexed by the vertices of T^b. The* Branching Random Walk (BRW) on T^b of depth n with step distribution Z *is the family of random variables $\{\phi_x^{T^b} : x \in L_n\}$ where for $x = (x_1, \dots, x_n) \in L_n$ we set*

$$\phi_x^{T^b} := \sum_{k=0}^{n-1} Z_{(x_1,\dots,x_k)}, \tag{7.15}$$

with the $k = 0$ term corresponding to the root value $Z_\varnothing$.

We write $n-1$ in (7.15) to ensure that the restriction of $z \mapsto \phi_z^{T^b} - \phi_x^{T^b}$ to the subtree of T^b rooted at x is independent of $\phi_x^{T^b}$ with the same law (after proper relabeling) as ϕ^{T^b}. This is, in fact, a statement of the Gibbs-Markov property for the BRW.

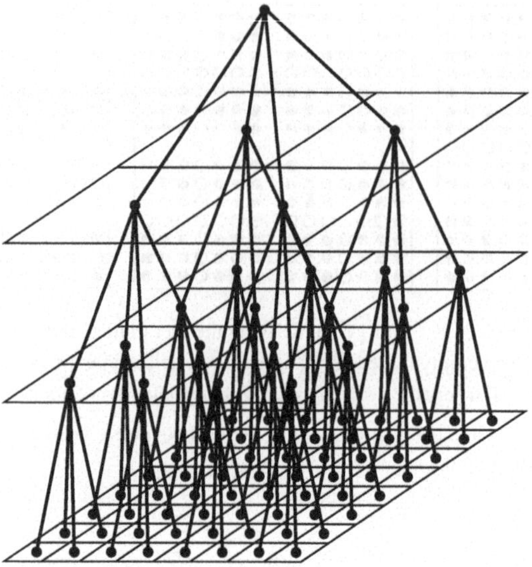

Fig. 12 A picture indicating how the values of the BRW on T^4 are naturally interpreted as a field on $\mathbb{Z}^2$. There is a close link to the Gibbs-Markov decomposition of the DGFF from (7.11)

The specific case of interest for us is the *Gaussian Branching Random Walk* where we take Z normal. The value of the BRW at a given point $x \in L_n$ is then very much like the last line in (7.11)—the sum of n independent Gaussians along the unique path from the root to x. As already noted, the correspondence is not perfect because of the more subtle covariance structure of the DGFF compared to the BRW and also because L_n has more active vertices than V_{2^n}. Still, we can use this fruitfully to get:

Lemma 7.5 (Domination of DGFF by BRW) *Consider a BRW ϕ^{T^4} on T^4 with step distribution $\mathcal{N}(0,1)$ and identify V_N for $N := 2^n$ with a subset of L_n as above. There is $c > 0$ such that for each $n \ge 1$ and each $x, y \in L_n$,*

$$E\big([h_x^{V_N} - h_y^{V_N}]^2\big) \le c + (g \log 2)\, E\Big(\big[\phi_x^{T^4} - \phi_y^{T^4}\big]^2\Big). \tag{7.16}$$

In particular, there is $k \in \mathbb{N}$ such that for each $n \ge 1$ (and $N := 2^n$),

$$E\Big(\max_{x \in V_N} h_x^{V_N}\Big) \le \sqrt{g \log 2}\, E\Big(\max_{x \in L_{n+k}} \phi_x^{T^4}\Big). \tag{7.17}$$

Proof. Since $V \mapsto E([h_x^V - h_y^V]^2)$ is non-decreasing under the set inclusion, the representation of the Green function from Lemma 1.19 along with the asymptotic for the potential kernel from Lemma 1.21 shows that, for some constant $\tilde{c} > 0$ and all distinct $x, y \in L_n$,

$$E\big([h_x^{V_N} - h_y^{V_N}]^2\big) \le \tilde{c} + 2g \log |x - y|. \tag{7.18}$$

Denoting by $d_n(x, y)$ the *ultrametric distance* between $x, y \in L_n$, which is defined as the graph-theoretical distance on T^b from x to the nearest common ancestor with y, from (7.15) we infer

$$E\Bigl(\bigl[\phi_x^{T^4} - \phi_y^{T^4}\bigr]^2\Bigr) = 2[d_n(x, y) - 1] \tag{7.19}$$

for any two distinct $x, y \in L_n$. We now pose: □

Exercise 7.6 *There is $\tilde{c}' \in (0, \infty)$ such that for each $n \geq 1$ and each $x, y \in L_n$,*

$$|x - y| \leq \tilde{c}'\, 2^{d_n(x,y)}. \tag{7.20}$$

Combining this with (7.18)–(7.19), we then get (7.16).

To get (7.17), let $k \in \mathbb{N}$ be so large that c in (7.16) obeys $c \leq 2k(g \log 2)$. Now, for each $x = (x_1, \dots, x_n) \in L_n$ let $\theta(x) := (x_1, \dots, x_n, 1, \dots, 1) \in L_{n+k}$. Then (7.16) implies

$$E\bigl([h_x^{V_N} - h_y^{V_N}]^2\bigr) \leq (g \log 2)\, E\Bigl(\bigl[\phi_{\theta(x)}^{T^4} - \phi_{\theta(y)}^{T^4}\bigr]^2\Bigr), \qquad x, y \in L_n. \tag{7.21}$$

The Sudakov-Fernique inequality then gives

$$E\Bigl(\max_{x \in V_N} h_x^{V_N}\Bigr) \leq \sqrt{g \log 2}\; E\Bigl(\max_{x \in L_n} \phi_{\theta(x)}^{T^4}\Bigr). \tag{7.22}$$

The claim follows by extending the maximum on the right from $\theta(L_n)$ to all vertices in L_{n+k}. □

7.3 Maximum of Gaussian Branching Random Walk

In order to use Lemma 7.5 to bound the expected maximum of the DGFF, we need a good control of the expected maximum of the BRW. This is a classical subject with strong connections to large deviation theory. (Indeed, as there are b^n branches of the tree, the maximum will be determined by events whose probability decays exponentially with n. See, e.g., Zeitouni's notes [126].) For Gaussian BRWs, we can rely on explicit calculations and so the asymptotic is completely explicit as well:

Theorem 7.7 (Maximum of Gaussian BRW) *For $b \geq 2$, let $\{\phi_x^{T^b} : x \in T^b\}$ be the Branching Random Walk on b-ary tree with step distribution $\mathcal{N}(0, 1)$. Then*

$$E\Bigl(\max_{x \in L_n} \phi_x^{T^b}\Bigr) = \sqrt{2 \log b}\; n - \frac{3}{2\sqrt{2 \log b}} \log n + O(1), \tag{7.23}$$

where $O(1)$ is a quantity that is bounded uniformly in $n \geq 1$. Moreover,

$$\Bigl\{\max_{x \in L_n} \phi_x^{T^b} - E(\max_{x \in L_n} \phi_x^{T^b}) \colon n \geq 1\Bigr\} \tag{7.24}$$

is tight.

Let us henceforth abbreviate the quantity on the right of (7.23) as

$$\widetilde{m}_n := \sqrt{2\log b}\, n - \frac{3}{2\sqrt{2\log b}} \log n\,. \tag{7.25}$$

The proof starts by showing that the maximum exceeds $\widetilde{m}_n - O(1)$ with a uniformly positive probability. This is achieved by a second moment estimate of the kind we employed for the intermediate level sets of the DGFF. However, as we are dealing with the absolute maximum, a truncation is necessary. Thus, for $x = (x_1, \dots, x_n) \in L_n$, let

$$G_n(x) := \bigcap_{k=0}^{n-1} \Bigl\{ \phi^{T^b}_{(x_1,\dots,x_{k-1})} \le \frac{k}{n}\widetilde{m}_n + 2 \Bigr\} \tag{7.26}$$

be the "good" event that curbs the growth the BRW on the unique path from the root to x. Now define

$$\Gamma_n := \bigl\{ x \in L_n \colon \phi^{T^b}_x \ge \widetilde{m}_n,\ G_n(x) \text{ occurs} \bigr\} \tag{7.27}$$

as the analogue of the truncated level set $\widehat{\Gamma}^{D,M}_N(b)$ from our discussion of intermediate levels of the DGFF. We now claim:

Lemma 7.8 *For the setting as above,*

$$\inf_{n\ge 1} E|\Gamma_n| > 0 \tag{7.28}$$

while

$$\sup_{n\ge 1} E\bigl(|\Gamma_n|^2\bigr) < \infty. \tag{7.29}$$

Let us start with the first moment calculations:

Proof of (7.28). Fix $x \in L_n$ and, for $k = 1, \dots, n$, abbreviate $Z_k := Z_{(x_1,\dots,x_{k-1})}$ (with $Z_1 := Z_\varnothing$). Set

$$S_k := Z_1 + \dots + Z_k, \quad k = 1, \dots, n. \tag{7.30}$$

Then

$$P\bigl(\phi^{T^b}_x \ge \widetilde{m}_n,\ G_n(x) \text{ occurs}\bigr) = P\Biggl(\{S_n \ge \widetilde{m}_n\} \cap \bigcap_{k=1}^{n} \Bigl\{ S_k \le \frac{k}{n}\widetilde{m}_n + 2 \Bigr\} \Biggr). \tag{7.31}$$

In what follows we will make frequent use of: □

Exercise 7.9 *Prove that, for $Z_1, \dots, Z_n$ i.i.d. normal and any $a \in \mathbb{R}$,*

$$P\left((Z_1,\dots,Z_n)\in\cdot\,\Big|\,\sum_{i=1}^n Z_i=a\right)=P\left(\Big(Z_1+\frac{a}{n},\dots,Z_n+\frac{a}{n}\Big)\in\cdot\,\Big|\,\sum_{i=1}^n Z_i=0\right). \tag{7.32}$$

To see this in action, denote

$$\mu_n(\mathrm{d}s):=P(S_n-\widetilde{m}_n\in\mathrm{d}s) \tag{7.33}$$

and use (7.32) to express the probability in (7.31) as

$$\int_0^\infty \mu_n(\mathrm{d}s)P\left(\bigcap_{k=1}^n\Big\{S_k\le-\frac{k}{n}s+2\Big\}\,\Big|\,S_n=0\right). \tag{7.34}$$

As a lower bound, we may restrict the integral to $s\in[0,1]$ which yields $\frac{k}{n}s\le 1$. Realizing Z_k as the increment of the standard Brownian motion on $[k-1,k)$, the giant probability on the right is bounded from below by the probability that the standard Brownian motion on $[0,n]$, conditioned on $B_n=0$, stays below 1 for all times in $[0,n]$. For this we observe:

Exercise 7.10 *Let $\{B_t\colon t\ge 0\}$ be the standard Brownian motion started from 0. Prove that for all $a,b>0$ and all $r>0$,*

$$P^a\Big(B_t\ge 0\colon t\in[0,r]\,\Big|\,B_r=b\Big)=1-\exp\Big\{-2\frac{ab}{r}\Big\}. \tag{7.35}$$

Invoking (7.35) with $a,b:=1$ and $r:=n$ and applying the shift invariance of the Brownian motion, the giant probability in (7.34) is at most $1-\mathrm{e}^{-2/n}$. A calculation shows that, for some constant $c>0$,

$$\mu_n([0,1])\ge\frac{c}{\sqrt{n}}\,\mathrm{e}^{-\frac{\widetilde{m}_n^2}{2n}}=c\,\mathrm{e}^{O(n^{-1}\log n)}b^{-n}n \tag{7.36}$$

thanks to our choice of $\widetilde{m}_n$. The product $n(1-\mathrm{e}^{-2/n})$ is uniformly positive and so we conclude that the probability in (7.31) is at least a constant times b^{-n}. Since $|L_n|=b^n$, summing over $x\in L_n$ we get (7.28). □

We remark that (7.36) (and later also (7.48)) is exactly what determines the precise constant in the subleading term in (7.25). Next we tackle the second moment estimate which is somewhat harder:

Proof of (7.29). Pick distinct $x,y\in L_n$ and let $k\in\{1,\dots,n-1\}$ be such that the paths from the root to these vertices have exactly k vertices (including the root) in common. Let $S_1,\dots,S_n$ be the values of the BRW on the path to x and let $S_1',\dots,S_n'$ be the values on the path to y. Then $S_i=S_i'$ for $i=1,\dots,k$ while

$$\{S_{k+j}-S_k\colon j=1,\dots,n-k\}\quad\text{and}\quad\{S_{k+j}'-S_k'\colon j=1,\dots,n-k\} \tag{7.37}$$

are independent and each equidistributed to $S_1, \ldots, S_{n-k}$. Denoting

$$\mu_{n,k}(\mathrm{d}s) = P\left(S_k - \frac{k}{n}\widetilde{m}_n \in \mathrm{d}s\right), \tag{7.38}$$

conditioning on $S_k - \frac{k}{n}\widetilde{m}_n$ then implies

$$P\Big(\phi_x^{T^b} \vee \phi_y^{T^b} \geq \widetilde{m}_n,\ G_n(x) \cap G_n(y) \text{ occurs}\Big) = \int_{-\infty}^{2} \mu_{n,k}(\mathrm{d}s)\, f_k(s)\, g_{k,n-k}(s)^2, \tag{7.39}$$

where

$$f_k(s) := P\left(\bigcap_{j=1}^{k}\left\{S_j \leq \frac{j}{n}\widetilde{m}_n + 2\right\} \,\middle|\, S_k - \frac{k}{n}\widetilde{m}_n = s\right) \tag{7.40}$$

and

$$g_{k,r}(s) := P\left(\bigcap_{j=1}^{r}\left\{S_j \leq \frac{j}{n}\widetilde{m}_n + 2 - s\right\} \cap \left\{S_r \geq \frac{r}{n}\widetilde{m}_n - s\right\}\right). \tag{7.41}$$

We now need good estimates on both f_k and $g_{k,r}$. For this we will need a version of Exercise 7.10 for random walks:

Exercise 7.11 *(Ballot problem) Let $S_k := Z_1 + \cdots + Z_k$ be the above Gaussian random walk. Prove that there is $c \in (0, \infty)$ such that for any $a \geq 1$ and any $n \geq 1$,*

$$P\left(\bigcap_{k=1}^{n-1}\{S_k \leq a\} \,\middle|\, S_n = 0\right) \leq c\frac{a^2}{n}. \tag{7.42}$$

In what follows we will use c to denote a positive constant whose meaning may change line to line.

Concerning an upper bound on f_k, Exercise 7.9 gives

$$f_k(s) = P\left(\bigcap_{j=1}^{k}\left\{S_j \leq 2 - \frac{j}{k}s\right\} \,\middle|\, S_k = 0\right). \tag{7.43}$$

Exercise 7.11 (and $s \leq 2$) then yields

$$f_k(s) \leq c\frac{1+s^2}{k}. \tag{7.44}$$

As for $g_{k,r}$, we again invoke Exercise 7.9 to write

$$g_{k,r}(s) = \int_{-s}^{2-s} \mu_{n,r}(\mathrm{d}u)\, P\left(\bigcap_{j=1}^{r} \left\{ S_j \le 2 - s - \frac{j}{r} u \right\} \,\Big|\, S_r = 0 \right). \tag{7.45}$$

Exercise 7.11 (and $s \le 2$) again gives

$$g_{k,r}(s) \le c\, \frac{1+s^2}{r}\, P\left(S_r \ge \frac{r}{n} \widetilde{m}_n - s \right). \tag{7.46}$$

In order to plug this into the integral in (7.39), we invoke the standard Gaussian estimate (2.2) and the fact that $\widetilde{m}_n/n$ is uniformly positive and bounded to get, for all $k = 1, \dots, n-1$ and all $s \le 2$,

$$\begin{aligned}
\mu_{n,k}(\mathrm{d}s) P\left(S_{n-k} \ge \frac{n-k}{n} \widetilde{m}_n - s \right)^2 \\
&\le \frac{c}{\sqrt{k}}\, \mathrm{e}^{-\frac{(\frac{k}{n}\widetilde{m}_n + s)^2}{2k}} \left[\frac{1}{\sqrt{n-k}}\, \mathrm{e}^{-\frac{(\frac{n-k}{n}\widetilde{m}_n - s)^2}{2(n-k)}} \right]^2 \mathrm{d}s \\
&\le \frac{c}{\sqrt{k}\,(n-k)}\, \mathrm{e}^{-\frac{1}{2}(\widetilde{m}_n/n)^2(2n-k) + (\widetilde{m}_n/n)s}\, \mathrm{d}s\,.
\end{aligned} \tag{7.47}$$

(When $\frac{n-k}{n}\widetilde{m}_n - s$ is not positive, which for n large happens only when $\log b \le 2$, $k = n-1$ and s is close to 2, the probability is bounded simply by one.) The explicit form of $\widetilde{m}_n$ then gives

$$\mathrm{e}^{-\frac{1}{2}(\widetilde{m}_n/n)^2(2n-k)} \le c b^{-(2n-k)} n^{3 - \frac{3}{2}k/n}. \tag{7.48}$$

The exponential factor in s in (7.47) ensures that the integral in (7.39), including all s-dependent terms arising from (7.44) and (7.46), is bounded. Collecting the denominators from (7.44) and (7.46), we get that

$$P\left(\phi_x^{T^b} \vee \phi_y^{T^b} \ge \widetilde{m}_n,\ G_n(x) \cap G_n(y) \text{ occurs} \right) \le c b^{-(2n-k)} \frac{n^{3-\frac{3}{2}k/n}}{k^{3/2}(n-k)^3} \tag{7.49}$$

holds whenever $x, y \in L_n$ are distinct and k is as above.

The number of distinct pairs $x, y \in L_n$ with the same k is b^{2n-k}. Splitting off the term corresponding to $x = y$, from (7.49) we obtain

$$E\left(|\Gamma_n|^2\right) \le E|\Gamma_n| + c \sum_{k=1}^{n-1} \frac{n^{3-\frac{3}{2}k/n}}{k^{3/2}(n-k)^3}. \tag{7.50}$$

For $k < n/2$ the expression under the sum is bounded by $8k^{-3/2}$ which is summable on all $k \ge 1$. For the complementary k we use $k^{-3/2} \le \sqrt{8} n^{-3/2}$ and then change variables to $j := n - k$ to get

$$\sum_{n/2\le k<n} \frac{n^{3-\frac{3}{2}k/n}}{k^{3/2}(n-k)^3} \le \sqrt{8} \sum_{1\le j\le n/2} \frac{e^{\frac{3}{2}(j/n)\log n}}{j^3}. \tag{7.51}$$

Using ℓ to denote the unique integer such that $\frac{n}{\log n}\ell \le j < \frac{n}{\log n}(\ell+1)$, the sum on the right is further bounded by

$$\sum_{j\ge 1} \frac{e^{3/2}}{j^3} + 2\frac{(\log n)^2}{n^2} \sum_{1\le \ell<\log n} \frac{e^{\frac{3}{2}(\ell+1)}}{\ell^3}. \tag{7.52}$$

The first sum is convergent and, by $e^{\frac{3}{2}(\ell+1)} \le cn^{3/2}$ for $\ell < \log n$, the second part of the expression is at most order $(\log n)^2/\sqrt{n}$. Using this in (7.50) we get $E(|\Gamma_n|^2) \le E|\Gamma_n| + c$. The Cauchy–Schwarz inequality and some algebra then show $E(|\Gamma_n|^2) \le 1 + \sqrt{c} + c$ thus proving (7.29). □

As a consequence of Lemma 7.8 we get:

Corollary 7.12 *Using the notation $\widetilde{m}_n$ from* (7.25),

$$\inf_{n\ge 1} P\Big(\max_{x\in L_n} \phi_x^{T^b} \ge \widetilde{m}_n\Big) > 0. \tag{7.53}$$

Proof. The probability in (7.53) is bounded below by $P(|\Gamma_n| > 0)$ which by (2.13) is bounded from below by $(E|\Gamma_n|)^2/E(|\Gamma_n|^2)$. Thanks to (7.28)–(7.29), this ratio is positive uniformly in $n \ge 1$. □

7.4 Bootstrap to Exponential Tails

Our next task in this lecture is to boost the uniform lower bound (7.53) to an exponential tail estimate. Our method of proof will for convenience be restricted to $b > 2$ (remember that we are interested in $b = 4$) and so this is what we will assume in the statement:

Lemma 7.13 (Lower tail) *For each integer $b > 2$ there is $a = a(b) > 0$ such that*

$$\sup_{n\ge 1} P\Big(\max_{x\in L_n} \phi_x^{T^b} < \widetilde{m}_n - t\Big) \le \frac{1}{a}e^{-at}, \quad t > 0. \tag{7.54}$$

In particular, "$\ge$" holds in (7.23).

Proof. The proof will be based on a percolation argument. Recall that the threshold for site percolation on T^b is $p_c(b) = 1/b$. (This is also the survival threshold of a branching process with offspring distribution $\mathrm{Bin}(b, p)$.) Since $P(Z_x \ge 0) = 1/2$, for any $b > 2$ there is $\epsilon > 0$ such that the set $\{x \in T^b \colon Z_x \ge \epsilon\}$ contains an infinite connected component a.s. We denote by $\mathcal{C}$ the one closest to the origin (breaking ties using an arbitrary a priori ordering of the vertices).

Noting that $\mathcal{C}$, if viewed from the point closest to the origin, contains a supercritical branching process that survives forever, the reader will surely be able to solve:

Exercise 7.14 *Show that there are $\theta > 1$ and $c > 0$ such that for all $r \geq 1$,*

$$P\Bigl(\exists n \geq r \colon |\mathcal{C} \cap L_n| < \theta^n\Bigr) \leq \mathrm{e}^{-cr} \tag{7.55}$$

Writing again c for a generic positive constant, we claim that this implies

$$P\Bigl(|\{x \in L_k \colon \phi_x^{T^b} \geq 0\}| < \theta^k\Bigr) \leq \mathrm{e}^{-ck}, \quad k \geq 1. \tag{7.56}$$

Indeed, a crude first moment estimate shows

$$P\Bigl(\min_{x \in L_r} \phi_x^{T^b} \leq -2\sqrt{\log b}\, r\Bigr) \leq cb^{-r}, \quad r \geq 0. \tag{7.57}$$

Taking $r := \delta n$, with $\delta \in (0, 1)$, on the event that $\min_{x \in L_r} \phi_x^{T^b} > -2\sqrt{\log b}\, r$ and $\mathcal{C} \cap L_r \neq \emptyset$, we then have

$$\phi_x^{T^b} \geq -2\sqrt{\log b}\, \delta n + \epsilon(n - \delta n), \qquad x \in \mathcal{C} \cap L_n. \tag{7.58}$$

Assuming $\delta > 0$ is small enough so that $\epsilon(1 - \delta) > 2\sqrt{\log b}\, \delta$, this gives $\phi_x^{T^b} \geq 0$ for all $x \in \mathcal{C} \cap L_n$. Hence (7.56) follows from (7.55) and (7.57).

Moving to the proof of (7.54), fix $t > 0$ and let k be the largest integer less than n such that $\widetilde{m}_n - t \leq \widetilde{m}_{n-k}$. Denote $A_k := \{x \in L_k \colon \phi_x^{T^b} \geq 0\}$. On the event in (7.54), the maximum of the BRW of depth $n - k$ started at any vertex in A_k must be less than $\widetilde{m}_{n-k}$. Conditional on A_k, this has probability at most $(1 - q)^{|A_k|}$, where q is the infimum in (7.53). On the event that $|A_k| \geq \theta^k$, this decays double exponentially with k and so the probability in (7.54) is dominated by that in (7.56). The claim follows by noting that $t \approx \sqrt{2\log b}\, k$. □

With the lower-tail settled, we can address the upper bound as well:

Lemma 7.15 (Upper tail) *For each $b > 0$ there is $\tilde{a} = \tilde{a}(b) > 0$ such that*

$$\sup_{n \geq 1} P\Bigl(\max_{x \in L_n} \phi_x^{T^b} > \widetilde{m}_n + t\Bigr) \leq \frac{1}{\tilde{a}} \mathrm{e}^{-\tilde{a}t}, \quad t > 0. \tag{7.59}$$

Proof. The continuity of the involved Gaussian random variables ensures that the maximum occurs at a unique $x = (x_1, \ldots, x_n) \in L_n$ a.s. Write $Z_k := Z_{(x_1,\ldots,x_{k-1})}$ (with $Z_1 := Z_\varnothing$) and recall the notation S_k from (7.30). Note that each vertex $(x_1, \ldots, x_k)$ on the path from the root to x has $b - 1$ "children" $y_1, \ldots, y_{b-1}$ not lying on this path. Let $\widetilde{M}_\ell^{(i)}$ denote the maximum of the BRW of depth ℓ rooted at y_i and abbreviate

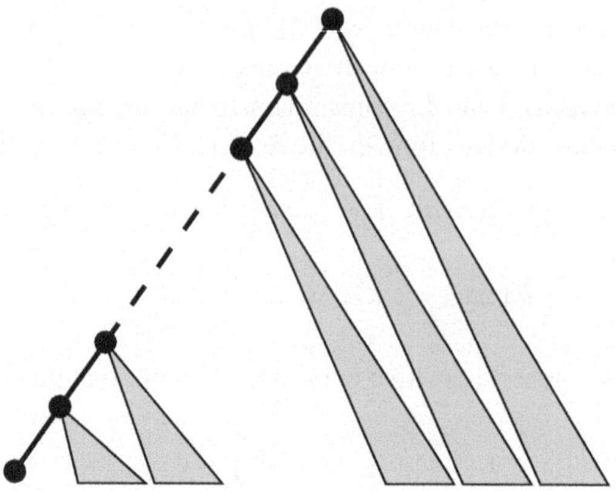

Fig. 13 The picture demonstrating the geometric setup for the representation in (7.61). The bullets mark the vertices on the path from the root (top vertex) to x (the vertex on the bottom left). The union of the relevant subtrees of these vertices are marked by shaded triangles. The maximum of the field in the subtrees of ℓth vertex on the path is the quantity in (7.60)

$$\widehat{M}_\ell := \max_{i=1,\dots,b-1} \widetilde{M}^{(i)}_{\ell-1}, \tag{7.60}$$

see Fig. 13. Since the maximum occurs at x and equals $\widetilde{m}_n + u$ for some $u \in \mathbb{R}$, we must have $S_k + \widehat{M}_{n-k} \le \widetilde{m}_n + u$ for all $k = 1, \dots, n-1$. The symmetries of the BRW then allow us to write the probability in (7.59) as

$$b^n \int_t^\infty \mu_n(\mathrm{d}u) P\left(\bigcap_{k=1}^{n-1} \{S_k + \widehat{M}_{n-k} \le \widetilde{m}_n + u\} \,\Big|\, S_n = \widetilde{m}_n + u \right), \tag{7.61}$$

where μ_n is the measure in (7.33) and where $\widehat{M}_1, \dots, \widehat{M}_n$ are independent of each other and of the random variables $Z_1, \dots, Z_n$ that define the S_k's.

We will now estimate (7.61) similarly as in the proof of Lemma 7.8. First, shifting the normals by their arithmetic mean, the conditional probability in the integral is recast as

$$P\left(\bigcap_{k=1}^{n-1} \left\{ S_k + \widehat{M}_{n-k} \le \frac{n-k}{n}(\widetilde{m}_n + u) \right\} \,\Big|\, S_n = 0 \right). \tag{7.62}$$

Letting $\theta_n(k) := (k \wedge (n-k))^{1/5}$, we readily check that

$$\frac{n-k}{n}\widetilde{m}_n \le \widetilde{m}_{n-k} + \theta_n(k), \quad k = 1, \dots, n, \tag{7.63}$$

as soon as n is sufficiently large. Introducing

$$\Theta_n := \max_{k=1,\dots,n} \big[\widetilde{m}_{n-k} - \widehat{M}_{n-k} - \theta_n(k)\big]_+, \tag{7.64}$$

the probability in (7.62) is thus bounded from above by

$$P\left(\bigcap_{k=1}^{n-1}\Big\{S_k \le \Theta_n + 2\theta_n(k) + u\Big\} \,\Big|\, S_n = 0\right). \tag{7.65}$$

We now observe:

Exercise 7.16 *(Inhomogeneous ballot problem) Let $S_k := Z_1 + \dots + Z_k$ be the above Gaussian random walk. Prove that there is $c \in (0, \infty)$ such that for any $a \ge 1$ and any $n \ge 1$,*

$$P\left(\bigcap_{k=1}^{n-1}\Big\{S_k \le a + 2\theta_n(k)\Big\} \,\Big|\, S_n = 0\right) \le c\frac{a^2}{n}. \tag{7.66}$$

(This is in fact quite hard. Check the Appendix of [27] for ideas and references.)

Noting that Θ_n is independent of $Z_1, \dots, Z_n$, Exercise 7.16 bounds the probability in (7.65) by a constant times $n^{-1}E([\Theta_n + u]^2)$. The second moment of Θ_n is bounded uniformly in $n \ge 1$ because, by Lemma 7.13 and a union bound,

$$P(\Theta_n > u) \le \sum_{k=1}^{n} \mathrm{e}^{-a(\theta_n(k)+u)}, \quad u > 0. \tag{7.67}$$

The probability in (7.65) is thus at most a constant times $(1 + u^2)/n$. Since

$$\mu_n\big([u, u+1]\big) \le c\,\mathrm{e}^{-(\widetilde{m}_n/n)u}\, nb^{-n}, \quad u \ge 0, \tag{7.68}$$

the claim follows by a routine calculation. □

Remark 7.17 Exercise 7.16 is our first excursion to the area of "random walks above slowly-varying curves" or "Inhomogenous Ballot Theorems" which we will encounter several times in these notes. We will not supply detailed proofs of these estimates as these are quite technical and somewhat detached from the main theme of these notes. The reader is encouraged to consult Bramson's seminal work [35] as well as the Appendix of [27] or the recent posting by Cortines, Hartung and Louidor [49] for a full treatment.

We now quickly conclude:

Proof of Theorem 7.7. Combining Lemmas 7.13–7.15, the maximum has exponential tails away from $\widetilde{m}_n$, uniformly in $n \ge 1$. This yields the claim. □

Notice that the underlying idea of the previous proof is to first establish a bound on the lower tail of the maximum and then use it (via a bound on the second moment of Θ_n) to control the upper tail. A similar strategy, albeit with upper and lower tails interchanged, will be used in the next lecture to prove tightness of the DGFF maximum.

Lecture 8: Tightness of DGFF Maximum

We are now ready to tackle the tightness of the DGFF maximum stated in Theorem 7.3. The original proof due to Bramson and Zeitouni [37] was based on comparisons with the so called modified Branching Random Walk. We bypass this by proving tightness of the upper tail directly using a variation of the Dekking–Host argument and controlling the lower tail via a concentric decomposition of the DGFF. This brings us closer to what we have done for the Branching Random Walk. The concentric decomposition will be indispensable later as well; specifically, in the analysis of the local structure of nearly-maximal local maxima and the proof of distributional convergence of the DGFF maximum.

8.1 Upper Tail of DGFF Maximum

Recall the notation m_N from (7.9) and $\widetilde{m}_n$ from (7.25). As is easy to check, for $b := 4$ and $N := 2^n$ we have

$$\sqrt{g\log 2}\,\widetilde{m}_n = m_N + O(1) \tag{8.1}$$

and so (7.17) and (7.23) yield $EM_N \le m_N + O(1)$. Unfortunately, this does not tell us much by itself (indeed, the best type of bound we can extract from this is that $P(M_N > 2m_N)$ is at most about a half.) Notwithstanding, the argument can be enhanced to yield tightness of the upper tail of M_N:

Lemma 8.1 (Upper tail tightness) *We have*

$$\sup_{N\ge1} E\bigl((M_N - m_N)_+\bigr) < \infty. \tag{8.2}$$

For the proof we will need the following general inequality:

Lemma 8.2 (Between Slepian and Sudakov-Fernique) *Suppose X and Y are centered Gaussians on $\mathbb{R}^n$ such that*

$$E\bigl((X_i - X_j)^2\bigr) \le E\bigl((Y_i - Y_j)^2\bigr), \quad i,j = 1,\dots,n \tag{8.3}$$

and

$$E(X_i^2) \le E(Y_i^2), \quad i = 1,\dots,n. \tag{8.4}$$

Abbreviate

$$M_X := \max_{i=1,\dots,n} X_i \quad \textit{and} \quad M_Y := \max_{i=1,\dots,n} Y_i \tag{8.5}$$

and let $M'_Y \overset{\text{law}}{=} M_Y$ *be such that* $M'_Y \perp\!\!\!\perp M_Y$*. Then*

$$E\bigl((M_X - EM_Y)_+\bigr) \le E\bigl(\max\{M_Y, M'_Y\}\bigr) - EM_Y. \tag{8.6}$$

Proof. Let Y' be a copy of Y and assume X, Y, Y' are realized as independent on the same probability space. Define random vectors $Z, \widetilde{Z} \in \mathbb{R}^{2n}$ as

$$Z_i := X_i \quad \text{and} \quad Z_{n+i} := Y'_i, \quad i = 1, \dots, n \tag{8.7}$$

and

$$\widetilde{Z}_i := Y_i \quad \text{and} \quad \widetilde{Z}_{n+i} := Y'_i, \quad i = 1, \dots, n. \tag{8.8}$$

Since $E((X_i - Y'_j)^2) = E(X_i^2) + E((Y'_j)^2)$, from (8.3)–(8.4) we readily get

$$E\bigl((Z_i - Z_j)^2\bigr) \le E\bigl((\widetilde{Z}_i - \widetilde{Z}_j)^2\bigr), \quad i, j = 1, \dots, 2n. \tag{8.9}$$

Writing $M'_Y := \max_{i=1,\dots,n} Y'_i$, from the Sudakov-Fernique inequality we infer

$$E\bigl(\max\{M_X, M'_Y\}\bigr) \le E\bigl(\max\{M_Y, M'_Y\}\bigr). \tag{8.10}$$

Invoking $(a-b)_+ = \max\{a,b\} - b$ and using Jensen's inequality to pass the expectation over Y' inside the positive-part function, the claim follows. □

Proof of Lemma 8.1. Let n be the least integer such that $N \le 4^n$, assume that V_N is naturally embedded into L_n and consider the map $\theta\colon L_n \to L_{n+k}$ for k as in Lemma 7.5. Assume h^{V_N} and ϕ^{T^4} are realized independently on the same probability space, denote

$$\widetilde{M}_n := \sqrt{g \log 2}\, \max_{x \in L_n} \phi_x^{T^4}, \tag{8.11}$$

and observe that

$$\sqrt{g \log 2}\, \max_{x \in L_n} \phi_{\theta(x)}^{T^4} \le M_{n+k} \tag{8.12}$$

and

$$\sqrt{g \log 2}\, E\Bigl(\max_{x \in L_n} \phi_{\theta(x)}^{T^4}\Bigr) \ge E\widetilde{M}_n, \tag{8.13}$$

where the second inequality follows by the same reasoning as (7.6) in the Dekking–Host argument. In light of (7.21) and

$$E\bigl([h_x^{V_N}]^2\bigr) \le (g \log 2)\, E\bigl([\phi_{\theta(x)}^{T^4}]^2\bigr), \qquad x \in L_n, \tag{8.14}$$

(proved by the same computation as (7.21)) the conditions (8.3)–(8.4) are satisfied for $X := h^{V_N}$ and $Y := \sqrt{g\log 2}\,\phi^{T^4}_{\theta(\cdot)}$ (both indexed by L_n). Lemma 8.2 along with (8.12)–(8.13) and downward monotonicity of $b \mapsto (a-b)_+$ show

$$E\big((M_N - E\widetilde{M}_{n+k})_+\big) \le E\big(\max\{\widetilde{M}_{n+k}, \widetilde{M}'_{n+k}\}\big) - E(\widetilde{M}_n), \tag{8.15}$$

where

$$\widetilde{M}'_{n+k} \overset{\text{law}}{=} \widetilde{M}_{n+k} \quad \text{with} \quad \widetilde{M}'_{n+k} \perp\!\!\!\perp \widetilde{M}_{n+k}. \tag{8.16}$$

The maximum on the right-hand side is now dealt with as in the Dekking–Host argument (see the proof of Lemma 7.1). Indeed, the definition of the BRW gives

$$M_{n+k+1} \overset{\text{law}}{\ge} Z + \max\{\widetilde{M}_{n+k}, \widetilde{M}'_{n+k}\}, \tag{8.17}$$

where $Z \overset{\text{law}}{=} \mathcal{N}(0, g\log 2)$ is independent of $\widetilde{M}_{n+k}$ and $\widetilde{M}'_{n+k}$. It follows that

$$E\big((M_N - E\widetilde{M}_{n+k})_+\big) \le E(\widetilde{M}_{n+k+1}) - E(\widetilde{M}_n). \tag{8.18}$$

By Theorem 7.7 and (8.1), both the right-hand side and $E\widetilde{M}_{n+k} - m_N$ are bounded uniformly in $N \ge 1$, thus proving the claim. □

Once we know that $(M_N - m_N)_+$ cannot get too large, we can bootstrap this to an exponential upper tail, just as in Lemma 7.15 for the BRW:

Lemma 8.3 *There are $\tilde a > 0$ and $t_0 > 0$ such that*

$$\sup_{N\ge 1} P\big(M_N \ge m_N + t\big) \le \mathrm{e}^{-\tilde a t}, \qquad t \ge t_0. \tag{8.19}$$

Proof. Lemma 8.1 and the Markov inequality ensure that, for some $r > 0$,

$$\inf_{N\ge 1} P\big(M_N \le m_N + r\big) \ge \frac{1}{2}. \tag{8.20}$$

Fix an even integer $K \ge 1$ and consider the DGFF in V_{3KN}. Identify K^2 disjoint translates of V_{3N} inside V_{3KN} such that any pair of adjacent translates is separated by a line of sites. Denote these translates $V_{3N}^{(i)}$, with $i = 1, \dots, K^2$, and abusing our earlier notation slightly, write $V_{3KN}^\circ := \bigcup_{i=1}^{K^2} V_{3N}^{(i)}$. Moreover, let $V_N^{(i)}$ be a translate of V_N centered at the same point as $V_{3N}^{(i)}$; see Fig. 14. Using the Gibbs-Markov decomposition, we then have

$$M_{3KN} \overset{\text{law}}{\ge} \max_{i=1,\dots,K^2} \max_{x\in V_N^{(i)}} \Big(h_x^{V_{3N}^{(i)}} + \varphi_x^{V_{3KN}, V_{3KN}^\circ}\Big). \tag{8.21}$$

Consider the event

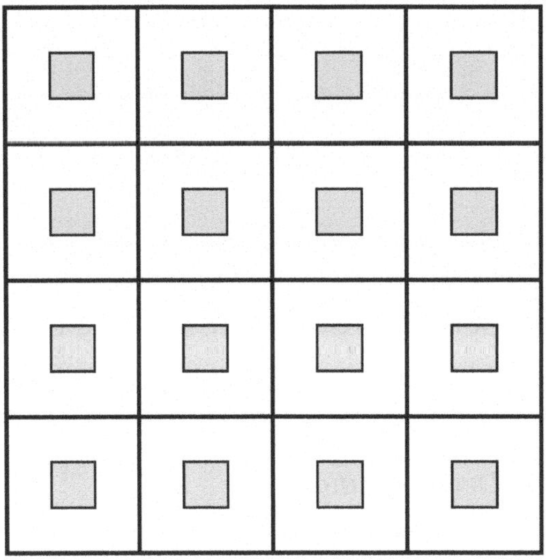

Fig. 14 The geometric setup underlying the proof of Lemma 8.3. The large square marks the domain V_{3KN} for $K := 4$. The shaded squares are the translates $V_N^{(i)}$, $i = 1, \dots, K^2$, of V_N

$$A_K := \Bigl\{\#\Bigl\{i \in \{1, \dots, K^2\}\colon \min_{x \in V_N^{(i)}} \varphi_x^{V_{3KN}, V_{3KN}^\circ} \ge -\sqrt{t}\sqrt{\log K}\Bigr\} \ge K^2/2\Bigr\}. \tag{8.22}$$

Since $\mathrm{Var}(\varphi_x^{V_{3KN}, V_{3KN}^\circ}) \le c \log K$, a combination of Borell–TIS inequality with Fernique's majorization permits us to solve:

Exercise 8.4 *Prove that there are $a > 0$ and $t_0 \ge 0$ such that for all $t \ge t_0$,*

$$\sup_{N \ge 1} \max_{i=1,\dots,K} P\Bigl(\min_{x \in V_N^{(i)}} \varphi_x^{V_{3KN}, V_{3KN}^\circ} < -\sqrt{t}\sqrt{\log K}\Bigr) \le \mathrm{e}^{-at}. \tag{8.23}$$

Assuming $t \ge t_0$, the Markov inequality shows $P(A_K^{\mathrm{c}}) \le 2\mathrm{e}^{-at}$. The Gibbs-Markov decomposition (8.21) (and translation invariance of the DGFF) then yields

$$P\bigl(M_{3KN} \le m_{3KN} + r\bigr) \le 2\mathrm{e}^{-at} + P\Bigl(\max_{x \in V_N'} h_x^{V_{3N}} \le m_{KN} + r + \sqrt{t}\sqrt{\log K}\Bigr)^{K^2/2}, \tag{8.24}$$

where V_N' is the translate of V_N centered at the same point as V_{3N}. We now invoke the bound $P(X \le s) = 1 - P(X > s) \le \mathrm{e}^{-P(X > s)}$ along with Exercise 3.4 to replace $h^{V_{3N}}$ by $h^{V_N'}$ in the maximum over $x \in V_N'$ at the cost of another factor of $1/2$ popping in front of the probability. This turns (8.20) and (8.24) into

$$\frac{1}{2} - 2\mathrm{e}^{-at} \le \exp\Bigl\{-\frac{1}{4}K^2 P\bigl(M_N > m_{3NK} + r + \sqrt{t}\sqrt{\log K}\bigr)\Bigr\}. \tag{8.25}$$

Assuming t is so large that $2\mathrm{e}^{-at} \le 1/4$ and applying the inequality

$$m_{3KN} \le m_N + 2\sqrt{g}\log K, \tag{8.26}$$

this proves the existence of $c' > 0$ such that, for all even $K \ge 1$,

$$\sup_{N\ge 1} P\Bigl(M_N > m_N + r + 2\sqrt{g}\log K + \sqrt{t\log K}\Bigr) \le c' K^{-2}. \tag{8.27}$$

Substituting $t := c\log K$ for c large enough that $r + 2\sqrt{g}\log K + \sqrt{t\log K} \le t$ then gives (8.19). □

8.2 Concentric Decomposition

Although the above conclusions seem to be quite sharp, they are not inconsistent with M_N being concentrated at values much smaller than m_N. (Indeed, comparisons with the BRW will hardly get us any further because of the considerable defect in (7.20). This is what the modified BRW was employed in [37] for but this process is then much harder to study than the BRW.) To address this deficiency we now develop the technique of *concentric decomposition* that will be useful in several parts of these notes.

In order to motivate the definitions to come, note that, to rule out $M_N \ll m_N$, by Lemma 8.3 it suffices to show $EM_N \ge m_N + O(1)$. In the context of BRW, this was reduced (among other things) to calculating the asymptotic of a probability that for the DGFF takes the form

$$P\bigl(h^{D_N} \le m_N + t \,\big|\, h_0^{D_N} = m_N + t\bigr), \tag{8.28}$$

where we assumed that $0 \in D_N$. For the BRW it was useful (see Exercise 7.9) that the conditional event can be transformed into (what for the DGFF is) $h_0^{D_N} = 0$ at the cost of subtracting a suitable linear expression in $h_0^{D_N}$ from all fields. Such a reduction is possible here as well and yields:

Lemma 8.5 (Reduction to pinned DGFF) *Suppose $D_N \subseteq \mathbb{Z}^2$ is finite with $0 \in D_N$. Then for all $t \in \mathbb{R}$ and $s \ge 0$,*

$$\begin{aligned} P\bigl(h^{D_N} \le m_N + t + s \,\big|\, h_0^{D_N} = m_N + t\bigr) \\ = P\bigl(h^{D_N} \le (m_N + t)(1 - \mathfrak{g}^{D_N}) + s \,\big|\, h_0^{D_N} = 0\bigr), \end{aligned} \tag{8.29}$$

where $\mathfrak{g}^{D_N} \colon \mathbb{Z}^2 \to [0,1]$ is discrete-harmonic on $D_N \smallsetminus \{0\}$ with $\mathfrak{g}^{D_N}(0) = 1$ and $\mathfrak{g}^{D_N} = 0$ on D_N^{c}. In particular, the probability is non-decreasing in s and t.

Proof. The Gibbs-Markov decomposition of h^{D^N} reads

$$h^{D_N} \overset{\text{law}}{=} h^{D_N \smallsetminus \{0\}} + \varphi^{D_N, D_N \smallsetminus \{0\}}. \tag{8.30}$$

Now $\varphi^{D_N, D_N \smallsetminus \{0\}}$ has the law of the discrete-harmonic extension of h^{D_N} on $\{0\}$ to $D_N \smallsetminus \{0\}$. This means $\varphi^{D_N, D_N \smallsetminus \{0\}} = \mathfrak{g}^{D_N} h^{D_N}(0)$. Using this, the desired probability can be written as

$$P\big(h^{D_N \smallsetminus \{0\}} \le (m_N + t)(1 - \mathfrak{g}^{D_N}) + s\big). \tag{8.31}$$

The claim then follows from the next exercise. □

Exercise 8.6 *(Pinning is conditioning) For any finite* $D \subseteq \mathbb{Z}^2$ *with* $0 \in D$,

$$(h^D \mid h_0^D = 0) \overset{\text{law}}{=} h^{D \smallsetminus \{0\}}. \tag{8.32}$$

The conditioning the field to be zero is useful for the following reason:

Exercise 8.7 *(Pinned field limit) Prove that, for* $0 \in D_n \uparrow \mathbb{Z}^2$,

$$h^{D_N \smallsetminus \{0\}} \underset{N \to \infty}{\overset{\text{law}}{\longrightarrow}} h^{\mathbb{Z}^2 \smallsetminus \{0\}} \tag{8.33}$$

in the sense of finite dimensional distributions.

Let us now inspect the event $h^{D_N} \le m_N(1 - \mathfrak{g}^{D_N})$ in (8.29)—with t and s dropped for simplicity. The following representation using the Green function G^{D_N} will be useful

$$m_N\big(1 - \mathfrak{g}^{D_N}(x)\big) = m_N \frac{G^{D_N}(0,0) - G^{D_N}(0,x)}{G^{D_N}(0,0)}. \tag{8.34}$$

Now $m_N = 2\sqrt{g} \log N + o(\log N)$ while (for 0 deep inside D_N) for the Green function we get $G^{D_N}(0,0) = g \log N + O(1)$. With the help of the relation between the Green function and the potential kernel $\mathfrak{a}$ as well as the large-scale asymptotic form of $\mathfrak{a}$ (see Lemmas 1.19 and 1.21) we then obtain

$$m_N\big(1 - \mathfrak{g}^{D_N}(x)\big) = \frac{2}{\sqrt{g}} \mathfrak{a}(x) + o(1) = 2\sqrt{g} \log|x| + O(1). \tag{8.35}$$

The event $\{h^{D_N} \le m_N(1 - \mathfrak{g}^{D_N})\} \cap \{h_0^{D_N} = 0\}$ thus forces the field to stay below the logarithmic cone $x \mapsto 2\sqrt{g} \log|x| + O(1)$. Notice that this is the analogue of the event that the BRW on *all subtrees* along the path to the maximum stay below a linear curve; see (7.61).

In order to mimic our earlier derivations for the BRW, we need to extract as much independence from the DGFF as possible. The Gibbs-Markov property is the right tool to use here. We will work with a decomposition over a sequence of domains defined (abusing our earlier notation) by

$$\Delta^k := \begin{cases} \{x \in \mathbb{Z}^2 \colon |x|_\infty < 2^k\}, & \text{if } k = 0, \dots, n-1, \\ D_N, & \text{if } k = n, \end{cases} \tag{8.36}$$

where n is the largest integer such that $\{x \in \mathbb{Z}^2 \colon |x|_\infty \le 2^{n+1}\} \subseteq D_N$; see Fig. 9 in Sect. 4.5. The Gibbs-Markov property now gives

$$h^{D_N} = h^{\Delta^n} \overset{\text{law}}{=} h^{\Delta^{n-1}} + h^{\Delta^n \smallsetminus \overline{\Delta^{n-1}}} + \varphi^{\Delta^n, \Delta^n \smallsetminus \partial \Delta^{n-1}}, \tag{8.37}$$

where ∂D stands for the set of vertices on the external boundary of $D \subseteq \mathbb{Z}^2$ and $\overline{D} := D \cup \partial D$. This relation can be iterated to yield:

Lemma 8.8 *For the setting as above,*

$$h^{D_N} \overset{\text{law}}{=} \sum_{k=0}^{n} (\varphi_k + h'_k) \tag{8.38}$$

where all the fields on the right are independent with

$$\varphi_k \overset{\text{law}}{=} \varphi^{\Delta^k, \Delta^k \smallsetminus \partial \Delta^{k-1}} \quad \text{and} \quad h'_k \overset{\text{law}}{=} h^{\Delta^k \smallsetminus \overline{\Delta^{k-1}}} \tag{8.39}$$

for $k = 1, \dots, n$ *and*

$$\varphi_0 \overset{\text{law}}{=} h^{\{0\}} \quad \text{and} \quad h'_0 = 0. \tag{8.40}$$

Proof. Apply induction to (8.37) while watching for the provisos at $k = 0$. □

The representation (8.38) is encouraging in that it breaks h^{D_N} into the sum of independent contributions of which one (the φ_k's) is "smooth" and timid and the other (the h'_k's) is, while "rough" and large, localized to the annulus $\Delta^k \smallsetminus \overline{\Delta^{k-1}}$. In order to make the correspondence with the BRW closer, we need to identify an analogue of the Gaussian random walk from (7.30) in this expression. Here we use that, since φ_k is harmonic on $\Delta^k \smallsetminus \partial \Delta^{k-1}$, its typical value is well represented by its value at the origin. This gives rise to:

Proposition 8.9 (Concentric decomposition of DGFF) *For the setting as above,*

$$h^{D_N} \overset{\text{law}}{=} \sum_{k=0}^{n} \Big((1 + \mathfrak{b}_k)\varphi_k(0) + \chi_k + h'_k \Big), \tag{8.41}$$

where all the random objects in $\bigcup_{k=0}^{n} \{\varphi_k(0), \chi_k, h'_k\}$ *are independent of one another with the law of* $\varphi_k(0)$ *and* h'_k *as in (8.39)–(8.40) and with*

$$\chi_k(\cdot) \overset{\text{law}}{=} \varphi_k(\cdot) - E\big(\varphi_k(\cdot)\,\big|\,\sigma(\varphi_k(0))\big). \tag{8.42}$$

The function $\mathfrak{b}_k\colon \mathbb{Z}^2 \to \mathbb{R}$ *is defined by*

$$\mathfrak{b}_k(x) := \frac{E\big([\varphi_k(x) - \varphi_k(0)]\varphi_k(0)\big)}{E\big(\varphi_k(0)^2\big)}. \tag{8.43}$$

Proof. Define χ_k from φ_k by the right-hand side of (8.42). Then χ_k and $\varphi_k(0)$ are uncorrelated and, being Gaussian, independent. Moreover, the fact that conditional expectation is a projection in L^2 ensures that $E\big(\varphi_k(\cdot)\,\big|\,\sigma(\varphi_k(0))\big)$ is a linear function of $\varphi_k(0)$. The fact that these fields have zero mean then implies

$$E\big(\varphi_k(x)\,\big|\,\sigma(\varphi_k(0))\big) = \mathfrak{f}_k(x)\varphi_k(0) \tag{8.44}$$

for some deterministic $\mathfrak{f}_k\colon \mathbb{Z}^2 \to \mathbb{R}$. A covariance computation shows $\mathfrak{f}_k = 1 + \mathfrak{b}_k$. Substituting

$$\varphi_k = (1 + \mathfrak{b}_k)\varphi_k(0) + \chi_k, \tag{8.45}$$

which, we note, includes the case $k = 0$, into (8.38) then gives the claim. □

8.3 Bounding the Bits and Pieces

One obvious advantage of (8.41) is that it gives us a representation of DGFF as the sum of *independent*, and reasonably localized, objects. However, in order to make use of this representation, we need estimates on the sizes of these objects as well. The various constants in the estimates that follow will depend on the underlying set D_N but only via the smallest $k_1 \in \mathbb{N}$ such that

$$D_N \subseteq \{x \in \mathbb{Z}^2\colon |x|_\infty \le 2^{n+1+k_1}\} \tag{8.46}$$

with n as above. We thus assume this k_1 to be fixed; all estimates are then uniform in the domains satisfying (8.46). We begin with the $\varphi_k(0)$'s:

Lemma 8.10 *For each* $\epsilon > 0$ *there is* $k_0 \ge 0$ *such that for all* $n \ge k_0 + 1$,

$$\max_{k=k_0,\dots,n-1} \Big|\mathrm{Var}\big(\varphi_k(0)\big) - g\log 2\Big| < \epsilon \tag{8.47}$$

Moreover, $\mathrm{Var}(\varphi_k(0))$ *is bounded away from zero and infinity by positive constants that depend only on* k_1 *from* (8.46).

Proof (sketch). For $k < n$ large, $\varphi_k(0)$ is close in law to the value at zero of the continuum binding field $\Phi^{B_2, B_2 \smallsetminus \partial B_1}$, where $B_r := [-r, r]^2$. A calculation shows $\mathrm{Var}(\Phi^{B_2, B_2 \smallsetminus \partial B_1}(0)) = g\log 2$. □

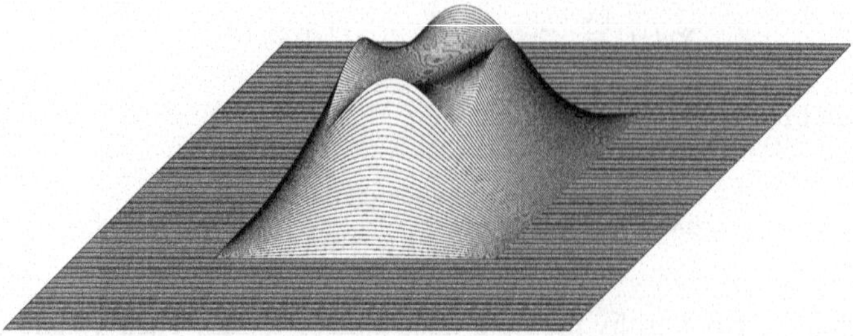

Fig. 15 A plot of $\mathfrak{b}_k$ on the set Δ^{k+1} for k large. The function equals -1 outside Δ^k and vanishes at the origin. It is discrete harmonic on $\Delta^k \smallsetminus \partial\Delta^{k-1}$

Lemma 8.11 *The function $\mathfrak{b}_k$ is bounded uniformly in k, is discrete-harmonic on $\Delta^k \smallsetminus \partial\Delta^{k-1}$ and obeys*

$$\mathfrak{b}_k(0) = 0 \quad \text{and} \quad \mathfrak{b}_k(\cdot) = -1 \text{ on } \mathbb{Z}^2 \smallsetminus \Delta^k. \tag{8.48}$$

Moreover, there is $c > 0$ such that for all $k = 0, \dots, n$,

$$\bigl|\mathfrak{b}_k(x)\bigr| \le c\frac{\text{dist}(0, x)}{\text{dist}(0, \partial\Delta^k)}, \qquad x \in \Delta^{k-2}. \tag{8.49}$$

Proof (*sketch*). The harmonicity of $\mathfrak{b}_k$ follows from harmonicity of φ_k. The overall boundedness is checked by representing $\mathfrak{b}_k$ using the covariance structure of φ_k. The bound (8.49) then follows from uniform Lipschitz continuity of the (discrete) Poisson kernels in square domains. See Fig. 15. □

Lemma 8.12 *For $k = 0, \dots, n$ and $\ell = 0, \dots, k-2$,*

$$E\Bigl(\max_{x\in\Delta^\ell}\bigl|\chi_k(x)\bigr|\Bigr) \le c2^{\ell-k} \tag{8.50}$$

and

$$P\biggl(\Bigl|\max_{x\in\Delta^\ell}\chi_k(x) - E\max_{x\in\Delta^\ell}\chi_k(x)\Bigr| > \lambda\biggr) \le \mathrm{e}^{-c4^{k-\ell}\lambda^2}. \tag{8.51}$$

Proof (*idea*). These follow from Fernique majorization and Borell–TIS inequality and Lipschitz property of the covariances of φ_k (which extend to χ_k). □

The case of $\ell = k-1, k$ has intentionally been left out of Lemma 8.12 because χ_k ceases to be regular on and near $\partial\Delta^{k-1}$, being essentially equal (in law) to the DGFF there; see Fig. 16. Combining χ_k with h'_k and χ_{k+1} we get:

Lemma 8.13 (Consequence of upper-tail tightness of M_N) *There exists $a > 0$ such that each $k = 1, \dots, n$ and each $t \ge 1$,*

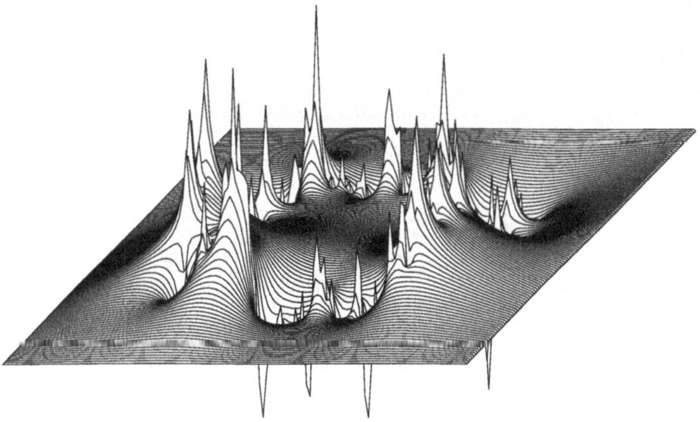

Fig. 16 A plot of a sample of χ_k on Δ^k for $k := 7$. The sample function is discrete harmonic on $\Delta^k \smallsetminus \partial\Delta^{k-1}$ but quite irregular on and near $\partial\Delta^{k-1}$

$$P\Bigl(\max_{x\in\Delta^k\smallsetminus\Delta^{k-1}}\bigl[\chi_{k+1}(x)+\chi_k(x)+h'_k(x)\bigr]-m_{2^k}\ge t\Bigr)\le \mathrm{e}^{-at}. \tag{8.52}$$

Proof (*sketch*). Recalling how the concentric decomposition was derived,

$$\varphi_{k+1}+\varphi_k+h'_k \overset{\text{law}}{=} h^{\Delta^k} \quad \text{on } \Delta^k\smallsetminus\Delta^{k-1}. \tag{8.53}$$

Lemma 8.3 along with the first half of Exercise 3.4 show that this field has exponential upper tail above m_{2^k}. But this field differs from the one in the statement by the term $(1+\mathfrak{b}_k)\varphi_k(0)+(1+\mathfrak{b}_{k+1})\varphi_{k+1}(0)$ which has even a Gaussian tail. The claim follows from a union bound. □

Remark 8.14 Once we prove the full tightness of the DGFF maximum, we will augment (8.52) to an estimate on the maximal *absolute value* of the quantity in (8.52), see Lemma 11.4. However, at this point we claim only a bound on the upper tail in (8.52).

In light of $\chi_k(0)=0$, $h'_k(0)=0$ and $\mathfrak{b}_k(0)=0$ we have $h_0^{\Delta^n}=\sum_{k=0}^n\varphi_k(0)$. This leads to a representation of the field at the prospective maximum by the $(n+1)$st member of the sequence

$$S_k := \sum_{\ell=0}^{k-1}\varphi_\ell(0)\,, \tag{8.54}$$

which we think of as an analogue of the random walk in (7.30) albeit this time with time inhomogeneous steps. The observation

$$h_0^{\Delta^n}=0 \quad\Leftrightarrow\quad S_{n+1}=0 \tag{8.55}$$

along with the fact that, for any $k = 0, \dots, n-1$, neither $\mathfrak{b}_k$ nor the laws of φ_k and h'_k depend on n and D_N then drive:

Exercise 8.15 *(Representation of pinned DGFF) The DGFF on $\mathbb{Z}^2 \smallsetminus \{0\}$ can be represented as the a.s.-convergent sum*

$$h^{\mathbb{Z}^2 \smallsetminus \{0\}} \overset{\text{law}}{=} \sum_{k=0}^{\infty} \Bigl(\varphi_k(0)\mathfrak{b}_k + \chi_k + h'_k \Bigr), \tag{8.56}$$

where the objects on the right are independent with the laws as above for the sequence $\{\Delta^k : k \geq 0\}$ from (8.36) *with $n := \infty$. [Hint: Use Exercise 8.7.]*

We remark that, besides the connection to the random walk from (7.30) associated with the Gaussian BRW, the random walk in (8.54) can also be thought of as an analogue of circle averages of the CGFF; see Exercise 1.30. As we will show next, this random walk will by and large determine the behavior of the DGFF in the vicinity of a point where the field has a large local maximum.

8.4 Random Walk Representation

We will now move to apply the concentric decomposition in the proof of the lower bound on EM_N. A key technical step in this will be the proof of:

Proposition 8.16 *For all $\epsilon \in (0, 1)$ there is $c = c(\epsilon) > 1$ such that for all naturals $N > 2$ and all sets $D_N \subseteq \mathbb{Z}^2$ satisfying*

$$[-\epsilon N, \epsilon N]^2 \cap \mathbb{Z}^2 \subseteq D_N \subseteq [-\epsilon^{-1} N, \epsilon^{-1} N]^2 \cap \mathbb{Z}^2 \tag{8.57}$$

we have

$$P\bigl(h^{D_N} \leq m_N \,\big|\, h_0^{D_N} = m_N \bigr) \geq \frac{c^{-1}}{\log N} \bigl[1 - 2P(M_N > m_N - c) \bigr], \tag{8.58}$$

where, abusing our earlier notation, $M_N := \max_{x \in D_N} h_x^{D_N}$.

In order to prove this, we will need to control the growth of the various terms on the right-hand side of (8.41). This will be achieved using a single *control variable K* that we define next:

Definition 8.17 *(Control variable) For k, ℓ, n positive integers denote*

$$\theta_{n,k}(\ell) := \bigl[\log(k \vee (\ell \wedge (n - \ell))) \bigr]^2. \tag{8.59}$$

Then define K as the smallest $k \in \{1, \dots, \lfloor \frac{n}{2} \rfloor\}$ such that for all $\ell = 0, \dots, n$:

(1) $|\varphi_\ell(0)| \le \theta_{n,k}(\ell)$,
(2) *for all* $r = 1, \dots, \ell - 2$,

$$\max_{x\in\Delta^r} \bigl|\chi_\ell(x)\bigr| \le 2^{(r-\ell)/2}\theta_{n,k}(\ell)\,, \tag{8.60}$$

(3)

$$\max_{x\in\Delta^\ell\smallsetminus\Delta^{\ell-1}} \bigl[\chi_\ell(x) + \chi_{\ell+1}(x) + h'_\ell(x) - m_{2^\ell}\bigr] \le \theta_{n,k}(\ell)\,. \tag{8.61}$$

If no such K *exists, we set* $K := \lfloor\frac{n}{2}\rfloor + 1$. *We call* K *the control variable.*

Based on the above lemmas, one readily checks:

Exercise 8.18 *For some* $c, c' > 0$, *all* $n \ge 1$ *and all* $k = 1, \dots, \lfloor\frac{n}{2}\rfloor + 1$,

$$P(K = k) \le c'\,\mathrm{e}^{-c(\log k)^2}, \qquad k \ge 1. \tag{8.62}$$

As we will only care to control events up to probabilities of order $1/n$, this permits us to disregard the situations when K is at least n^ϵ, for any $\epsilon > 0$. Unfortunately, for smaller k we will need to control the growth of the relevant variables on the background of events whose probability is itself small (the said order $1/n$). The key step is to link the event in Proposition 8.16 to the behavior of the above random walk. This is the content of:

Lemma 8.19 (Reduction to a random walk event) *Assume* h^{D_N} *is realized as the sum in* (8.41). *There is a numerical constant* $C > 0$ *such that, uniformly in the above setting, the following holds for each* $k = 0, \dots, n$:

$$\begin{aligned}\{h_0^{D_N} = 0\} \cap \bigl\{h^{D_N} \le m_N(1 - \mathfrak{g}^{D_N}) \text{ on } \Delta^k \smallsetminus \Delta^{k-1}\bigr\}&\\ \supseteq \{S_{n+1} = 0\} \cap \bigl\{S_k \ge C[1 + \theta_{n,K}(k)]\bigr\}&.\end{aligned} \tag{8.63}$$

Proof. Fix k as above and let $x \in \Delta^k \smallsetminus \Delta^{k-1}$. In light of (8.54)–(8.55), on the event $\{h_0^{D_N} = 0\}$ we can drop the "1" in the first term on the right-hand side of (8.41) without changing the result. Noting that $\mathfrak{b}_\ell(x) = -1$ for $\ell < k$, on this event we then get

$$\begin{aligned}h_x^{D_N} - m_{2^k} = -S_k + \sum_{\ell=k}^{n} \mathfrak{b}_\ell(x)\varphi_\ell(0)&\\ + \left(\sum_{\ell=k+2}^{n} \chi_\ell(x)\right) + \bigl[\chi_{k+1}(x) + \chi_k(x) + h'_k(x) - m_{2^k}\bigr]&.\end{aligned} \tag{8.64}$$

The bounds in the definition of the control variable permit us to estimate all terms after $-S_k$ by $C\theta_{n,K}(k)$ from above, for $C > 0$ a numerical constant independent of k and x. Adjusting C if necessary, (8.34) along with the approximation of the Green function using the potential kernel and invoking the downward monotonicity of $N \mapsto \frac{\log\log N}{\log N}$ for N large enough shows

$$m_N\bigl(1-\mathfrak{g}^{D_N}(x)\bigr) \ge m_{2^k} - C. \tag{8.65}$$

Hence,

$$m_N(1-\mathfrak{g}^{D_N}) - h^{D_N}(x) \ge m_{2^k} - h^{D_N}(x) - C \ge S_k - C[1+\theta_{n,K}(k)]. \tag{8.66}$$

This, along with (8.55), now readily yields the claim. □

We are ready to give:

Proof of Proposition 8.16. We first use Lemma 8.5 to write

$$P\bigl(h^{D_N} \le m_N \,\big|\, h_0^{D_N} = m_N\bigr) = P\bigl(h^{D_N \smallsetminus \{0\}} \le m_N(1-\mathfrak{g}^{D_N})\bigr). \tag{8.67}$$

Next pick a $k \in \{1,\dots,\lfloor n/2\rfloor\}$ and note that, by the FKG inequality for the DGFF (see Proposition 5.22),

$$P\bigl(h^{D_N \smallsetminus \{0\}} \le m_N(1-\mathfrak{g}^{D_N})\bigr) \ge P(A^1_{n,k})P(A^2_{n,k})P(A^3_{n,k}), \tag{8.68}$$

where we used that

$$\begin{aligned} A^1_{n,k} &:= \bigl\{h^{D_N \smallsetminus \{0\}} \le m_N(1-\mathfrak{g}^{D_N}) \text{ on } \Delta^k\bigr\} \\ A^2_{n,k} &:= \bigl\{h^{D_N \smallsetminus \{0\}} \le m_N(1-\mathfrak{g}^{D_N}) \text{ on } \Delta^{n-k} \smallsetminus \Delta^k\bigr\} \\ A^3_{n,k} &:= \bigl\{h^{D_N \smallsetminus \{0\}} \le m_N(1-\mathfrak{g}^{D_N}) \text{ on } \Delta^n \smallsetminus \Delta^{n-k}\bigr\} \end{aligned} \tag{8.69}$$

are increasing events. We will now estimate the three probabilities on the right of (8.68) separately.

First we observe that, for any k fixed,

$$\inf_{n\ge 1} P(A^1_{n,k}) > 0 \tag{8.70}$$

in light of the fact that $h^{D_N \smallsetminus \{0\}}$ tends in law to $h^{\mathbb{Z}^2 \smallsetminus \{0\}}$ (see Exercise 8.7), while $m_N(1-\mathfrak{g}^{D_N})$ tends to $\frac{2}{\sqrt{g}}\alpha$, see (8.35). For $A^3_{n,k}$ we note (similarly as in (8.65)) that, for some c depending only on k,

$$m_N(1-\mathfrak{g}^{D_N}) \ge m_N - c \quad \text{on } \Delta^n \smallsetminus \Delta^{n-k}. \tag{8.71}$$

Denoting $M_N := \max_{x\in D_N} h_x^{D_N}$ and $M_N^0 := \max_{x\in D_N} h_x^{D_N \smallsetminus \{0\}}$, Exercise 3.3 then yields

$$P(A^3_{n,k}) \ge P\bigl(M_N^0 \le m_N - c\bigr) \ge 1 - 2P\bigl(M_N > m_N - c\bigr). \tag{8.72}$$

We may and will assume the right-hand side to be positive in what follows, as there is nothing to prove otherwise.

It remains to estimate $P(A^2_{n,k})$. Using Lemma 8.19 and the fact that $k \mapsto \theta_{n,k}(\ell)$ is non-decreasing, we bound this probability as

$$\begin{aligned} P(A^2_{n,k}) &\ge P\left(\{K \le k\} \cap \bigcap_{\ell=k+1}^{n-k-1} \big\{S_\ell \ge C[1+\theta_{n,k}(\ell)]\big\} \,\Big|\, S_{n+1}=0\right) \\ &\ge P\left(\bigcap_{\ell=k+1}^{n-k-1} \big\{S_\ell \ge C[1+\theta_{n,k}(\ell)]\big\} \cap \bigcap_{\ell=1}^{n}\{S_\ell \ge -1\} \,\Big|\, S_{n+1}=0\right) \\ &\qquad - P\left(\{K > k\} \cap \bigcap_{\ell=1}^{n}\{S_\ell \ge -1\} \,\Big|\, S_{n+1}=0\right). \end{aligned} \tag{8.73}$$

To estimate the right-hand side, we invoke the following lemmas:

Lemma 8.20 (Entropic repulsion) *There is a constant $c_1 > 0$ such that for all $n \ge 1$ and all $k = 1, \dots, \lfloor n/2 \rfloor$*

$$P\left(\bigcap_{\ell=k+1}^{n-k-1} \big\{S_\ell \ge C[1+\theta_{n,k}(\ell)]\big\} \,\Big|\, \bigcap_{\ell=1}^{n}\{S_\ell \ge -1\} \cap \{S_{n+1}=0\}\right) \ge c_1. \tag{8.74}$$

Lemma 8.21 *There is $c_2 > 0$ such that for all $n \ge 1$ and all $k = 1, \dots, \lfloor n/2 \rfloor$,*

$$P\left(\{K > k\} \cap \bigcap_{\ell=1}^{n}\{S_\ell \ge -1\} \,\Big|\, S_{n+1}=0\right) \le \frac{1}{n}\mathrm{e}^{-c_2(\log k)^2}. \tag{8.75}$$

As noted before, we will not supply a proof of these lemmas as that would take us on a detour into the area of "Inhomogenous Ballot Theorems;" instead, the reader is asked to consult [27] where these statements are given a detailed proof. We refer to Lemma 8.20 using the term "entropic repulsion" in reference to the following observation from statistical mechanics: An interface near a barrier appears to be pushed away as that increases the entropy of its available fluctuations.

Returning to the proof of Proposition 8.16, we note that Lemmas 8.20 and 8.21 reduce (8.73) to a lower bound on the probability of $\bigcap_{\ell=1}^n\{S_\ell \ge -1\}$ conditional on $S_{n+1} = 0$. We proceed by embedding the random walk into a path of the standard Brownian motion $\{B_t : t \ge 0\}$ via

$$S_k := B_{t_k} \quad \text{where} \quad t_k := \mathrm{Var}(S_k) = \sum_{\ell=0}^{k-1} \mathrm{Var}\big(\varphi_\ell(0)\big). \tag{8.76}$$

This readily yields

$$P\left(\bigcap_{\ell=1}^{n}\{S_\ell \ge -1\}\,\Big|\, S_{n+1}=0\right) \ge P^0\Big(B_t \ge -1\colon t\in[0,t_{n+1}]\,\Big|\, B_{t_{n+1}}=0\Big). \tag{8.77}$$

Lemma 8.10 ensures that t_{n+1} grows proportionally to n and the Reflection Principle then bounds the last probability by c_3/n for some $c_3>0$ independent of n; see Exercise 7.10. Lemmas 8.20–8.21 then show

$$P(A_{n,k}^2) \ge \frac{c_1c_3}{n} - \frac{1}{n}\mathrm{e}^{-c_2(\log k)^2}. \tag{8.78}$$

For k sufficiently large, this is at least a constant over n. Since $n\approx \log_2 N$, we get (8.58) from (8.68), (8.70) and (8.72). □

8.5 Tightness of DGFF Maximum: Lower Tail

We will now harvest the fruits of our hard labor in the previous sections and prove tightness of the maximum of DGFF. First we claim:

Lemma 8.22 *For the DGFF in V_N, we have*

$$\inf_{N\ge1} P\big(M_N\ge m_N\big)>0. \tag{8.79}$$

Proof. We may and will assume $N\ge 10$ without loss of generality. Let $V'_{N/2}$ denote the square of side $\lfloor N/2\rfloor$ centered (roughly) at the same point as V_N. For each $x\in V'_{N/2}$ and denoting $D_N:=-x+V_N$, the translation invariance of the DGFF gives

$$\begin{aligned}P\big(h_x^{V_N}\ge m_N,\, h^{V_N}\le h_x^{V_N}\big) &= P\big(h_0^{D_N}\ge m_N,\, h^{D_N}\le h_0^{D_N}\big)\\ &=\int_0^\infty P\big(h_0^{D_N}-m_N\in \mathrm{d}s\big)\,P\big(h^{D_N}\le m_N+s\,\big|\,h_0^{D_N}=m_N+s\big).\end{aligned} \tag{8.80}$$

Rewriting the conditional probability using the DGFF on $D_N\smallsetminus\{0\}$, the monotonicity in Lemma 8.5 along with Proposition 8.16 show, for any $s\ge0$,

$$\begin{aligned}&P\big(h^{D_N}\le m_N+s\,\big|\,h_0^{D_N}=m_N+s\big)\\ &\qquad\ge P\big(h^{D_N}\le m_N\,\big|\,h_0^{D_N}=m_N\big)\ge \frac{c^{-1}}{\log N}\big[1-2P(M_N\le m_N-c)\big].\end{aligned} \tag{8.81}$$

Plugging this in (8.80) yields

$$P\big(h_x^{V_N} \geq m_N,\, h^{V_N} \leq h_x^{V_N}\big) \geq \frac{c^{-1}}{\log N} P\big(h_x^{V_N} \geq m_N\big)\big[1 - 2P(M_N \leq m_N - c)\big]. \tag{8.82}$$

If $P(M_N \leq m_N - c) > 1/2$, we may skip directly to (8.85). Otherwise, invoking

$$P\big(h_x^{V_N} \geq m_N\big) \geq c_1(\log N)N^{-2} \tag{8.83}$$

with some $c_1 > 0$ uniformly for all $x \in V'_{N/2}$, summing (8.82) over $x \in V'_{N/2}$ and using that $|V'_{N/2}|$ has order N^2 vertices shows

$$P(M_N \geq m_N) \geq c_2\big[1 - 2P(M_N > m_N - c)\big] \tag{8.84}$$

for some $c_2 > 0$. As $c > 0$, this implies

$$P(M_N > m_N - c) \geq c_2/(1 + 2c_2). \tag{8.85}$$

This is almost the claim except for the constant c in the event.

Consider the translate $V'_{N/2}$ of $V_{\lfloor N/2 \rfloor}$ centered at the same point as V_N and use the Gibbs-Markov property to write h^{V_N} as $h^{V'_{N/2}} + \varphi^{V_N, V'_{N/2}}$. Denoting $M'_{N/2} := \max_{x \in V'_{N/2}} h_x^{V'_{N/2}}$, a small variation on Exercise 3.4 shows

$$P(M_N \geq m_N) \geq P(M'_{N/2} > m_{N/2} - c) \min_{x \in V'_{N/2}} P\big(\varphi_x^{V_N, V'_{N/2}} \geq c + m_N - m_{N/2}\big). \tag{8.86}$$

Since $m_N - m_{N/2}$ remains bounded as $N \to \infty$ and

$$\inf_{N \geq 10} \min_{x \in V'_{N/2}} \operatorname{Var}\big(\varphi_x^{V_N, V'_{N/2}}\big) > 0\,, \tag{8.87}$$

the minimum in (8.86) is positive uniformly in $N \geq 10$. The claim follows from (8.85). □

As our final step, we boost (8.79) to a bound on the lower tail:

Lemma 8.23 (Tightness of lower tail) *There is $a > 0$ and $t_0 > 0$ such that*

$$\sup_{N \geq 1} P\big(M_N < m_N - t\big) \leq \mathrm{e}^{-at}, \quad t > t_0. \tag{8.88}$$

Before we delve into the proof, let us remark that the bound (8.88) is not sharp even as far its overall structure is concerned. Indeed, Ding and Zeitouni [58] showed that the lower tails of M_N are in fact doubly exponential. However, the proof of the above is easier and fully suffices for our later needs.

Proof of Lemma 8.23. Consider the setting as in the proof of Lemma 8.3. The reasoning leading up to (8.24) plus the arguments underlying (8.25) show

$$P\big(M_{3KN} < m_{3KN} - t\big) \le 2\mathrm{e}^{-at} + \mathrm{e}^{-\frac{1}{4}K^2 P(M_N \ge m_{3NK} - t + \sqrt{t}\,\sqrt{\log K})}. \tag{8.89}$$

Now link K to t by setting $t := c\log K$ with $c > 0$ so large that $m_{3NK} - t + \sqrt{t}\,\sqrt{\log K} \le m_N$. With the help of (8.79), the second term on the right of (8.89) is doubly exponentially small in t, thus proving the claim (after an adjustment of a) for $N \in (3K)\mathbb{N}$. The claim follows from the next exercise. □

Exercise 8.24 *Use a variation of Exercise 3.4 to show that if c relating K to t is chosen large enough, then* (8.88) *for $N \in (3K)\mathbb{N}$ extends to a similar bound for all $N \in \mathbb{N}$. Hint: Note that* $\max_{0\le r<3K}[m_{3Kn+r} - m_{3Kn}] \le 2\sqrt{g}\,\log(3K)$.

From here we finally get:

Proof of Theorem 7.3. The tightness of $\{M_N - m_N : N \ge 1\}$ follows from Lemmas 8.3 and 8.23. Alternatively, use these lemmas to infer $EM_N = m_N + O(1)$ and then apply Lemma 7.1. □

Lecture 9: Extremal Local Extrema

Having established tightness of the DGFF maximum, we turn our attention to the structure of the extremal level sets—namely, the sets of vertices where the DGFF takes values within order unity of the absolute maximum. As for the intermediate levels, we cast these in the form of a point process. In this lecture we state the main result proved in a sequence of joint papers with Louidor [24, 25, 27] and then discuss one of the main ingredients of the proof: Liggett's theory of invariant measures for point processes evolving (or Dysonized) by independent Markov chains. This characterizes the subsequential limits of the resulting point process; uniqueness (and thus existence) of the limit will be addressed later.

9.1 Extremal Level Sets

In the previous lectures we showed that the maximum of the DGFF in V_N is tight around the sequence m_N defined in (7.9). By Exercise 3.4, this applies to any sequence $\{D_N : N \ge 1\}$ of admissible discretizations of a continuum domain $D \in \mathfrak{D}$. Once the tightness of the maximum is in place, additional conclusions of interest can be derived concerning the structure of the *extremal level set*

$$\Gamma_N^D(t) := \big\{x \in D_N : h_x^{D_N} \ge m_N - t\big\}. \tag{9.1}$$

We will need two theorems, both proved originally in Ding and Zeitouni [58]:

Theorem 9.1 (Size of extremal level set) *There are constants $c, C \in (0,\infty)$ such that*

$$\lim_{t\to\infty}\liminf_{N\to\infty} P\bigl(e^{ct} \le |\Gamma_N^D(t)| \le e^{Ct}\bigr) = 1. \tag{9.2}$$

Theorem 9.2 (Geometry of extremal level set) *For each $t \in \mathbb{R}$,*

$$\lim_{r\to\infty}\limsup_{N\to\infty} P\left(\exists x, y \in \Gamma_N^D(t)\colon r < |x-y| < \frac{N}{r}\right) = 0\,. \tag{9.3}$$

The upshot of these theorems is the following picture: *With high probability, the extremal level set $\Gamma_N^D(t)$ is the union of a finite number of "islands" of bounded size at distances of order N*. For this reason, we like to think of Theorems 9.1–9.2 as tightness results; the former for the size and the latter for the spatial distribution of $\Gamma_N^D(t)$. Our proofs of these theorems will follow a different route than [58] and are therefore postponed to Lecture 12.

Our focus on scaling limits naturally suggests considering subsequential distributional limits of the scaled version of $\Gamma_N^D(t)$, namely, the set $\{x/N\colon x \in \Gamma_N^D(t)\}$ regarded as a random subset of the continuum domain D. As is common in the area of extreme-order statistics, these are best encoded in terms of the empirical point measure on $D \times \mathbb{R}$ defined by

$$\eta_N^D := \sum_{x\in D_N} \delta_{x/N} \otimes \delta_{h_x^{D_N} - m_N}\,. \tag{9.4}$$

Indeed, we have

$$\eta_N^D\bigl(D \times [-t,\infty)\bigr) = \bigl|\Gamma_N^D(t)\bigr| \tag{9.5}$$

and so η_N^D definitely captures at least the size of the level set. In addition, η_N^D also keeps track of how the points are distributed in the level set and what the value of the field is there, very much like the measures discussed in Lectures 2–5. However, unlike for the intermediate level sets, no normalization needs to be imposed here thanks to Theorem 9.1.

Notwithstanding all that was just said, η_N^D are actually not the most natural *primary* objects to work with. This is because each high value of the DGFF comes along with a whole *cluster* of high values; namely, the values of the field at the points of $\mathbb{Z}^2$ in close vicinity thereof. In the limit $N \to \infty$, this whole cluster collapses to a single spatial location; see Fig. 17. As the field-values within a single cluster are heavily correlated (and remain so through $N \to \infty$ limit) depending on their relative lattice positions, it is advantageous to track them at the (original) lattice scale; i.e., without scaling space by N.

Thus, denoting (with a slight abuse of our earlier notation) by

$$\Lambda_r(x) := \bigl\{y \in \mathbb{Z}^2\colon |x-y| < r\bigr\} \tag{9.6}$$

the r-neighborhood of x in (say) Euclidean norm, instead of η_N^D we will consider its *structured* version

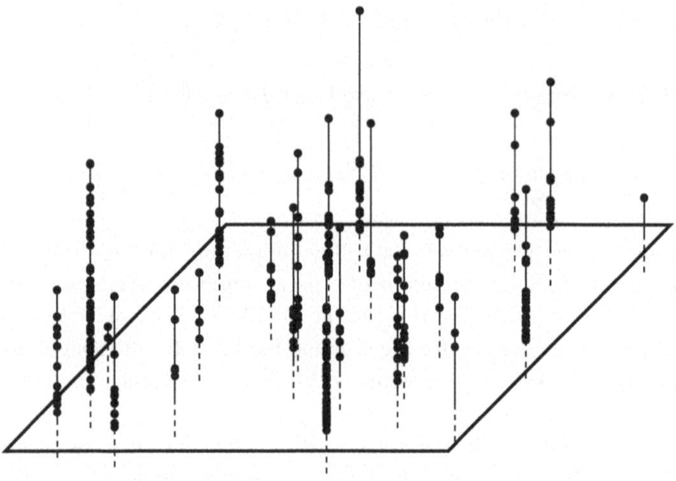

Fig. 17 A sample of the limit process obtained from η_N^D in (9.4) in the limit $N \to \infty$. Each "leading" local maximum is accompanied by a cluster of points that, in the limit, all line up over the same spatial location

$$\eta_{N,r}^D := \sum_{x\in D_N} 1_{\{h_x^{D_N}=\max_{y\in\Lambda_r(x)} h_y^{D_N}\}}\,\delta_{x/N}\otimes\delta_{h_x^{D_N}-m_N}\otimes\delta_{\{h_x^{D_N}-h_{x+z}^{D_N}\colon z\in\mathbb{Z}^2\}}\,. \tag{9.7}$$

This measure picks one reference point in each "cluster"—namely, the "r-local maximum" which is a point x where h^{D_N} dominates all values in $\Lambda_r(x)$—and records the scaled position and reduced field-value at this point along with the "shape" of the field thereabout.

Let us write PPP(μ) to denote the Poisson point process with (σ-finite) intensity measure μ. Abbreviate $\overline{\mathbb{R}} := \mathbb{R}\cup\{+\infty,-\infty\}$. Our main result on the structured extremal point processes (9.7) is the following theorem, derived jointly with O. Louidor in [24, 25, 27]:

Theorem 9.3 (DGFF extremal process) *There is a probability measure ν on $[0,\infty)^{\mathbb{Z}^2}$ and, for each $D\in\mathfrak{D}$, a random Borel measure Z^D on D with $Z^D(D)\in(0,\infty)$ a.s. such that the following holds for any sequence $\{D_N\colon N\ge1\}$ of admissible approximations of D and any sequence $\{r_N\colon N\ge1\}$ of numbers with $r_N\to0$ and $N/r_N\to\infty$:*

$$\eta_{N,r_N}^D \;\underset{N\to\infty}{\overset{\text{law}}{\longrightarrow}}\; \text{PPP}\bigl(Z^D(\mathrm{d}x)\otimes \mathrm{e}^{-\alpha h}\mathrm{d}h\otimes\nu(\mathrm{d}\phi)\bigr), \tag{9.8}$$

where the convergence in law is with respect to the vague convergence of Radon measures on $\overline{D}\times(\mathbb{R}\cup\{\infty\})\times\overline{\mathbb{R}}^{\mathbb{Z}^d}$ and $\alpha := 2/\sqrt{g}$.

A couple of remarks are in order:

(1) By Exercise 2.8 with $\mathscr{X} := \overline{D} \times (\mathbb{R} \cup \{\infty\}) \times \overline{\mathbb{R}}^{\mathbb{Z}^d}$, the statement (9.8) is equivalent to the distributional convergence of the sequence of real-valued random variables $\{\langle \eta^D_{N,r_N}, f \rangle : N \geq 1\}$ for all functions $f \in C_c(\mathscr{X})$, or even just such f's that depend only on a finite number of the coordinates in the third (i.e.,"cluster") variable.
(2) The PPP process with a random intensity is to be understood as a random sample from a class of Poisson processes. To generate PPP(μ) for μ random, we thus first sample μ and then generate the Poisson process conditional on μ. The term *Cox process* is sometimes used for this object as well.
(3) Somewhat surprisingly, the expression we saw in the limit of the intermediate level sets appears in (9.8) as well, albeit now as the intensity measure of a Poisson process. See Conjectures 16.5 and 16.7 for some explanation.
(4) Using local maxima as the reference points in (9.7) is the most natural choice although other choices might presumably work as well. (For instance, one could take x to be a sample from a suitably normalized measure $z \mapsto e^{\beta h^{D_N}_z}$ on $\Lambda_r(y)$ for y being an r-local maximum of h^{D_N}.)

Let us elaborate on remark (2) above by offering a more explicit way of sampling the limit process in (9.8). Writing $Z^D(\cdot)$ as $Z^D(D)\widehat{Z}^D(\cdot)$, where

$$\widehat{Z}^D(A) := \frac{Z^D(A)}{Z^D(D)}, \tag{9.9}$$

the limit process in (9.8) can be sampled as follows: Let $\{h'_i : i \in \mathbb{N}\}$ be points in a sample of the Poisson point process $\mathbb{R}$ with (Gumbel) intensity $e^{-\alpha h} dh$. Given an independent sample of the Z^D measure, set

$$h_i := h'_i + \alpha^{-1} \log Z^D(D), \quad i \in \mathbb{N}, \tag{9.10}$$

and, conditionally on Z^D, sample independently:

(1) $\{x_i : i \in \mathbb{N}\} \overset{\text{law}}{=}$ i.i.d. with law $\widehat{Z}^D$, and
(2) $\{\phi_i : i \in \mathbb{N}\} \overset{\text{law}}{=}$ i.i.d. with law ν.

The point measure $\sum_{i \in \mathbb{N}} \delta_{x_i} \otimes \delta_{h_i} \otimes \delta_{\phi_i}$ then has the law on the right-hand side of (9.8).

We remark that a conclusion analogous to (9.8) is known to hold also for the Branching Brownian Motion. This is thanks to the work of McKean [93], Bramson [35, 36], Lalley and Sellke [84] culminating in Arguin, Bovier and Kistler [11–13], Aïdekon, Berestycki, Brunet and Shi [8] and Bovier and Hartung [34]; see also the review by Bovier [33]. The corresponding problem was solved for the DGFF in $d \geq 3$ as well (Chiarini, Cipriani and Hazra [42–44]) although (since these DGFFs are no longer log-correlated) there the limit process has a non-random intensity. In fact, the same applies to the field of i.i.d. standard normals.

9.2 Distributional Invariance

The proof of Theorem 9.3 will take the total of four lectures. (The actual proof appears in Sects. 10.2 and 11.3 with all needed technicalities settled only in Lecture 12.) The basic strategy will be very much like that used for the intermediate level sets although, chronologically, these ideas started with extremal level sets. Namely, we extract a subsequential limit of the processes of interest and then derive enough properties to identify the limit law uniquely.

We start by noting a tightness statement:

Exercise 9.4 *Let $f \in C_c(\overline{D} \times (\mathbb{R} \cup \{\infty\}) \times \overline{\mathbb{R}}^{\mathbb{Z}^2})$. Use Theorems 9.1–9.2 to show that the sequence of random variables $\{\langle \eta^D_{N,r_N}, f\rangle \colon N \geq 1\}$ is tight.*

The use of subsequential convergence is then enabled by:

Exercise 9.5 *Suppose $\mathscr{X}$ is a locally compact, separable Hausdorff space and let η_N be a sequence of random (non-negative) integer-valued Radon measures on $\mathscr{X}$. Assume $\{\langle \eta_N, f\rangle \colon N \geq 1\}$ is tight for each $f \in C_c(\mathscr{X})$. Prove that there is a sequence $N_k \to \infty$ such that $\langle \eta_{N_k}, f\rangle$ converges in law to a random variable that takes the form $\langle \eta, f\rangle$ for some random Radon measure η on $\mathscr{X}$. Prove that η is integer valued.*

In this lecture we go only part of the way to the full proof of Theorem 9.3 by showing that any (subsequential) limit of the measures $\{\eta^D_{N,r_N} \colon N \geq 1\}$ must take the form on the right of (9.8). We will work only with the first two coordinates of the process for a while so we will henceforth regard η^D_{N,r_N} as a random measure on $D \times \mathbb{R}$. The main focus of this lecture is thus:

Theorem 9.6 (Poisson structure of subsequential limits) *Any weak subsequential limit η^D of the processes $\{\eta^D_{N,r_N} \colon N \geq 1\}$ restricted to just the first two coordinates takes the form*

$$\eta^D \overset{\text{law}}{=} \text{PPP}\big(Z^D(\mathrm{d}x) \otimes \mathrm{e}^{-\alpha h}\mathrm{d}h\big) \tag{9.11}$$

for some random Borel measure Z^D on $\overline{D}$ with $Z^D(\overline{D}) \in (0, \infty)$ a.s.

As we will demonstrate, this arises from the fact that *every* subsequential limit measure η^D has the following distributional symmetry:

Proposition 9.7 (Invariance under Dysonization) *For any η^D as above (projected on the first two coordinates) and any function $f \in C_c(\overline{D} \times \mathbb{R})$, we have*

$$E\big(\mathrm{e}^{-\langle \eta, f\rangle}\big) = E\big(\mathrm{e}^{-\langle \eta, f_t\rangle}\big), \quad t > 0, \tag{9.12}$$

where

$$f_t(x, h) := -\log E^0\big(\mathrm{e}^{-f(x, h + B_t - \frac{\alpha}{2}t)}\big) \tag{9.13}$$

with $\{B_t \colon t \geq 0\}$ denoting the standard Brownian motion.

Let us pause to explain why we refer to this as "invariance under Dysonization." Exercises 9.4–9.5 ensure that η^D is a point measure, i.e.,

$$\eta^D = \sum_{i\in\mathbb{N}} \delta_{x_i} \otimes \delta_{h_i}. \tag{9.14}$$

Given a collection $\{B_t^{(i)} : t \geq 0\}_{i=1}^{\infty}$ of independent standard Brownian motions, we then set

$$\eta_t^D := \sum_{i\in\mathbb{N}} \delta_{x_i} \otimes \delta_{h_i + B_t^{(i)} - \frac{\alpha}{2}t}. \tag{9.15}$$

Of course, for $t > 0$ this may no longer be a "good" point measure as we cannot a priori guarantee that $\eta_t^D(C) < \infty$ for any compact set. Nonetheless, we can use Tonelli's theorem to perform the following computations:

$$\begin{aligned} E\big(\mathrm{e}^{-\langle \eta_t, f\rangle}\big) &= E\left(\prod_{i\in\mathbb{N}} \mathrm{e}^{-f(x_i, h_i + B_t^{(i)} - \frac{\alpha}{2}t)}\right) \\ &= E\left(\prod_{i\in\mathbb{N}} \mathrm{e}^{-f_t(x_i, h_i)}\right) = E\big(\mathrm{e}^{-\langle \eta, f_t\rangle}\big), \end{aligned} \tag{9.16}$$

where in the middle equality we used the Bounded Convergence Theorem to pass the expectation with respect to each Brownian motion inside the infinite product. Proposition 9.7 then tells us

$$E\big(\mathrm{e}^{-\langle \eta_t, f\rangle}\big) = E\big(\mathrm{e}^{-\langle \eta, f\rangle}\big), \quad t \geq 0, \tag{9.17}$$

and, since this holds for all f as above,

$$\eta_t \overset{\text{law}}{=} \eta, \quad t \geq 0. \tag{9.18}$$

Thus, attaching to each point of η^D an independent diffusion $t \mapsto B_t - \frac{\alpha}{2}t$ in the second coordinate preserves the law of η^D. We call this operation "Dysonization" in analogy to Dyson's proposal [70] to consider dynamically-evolving random matrix ensembles; i.e., those where the static matrix entries are replaced by stochastic processes.

Proof of Proposition 9.7 *(main idea).* The proof is based on the following elementary computation. Writing h for the DGFF on D_N, let h', h'' be independent copies of h. For any $s \in [0, 1]$, we can then realize h as

$$h = \sqrt{1-s}\, h' + \sqrt{s}\, h''. \tag{9.19}$$

Choosing $s := t/(g \log N)$, we get

$$h = \sqrt{1 - \frac{t}{g \log N}}\, h' + \frac{\sqrt{t}}{\sqrt{g \log N}}\, h''. \tag{9.20}$$

Now pick x at or near a local maximum of h' of value $m_N + O(1)$. Expanding the first square-root into the first-order Taylor polynomial and using that $h'_x = 2\sqrt{g}\,\log N + O(\log\log N)$ yields

$$\begin{aligned} h_x &= h'_x - \frac{1}{2}\frac{t}{g \log N}\, h'_x + \frac{\sqrt{t}}{\sqrt{g \log N}}\, h''_x + O\Bigl(\frac{1}{\log N}\Bigr) \\ &= h'_x - \frac{t}{\sqrt{g}} + \frac{\sqrt{t}}{\sqrt{g \log N}}\, h''_x + O\Bigl(\frac{\log\log N}{\log N}\Bigr). \end{aligned} \tag{9.21}$$

As noted in (1.24), the covariance of the second field on the right of (9.20) satisfies

$$\operatorname{Cov}\Bigl(\frac{\sqrt{t}}{\sqrt{g \log N}}\, h''_x, \frac{\sqrt{t}}{\sqrt{g \log N}}\, h''_y\Bigr) = \begin{cases} t + o(1), & \text{if } |x-y| < r_N, \\ o(1), & \text{if } |x-y| > N/r_N. \end{cases} \tag{9.22}$$

Thus, the second field behaves as a *random constant* (with the law of $\mathcal{N}(0,t)$) on the whole island of radius r_N around x, and these random constants on distinct islands can be regarded as more or less independent.

The technical part of the proof (which we skip) requires showing that, if x is a local maximum of h'_x at height $m_N + O(1)$, the *gap* between h'_x and the second largest value in an r_N-neighborhood of x stays positive with probability tending to 1 as $N \to \infty$. Once this is known, the errors in all approximations can be fit into this gap and x will also be the local maximum of the field h—and so the locations of the relevant local maxima of h and h' coincide with high probability.

The values of the field $\frac{\sqrt{t}}{\sqrt{g \log N}}\, h''_x$ for x among the local maxima of h (or h') then act as independent copies of $\mathcal{N}(0,t)$, and so they can be realized as values of independent Brownian motions at time t. The drift term comes from the shift by $t/\sqrt{g}$ by noting that $1/\sqrt{g} = \alpha/2$. □

The full proof of Proposition 9.7 goes through a series of judicious approximations for which we refer the reader to [25]. One of them addresses the fact that f_t no longer has compact support. We pose this as:

Exercise 9.8 *Show that for each $f \in C_c(\overline{D} \times \mathbb{R})$, each $t > 0$ and each $\epsilon > 0$ there is $g \in C_c(\overline{D} \times \mathbb{R})$ such that*

$$\limsup_{N\to\infty} P\Bigl(\bigl|\langle \eta^D_{N,r_N}, f_t\rangle - \langle \eta^D_{N,r_N}, g\rangle\bigr| > \epsilon\Bigr) < \epsilon. \tag{9.23}$$

Hint: Use Theorem 9.1.

We will also revisit the "gap estimate" in full detail later; see Exercise 11.10.

9.3 Dysonization-Invariant Point Processes

Our next task is to use the distributional invariance articulated in Proposition 9.7 to show that the law of η^D must take the form in Theorem 9.6. A first natural idea is to look at the moments of η^D, i.e., measures μ_n on $\mathbb{R}^n$ of the form

$$\mu_n(A_1 \times \cdots \times A_n) := E\big[\eta^D(D \times A_1) \dots \eta^D(D \times A_n)\big]. \tag{9.24}$$

As is readily checked, by the smoothness of the kernel associated with $t \mapsto B_t - \frac{\alpha}{2}t$, these would have a density $f(x_1, \dots, x_n)$ with respect to the Lebesgue measure on $\mathbb{R}^n$ and this density would obey the PDE

$$\alpha(1, \dots, 1) \cdot \nabla f + \Delta f = 0. \tag{9.25}$$

This seems promising since all full-space solutions of this PDE can be classified. Unfortunately, the reasoning fails right at the beginning because, as it turns out, all positive-integer moments of η^D are infinite. We thus have to proceed using a different argument. Fortunately, a 1978 paper of Liggett [87] does all what we need to do, so this is what we will discuss next.

Liggett's interest in [87] was in *non-interacting* particle systems which he interpreted as families of independent Markov chains. His setting is as follows: Let $\mathscr{X}$ be a locally compact, separable Hausdorff space and, writing $\mathcal{B}(\mathscr{X})$ for the class of Borel sets in $\mathscr{X}$, let $\mathsf{P}\colon \mathscr{X} \times \mathcal{B}(\mathscr{X}) \to [0, 1]$ be a transition kernel of a Markov chain on $\mathscr{X}$. Denoting $\mathbb{N}^\star := \mathbb{N} \cup \{0\} \cup \{\infty\}$, we are interested in the evolution of point processes on $\mathscr{X}$, i.e., random elements of

$$\mathcal{M} := \big\{\mathbb{N}^\star\text{-valued Radon measures on } (\mathscr{X}, \mathcal{B}(\mathscr{X}))\big\}, \tag{9.26}$$

under the (independent) Markovian "dynamics."

Recall that being "Radon" means that the measure is inner and outer regular with a finite value on every compact subset of $\mathscr{X}$. Calling the measures in $\mathcal{M}$ "point processes" is meaningful in light of:

Exercise 9.9 *Every $\eta \in \mathcal{M}$ takes the form*

$$\eta = \sum_{i=1}^{N} \delta_{x_i}, \tag{9.27}$$

where $N \in \mathbb{N}^\star$ and where (for $N > 0$) $\{x_i : i = 1, \dots, N\}$ is a multiset of points from $\mathscr{X}$ such that $\eta(C) < \infty$ a.s. for every $C \subseteq \mathscr{X}$ compact.

To describe the above "dynamics" more precisely, given $\eta \in \mathcal{M}$ of the form (9.27), we consider a collection of independent Markov chains $\{X_n^{(i)} : n \geq 0\}_{i=1}^N$ such that

$$P\big(X_0^{(i)} = x_i\big) = 1, \quad i = 1, \dots, N. \tag{9.28}$$

Then we define

$$\eta_n := \sum_{i=1}^{N} \delta_{X_n^{(i)}}, \quad n \geq 0, \tag{9.29}$$

while noting that, due to the independence of the chains $X^{(i)}$, the law of this object does not depend on the initial labeling of the atoms of η.

Note that we cannot generally claim that $\eta_n \in \mathcal{M}$ for $n \geq 1$; in fact, easy counterexamples can be produced to the contrary whenever $N = \infty$. Nonetheless, for processes whose law is *invariant* under the above time evolution $\eta \mapsto \{\eta_n : n \geq 0\}$, i.e., those in

$$\mathscr{I} := \big\{\eta\colon \text{ random element of } \mathcal{M} \text{ such that } \eta_1 \overset{\text{law}}{=} \eta\big\}, \tag{9.30}$$

this is guaranteed *ex definitio* in light of

$$\eta \in \mathcal{M} \quad \text{and} \quad \eta_1 \overset{\text{law}}{=} \eta \quad \Rightarrow \quad \eta_n \in \mathcal{M} \text{ a.s. } \forall n \geq 0. \tag{9.31}$$

Let P^n denote the nth (convolution) power of the kernel P defined, inductively, by $\mathsf{P}^{n+1}(x,\cdot) = \int_{\mathscr{X}} \mathsf{P}(x,\mathrm{d}y)\mathsf{P}^n(y,\cdot)$. The starting point of Liggett's paper is the following observation which, although attributed to folklore, is still quite ingenious:

Theorem 9.10 (Folklore theorem) *Suppose* P *obeys the following "strong dispersivity" assumption*

$$\forall C \subseteq \mathscr{X} \text{ compact:} \quad \sup_{x \in \mathscr{X}} \mathsf{P}^n(x, C) \underset{n\to\infty}{\longrightarrow} 0\,. \tag{9.32}$$

Then the set of invariant measures $\mathscr{I}$ *defined in* (9.30) *is given as*

$$\mathscr{I} = \big\{\text{PPP}(M)\colon M = \text{ (random) Radon measure on } \mathscr{X} \text{ s.t. } M \overset{\text{law}}{=} M\mathsf{P}\big\}, \tag{9.33}$$

where $M\mathsf{P}(\cdot) := \int M(\mathrm{d}x)\mathsf{P}(x,\cdot)$.

We refer to the condition (9.32) using the physics term "dispersivity" as that is often used to describe the rate of natural spread (a.k.a. dispersion) of waves or wave packets in time. Note that this condition rules out existence of stationary *distributions* for the Markov chain.

Before we prove Theorem 9.10, let us verify its easier part—namely, that all the Poisson processes PPP(M) with $M \overset{\text{law}}{=} M\mathsf{P}$ lie in $\mathscr{I}$. This follows from:

Lemma 9.11 *Let* M *be a Radon measure on* $\mathscr{X}$. *Then*

$$\eta \overset{\text{law}}{=} \text{PPP}(M) \quad \Rightarrow \quad \eta_n \overset{\text{law}}{=} \text{PPP}(M\mathsf{P}^n), \quad \forall n \geq 0. \tag{9.34}$$

Proof. We start by noting that η being a sample from PPP(M) for a given Radon measure M is equivalent to saying that, for every $f \in C_c(\mathscr{X})$ with $f \geq 0$,

$$E\bigl(\mathrm{e}^{-\langle \eta, f\rangle}\bigr) = \exp\biggl\{-\int M(\mathrm{d}x)(1-\mathrm{e}^{-f(x)})\biggr\}. \tag{9.35}$$

The argument in (9.16) shows

$$E\bigl(\mathrm{e}^{-\langle \eta_n, f\rangle}\bigr) = E\bigl(\mathrm{e}^{-\langle \eta, f_n\rangle}\bigr), \tag{9.36}$$

where

$$f_n(x) := -\log\bigl[(\mathsf{P}^n\mathrm{e}^{-f})(x)\bigr]. \tag{9.37}$$

This equivalently reads as $\mathrm{e}^{-f_n} = \mathsf{P}^n\mathrm{e}^{-f}$ and so from (9.35) for f replaced by f_n and the fact that $\mathsf{P}^n 1 = 1$ we get

$$E\bigl(\mathrm{e}^{-\langle \eta, f_n\rangle}\bigr) = \exp\biggl\{-\int M(\mathrm{d}x)\,\mathsf{P}^n(1-\mathrm{e}^{-f})(x)\biggr\}. \tag{9.38}$$

Tonelli's theorem identifies the right-hand side with that of (9.35) for M replaced by $M\mathsf{P}^n$. Since (9.35) characterizes PPP(M), the claim follows from (9.36). □

We are now ready to give:

Proof of Theorem 9.10. In light of Lemma 9.11 we only need to verify that every element of $\mathscr{I}$ takes the form PPP(M) for some M satisfying $M \overset{\text{law}}{=} M\mathsf{P}$. Let $\eta \in \mathscr{I}$ and pick $f \in C_c(\mathscr{X})$ with $f \geq 0$. Since e^{-f} equals 1 outside a compact set, the strong dispersivity condition (9.32) implies

$$\sup_{x\in\mathscr{X}} \bigl|(\mathsf{P}^n\mathrm{e}^{-f})(x) - 1\bigr| \underset{n\to\infty}{\longrightarrow} 0. \tag{9.39}$$

Recalling the definition of $f_n(x)$ from (9.37), this and $\mathsf{P}^n 1 = 1$ yields the existence of an $\epsilon_n \downarrow 0$ such that

$$(1-\epsilon_n)\mathsf{P}^n(1-\mathrm{e}^{-f})(x) \leq f_n(x) \leq (1+\epsilon_n)\mathsf{P}^n(1-\mathrm{e}^{-f})(x), \quad x \in \mathscr{X}. \tag{9.40}$$

By the Intermediate Value Theorem there is a (random) $\widetilde{\epsilon}_n \in [-\epsilon_n, \epsilon_n]$ such that

$$\langle \eta, f_n\rangle = (1+\widetilde{\epsilon}_n)\bigl\langle \eta, \mathsf{P}^n(1-\mathrm{e}^{-f})\bigr\rangle. \tag{9.41}$$

Denoting $\eta\mathsf{P}^n(\cdot) := \int \eta(\mathrm{d}x)\mathsf{P}^n(x,\cdot)$, from $\eta_n \overset{\text{law}}{=} \eta$ and (9.36) we get

$$E\bigl(\mathrm{e}^{-\langle \eta, f\rangle}\bigr) = E\bigl(\mathrm{e}^{-(1+\widetilde{\epsilon}_n)\langle \eta\mathsf{P}^n, (1-\mathrm{e}^{-f})\rangle}\bigr). \tag{9.42}$$

Noting that every $g \in C_c(\mathscr{X})$ can be written as $g = \lambda(1 - e^{-f})$ for some $f \in C_c(\mathscr{X})$ and $\lambda > 0$, we now ask the reader to solve:

Exercise 9.12 *Prove that* $\{\langle \eta \mathsf{P}^n, g\rangle : n \geq 0\}$ *is tight for all* $g \in C_c(\mathscr{X})$.

Using this along with Exercise 9.5 permits us to extract a subsequence $\{n_k : k \geq 1\}$ with $n_k \to \infty$ and a random Borel measure M on $\mathscr{X}$ such that

$$\langle \eta \mathsf{P}^{n_k}, g\rangle \xrightarrow[k\to\infty]{\text{law}} \langle M, g\rangle, \quad g \in C_c(\mathscr{X}). \tag{9.43}$$

As $\widetilde{\epsilon}_n \to 0$ in L^∞, from (9.42)–(9.43) we conclude

$$E\bigl(e^{-\langle \eta, f\rangle}\bigr) = E\bigl(e^{-\langle M, (1-e^{-f})\rangle}\bigr), \quad f \in C_c(\mathscr{X}). \tag{9.44}$$

A comparison with (9.35) proves that $\eta \overset{\text{law}}{=} \text{PPP}(M)$. Replacing n by $n+1$ in (9.42) shows that (9.44) holds with M replaced by $M\mathsf{P}$. From (9.44) we infer that $M\mathsf{P}$ is equidistributed to M. □

9.4 Characterization of Subsequential Limits

By Proposition 9.7, for the case at hand—namely, the problem of the extremal process of the DGFF restricted to first two coordinates—the relevant setting is $\mathscr{X} := \overline{D} \times \mathbb{R}$ and, fixing any $t > 0$, the transition kernel

$$\mathsf{P}_t\bigl((x,h), A\bigr) := P^0\bigl((x, h + B_t - \tfrac{\alpha}{2}t) \in A\bigr). \tag{9.45}$$

We leave it to the reader to verify:

Exercise 9.13 *Prove that (for all $t > 0$) the kernel* P_t *has the strong dispersivity property* (9.32).

Hence we get:

Corollary 9.14 *Every subsequential limit η^D of processes $\{\eta^D_{N,r_N} : N \geq 1\}$ (projected on the first two coordinates and taken with respect to the vague topology on Random measures on $\overline{D} \times \mathbb{R}$) takes the form*

$$\eta^D \overset{\text{law}}{=} \text{PPP}(M), \tag{9.46}$$

where $M = M(\mathrm{d}x\mathrm{d}h)$ is a Radon measure on $\overline{D} \times \mathbb{R}$ such that

$$M\mathsf{P}_t \overset{\text{law}}{=} M, \quad t > 0. \tag{9.47}$$

Proof. Just combine Exercise 9.13 with Theorem 9.10. □

Moving back to the general setting of the previous section, Theorem 9.10 reduces the problem of classifying the invariant measures on point processes evolving by independent Markov chains to a question involving a *single* Markov chain only:

$$\text{Characterize (random) Radon measures } M \text{ on } \mathscr{X} \text{ satisfying } M\mathsf{P} \overset{\text{law}}{=} M.$$

Note that if M is a random sample from the invariant measures for the Markov chain P, then $M\mathsf{P} = M$ a.s. and so, in particular, $M\mathsf{P} \overset{\text{law}}{=} M$. This suggests we recast the above question as:

$$\text{When does } M\mathsf{P} \overset{\text{law}}{=} M \text{ imply } M\mathsf{P} = M \text{ a.s.?}$$

In his 1978 paper, Liggett [87] identified a number of examples when this is answered in the affirmative. The salient part of his conclusions is condensed into:

Theorem 9.15 (Cases when $M\mathsf{P} \overset{\text{law}}{=} M$ implies $M\mathsf{P} = M$ a.s.) *Let M be a random Radon measure on $\mathscr{X}$. Suppose that*

(1) either P is an irreducible, aperiodic, Harris recurrent Markov chain, or
(2) P is a random walk on an Abelian group such that $\mathsf{P}(0, \cdot)$, where 0 is the identity, is not supported on a translate of a proper closed subgroup.

Then $M\mathsf{P} \overset{\text{law}}{=} M$ implies $M\mathsf{P} = M$ a.s.

Liggett's proof of Theorem 9.15 uses sophisticated facts from the theory of Markov chains and/or random walks on Abelian groups and so we will not reproduce it in full generality here. Still, the reader may find it instructive to solve:

Exercise 9.16 *Find examples of Markov chains that miss just one of the three required attributes in (1) and for which the conclusion of the theorem fails.*

Returning back to the specific case of extremal point process associated with the DGFF, here the first alternative in Theorem 9.15 does not apply as our Markov chain—namely, the Brownian motion with a constant negative drift evaluated at integer multiples of some $t > 0$—is definitely not Harris recurrent. Fortunately, the second alternative does apply and hence we get:

Corollary 9.17 *For each $t > 0$, any M satisfying* (9.47) *obeys $M\mathsf{P}_t = M$ a.s.*

We will provide an independent proof of this corollary, and thus also highlight the main ideas behind the second part of Theorem 9.15, by showing:

Lemma 9.18 *Any M satisfying* (9.47) *takes the form*

$$M(\mathrm{d}x\mathrm{d}h) \overset{\text{law}}{=} Z^D(\mathrm{d}x) \otimes \mathrm{e}^{-\alpha h}\mathrm{d}h + \widetilde{Z}^D(\mathrm{d}x) \otimes \mathrm{d}h, \tag{9.48}$$

where $(Z^D, \widetilde{Z}^D)$ is a pair of random Borel measures on $\overline{D}$.

Proof. Let M be a random Radon measure on $\overline{D} \times \mathbb{R}$ with $M\mathsf{P}_t \overset{\text{law}}{=} M$ and let $A \subseteq \overline{D}$ be a Borel set. Since the Markov kernel P_t does not move the spatial coordinate of the process, we project that coordinate out by considering the measure

$$M^A(C) := M(A \times C), \quad C \subseteq \mathbb{R} \text{ Borel}, \tag{9.49}$$

and the kernel

$$\mathsf{Q}_t(h, C) := P^0\left(h + B_t - \tfrac{\alpha}{2}t \in C\right). \tag{9.50}$$

Note that the only invariant sets for Q_t are $\emptyset$ and $\mathbb{R}$ and so Q_t falls under alternative (2) in Theorem 9.15. We will nevertheless give a full proof in this case.

By assumption, the sequence $\{M^A\mathsf{Q}_t^n : n \geq 0\}$ is stationary with respect to the left shift. The Kolmogorov Extension Theorem (or weak-limit arguments) then permits us to embed it into a *two-sided* stationary sequence $\{M_n^A : n \in \mathbb{Z}\}$ such that

$$M_n^A \overset{\text{law}}{=} M^A \quad \text{and} \quad M_{n+1}^A = M_n^A\mathsf{Q}_t \text{ a.s.} \tag{9.51}$$

hold for each $n \in \mathbb{Z}$. This makes $\{M_n^A : n \in \mathbb{Z}\}$ a random instance of:

Definition 9.19 *(Entrance law) Given a Markov chain on S with transition kernel P, a family of measures $\{\pi_n : n \in \mathbb{Z}\}$ on S satisfying*

$$\pi_{n+1} = \pi_n\mathsf{P}, \quad n \in \mathbb{Z}. \tag{9.52}$$

is called an entrance law.

From $\mathsf{Q}_t(h, \cdot) \ll \text{Leb}$ we infer $M_n^A \ll \text{Leb}$ for each $n \in \mathbb{Z}$. Hence, there are (random) densities $h \mapsto f(n, h)$ such that $M_n^A(\mathrm{d}h) = f(n, h)\mathrm{d}h$. Denoting by

$$k_t(h) := \frac{1}{\sqrt{2\pi t}}\,\mathrm{e}^{-\frac{h^2}{2t}} \tag{9.53}$$

the probability density of $\mathcal{N}(0, t)$, the second relation in (9.51) then reads

$$f(n + 1, h) = f(n, \cdot) \star k_t\left(h + \tfrac{\alpha}{2}t\right), \tag{9.54}$$

where $\star$ denotes the convolution of functions over $\mathbb{R}$. Strictly speaking, this relation holds only for Lebesgue a.e. h but this can be mended by solving:

Exercise 9.20 *Show that, for each $n \in \mathbb{N}$, $h \mapsto f(n, h)$ admits a continuous version such that $f(n, h) \in (0, \infty)$ for each $h \in \mathbb{R}$. The identity (9.54) then holds for all $h \in \mathbb{R}$.*

A key observation underlying Liggett's proof of (2) in Theorem 9.15 is then:

Exercise 9.21 *Let X be a random variable on an Abelian group S. Prove that every entrance law $\{\pi_n : n \in \mathbb{Z}\}$ for the random walk on S with step distribution X is a stationary measure for the random walk on $\mathbb{Z} \times S$ with step distribution $(1, X)$.*

This is relevant because it puts us in the setting to which the *Choquet–Deny theory* applies (see Choquet and Deny [45] or Deny [54]) with the following conclusion: Every $f\colon \mathbb{Z}\times\mathbb{R}\to(0,\infty)$ that obeys (9.54) takes the form

$$f(n,h)=\int \lambda(\kappa)^n\,\mathrm{e}^{\kappa h}\,\nu(\mathrm{d}\kappa) \tag{9.55}$$

for some Borel measure ν on $\mathbb{R}$ and for

$$\lambda(\kappa):=\mathrm{e}^{\frac12\kappa(\kappa+\alpha)t}\,. \tag{9.56}$$

One important ingredient that goes in the proof is:

Exercise 9.22 *Prove that the random walk on (the Abelian group) $\mathbb{Z}\times\mathbb{R}$ with step distribution $(1,\mathcal{N}(t,-\frac{\alpha}{2}t))$ has no non-trivial closed invariant subgroup.*

Underlying the Choquet–Deny theory is Choquet's Theorem, which states that every compact, convex subset of a Banach space is the closed convex hull of its extreme points—i.e., those that cannot be written as a convex combination of distinct points from the set. It thus suffices to classify the extreme points:

Exercise 9.23 *Observe that if f lies in*

$$\mathcal{C}:=\Bigl\{f\colon \mathbb{Z}\times\mathbb{R}\to(0,\infty)\colon (9.54)\ \textit{holds},\ f(0,0)=1\Bigr\}, \tag{9.57}$$

then

$$f_s(n,h):=\frac{f(n-1,h+s)}{f(-1,s)} \tag{9.58}$$

obeys $f_s\in\mathcal{C}$ for each $s\in\mathbb{R}$. Use this to prove that every extreme $f\in\mathcal{C}$ takes the form $f(n,h)=\lambda(\kappa)^n\mathrm{e}^{\kappa h}$ for some $\kappa\in\mathbb{R}$ and $\lambda(\kappa)$ as in (9.56).

With (9.55)–(9.56) in hand, elementary integration (and Tonelli's theorem) gives

$$M_n^A([-1,1])=\int_{[-1,1]} f(n,h)\,\mathrm{d}h=\int \lambda(\kappa)^n\,\frac{\sinh(\kappa)}{\kappa}\,\nu(\mathrm{d}\kappa)\,, \tag{9.59}$$

where, in light of (9.55), ν is determined by (and thus a function of) the realization of M^A. Since $\{M_n^A\colon n\in\mathbb{Z}\}$ is stationary, the last integral in (9.59) cannot diverge to infinity as $n\to\infty$ or $n\to-\infty$. This means that ν must be supported on $\{\kappa\in\mathbb{R}\colon \lambda(\kappa)=1\}$ which, via (9.56), forces ν to be of the form

$$\nu=X^A\delta_{-\alpha}+Y^A\delta_0 \tag{9.60}$$

for some non-negative X^A and Y^A. Hence we get

$$M^A(\mathrm{d}h)=X^A\mathrm{e}^{-\alpha h}\mathrm{d}h+Y^A\mathrm{d}h. \tag{9.61}$$

But $A \mapsto M^A$ is a Borel measure and so $Z^D(A) := X^A$ and $\widetilde{Z}^D(A) := Y^A$ defines two random Borel measures for which (9.48) holds. □

Exercise 9.24 *A technical caveat in the last argument is that* (9.61) *holds only a.s. with the null set depending possibly on A. Prove that, thanks to the fact that the Borel sets in* $\mathbb{R}$ *are countably generated, the conclusion does hold as stated. (Remember to resolve all required measurability issues.)*

We are now ready to establish the desired Poisson structure for any weak subsequential limit of the processes $\{\eta^D_{N,r_N} : N \ge 1\}$:

Proof of Theorem 9.6. First we note that, if $f \in C_c(\overline{D} \times (\mathbb{R} \cup \{\infty\}))$ is such that $f \ge 0$ and $\operatorname{supp}(f) \subseteq \overline{D} \times [t, \infty]$, then

$$\Bigl\{\max_{x \in D_N} h^{D_N}_x < m_N + t\Bigr\} = \Bigl\{\eta^D_{N,r_N}\bigl(D \times [t, \infty)\bigr) = 0\Bigr\} \tag{9.62}$$

implies

$$P\bigl(\langle \eta^D_{N,r_N}, f\rangle > 0\bigr) \le P\Bigl(\max_{x \in D_N} h^{D_N}_x \ge m_N + t\Bigr). \tag{9.63}$$

The upper-tail tightness of the centered maximum (cf Lemma 8.3) shows that the right-hand side tends to zero as $N \to \infty$ and $t \to \infty$. As a consequence we get that every subsequential limit of $\{\eta^D_{N,r_N} : N \ge 1\}$ (in the first two coordinates) is concentrated on $\overline{\mathbb{D}} \times \mathbb{R}$ and so may focus on convergence in this space.

Consider a sequence $N_k \to \infty$ such that $\eta^D_{N_k,r_{N_k}}$ converges in law with respect to the vague topology on the space of Radon measures on $\overline{D} \times \mathbb{R}$. By Corollary 9.14 and Lemma 9.18, the limit is then of the form PPP(M) for some Radon measure M on $\overline{D} \times \mathbb{R}$ of the form (9.48). The probability that PPP(M) has no points in the set A is $\mathrm{e}^{-M(A)}$. Standard approximation arguments and (9.62) then show

$$P\Bigl(\max_{x \in D_{N_k}} h^{D_{N_k}}_x < m_{N_k} + t\Bigr) \underset{k \to \infty}{\longrightarrow} E\Bigl(\mathrm{e}^{-M(\overline{D} \times [t,\infty))}\Bigr). \tag{9.64}$$

(The convergence a priori holds only for a dense set of t's but, since we already know that M has a density in the h variable, it extends to all t.) The upper-tail tightness of the maximum bounds the left-hand side by $1 - \mathrm{e}^{-\tilde{a}t}$ from below once $t > t_0$. Taking $t \to \infty$ thus yields

$$M\bigl(\overline{D} \times [t, \infty)\bigr) \underset{t \to \infty}{\longrightarrow} 0 \text{ a.s.} \tag{9.65}$$

which forces $\widetilde{Z}^D(\overline{D}) = 0$ a.s.

To get that $Z^D(\overline{D}) > 0$ a.s. we instead invoke the lower-tail tightness of the maximum (cf Lemma 8.23), which bounds the left-hand side of (9.64) by e^{at} from above once t is large negative. The right-hand side in turn equals

$$E\bigl(\mathrm{e}^{-\alpha^{-1}\mathrm{e}^{-\alpha t} Z^D(\overline{D})}\bigr), \tag{9.66}$$

which tends to zero as $t \to -\infty$ only if $Z^D(\overline{D}) > 0$ a.s. □

We will continue the proof of Theorem 9.3 in the upcoming lectures. To conclude the present lecture we note that, although the subject of entrance laws has been studied intensely (e.g., by Cox [50], the aforementioned paper of Liggett [87] and also the diploma thesis of Secci [111]), a number of interesting open questions remain; particularly, for transient Markov chains. Ruzmaikina and Aizenman [109] applied Liggett's theory to classify quasi-stationary states for competing particle systems on the real line where the limiting distribution ends up to be also of Gumbel extreme-order type.

Lecture 10: Nailing the Intensity Measure

In this lecture we continue the proof of extremal point process convergence in Theorem 9.3. We first observe that the intensity measure associated with a subsequential limit of the extremal point process is in close correspondence with the limit distribution of the DGFF maximum. The existence of the latter limit is supplied by a theorem of Bramson, Ding and Zeitouni [38] which we will prove in Lecture 12, albeit using a different technique. Next we state properties that link the intensity measure in various domains; e.g., under the restriction to a subdomain (Gibbs-Markov) and conformal maps between domains. These give rise to a set of conditions that ultimately identify the intensity measure uniquely and, in fact, link them to a version of the critical Liouville Quantum Gravity.

10.1 Connection to the DGFF Maximum

On the way to the proof of Theorem 9.3 we have so far shown that any *subsequential* limit of the measures $\{\eta^D_{N,r_N} : N \geq 1\}$, restricted to the first two coordinates, is a Poisson Point Process on $D \times \mathbb{R}$ with intensity

$$Z^D(\mathrm{d}x) \otimes \mathrm{e}^{-\alpha h}\mathrm{d}h \tag{10.1}$$

for some random Borel measure Z^D on D. Our next task is to prove the existence of the limit. Drawing on the proofs for the intermediate level sets, a natural strategy would be to identify Z^D through its properties uniquely. Although this strategy now seems increasingly viable, it was not available at the time when these results were first derived (see the remarks at the end of this lecture). We thus base our proof of uniqueness on the connection with the DGFF maximum, along the lines of the original proof in [25].

We start by formally recording an observation used in the last proof of the previous lecture:

Lemma 10.1 *Suppose $N_k \to \infty$ is such that*

$$\eta^D_{N_k, r_{N_k}} \underset{k\to\infty}{\overset{\text{law}}{\longrightarrow}} \text{PPP}\bigl(Z^D(\mathrm{d}x) \otimes \mathrm{e}^{-\alpha h}\mathrm{d}h\bigr). \tag{10.2}$$

Then for each $t \in \mathbb{R}$,

$$P\Bigl(\max_{x\in D_{N_k}} h_x^{D_{N_k}} < m_{N_k} + t\Bigr) \underset{k\to\infty}{\longrightarrow} E\Bigl(\mathrm{e}^{-Z^D(D)\alpha^{-1}\mathrm{e}^{-\alpha t}}\Bigr) \tag{10.3}$$

Proof. Apply (9.64) along with the known form of the intensity measure. □

Hence we get:

Corollary 10.2 *If $\max_{x\in D_N} h_x^{D_N} - m_N$ converges in distribution, then the law of $Z^D(D)$ is the same for every subsequential limit of $\{\eta^D_{N,r_N} : N \geq 1\}$.*

Proof. The function $t \mapsto \alpha^{-1}\mathrm{e}^{-\alpha t}$ sweeps through $(0, \infty)$ as t varies through $\mathbb{R}$. The limit distribution of the maximum thus determines the Laplace transform of the random variable $Z^D(D)$ which in turn determines the law of $Z^D(D)$. □

In the original argument in [25], the premise of Corollary 10.2 was supplied by the main result of Bramson, Ding and Zeitouni [38] (which conveniently appeared while the first version of [25] was being written):

Theorem 10.3 (Convergence of DGFF maximum) *As $N \to \infty$, the centered maximum $\max_{x\in V_N} h_x^{V_N} - m_N$ of the DGFF in $V_N := (0, N)^2 \cap \mathbb{Z}^2$ converges in law to a non-degenerate random variable.*

The proof of Theorem 10.3 in [38] is (similarly as the proof of tightness in [37]) based on an intermediate process between the BRW and the DGFF called the modified Branching Random Walk. In Lecture 12 we will give a different proof that instead relies on the concentric decomposition and entropic-repulsion techniques encountered already in our proof of tightness. Our proof of Theorem 10.3 runs somewhat logically opposite to Corollary 10.2 as it yields directly the uniqueness of the Z^D measure; the limit of the DGFF maximum then follows from Lemma 10.1. The following text highlights some of the main ideas shared by both proofs.

Pick $N, K \in \mathbb{N}$ large and consider the DGFF on V_{KN}. Building towards the use of the Gibbs-Markov decomposition, identify within V_{KN} the collection of translates $\{V_N^{(i)} : i = 1, \dots, K^2\}$ of V_N separated by "lines of sites" in-between; cf Fig. 18. Realizing the DGFF $h^{V_{KN}}$ on V_{KN} by way of a binding field and K^2 independent copies of the DGFF on V_N's, i.e.,

$$h^{V_{KN}} := h^{V_{KN}^\circ} + \varphi^{V_{KN}, V_{KN}^\circ} \quad \text{with} \quad h^{V_{KN}^\circ} \perp\!\!\!\perp \varphi^{V_{KN}, V_{KN}^\circ}, \tag{10.4}$$

where $V_{KN}^\circ := \bigcup_{i=1}^{K^2} V_N^{(i)}$, we will now study what the occurrence of the absolute maximum of $h^{V_{KN}}$ at some $x \in V_N^{(i)}$ means for the DGFF on $V_N^{(i)}$.

We first observe that we may safely ignore the situations when x falls near the said "lines of sites." Indeed, in Lecture 12 we will prove:

Lemma 10.4 *There is a constant $c > 0$ such that for all $N \geq 1$, all t with $0 \leq |t| \leq (\log N)^{1/5}$ and all non-empty sets $A \subseteq D_N$,*

$$P\bigl(\max_{x\in A} h_x^{D_N} \geq m_N + t\bigr) \leq c(1+t^2)\mathrm{e}^{-\alpha t}\frac{|A|}{N^2}\left[\log\left(1+\frac{N^2}{|A|}\right)\right]^5. \tag{10.5}$$

Remark 10.5 We note that, although we expect the fifth power to be suboptimal, the example of A being a singleton shows that a logarithmic correction cannot be eliminated completely. The prefactor t^2 is suboptimal as well as it can be omitted for $t < 0$ and reduced to t for $t > 0$; see Theorem 10.7 below.

Next we note that, once x is at least ϵN away from the boundaries of boxes $V_N^{(i)}$, the binding field $\varphi^{V_{KN},V_{KN}^\circ}$ is well behaved. This might suggest that we could replace the whole binding field by its value at the center x_i of $V_N^{(i)}$. However, this is correct only to the leading order as the field

$$x \mapsto \varphi_x^{V_{KN},V_{KN}^\circ} - \varphi_{x_i}^{V_{KN},V_{KN}^\circ} \quad \text{for } x \text{ with } \operatorname{dist}(x,(V_N^{(i)})^{\mathrm{c}}) > \epsilon N, \tag{10.6}$$

retains variance of order unity (after all, it scales to a non-trivial Gaussian field with smooth, but non-constant, sample paths). Notwithstanding, the covariances at the center points satisfy

$$\operatorname{Cov}\Bigl(\varphi_{x_i}^{V_{KN},V_{KN}^\circ}, \varphi_{x_j}^{V_{KN},V_{KN}^\circ}\Bigr) = \operatorname{Cov}\bigl(h_{x_i}^{V_{KN}}, h_{x_j}^{V_{KN}}\bigr) - \operatorname{Cov}\bigl(h_{x_i}^{V_N^{(i)}}, h_{x_j}^{V_N^{(j)}}\bigr) \tag{10.7}$$

from which, checking the cases $i = j$ and $i \neq j$ separately, we get

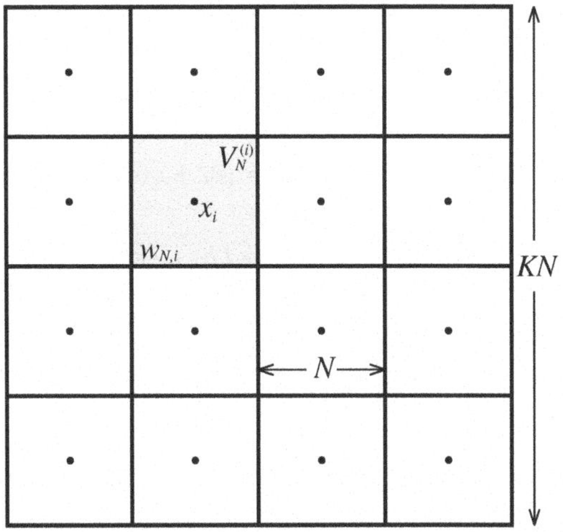

Fig. 18 The decomposition of V_{NK} into K^2 translates of V_N. The center of (the shaded box) $V_N^{(i)}$ is denoted by x_i; the lower-left corner thereof by $w_{N,i}$

$$\mathrm{Cov}\Bigl(\varphi_{x_i}^{V_{KN},V_{KN}^\circ},\varphi_{x_j}^{V_{KN},V_{KN}^\circ}\Bigr)=g\log\left(\frac{KN}{|x_i-x_j|\vee N}\right)+O(1). \tag{10.8}$$

In particular, $\{\varphi_{x_i}^{V_{KN},V_{KN}^\circ}\colon i=1,\dots,K^2\}$ behaves very much like the DGFF in V_K. This is further corroborated by solving:

Exercise 10.6 *Prove that for $c>0$ small enough,*

$$\sup_{N\ge1}P\Bigl(\max_{i=1,\dots,K^2}\varphi_{x_i}^{V_{KN},V_{KN}^\circ}>2\sqrt{g}\log K-c\log\log K\Bigr)\underset{K\to\infty}{\longrightarrow}0. \tag{10.9}$$

The factor $c\log\log K$ is quite important because, ignoring the variations of the field in (10.6), the assumption $h_x^{V_{KN}}\ge m_{KN}+t$ for $x\in V_N^{(i)}$ at least ϵN away from the boundary of the box implies (on the complementary event to (10.9))

$$\begin{aligned}h_x^{V_N^{(i)}}&\ge m_{KN}-2\sqrt{g}\log K+c\log\log K+t\\&\ge m_N+c\log\log K+t+O(1),\end{aligned} \tag{10.10}$$

i.e., (for K large) $h^{V_N^{(i)}}$ takes an *unusually high* value at x. Since the binding field has a well-defined (and smooth) scaling limit, it thus appears that, to prove convergence in law for the maximum the DGFF in a large box V_{KN} as $N\to\infty$, it suffices to prove the convergence in law of the maximum in V_N conditioned to exceed m_N+t, in the limits as $N\to\infty$ followed by $t\to\infty$.

What we just described will indeed be our strategy except that, since the spatial variations of the field (10.6) remain non-trivial in the said limits, we have to control *both* the value of the maximum *and* the position of the maximizer of the DGFF in V_N. This is the content of the following theorem proved originally (albeit in a different form) as [38, Proposition 4.1]:

Theorem 10.7 (Limit law for large maximum) *There is $c_\star>0$ such that*

$$P\bigl(\max_{x\in V_N}h_x^{V_N}\ge m_N+t\bigr)=\bigl[c_\star+o(1)\bigr]te^{-\alpha t}, \tag{10.11}$$

where $o(1)\to0$ in the limit $N\to\infty$ followed by $t\to\infty$. Furthermore, there is a continuous function $\psi\colon(0,1)^2\to[0,\infty)$ such that for all $A\subseteq(0,1)^2$ open,

$$P\Bigl(N^{-1}\operatorname*{argmax}_{V_N}h^{V_N}\in A\,\Big|\,\max_{x\in V_N}h_x^{V_N}\ge m_N+t\Bigr)=o(1)+\int_A\psi(x)\mathrm{d}x, \tag{10.12}$$

where $o(1)\to0$ in the limit $N\to\infty$ followed by $t\to\infty$.

Our proof of Theorem 10.7 will come in Lecture 12, along with the proof of Theorem 10.3 and the finishing touches for the proof of Theorem 9.3. We will now comment on how Theorem 10.7 is used in the proof of Theorem 10.3 in [38]. A key point to observe is that (except for the $o(1)$ term) the right-hand side of (10.12) is independent of t; the position of the conditional maximum thus becomes *asymptotically independent* of its value. This suggest that we define an auxiliary process of triplets of independent random variables,

$$\big\{(\wp_i, h_i, X_i)\colon i = 1, \dots, K^2\big\}, \tag{10.13}$$

which encode the limiting centered values and scaled positions of the excessive maxima in the boxes $V_N^{(i)}$, as follows: Fix a "cutoff scale"

$$t_K := (c/2) \log\log K \tag{10.14}$$

with c as in (10.9). Noting that ψ is a probability density on $(0, 1)^2$, set

(1) $\wp_i \in \{0, 1\}$ with $P(\wp_i = 1) := c_\star\, t_K\, \mathrm{e}^{-\alpha t_K}$,
(2) $h_i \ge 0$ with $P(h_i \ge t) := \frac{t_K + t}{t_K}\, \mathrm{e}^{-\alpha t}$ for all $t \ge 0$, and
(3) $X_i \in [0, 1]^2$ with $P\big(X_i \in A\big) := \int_A \psi(x) \mathrm{d}x$.

The meaning of $\wp_i$ is that $\wp_i = 1$ designates that the maximum of the DGFF in $V_N^{(i)}$ is at least $m_N + t_K$. The excess of this maximum above $m_N + t_K$ is then distributed as h_i while the position of the maximizer (scaled and shifted to a unit box) is distributed as X_i. See (10.11)–(10.12).

Writing $w_{K,i}$ for the lower-left corner of $V_N^{(i)}$, from Theorem 9.2 (which ensures that only one value per $V_N^{(i)}$ needs to be considered) it follows that the absolute maximum of $h^{V_{KN}} - m_{NK}$ is for N large well approximated by

$$2\sqrt{g}\, \log K + t_K + \max\Big\{h_i + \varphi^{V_{KN}, V_{KN}^\circ}_{w_{N,i} + \lfloor N X_i \rfloor} \colon i = 1, \dots, K^2,\ \wp_i = 1\Big\}, \tag{10.15}$$

where the very first term arises as the $N \to \infty$ limit of $m_{NK} - m_N$ and where the binding field is independent of the auxiliary process. The choice of t_K guarantees that, for N large, $\{i = 1, \dots, K^2\colon \wp_i = 1\} \neq \emptyset$ with high probability.

Next we note that all N-dependence in (10.15) now comes through the binding field which, for K fixed and $N \to \infty$, converges to

$$\Phi^{S, S_K^\circ} := \mathcal{N}(0, C^{S, S_K^\circ})\,, \tag{10.16}$$

where

$$S := (0, 1)^2 \quad\text{and}\quad S^\circ := \bigcup_{i=1}^{K^2} [\hat{w}_i + (0, 1/K)^2] \tag{10.17}$$

with $\hat{w}_i$ denoting the position of the lower-left corner of the box that, for finite N, has the lower-left corner at $w_{N,i}$. It follows that (10.15) approximates *every* subsequential limit of the absolute maximum by

$$2\sqrt{g}\log K + t_K + \max\Big\{h_i + \Phi^{S,S_K^\circ}(\hat{w}_i + X_i/K) : i = 1,\dots,K^2,\ \wp_i = 1\Big\}, \tag{10.18}$$

with all errors washed out when $K \to \infty$. The maximum thus converges in law to the $K \to \infty$ limit of the random variable in (10.18) (whose existence is a by-product of the above reasoning).

10.2 Gumbel Convergence

Moving back to the convergence of the two-coordinate extremal process, let us explain on how the above is used to prove the uniqueness of the subsequential limit of $\{\eta^D_{N,r_N} : N \ge 1\}$. First off, all that was stated above for square boxes V_N extends readily to any admissible sequence D_N of lattice approximations of $D \in \mathfrak{D}$. Defining, for $A \subseteq D$ with a non-empty interior,

$$h^{D_N}_{A,\star} := \max_{\substack{x \in D_N \\ x/N \in A}} h^{D_N}_x. \tag{10.19}$$

the methods underlying the proof of Theorem 10.3 also give:

Lemma 10.8 *Let $A_1,\dots,A_k \subseteq D$ be disjoint open sets. Then the joint law of*

$$\big\{h^{D_N}_{A_i,\star} - m_N : i = 1,\dots,k\big\} \tag{10.20}$$

admits a non-degenerate weak limit as $N \to \infty$.

This lemma was originally proved as [25, Theorem 5.1]; we will prove it in Lecture 12 via the uniqueness of the limiting Z^D measure. This will be aided by:

Lemma 10.9 *For any subsequential limit $\eta^D = \text{PPP}(Z^D(\mathrm{d}x) \otimes \mathrm{e}^{-\alpha h}\mathrm{d}h)$ of $\{\eta^D_{N,r_N} : N \ge 1\}$, any disjoint open sets $A_1,\dots,A_k \subseteq D$ and any $t_1,\dots,t_k \in \mathbb{R}$,*

$$P\big(h^{D_N}_{A_i,\star} < m_N + t_i : i = 1,\dots,k\big) \underset{N\to\infty}{\longrightarrow} E\big(\mathrm{e}^{-\sum_{i=1}^k Z^D(A_i)\alpha^{-1}\mathrm{e}^{-\alpha t_i}}\big). \tag{10.21}$$

Proof (sketch). The event on the left-hand side can be written as $\langle \eta^D_{N,r_N}, f\rangle = 0$, where $f := \sum_{i=1}^k 1_{A_i} \otimes 1_{[t_i,\infty)}$. Since the right-hand side equals the probability that $\langle \eta^D, f\rangle = 0$ the claim follows by approximating f by bounded continuous functions with compact support in $\overline{D} \times (\mathbb{R} \cup \{\infty\})$. The next exercise helps overcome (unavoidable) boundary issues. □

Exercise 10.10 *Use Lemma 10.4 to prove that, for every subsequential limit η^D of the processes of interest, the associated Z^D measure obeys*

$$\forall A \subseteq \overline{D} \text{ Borel}:\quad \text{Leb}(A) = 0 \ \Rightarrow\ Z^D(A) = 0 \quad a.s. \tag{10.22}$$

In addition, we get $Z^D(\partial D) = 0$ *a.s. so* Z^D *is concentrated on* D. *(You should not need to worry about whether* $\mathrm{Leb}(\partial D) = 0$ *for this.)*

These observations permit us to give:

Proof of Theorem 9.3 *(first two coordinates).* Lemmas 10.8 and 10.9 imply that the joint law of $(Z^D(A_1), \dots, Z^D(A_k))$ is the same for every subsequential limit η^D of our processes of interest. This means that we know the law of $\langle Z^D, f\rangle$ for any f of the form $f = \sum_{i=1}^k a_i 1_{A_i}$ with A_i open disjoint and $a_1, \dots, a_k > 0$. Every bounded and continuous f can be approximated by a function of the form $\sum_{i=1}^k a_i 1_{\{a_{i-1} \le f < a_i\}}$ with $\mathrm{Leb}(f = a_i) = 0$ for every $i = 1, \dots, k$. With the help of Exercise 10.10 we then get uniqueness of the law of $\langle Z^D, f\rangle$ for any $f \in C_c(D \times \mathbb{R})$. Thanks to Exercise 10.10, this identifies the law of Z^D on $\overline{D}$ uniquely. That $Z^D(D) \in (0, \infty)$ a.s. was already shown in the proof of Theorem 9.6. □

The structure of the limit process gives rise to interesting formulas that are worth highlighting at this point. First we ask the reader to solve:

Exercise 10.11 *(Connection to Gumbel law) Let* $\{(x_i, h_i) \colon i \in \mathbb{Z}\}$ *label the points in a sample of* $\mathrm{PPP}(Z^D(\mathrm{d}x) \otimes \mathrm{e}^{-\alpha h}\mathrm{d}h)$. *Show that*

$$\sum_{i \in \mathbb{N}} \delta_{h_i - \alpha^{-1}\log Z^D(D)} \overset{\text{law}}{=} \mathrm{PPP}(\mathrm{e}^{-\alpha h}\mathrm{d}h). \tag{10.23}$$

In particular, the absolute maximum $\max_{i \in \mathbb{N}} h_i$ *has the law of a randomly shifted Gumbel random variable. See also* (9.10).

By (10.23), the *gap* between the ith and $i+1$st local maximum are distributed as those in $\mathrm{PPP}(\mathrm{e}^{-\alpha h}\mathrm{d}h)$. (This will largely be responsible for the simple form of the limit in Corollary 11.22.) We remark that proving convergence of the maximum to a randomly-shifted Gumbel random variable has been one of the holy grails of the subject area (stimulated by the corresponding result for the Branching Brownian Motion; cf McKean [93] and Lalley and Sellke [84]). For the so called star-scale invariant Gaussian fields (in any spatial dimension), this was proved by Madaule [91] very soon after [25, 38] were posted. A corresponding result for general Branching Random Walks is due to Aïdekon [7].

Another, perhaps more interesting, aspect to showcase is:

Lemma 10.12 *Let* X_N *be the (a.s.-unique) vertex where* $h^{D_N}_{X_N} = \max_{x \in D_N} h^{D_N}_x$. *Then for any* $A \subseteq D$ *open with* $\mathrm{Leb}(\partial A) = 0$ *and any* $t \in \mathbb{R}$,

$$P\Bigl(\tfrac{1}{N}X_N \in A,\, \max_{x \in D_N} h^{D_N}_x < m_N + t\Bigr) \underset{N\to\infty}{\longrightarrow} E\bigl(\widehat{Z}^D(A)\mathrm{e}^{-\alpha^{-1}\mathrm{e}^{-\alpha t} Z^D(D)}\bigr), \tag{10.24}$$

where $\widehat{Z}$ *is the measure from* (9.9).

Proof. Let $A \subseteq D$ be open with $\mathrm{Leb}(\partial A) = 0$ (which implies $Z^D(\partial A) = 0$ a.s.). Lemma 10.8 along with a continuity argument based on Exercise 10.10 show

$$\left(h^{D_N}_{A,\star} - m_N,\, h^{D_N}_{A^c,\star} - m_N\right) \underset{N\to\infty}{\overset{\text{law}}{\longrightarrow}} (h^\star_A, h^\star_{A^c})\,. \tag{10.25}$$

The continuity of the law of the DGFF yields

$$\begin{aligned} P\Bigl(\tfrac{1}{N}X_N \in A,\, \max_{x\in D_N} h^{D_N}_x < m_N + t\Bigr) &= P\bigl(h^{D_N}_{A^c,\star} < h^{D_N}_{A,\star},\, h^{D_N}_{A,\star} - m_N < t\bigr) \\ &= P\bigl(h^{D_N}_{A^c,\star} \le h^{D_N}_{A,\star},\, h^{D_N}_{A,\star} - m_N \le t\bigr). \end{aligned} \tag{10.26}$$

Since the event in the first line is open while that in the second line is closed, and the limit distribution are continuous thanks to Theorem 9.6 and Exercise 10.13 below, the standard facts about the convergence in law imply

$$P\Bigl(\tfrac{1}{N}X_N \in A,\, \max_{x\in D_N} h^{D_N}_x < m_N + t\Bigr) \underset{N\to\infty}{\longrightarrow} P\bigl(h^\star_{A^c} < h^\star_A,\, h^\star_A < t\bigr). \tag{10.27}$$

Now we invoke:

Exercise 10.13 *Let $A \subseteq D$ be open with* $\text{Leb}(\partial A) = 0$. *Prove that*

$$h^\star_A \overset{\text{law}}{=} \inf\bigl\{t \in \mathbb{R} : \eta^D(A \times [t,\infty)) = 0\bigr\} \tag{10.28}$$

and that this in fact applies jointly to $(h^\star_A, h^\star_{A^c})$.

This means that we can now rewrite the probability on the right of (10.27) in terms of the maximal points in the sample of η^D. For $\eta = \text{PPP}(M \otimes \mathrm{e}^{-\alpha h}\mathrm{d}h)$ with a fixed M, the point with the maximal h-value in A has probability density $\alpha^{-1}\mathrm{e}^{-\alpha h}M(A)\mathrm{e}^{-\alpha^{-1}\mathrm{e}^{-\alpha h}M(A)}$ with respect to the Lebesgue measure, while the probability that no point in A^c has h-value above h is $\mathrm{e}^{-\alpha^{-1}\mathrm{e}^{-\alpha h}M(A^c)}$. The underlying Poisson structure (conditional on Z^D) therefore gives

$$P\bigl(h^\star_{A^c} < h^\star_A,\, h^\star_A < t\bigr) = \int_{-\infty}^{t} \alpha^{-1}\mathrm{e}^{-\alpha h}\, E\Bigl(Z^D(A)\mathrm{e}^{-\alpha^{-1}\mathrm{e}^{-\alpha h}Z^D(D)}\Bigr)\mathrm{d}h. \tag{10.29}$$

The result now follows by integration (and Tonelli's Theorem). □

Hereby we get:

Corollary 10.14 *The measure $A \mapsto E(\widehat{Z}^D(A))$ is the $N \to \infty$ weak limit of the marginal law of the (a.s.-unique) maximizer of h^{D_N} scaled by N.*

10.3 Properties of Z^D-Measures

Once the convergence issue has been settled (modulo proofs deferred till later), the next natural question is: What is Z^D? Or, more precisely: Can the law of Z^D be independently characterized? As was the case of intermediate level sets, although the laws of $\{Z^D : D \in \mathfrak{D}\}$ are defined individually, they are very much interrelated. The following theorem articulates these relations explicitly:

Theorem 10.15 (Properties of Z^D-measures) *The family of random measures $\{Z^D : D \in \mathfrak{D}\}$ satisfies the following:*

(1) $Z^D(A) = 0$ *a.s. when* $A = \partial D$ *and for any* $A \subseteq D$ *with* $\mathrm{Leb}(A) = 0$,
(2) for any $a \in \mathbb{C}$ *and any* $b > 0$,

$$Z^{a+bD}(a + b\mathrm{d}x) \overset{\text{law}}{=} b^4 Z^D(\mathrm{d}x), \tag{10.30}$$

(3) if $D \cap \widetilde{D} = \emptyset$, *then*

$$Z^{D\cup\widetilde{D}}(\mathrm{d}x) \overset{\text{law}}{=} Z^D(\mathrm{d}x) + Z^{\widetilde{D}}(\mathrm{d}x), \quad Z^D \perp\!\!\!\perp Z^{\widetilde{D}}, \tag{10.31}$$

(4) if $\widetilde{D} \subseteq D$ *and* $\mathrm{Leb}(D \smallsetminus \widetilde{D}) = 0$, *then for* $\Phi^{D,\widetilde{D}} := \mathcal{N}(0, C^{D,\widetilde{D}})$ *and* $\alpha := 2/\sqrt{g}$,

$$Z^D(\mathrm{d}x) \overset{\text{law}}{=} Z^{\widetilde{D}}(\mathrm{d}x)\,\mathrm{e}^{\alpha\Phi^{D,\widetilde{D}}(x)}, \quad Z^{\widetilde{D}} \perp\!\!\!\perp \Phi^{D,\widetilde{D}}, \tag{10.32}$$

(5) there is $\hat{c} \in (0, \infty)$ *such that for all open* $A \subseteq D$,

$$\lim_{\lambda\downarrow 0} \frac{E(Z^D(A)\mathrm{e}^{-\lambda Z^D(D)})}{\log(1/\lambda)} = \hat{c} \int_A r_D(x)^2\,\mathrm{d}x, \tag{10.33}$$

where $r_D(x)$ *is the conformal radius of* D *at* x.

Before we get to the proof of this theorem, let us make a few remarks. First off, with the exception of (5), these properties are shared by the whole family of measures $\{Z^D_\lambda : D \in \mathfrak{D}\}$ introduced in Lectures 2–5. The condition (5) is different than for the measures Z^D_λ because (by (5)) $EZ^D(A) = \infty$ for any non-empty open $A \subseteq D$. Incidentally, this is also what stands in the way of proving uniqueness of the law of Z^D by the argument underlying Proposition 4.11.

Proof of Theorem 10.15*(1) and (3)*. Part (1) is proved by way of reference to Exercise 10.10. For (3) we just observe that the DGFF on domains separated by at least two lattice steps are independent. (This is one place where the first condition (1.26) of admissible approximations is needed.) □

Next we will establish the *Gibbs-Markov property*:

Proof of Theorem 10.15*(4)*. Most of the argument can be borrowed from the proof of Proposition 4.9 on the Gibbs-Markov property for the measures arising from the intermediate level sets. Let $\widetilde{D}$ and D as in the statement, pick $f \in C_c(\widetilde{D} \times \mathbb{R})$ with $f \geq 0$ and recall the notation

$$f_\Phi(x, h) := f\big(x,\, h + \Phi^{D,\widetilde{D}}(x)\big)\,. \tag{10.34}$$

Writing η^D and $\eta^{\widetilde{D}}$ for the limit processes in the respective domains, the argument leading up to (4.27) then ensures

$$\langle \eta^D, f\rangle \overset{\text{law}}{=} \langle \eta^{\widetilde{D}}, f_\Phi\rangle, \quad \Phi^{D,\widetilde{D}} \perp\!\!\!\perp \eta^{\widetilde{D}}. \tag{10.35}$$

The Poisson nature of the limit process then gives, via a routine change of variables,

$$\begin{aligned} E\bigl(\mathrm{e}^{-\langle \eta^{\widetilde{D}}, f_\Phi\rangle}\bigr) &= E\left(\exp\left\{-\int Z^{\widetilde{D}}(\mathrm{d}x)\mathrm{e}^{-\alpha h}\mathrm{d}h\,\bigl(1-\mathrm{e}^{-f(x,h+\Phi^{D,\widetilde{D}}(x))}\bigr)\right\}\right) \\ &= E\left(\exp\left\{-\int Z^{\widetilde{D}}(\mathrm{d}x)\mathrm{e}^{\alpha\Phi^{D,\widetilde{D}}(x)}\mathrm{e}^{-\alpha h}\mathrm{d}h\,\bigl(1-\mathrm{e}^{-f(x,h)}\bigr)\right\}\right). \end{aligned} \tag{10.36}$$

The claim follows by comparing this with the expression one would get for the Laplace transform of $\langle \eta^D, f\rangle$. □

With the help of the Gibbs-Markov property we then readily solve:

Exercise 10.16 *(Support of Z^D and its non-atomicity) Prove that Z^D charges every non-empty open subset of D with probability one. In particular, we have* supp $Z^D = \overline{D}$ *a.s. Prove also that Z^D is a.s. non-atomic. [Hint: Recall that Theorem 9.2 rules out nearby occurrence of near-maximal local maxima.]*

Next let us address the behavior under shifts and dilations:

Proof of Theorem 10.15*(2) (sketch)*. The proof will require working with approximations of the given domain. We thus first ask the reader to solve:

Exercise 10.17 *Let $D \in \mathfrak{D}$ and assume that $\{D^n : n \ge 1\} \in \mathfrak{D}$ are such that $D^n \uparrow D$ with $C^{D,D^n}(x,y) \to 0$ locally uniformly on D. Then*

$$Z^{D^n}(\mathrm{d}x) \underset{n\to\infty}{\overset{\text{law}}{\longrightarrow}} Z^D(\mathrm{d}x). \tag{10.37}$$

Hint: Use the Gibbs-Markov property.

Thanks to this exercise we may assume that $a \in \mathbb{Q}^2$ and $b \in \mathbb{Q}$. As aN and bN will then be integer for an infinite number of N's, the existence of the limit permits us to even assume that $a \in \mathbb{Z}^2$ and $b \in \mathbb{N}$. The invariance of the law of Z^D under integer-valued shifts is a trivial consequence of the corresponding invariance of the DGFF. Concerning the behavior under scaling, here we note that if $\{D_N : N \ge 1\}$ approximates D, then $\{D_{bN} : N \ge 1\}$ approximates bD. The only item to worry about is the centering sequence for which we get

$$m_{bN} - m_N = 2\sqrt{g}\,\log b + o(1). \tag{10.38}$$

Following this change through the limit procedure yields

$$Z^{bD}(b\,\mathrm{d}x) \overset{\text{law}}{=} \mathrm{e}^{\alpha 2\sqrt{g}\,\log b} Z^D(\mathrm{d}x). \tag{10.39}$$

The claim follows by noting that $\alpha 2\sqrt{g} = 4$. □

Proof of Theorem 10.15(5). The starting point is an extension of Theorem 10.7 to all $D \in \mathfrak{D}$. Considering two domains, D and $\widetilde{D}$, with $\widetilde{D} \subseteq D$ and $\widetilde{D}$ a square; i.e., $\widetilde{D} = a + (0, b)^2$ for some $a \in \mathbb{C}$ and $b > 0$. We will for a while assume that $\widetilde{D}$ is in fact a unit square (i.e., $b = 1$); by Exercise 10.17 this can be achieved by redefining N. Pick approximating domains $\{D_N\}$ and $\{\widetilde{D}_N\}$ respectively, and couple the DGFFs therein via the Gibbs-Markov property to get

$$h^{D_N} = h^{\widetilde{D}_N} + \varphi^{D_N,\widetilde{D}_N} \quad \text{where} \quad h^{\widetilde{D}_N} \perp\!\!\!\perp \varphi^{D_N,\widetilde{D}_N}\,. \tag{10.40}$$

We then claim the following intuitive fact:

Lemma 10.18 *Conditional on the maximum of h^{D_N} to exceed $m_N + t$, the position of the (a.s.-unique) maximizer will, with probability tending to one as $N \to \infty$ and $t \to \infty$, lie within $o(N)$ distance of the position of the maximizer of $h^{\widetilde{D}_N}$.*

We will will not prove this lemma here; instead, we refer the reader to [24, Proposition 5.2]. (The proof can also be gleaned from that for a closely-related, albeit far stronger, Lemma 11.18 that we give in Lecture 12.) We now extract the limit law of the scaled maximizer/centered maximum for both domains and write $(X_\star, h_\star)$ for the limiting pair for D and $(\widetilde{X}_\star, \widetilde{h}_\star)$ for the limiting pair for $\widetilde{D}$ (these exist by Lemma 10.12). Let $A \subseteq \widetilde{D}$ be a non-empty open set with $\overline{A} \subseteq \widetilde{D}$ and $\text{Leb}(\partial A) = 0$. Lemma 10.18 and Exercise 10.10 then give

$$P\big(X_\star \in A,\, h_\star > t\big) = \big(1 + o(1)\big) P\Big(\widetilde{X}_\star \in A,\, h_\star + \Phi^{D,\widetilde{D}}(\widetilde{X}_\star) > t\Big), \tag{10.41}$$

where $o(1) \to 0$ in the limit $t \to \infty$. Since $\widetilde{D}$ is a unit square and $\Phi^{D,\widetilde{D}}$ has uniformly continuous sample paths on A, a routine approximation argument combined with conditioning on $\Phi^{D,\widetilde{D}}$ and Theorem 10.7 show

$$t^{-1}e^{\alpha t} P\Big(\widetilde{X}_\star \in A,\, \widetilde{h}_\star + \Phi^{D,\widetilde{D}}(\widetilde{X}_\star) > t\Big) \underset{t\to\infty}{\longrightarrow} c_\star E\left(\int_A e^{\alpha\Phi^{D,\widetilde{D}}(x)}\psi(x)\mathrm{d}x\right). \tag{10.42}$$

Hereby we get

$$t^{-1}e^{\alpha t} P\big(X_\star \in A,\, h_\star > t\big) \underset{t\to\infty}{\longrightarrow} \int_A \psi^D(x)\mathrm{d}x\,, \tag{10.43}$$

where

$$\psi^D(x) := c_\star\psi(x)e^{\frac{1}{2}\alpha^2 C^{D,\widetilde{D}}(x,x)}, \quad x \in \widetilde{D}. \tag{10.44}$$

Note that, since $\widetilde{D}$ is a unit square, we get $\psi^{\widetilde{D}}(x) := c_\star\psi(x)$ directly from Theorem 10.7.

From $\frac{1}{2}\alpha^2 g = 2$ and the relation between $C^{D,\widetilde{D}}$ and the conformal radius (see Definition 1.24 and (4.2)) we then get

$$\frac{\psi^D(x)}{r_D(x)^2} = \frac{\psi^{\widetilde{D}}(x)}{r_{\widetilde{D}}(x)^2}, \quad x \in \widetilde{D}. \tag{10.45}$$

This was derived for $\widetilde{D}$ a unit square but the following exercise relaxes that:

Exercise 10.19 *Suppose that $\widetilde{D}$ is a unit square centered at 0. Show that, for any $b \geq 1$,* (10.43) *holds for a domain $D \in \mathfrak{D}$ with $\widetilde{D} \subseteq D$ if and only if it holds for domain bD. Moreover,*

$$\psi^{bD}(bx) = b^2 \psi^D(x). \tag{10.46}$$

Conclude that (10.45) *holds if $\widetilde{D}$ is a square of any size (of the form $a + (0, b)^2$).*

We now claim that, for all $\widetilde{D}, D \in \mathfrak{D}$,

$$\frac{\psi^D(x)}{r_D(x)^2} = \frac{\psi^{\widetilde{D}}(x)}{r_{\widetilde{D}}(x)^2}, \quad x \in D \cap \widetilde{D}. \tag{10.47}$$

Indeed, if $x \in D \cap \widetilde{D}$ then (as $D \cap \widetilde{D}$ is open) there is a square $\widetilde{D}' \subseteq D \cap \widetilde{D}$ containing x and so we get (10.47) by iterating (10.45). Using (10.47) for $\widetilde{D}$ a translate of D shows that $x \mapsto \psi^D(x)/r_D(x)^2$ is constant on D; one more use of (10.47) shows that this constant is the same for all admissible domains. So

$$\psi^D(x) = c\, r_D(x)^2, \quad x \in D, \tag{10.48}$$

for some $c > 0$ independent of D.

Having extended Theorem 10.7 to all $D \in \mathfrak{D}$, writing $\lambda := \alpha^{-1} e^{-\alpha t}$, (10.43), Corollary 10.14 and (10.24) then show

$$\frac{E(\widehat{Z}^D(A)[1 - e^{-\lambda Z^D(D)}])}{\lambda \log(1/\lambda)} \underset{\lambda \downarrow 0}{\longrightarrow} \alpha c_\star c \int_A r_D(x)^2 \mathrm{d}x \tag{10.49}$$

for any $A \subseteq D$ open with $\mathrm{Leb}(\partial A) = 0$. The claim follows with $\hat{c} := \alpha c_\star c$ from the next exercise. □

Exercise 10.20 *Prove that* (10.49) *implies* (10.33).

We note that, in Lecture 12, we will establish the explicit relation between the asymptotic density ψ^D and the square of the conformal radius directly.

10.4 Uniqueness up to Overall Constant

As our next item of business, we wish to explain that the properties listed in Theorem 10.15 actually determine the laws of the Z^D's uniquely.

Given an integer $K \geq 1$, consider a tiling of the plane by equilateral triangles of side-length K^{-1}. For a domain $D \in \mathfrak{D}$, let $T^1, \dots, T^{m_K}$ be the triangles entirely contained in D, cf Fig. 19. Abbreviate

$$\widetilde{D} := \bigcup_{i=1}^{m_K} T^i. \tag{10.50}$$

Given $\delta \in (0, 1)$, label the triangles so that $i = 1, \dots, n_K$, for some $n_K \leq m_K$, enumerate the triangles that are at least distance δ away from D^c. Define $T_\delta^1, \dots, T_\delta^{n_K}$ as the equilateral triangles of side length $(1 - \delta)K^{-1}$ that have the same orientation and centers as $T^1, \dots, T^{n_K}$, respectively. Recall that the oscillation of a function f on a set A is given by

$$\operatorname{osc}_A f := \sup_{x \in A} f(x) - \inf_{x \in A} f(x). \tag{10.51}$$

We then claim:

Theorem 10.21 *Consider a family $\{M^D : D \in \mathfrak{D}\}$ of random Borel measures satisfying (1–5) in Theorem 10.15 with some $\hat{c} \in (0, \infty)$. Define events $A_{K,R}^i$, $i = 1, \dots, n_K$ by*

$$A_{K,R}^i := \bigl\{\operatorname{osc}_{T_\delta^i} \Phi^{D,\widetilde{D}} \leq R\bigr\} \cap \bigl\{\max_{T_\delta^i} \Phi^{D,\widetilde{D}} \leq 2\sqrt{g} \log K - R\bigr\}. \tag{10.52}$$

Then for any $D \in \mathfrak{D}$, for $\widetilde{D}$ related to K as in (10.50) *and $\Phi^{D,\widetilde{D}} := \mathcal{N}(0, C^{D,\widetilde{D}})$, the random measure*

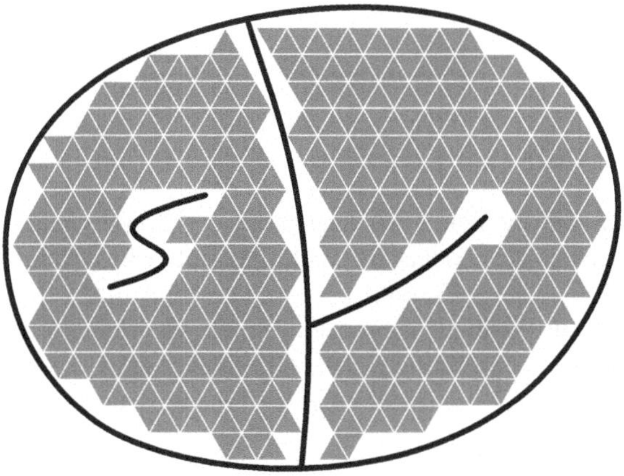

Fig. 19 A tiling of domain D by equilateral triangles of side-length K^{-1}

$$\alpha\hat{c}\, r_D(x)^2 \sum_{i=1}^{n_K} 1_{A^i_{K,R}} \bigl(\alpha \mathrm{Var}(\Phi^{D,\widetilde{D}}(x)) - \Phi^{D,\widetilde{D}}(x)\bigr)\, \mathrm{e}^{\alpha\Phi^{D,\widetilde{D}}(x) - \frac{1}{2}\alpha^2 \mathrm{Var}(\Phi^{D,\widetilde{D}}(x))}\, 1_{T^i_\delta}(x)\,\mathrm{d}x \tag{10.53}$$

tends in law to M^D in the limit as $K \to \infty$, $R \to \infty$ and $\delta \downarrow 0$ (in this order). This holds irrespective of the orientation of the triangular grid.

Before we delve into the proof we note that, by computations involving the covariance kernel $C^{D,\widetilde{D}}$, we get

$$\min_{i=1,\dots,n_K} \inf_{x\in T^i_\delta} \mathrm{Var}(\Phi^{D,\widetilde{D}}(x)) \ge g \log K - c \tag{10.54}$$

for some $c = c(\delta) > 0$ and all $K \ge 1$. Hence, for R sufficiently large (depending only on $\delta > 0$), $\alpha g = 2\sqrt{g}$ implies

$$\alpha \mathrm{Var}(\Phi^{D,\widetilde{D}}(x)) - \Phi^{D,\widetilde{D}}(x) > 0 \quad \text{on } A^i_{K,R}. \tag{10.55}$$

In particular, (10.53) is a positive measure. We also note that, by combining Theorems 10.15 and 10.21, we get:

Corollary 10.22 (Characterization of Z^D-measures) *The laws of the measures $\{Z^D\colon D \in \mathfrak{D}\}$ are determined uniquely by conditions (1–5) of Theorem 10.15.*

In order to avoid confusion about how the various results above depend on each other, we give the following disclaimer:

Remark 10.23 Although Corollary 10.22 may seem to produce the desired independent characterization of the Z^D measures by their underlying properties, we remind the reader that our derivation of these properties in Theorem 10.15 assumed (or came after proving, if we take Theorem 10.3 for granted) that the limit of $\{\eta^D_{N,r_N}: N \ge 1\}$ exists, which already entails some version of their uniqueness. We will amend this in Lecture 12 in our proofs of Theorems 10.7 and 10.3 which draw heavily on the proof of Theorem 10.21.

The main thrust of Theorem 10.21 is that it gives a representation of Z^D as the limit of the measures in (10.53) that are determined solely by the binding fields $\Phi^{D,\widetilde{D}}$ (where $\widetilde{D}$ depends on K). By Exercise 4.15, we may think of $\Phi^{D,\widetilde{D}}$ as the orthogonal projection of the CGFF onto the subspace of functions in $\mathrm{H}^1_0(D)$ that are harmonic in each of the triangles constituting $\widetilde{D}$. The representation by measures in (10.53) is thus akin to that of the Z^D_λ-measures by way of the measures in (4.28) except that instead of using

$$\mathrm{e}^{\beta\Phi^{D,\widetilde{D}}(x) - \frac{1}{2}\beta^2 \mathrm{Var}(\Phi^{D,\widetilde{D}}(x))} \tag{10.56}$$

as the density with respect to the Lebesgue measure, we use its *derivative*

$$-\frac{\mathrm{d}}{\mathrm{d}\beta}\,\mathrm{e}^{\beta\Phi^{D,\widetilde{D}}(x)-\frac{1}{2}\beta^2\mathrm{Var}(\Phi^{D,\widetilde{D}}(x))}\Big|_{\beta=\alpha}$$
$$=\bigl(\alpha\mathrm{Var}(\Phi^{D,\widetilde{D}}(x))-\Phi^{D,\widetilde{D}}(x)\bigr)\mathrm{e}^{\alpha\Phi^{D,\widetilde{D}}(x)-\frac{1}{2}\alpha^2\mathrm{Var}(\Phi^{D,\widetilde{D}}(x))}. \tag{10.57}$$

Note that the key fact underlying the derivations in Lecture 4 is still true: the expression in (10.57) is a martingale with respect to the filtration induced by the fields $\Phi^{D,\widetilde{D}}$ along nested subdivisions of D into finite unions of equilateral triangles (i.e., with K varying along powers of 2). However, the lack of positivity makes manipulations with this object considerably more difficult.

Proof of Theorem 10.21 (*sketch*). The proof is based on a number of relatively straightforward observations (and some technical calculations that we will mostly skip). Denote $\widetilde{D}_\delta := \bigcup_{i=1}^{n_K} T^i_\delta$ and, for $f \in C_c(D)$, let $f_\delta := f 1_{\widetilde{D}_\delta}$. Consider a family $\{M^D\colon D \in \mathfrak{D}\}$ of measures in the statement. Property (1) then ensures

$$\langle M^D, f_\delta\rangle \underset{\delta\downarrow 0}{\overset{\text{law}}{\longrightarrow}} \langle M^D, f\rangle\,, \tag{10.58}$$

and so we may henceforth focus on f_δ. Properties (3–4) then give

$$1_{\widetilde{D}_\delta}(x)M^D(\mathrm{d}x) \overset{\text{law}}{=} \sum_{i=1}^{n_K} \mathrm{e}^{\alpha\Phi^{D,\widetilde{D}}(x)} 1_{T^i_\delta}(x)\, M^{T^i}(\mathrm{d}x)\,, \tag{10.59}$$

with $M^{T^1},\dots,M^{T^{n_K}}$ and $\Phi^{D,\widetilde{D}}$ all independent. Let $x_1,\dots,x_{n_K}$ be the center points of the triangles $T^1,\dots,T^{n_K}$, respectively. A variation on Exercise (10.9) then shows

$$\limsup_{K\to\infty} P\Bigl(\max_{i=1,\dots,n_K} \Phi^{D,\widetilde{D}}(x_i) > 2\sqrt{g}\log K - c\log\log K\Bigr) = 0 \tag{10.60}$$

for some $c > 0$. The first technical claim is the content of:

Proposition 10.24 *For any* $\delta \in (0,1)$ *and any* $\epsilon > 0$,

$$\lim_{R\to\infty}\limsup_{K\to\infty} P\left(\sum_{i=1}^{n_K}\int_{T^i_\delta} M^{T^i}(\mathrm{d}x)\mathrm{e}^{\alpha\Phi^{D,\widetilde{D}}(x)}\,1_{\{\mathrm{osc}_{T^i_\delta}\Phi^{D,\widetilde{D}}>R\}} > \epsilon\right) = 0. \tag{10.61}$$

Proof (*main ideas*). We will not give a detailed proof for which we refer the reader to [24, Proposition 6.5]. Notwithstanding, we will give the main ideas and thus also explain why we have resorted to triangle partitions.

For triangle partitions the subspace of functions in $\mathsf{H}^1_0(D)$ that are piece-wise harmonic on $\widetilde{D}$ naturally decomposes into a direct sum of the space $\mathsf{H}^\triangle$ of functions that are *affine* on each T^i and its orthogonal complement $\mathsf{H}^\perp$ in $\mathsf{H}^1_0(D)$. The projection onto $\mathsf{H}^\perp$ can be controlled uniformly in K thanks to:

Exercise 10.25 *For D and $\widetilde{D}$ (which depends on K) as above, let $\Phi_K^{\perp}$ denote the orthogonal projection of $\Phi^{D,\widetilde{D}}$ onto $\mathsf{H}^{\perp}$. Show that, for each $\delta > 0$ small,*

$$\sup_{K\geq 1}\sup_{x\in\widetilde{D}_\delta} \mathrm{Var}\bigl(\Phi_K^{\perp}(x)\bigr) < \infty. \tag{10.62}$$

Hint: Use the observation (made already in Sheffield's review [114]) that

$$\Phi_K^{\triangle} := \Phi^{D,\widetilde{D}} - \Phi_K^{\perp} \tag{10.63}$$

is piece-wise affine and so it is determined by its values at the vertices of the triangles where it has the law of (a scaled) DGFF on the triangular lattice. See the proof of [24, Lemma 6.10].

The uniform bound (10.62) implies (via Borell–TIS inequality and Fernique majorization) a uniform Gaussian tail for the oscillation of $\Phi_K^{\perp}$. This gives

$$\epsilon_R := \sup_{K\geq 1}\max_{i=1,\dots,n_K}\sup_{x\in T_\delta^i} E\bigl(\mathrm{e}^{\alpha\Phi_K^{\perp}(x)}1_{\{\mathrm{osc}_{T_\delta^i}\Phi_K^{\perp}>R\}}\bigr) \underset{R\to\infty}{\longrightarrow} 0 \tag{10.64}$$

(see [24, Corollary 6.11] for details). To get the claim from this, let $M_{K,R}^{\perp}$ denote the giant sum in (the event in) (10.61). The independence of $\Phi_K^{\perp}$ and $\Phi_K^{\triangle}$ then gives, for each $\lambda > 0$ and for Y a centered normal independent of $\Phi_K^{\perp}$, $\Phi_K^{\triangle}$ and of M^{T^i}'s, and thus of $M_{K,R}^{\perp}$,

$$E\mathrm{e}^{-\lambda \mathrm{e}^{\alpha Y} M_{K,R}^{\perp}} \geq E\left(\exp\left\{-\lambda\sum_{i=1}^{n}\int_{T_\delta^i} M^{T^i}(\mathrm{d}x)\mathrm{e}^{\alpha\Phi_K^{\triangle}(x)+Y}\epsilon_R\right\}\right). \tag{10.65}$$

If $\mathrm{Var}(Y)$ exceeds the quantity in (10.62), Kahane's convexity inequality (cf Proposition 5.6) permits us to replace $\alpha\Phi_K^{\triangle}(x)+Y$ by $\alpha\Phi^{D,\widetilde{D}}+\frac{1}{2}\mathrm{Var}(Y)$ and wrap the result (as a further lower bound) into $E\mathrm{e}^{-\lambda c\epsilon_R M^D(D)}$, for some $c > 0$. As $\epsilon_R \to 0$ when $R\to\infty$ we conclude that $M_{K,R}^{\perp} \to 0$ in probability as $K\to\infty$ and $R\to\infty$. This gives (10.61). □

Returning to the proof of Theorem 10.21, (10.60)–(10.61) permit us to restrict attention only to those triangles where $A_{K,R}^i$ occurs. This is good because, in light of the piece-wise harmonicity of the sample paths of $\Phi^{D,\widetilde{D}}$, the containment in $A_{K,R}^i$ forces

$$x \mapsto \Phi^{D,\widetilde{D}}(x) - \Phi^{D,\widetilde{D}}(x_i) \tag{10.66}$$

to be bounded and uniformly Lipschitz on T_δ^i, for each $i = 1,\dots,_K$. Let $\mathcal{F}_{R,\beta,\delta}^T$, for $R,\beta,\delta > 0$, denote the class of continuous functions $\phi\colon T\to\mathbb{R}$ on an equilateral triangle T such that

$$\phi(x)\geq\beta \quad\text{and}\quad |\phi(x)-\phi(y)|\leq R|x-y|, \qquad x,y\in T_\delta. \tag{10.67}$$

For such functions, property (5) in Theorem 10.15 we assume for M^D yields:

Proposition 10.26 *Fix $\beta > 0$ and $R > 0$. For each $\epsilon > 0$ there are $\delta_0 > 0$ and $\lambda_0 > 0$ such that, for all $\lambda \in (0, \lambda_0)$, all $\delta \in (0, \delta_0)$ and all $f \in \mathcal{F}^T_{R,\beta,\delta}$,*

$$(1-\epsilon)\hat{c}\int_{T_\delta} f(x) r_T(x)^2\,\mathrm{d}x \le \frac{\log E(\mathrm{e}^{-\lambda M^T(f 1_{T_\delta})})}{\lambda\log\lambda} \le (1+\epsilon)\hat{c}\int_{T_\delta} f(x) r_T(x)^2\,\mathrm{d}x, \tag{10.68}$$

where $M^T(f 1_{T_\delta}) := \int_{T_\delta} M^T(\mathrm{d}x)\, f(x)$.

Proof (some ideas). This builds on Exercise 10.20 and a uniform approximation of f by functions that are piecewise constant on small subtriangles of T. Thanks to the known scaling properties of the Z^D measure (see Theorem 10.15(2)) and the conformal radius, it suffices to prove this just for a unit triangle. We refer the reader to [24, Lemma 6.8] for further details and calculations. □

As noted above, whenever $f \in \mathcal{F}_{R,\beta,\delta}$ and $A^i_{K,R}$ occurs, Proposition 10.26 may be applied (with perhaps slightly worse values of R and β) to the test function

$$x \mapsto f(x)\mathrm{e}^{\alpha(\Phi^{D,\widetilde{D}}(x)-\Phi^{D,\widetilde{D}}(x_i))} \tag{10.69}$$

because the harmonicity of $x \mapsto \Phi^{D,\widetilde{D}}(x)$ turns the uniform bound on oscillation into a Lipschitz property. Thus, for $\lambda := K^{-4}\mathrm{e}^{\alpha\Phi^{D,\widetilde{D}}(x_i)}$, on $A^i_{K,R}$ we then get

$$\begin{aligned} E\Bigl(\exp\Bigl\{-\mathrm{e}^{\alpha\Phi^{D,\widetilde{D}}(x_i)} M^{T^i}\bigl(f 1_{T^i_\delta}\mathrm{e}^{\alpha(\Phi^{D,\widetilde{D}}-\Phi^{D,\widetilde{D}}(x_i))}\bigr)\Bigr\}\Bigm|\Phi^{D,\widetilde{D}}\Bigr) \\ = \exp\Bigl\{\hat{c}(1+\tilde{\epsilon})\,\log\bigl(K^{-4}\mathrm{e}^{\alpha\Phi^{D,\widetilde{D}}(x_i)}\bigr)\int_{T^i_\delta}\mathrm{d}x\, f(x)\,\mathrm{e}^{\alpha\Phi^{D,\widetilde{D}}(x)}\, r_{T^i}(x)^2\Bigr\} \end{aligned} \tag{10.70}$$

for some random $\tilde{\epsilon} \in [-\epsilon, \epsilon]$ depending only on $\Phi^{D,\widetilde{D}}$, where we also used that, by (10.60) and $\alpha 2\sqrt{g} = 4$, we have $\lambda \le (\log K)^{-\alpha c}$ which tends to zero as $K \to \infty$.

Denote by $Z^D_{K,R,\delta}$ the measure in (10.53) and write $M^D_{K,R,\delta}$ for the expression on the right of (10.59) with the sum restricted to i where $A^i_{K,R}$ occurs. Note that, by (10.55) and standard (by now) estimates for $C^{D,\widetilde{D}}$,

$$\log(K^{-4}\mathrm{e}^{\alpha\Phi^{D,\widetilde{D}}(x_i)}) = \bigl(1+\tilde{\epsilon}'(x)\bigr)\alpha\bigl(\Phi^{D,\widetilde{D}}(x) - \alpha\mathrm{Var}(\Phi^{D,\widetilde{D}}(x))\bigr) \tag{10.71}$$

with $\tilde{\epsilon}'(x) \in [-\epsilon, \epsilon]$ for all $x \in T^i_\delta$, provided R is large enough. Recalling also that

$$r_{T^i}(x)^2 = r_D(x)^2\mathrm{e}^{-\frac{1}{2}\alpha^2\mathrm{Var}(\Phi^{D,\widetilde{D}}(x))}, \qquad x \in T^i, \tag{10.72}$$

from (10.70)–(10.71) we obtain

$$E\bigl(\mathrm{e}^{-(1+2\epsilon)Z^D_{K,R,\delta}(f)}\bigr) \le E\bigl(\mathrm{e}^{-M^D_{K,R,\delta}(f)}\bigr) \le E\bigl(\mathrm{e}^{-(1-2\epsilon)Z^D_{K,R,\delta}(f)}\bigr). \tag{10.73}$$

Since $M^D_{K,R,\delta}(f)$ tends in distribution to $M^D(f)$ in the stated limits, the law of $M^D(f)$ is given by the corresponding limit law of $Z^D_{K,R,\delta}(f)$, which must therefore exist as well. □

As an immediate consequence of Theorem 10.21, we then get:

Corollary 10.27 (Behavior under rigid rotations) *For each $a, b \in \mathbb{C}$,*

$$Z^{a+bD}(a + b\mathrm{d}x) \overset{\text{law}}{=} |b|^4 \, Z^D(\mathrm{d}x). \tag{10.74}$$

Proof. By Theorem 10.15(2), we just need to prove this for $a := 0$ and $|b| = 1$. This follows from Theorem 10.21 and the fact that both the law of $\Phi^{D,\widetilde{D}}$ and the conformal radius r_D are invariant under the rigid rotations of D (and $\widetilde{D}$). □

Somewhat more work is required to prove:

Theorem 10.28 (Behavior under conformal maps) *Let $f : D \to f(D)$ be a conformal bijection between admissible domains $D, f(D) \in \mathfrak{D}$. Then*

$$Z^{f(D)} \circ f(\mathrm{d}x) \overset{\text{law}}{=} \left|f'(x)\right|^4 Z^D(\mathrm{d}x). \tag{10.75}$$

In particular, the law of $r_D(x)^{-4} \, Z^D(\mathrm{d}x)$ is invariant under conformal maps.

This follows, roughly speaking, by the fact that the law of $\Phi^{D,\widetilde{D}}$ is conformally invariant and also by the observation that a conformal map is locally the composition of a dilation with a rotation. In particular, the triangles T^i map to near-triangles $f(T^i)$ with the deformation tending to zero with the size of the triangle. See the proof of [24, Theorem 7.2].

10.5 Connection to Liouville Quantum Gravity

We will now move to the question of direct characterization of the law of Z^D-measures. As we will show, Z^D has the law of a critical Liouville Quantum Gravity associated with the continuum GFF. These are versions of the measures from Lemma 2.17 for a critical value of parameter β which in our case corresponds to $\beta_{\mathrm{c}} = \alpha$. At this value, the martingale argument from Lemma 2.17 produces a vanishing measure and so additional manipulations are needed to obtain a non-trivial limit object.

One way to extract a non-trivial limit measure at β_{c} is to replace the exponential factor in (2.42) by its negative derivative; cf (10.56)–(10.57). The existence of the corresponding limit was proved in [66]. Another possibility is to re-scale the approximating measures so that a non-trivial limit is achieved. This goes by the name *Seneta–Heyde norming* as discussed in a 2014 paper by Duplantier, Rhodes, Sheffield and Vargas [67]. A technical advantage of the scaling over taking the derivative is the fact that Kahane's convexity inequality (Proposition 5.6) remains applicable in that case.

We will use approximating measures based on the *white-noise decomposition* of the CGFF. The specifics are as follows: For $\{B_t : t \geq 0\}$ the standard Brownian motion, let $p_t^D(x, y)$ be the transition density from x to y before exiting D. More precisely, letting $\tau_{D^c} := \inf\{t \geq 0 \colon B_t \notin D\}$ we have

$$p_t^D(x, y)\mathrm{d}y := P^x\big(B_t \in \mathrm{d}y,\ \tau_{D^c} > t\big). \tag{10.76}$$

Writing W for the white noise on $D \times (0, \infty)$ with respect to the Lebesgue measure, consider the Gaussian process $t, x \mapsto \varphi_t(x)$ defined by

$$\varphi_t(x) := \int_{D\times[\mathrm{e}^{-2t},\infty)} p_{s/2}^D(x, z) W(\mathrm{d}z\,\mathrm{d}s). \tag{10.77}$$

The Markov property of p^D along with reversibility give

$$\begin{aligned} \mathrm{Cov}\big(\varphi_t(x), \varphi_t(y)\big) &= \int_{D\times[\mathrm{e}^{-2t},\infty)} \mathrm{d}z \otimes \mathrm{d}s\ p_{s/2}^D(x, z) p_{s/2}^D(y, z) \\ &= \int_{[\mathrm{e}^{-2t},\infty)} \mathrm{d}s\ p_s^D(x, y) \underset{t\to\infty}{\longrightarrow} \widehat{G}^D(x, y) \end{aligned} \tag{10.78}$$

and so φ_t tends in law to the CGFF.

Define the random measure (and compare with (2.42))

$$\mu_t^{D,\alpha}(\mathrm{d}x) := \sqrt{t}\, 1_D(x)\, \mathrm{e}^{\alpha\varphi_t(x) - \frac{1}{2}\alpha^2 \mathrm{Var}[\varphi_t(x)]}\, \mathrm{d}x\,. \tag{10.79}$$

The key point to show that the scale factor $\sqrt{t}$ ensures that the $t \to \infty$ limit will produce a non-vanishing and yet finite limit object. This was proved in Duplantier, Rhodes, Sheffield and Vargas [67]:

Theorem 10.29 (Critical GMC measure) *There is a non-vanishing a.s.-finite random Borel measure $\mu_\infty^{D,\alpha}$ such that for every Borel set $A \subseteq D$,*

$$\mu_t^{D,\alpha}(A) \underset{t\to\infty}{\longrightarrow} \mu_\infty^{D,\alpha}(A), \quad \text{in probability.} \tag{10.80}$$

We call the measure $\mu_\infty^{D,\alpha}$ the *critical GMC* in D. See Fig. 20 for an illustration. It is a fact that $E\mu_\infty^{D,\alpha}(A) = \infty$ for any non-empty open subset of D so, unlike the subcritical cases $\beta = \lambda\alpha$ for $\lambda \in (0, 1)$, the overall normalization of this measure cannot be fixed by its expectation.

Alternative definitions of $\mu_\infty^{D,\alpha}$ have appeared in the meantime which, through the contributions of Rhodes and Vargas [105], Powell [104] and Junnila and Saksman [80], are now known to be all equal up to a multiplicative (deterministic) constant. The measure in (10.53) is yet another example of this kind, although the abovementioned uniqueness theorems do not apply to this case. Notwithstanding, one is able to call on Theorem 10.15 instead and get:

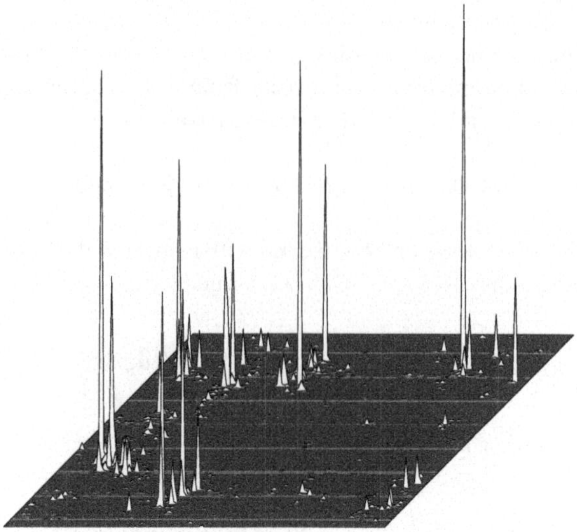

Fig. 20 A sample of critical Liouville quantum gravity measure over a square domain. Despite the occurrence of prominent spikes, the measure is non-atomic although just barely so as it is (conjecturally) supported on a set of zero Hausdorff dimension

Theorem 10.30 (Identification with critical LQG measure) *The family of measures*

$$r_D(x)^2\mu_\infty^{D,\alpha}(\mathrm{d}x),\quad D\in\mathfrak{D}\,, \tag{10.81}$$

constructed in Theorem 10.29 obeys conditions (1–5) of Theorem 10.15 for some $\hat{c}>0$. *In particular, there is a constant* $c\in(0,\infty)$ *such that*

$$Z^D(\mathrm{d}x)\overset{\text{law}}{=}c\,r_D(x)^2\,\mu_\infty^{D,\alpha}(\mathrm{d}x),\quad D\in\mathfrak{D}. \tag{10.82}$$

This appears as [24, Theorem 2.9]. The key technical challenge is to prove that the measure in (10.81) obeys the Laplace-transform asymptotic (10.33). This is based on a version of the concentric decomposition and approximations via Kahane's convexity inequality. We refer the reader to [24] for further details.

We conclude by reiterating that many derivations in the present lecture were originally motivated by the desire to prove directly the uniqueness of the subsequential limit of η^D_{N,r_N} (as we did for the intermediate level sets) based only on the easily obtainable properties of the Z^D-measure. This program will partially be completed in our proof of Theorem 12.16 although there we still rely on the Laplace transform tail (10.33) which also underpins the limit of the DGFF maximum. We believe that even this step can be bypassed and the reference to convergence of the DGFF maximum avoided; see Conjecture 16.3.

Lecture 11: Local Structure of Extremal Points

In this lecture we augment the conclusions obtained for the point processes associated with extremal local maxima to include information about the local behavior. The proofs are based on the concentric decomposition of the DGFF and entropic-repulsion arguments developed for the proof of tightness of the absolute maximum. Once this is done, we give a formal proof of our full convergence result from Theorem 9.3. A number of corollaries are presented that concern the cluster-process structure of the extremal level sets, a Poisson–Dirichlet limit of the Gibbs measure associated with the DGFF, the Liouville Quantum Gravity in the glassy phase and the freezing phenomenon.

11.1 Cluster at Absolute Maximum

Our focus in this lecture is on the local behavior of the field near its near-maximal local extrema. For reasons mentioned earlier, we will refer of these values as a *cluster*. As it turns out, all that will technically be required is the understanding of the cluster associated with the absolute maximum:

Theorem 11.1 (Cluster law) *Let $D \in \mathfrak{D}$ with $0 \in D$ and let $\{D_N : N \geq 1\}$ be an admissible sequence of approximating domains. Then for each $t \in \mathbb{R}$ and each function $f \in C_c(\overline{\mathbb{R}}^{\mathbb{Z}^2})$ depending only on a finite number of coordinates,*

$$E\Bigl(f(h_0^{D_N} - h^{D_N}) \Bigm| h_0^{D_N} = m_N + t,\, h^{D_N} \leq h_0^{D_N} \Bigr) \underset{N\to\infty}{\longrightarrow} E_\nu(f), \tag{11.1}$$

where ν is a probability measure on $[0,\infty)^{\mathbb{Z}^2}$ defined from $\phi :=$ DGFF on $\mathbb{Z}^2 \smallsetminus \{0\}$ via the weak limit

$$\nu(\cdot) := \lim_{r\to\infty} P\biggl(\phi + \frac{2}{\sqrt{g}}\mathfrak{a} \in \cdot \biggm| \phi_x + \frac{2}{\sqrt{g}}\mathfrak{a}(x) \geq 0\colon |x| \leq r \biggr) \tag{11.2}$$

in which (we recall) $\mathfrak{a}$ denotes the potential kernel on $\mathbb{Z}^2$.

The existence of the weak limit in (11.2) is part of the statement of the theorem. Remarkably, the convergence alone may be inferred by much softer means:

Exercise 11.2 *Let ν_r be the conditional measure on the right of* (11.2). *Prove that $r \mapsto \nu_r$ is stochastically increasing. [Hint: Under this type of conditioning, the same holds true for any strong-FKG measure.]*

The exercise shows that $r \mapsto \nu_r(A)$ is increasing on increasing events and so the limit in (11.2) exists for any event depending on a finite number of coordinates. The problem is that ν_r is a priori a measure on $[0,\infty]^{\mathbb{Z}^2}$ and the interpretation of the limit

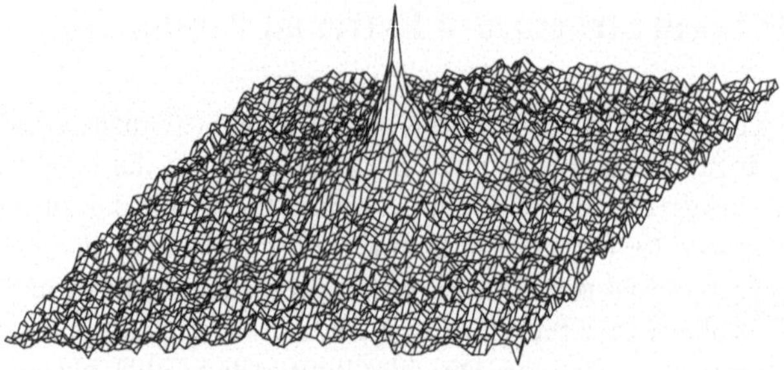

Fig. 21 A sample of the configuration of the DGFF in the vicinity of its (large) local maximum

as a distribution on $[0, \infty)^{\mathbb{Z}^2}$ requires a proof of tightness. This additional ingredient will be supplied by our proof of Theorem 11.1.

That the limit takes the form in (11.2) may be explained via a simple heuristic calculation. Indeed, by Lemma 8.5, conditioning the field h^{D_N} on $h_0^{D_N} = m_N + t$ effectively shifts the mean of $h_0^{D_N} - h_x^{D_N}$ by a quantity with the asymptotic

$$(m_N + t)\big(1 - \mathfrak{g}^{D_N}(x)\big) \underset{N\to\infty}{\longrightarrow} \frac{2}{\sqrt{g}}\mathfrak{a}(x). \tag{11.3}$$

A variance computation then shows that the law of $x \mapsto h_0^{D_N} - h_x^{D_N}$ tends, in the sense of finite-dimensional distributions, to $x \mapsto \phi_x + \frac{2}{\sqrt{g}}\mathfrak{a}(x)$, where ϕ is the pinned DGFF; see Fig. 21. The conditioning on the origin being the maximum then translates into the conditioning in (11.2). This would more or less suffice to prove the desired result were it not for the following fact:

Theorem 11.3 *There exists $c_\star \in (0, \infty)$ such that*

$$P\left(\phi_x + \frac{2}{\sqrt{g}}\mathfrak{a}(x) \geq 0\colon |x| \leq r\right) = \frac{c_\star}{\sqrt{\log r}}\big(1 + o(1)\big), \quad r \to \infty. \tag{11.4}$$

Since the right-hand side of (11.4) vanishes in the limit $r \to \infty$, the conditioning in (11.2) is increasingly singular and so it is hard to imagine that one could control the limit in (11.1) solely via manipulations based on weak convergence.

11.2 Random Walk Based Estimates

Our proof of Theorem 1.11 relies on the concentric decomposition of the DGFF developed in Sects. 8.2–8.4. As these sections were devoted to the proof of tightness of the lower tail of the maximum, we were not allowed to assume any bounds on the lower tails in estimates there. However, with the tightness settled, Lemma 8.23 augments Lemma 8.13 to a two-sided estimate:

Lemma 11.4 *There is $a > 0$ such that each $k = 1, \dots, n$ and each $t \ge 0$,*

$$P\biggl(\Bigl|\max_{x\in\Delta^k\smallsetminus\Delta^{k-1}}\bigl[\chi_{k+1}(x)+\chi_k(x)+h'_k(x)\bigr]-m_{2^k}\Bigr|\ge t\biggr)\le \mathrm{e}^{-at}. \tag{11.5}$$

This allows for control of the deviations of the field h^{D_N} from the random walk $-S_k$ in both directions which upgrades Lemma 8.19 to the form:

Lemma 11.5 (Reduction to random walk events) *Assume h^{D_N} is realized as the sum on the right of* (8.41). *There is a numerical constant $C > 0$ such that uniformly in the above setting, the following holds for each $k = 1, \dots, n$ and each $t \in \mathbb{R}$:*

$$\begin{aligned}\{S_{n+1}=0\}\cap\bigl\{S_k\ge R_K(k)+|t|\bigr\}&\\ \subseteq\{h_0^{D_N}=0\}\cap\bigl\{h^{D_N}\le(m_N+t)(1-\mathfrak{g}^{D_N})\text{ on }\Delta^k\smallsetminus\Delta^{k-1}\bigr\}&\\ \subseteq\{S_{n+1}=0\}\cap\bigl\{S_k\ge -R_K(k)-|t|\bigr\},&\end{aligned} \tag{11.6}$$

where K is the control variable from Definition 8.17 with the absolute value signs added around the quantity on the left of (8.61) *and*

$$R_k(\ell) := C[1+\theta_{n,k}(\ell)], \tag{11.7}$$

with $\theta_{n,k}$ as in (8.59) *and C as in Lemma 8.19. (We recall that, here and henceforth, n is the largest integer such that $\{x\in\mathbb{Z}^2\colon |x|_\infty\le 2^{n+1}\}\subseteq D_N$.)*

Proof. From (8.64) we get, for all $k = 0, \dots, n$ (and with $\Delta^{-1} := \emptyset$),

$$\begin{aligned}&\Bigl|\max_{x\in\Delta^k\smallsetminus\Delta^{k-1}}\bigl[h_x^{D_N}-(m_N+t)(1-\mathfrak{g}^{D_N}(x))\bigr]+S_k\Bigr|\\ &\qquad\le\sum_{\ell=k}^{n}\max_{x\in\Delta^k\smallsetminus\Delta^{k-1}}|\mathfrak{b}_\ell(x)||\varphi_\ell(0)|+\biggl(\sum_{\ell=k+2}^{n}\max_{x\in\Delta^k\smallsetminus\Delta^{k-1}}|\chi_\ell(x)|\biggr)\\ &\qquad\quad+\Bigl|\max_{x\in\Delta^k\smallsetminus\Delta^{k-1}}\bigl[\chi_{k+1}(x)+\chi_k(x)+h'_k(x)-m_{2^k}\bigr]\Bigr|\\ &\qquad\qquad+\max_{x\in\Delta^k\smallsetminus\Delta^{k-1}}\Bigl|m_N(1-\mathfrak{g}^{D_N}(x))-m_{2^k}\Bigr|+|t|.\end{aligned} \tag{11.8}$$

The definition of K then bounds the first three terms on the right by a quantity of order $1+\theta_{n,K}(k)$. For the second to last term, here instead of (8.65) we need:

Exercise 11.6 *There is $c > 0$ such that for all $n \ge 1$ and all $k = 0, \dots, n$,*

$$\max_{x\in\Delta^k\smallsetminus\Delta^{k-1}}\Bigl|m_N(1-\mathfrak{g}^{D_N}(x))-m_{2^k}\Bigr|\le c\bigl[1+\log(1+k\wedge(n-k))\bigr]. \tag{11.9}$$

The inclusions (11.6) follow readily from this. □

We will now use the random walk $\{S_1,\dots,S_n\}$ to control all important aspects of the conditional expectation in the statement of Theorem 11.1.

First note that the event $\bigcap_{k=1}^n\{S_k\ge -R_K(k)-|t|\}$ encases all of the events of interest and so we can use it as the basis for estimates of various undesirable scenarios. (This is necessary because the relevant events will have probability tending to zero proportionally to $1/n$.) In particular, we need to upgrade Lemma 8.21 to the form:

Lemma 11.7 *There are constants $c_1, c_2 > 0$ such that for all $n\ge 1$, all t with $0\le t\le n^{1/5}$ and all $k=1,\dots,\lfloor n/2\rfloor$,*

$$P\biggl(\{K>k\}\cap\bigcap_{\ell=1}^n\{S_\ell\ge -R_k(\ell)-t\}\,\bigg|\,S_{n+1}=0\biggr)\le c_1\frac{1+t^2}{n}\mathrm{e}^{-c_2(\log k)^2}. \tag{11.10}$$

Since the target decay is order-$1/n$, this permits us to assume $\{K\le k\}$ for k sufficiently large but independent of n whenever need arises in the forthcoming derivations. Lemma 8.20 then takes the form:

Lemma 11.8 (Entropic repulsion) *For each $t\ge 0$ there is $c>0$ such that for all $n\ge 1$ and all $k=1,\dots,\lfloor n/2\rfloor$*

$$P\biggl(\{S_k,S_{n-k}\ge k^{1/6}\}\cap\bigcap_{\ell=k+1}^{n-k-1}\{S_\ell\ge R_k(\ell)+t\}\,\bigg|\,\bigcap_{\ell=1}^{n}\{S_\ell\ge -R_k(\ell)-t\}\cap\{S_{n+1}=0\}\biggr)\ge 1-ck^{-\frac{1}{16}}. \tag{11.11}$$

We will not give formal proofs of Lemmas 11.7–11.8 here; instead we refer the reader to the paper [27, Sect. 4].

Consider the expectation in the statement of Theorem 11.1. Lemma 8.5 permits us to shift the conditioning event to $h_0^{D_N}=0$ at the cost of adding $(m_N+t)\mathfrak{g}^{D^N}$ to all occurrences of the field. Abbreviating

$$m_N(t,x):=(m_N+t)(1-\mathfrak{g}^{D_N}(x)), \tag{11.12}$$

the expectation in (11.1) can thus be written as

$$\frac{E\Bigl(f\bigl(m_N(t,\cdot)-h^{D_N}\bigr)1_{\{h^{D_N}\le m_N(t,\cdot)\}}\,\Big|\,h_0^{D_N}=0\Bigr)}{E\Bigl(1_{\{h^{D_N}\le m_N(t,\cdot)\}}\,\Big|\,h_0^{D_N}=0\Bigr)}. \tag{11.13}$$

We will control this ratio by deriving, separately, asymptotic expressions for the numerator and the denominator which (in light of Lemmas 11.4 and 11.7) will both decay as a constant over n. As both the numerator and the denominator have the same structure, it suffices to focus on the numerator. We claim:

Proposition 11.9 *For each $\epsilon > 0$ and each $t_0 > 0$ there is $k_0 \ge 1$ such that for all natural k with $k_0 \le k \le n^{1/6}$ and all $t \in [-t_0, t_0]$,*

$$\begin{aligned}\Biggl|E\Bigl(f\bigl(m_N(t,\cdot) - h^{D_N}\bigr)1_{\{h^{D_N} \le m_N(t,\cdot)\}}\Bigm| h_0^{D_N} = 0\Bigr) \\ - E\biggl(f\Bigl(\tfrac{2}{\sqrt{g}}\mathfrak{a} + \phi_k\Bigr)1_{\left\{\phi_k + \frac{2}{\sqrt{g}}\mathfrak{a} \ge 0 \text{ in } \Delta^k\right\}}1_{\{S_k, S_{n-k} \in [k^{1/6}, k^2]\}}\Bigl(\prod_{\ell=k}^{n-k} 1_{\{S_\ell \ge 0\}}\Bigr) \\ \times 1_{\{h^{D_N} \le m_N(t,\cdot) \text{ in } D_N \smallsetminus \Delta^{n-k}\}}\Bigm| h_0^{D_N} = 0\biggr)\Biggr| \le \frac{\epsilon}{n},\end{aligned} \tag{11.14}$$

where we used the shorthand

$$\phi_k(x) := h_0^{\Delta^k} - h_x^{\Delta^k}. \tag{11.15}$$

Proof (*sketch*). The proof is based on a sequence of replacements that gradually convert one expectation into the other. First we separate scales by noting that, given $k \in \{1, \dots, \lfloor n/2 \rfloor\}$, we can write the "hard" event in the expectation as the intersection of "inner", "middle" and "outer" events,

$$\begin{aligned}1_{\{h^{D_N} \le m_N(t,\cdot)\}} = 1_{\{h^{D_N} \le m_N(t,\cdot) \text{ in } \Delta^k\}} \\ \times 1_{\{h^{D_N} \le m_N(t,\cdot) \text{ in } \Delta^{n-k} \smallsetminus \Delta^k\}} 1_{\{h^{D_N} \le m_N(t,\cdot) \text{ in } D_N \smallsetminus \Delta^{n-k}\}}.\end{aligned} \tag{11.16}$$

Plugging this in we can also use Lemma 11.7 to insert the indicator of $\{K \le k\}$ into the expectation. The inclusions in (11.6) permit us to replace the "middle" event

$$\bigl\{h^{D_N} \le m_N(t,\cdot) \text{ in } \Delta^{n-k} \smallsetminus \Delta^k\bigr\} \tag{11.17}$$

by

$$\bigcap_{\ell=k}^{n-k}\{S_\ell \ge \pm(R_k(\ell) + |t|)\} \tag{11.18}$$

with the sign depending on whether we aim to get upper or lower bounds. Lemma 11.8 then tells us that the difference between these upper and lower bounds is negligible, and so we may further replace $\{S_\ell \ge \pm(R_k(\ell) + |t|)\}$ by $\{S_\ell \ge 0\}$. In addition, by Lemma 11.8 we may also assume $S_k, S_{n-k} \ge k^{1/6}$. The bounds $S_k, S_{n-k} \le k^2$ then arise (for k large enough) from the restriction to $\{K \le k\}$ and the inequalities in Definition 8.17.

At this point we have replaced the first expectation in (11.14) by $1+O(k^{-\frac{1}{16}})$-times the conditional expectation

$$E\Big(f\big(m_N(t,\cdot)-h^{D_N}\big)1_{\{K\le k\}}1_{\{h^{D_N}\le m_N(t,\cdot)\text{ in }\Delta^k\}}1_{\{S_k,S_{n-k}\in[k^{1/6},\,k^2]\}} \times\Bigg(\prod_{\ell=k}^{n-k}1_{\{S_\ell\ge 0\}}\Bigg)1_{\{h^{D_N}\le m_N(t,\cdot)\text{ in }D_N\smallsetminus\Delta^{n-k}\}}\,\Bigg|\,h_0^{D_N}=0\Big) \tag{11.19}$$

plus a quantity of order $\frac{1+t^2}{n}\text{e}^{-c_2(\log k)^2}$. Next we will use the continuity of f to replace $m_N(t,\cdot)$ in the argument of f by its limit value $\frac{2}{\sqrt{g}}\mathfrak{a}$. For this we assume that k is much larger than the diameter of the set of vertices that f depends on. Conditional on $h_0^{D_N}=0$ we then have

$$h_x^{D_N}=-\phi_k(x)+\sum_{\ell>k}\big[\mathfrak{b}_\ell(x)\varphi_\ell(0)+\chi_\ell\big],\quad x\in\Delta^k. \tag{11.20}$$

The bounds arising from the restriction $K\le k$ then let us replace h^{D_N} in the argument of f by $-\phi_k$.

It remains to deal with the indicator of the "inner" event

$$\big\{h^{D_N}\le m_N(t,\cdot)\text{ in }\Delta^k\big\} \tag{11.21}$$

which we want to replace by

$$\big\{\phi_k+\tfrac{2}{\sqrt{g}}\mathfrak{a}\ge 0\text{ in }\Delta^k\big\}. \tag{11.22}$$

Here continuity arguments cannot be used; instead, we have to show that entropic repulsion creates a sufficiently large gap between h^{D_N} and $m_N(t,\cdot)$ in the first expectation in (11.14), and between $\phi_k+\frac{2}{\sqrt{g}}\mathfrak{a}$ and zero in the second expectation, to absorb the sum on the right of (11.20) and the difference between $m_N(t,\cdot)$ and $\frac{2}{\sqrt{g}}\mathfrak{a}$ everywhere on Δ^k. We refer to [27, Lemma 4.22] (and Exercise 11.10 below) for details of this step.

Even after all the replacements above have been done, the quantity under expectation remains concentrated on $\bigcap_{\ell=1}^n\{S_\ell\ge -R_k(\ell)-|t|\}$. Lemma 11.8 then permits us to drop the restriction to $\{K\le k\}$ and get the desired result. □

The entropic repulsion step from [27, Lemma 4.22] relies on the following general fact which constitutes a nice exercise for the reader:

Exercise 11.10 *(Controlling the gap) Prove that, if ϕ is a field on a set Λ with the strong-FKG property, then for all $\delta>0$ and all $f\colon\Lambda\to\mathbb{R}$ (and writing $\{\phi\not\ge f\}$*

for $\{\phi(x) \ge f(x) \colon x \in \Lambda\}^c$*)*

$$P\bigl(\phi \not\ge f + \delta \,\big|\, \phi \ge f\bigr) \le \sum_{x\in\Lambda} P\bigl(\phi(x) < f(x) + \delta \,\big|\, \phi(x) \ge f(x)\bigr). \tag{11.23}$$

For $\phi(x) = \mathcal{N}(0, \sigma^2)$ *with* $\sigma^2 > 0$, *and* $f(x) \le 0$, *show also that*

$$P\bigl(\phi(x) \not\ge f(x) + \delta \,\big|\, \phi(x) \ge f(x)\bigr) \le \frac{2}{\sqrt{2\pi}} \frac{\delta}{\sigma} \le \frac{\delta}{\sigma}. \tag{11.24}$$

Thanks to (11.23)–(11.24), in order to show the existence of a gap it suffices to prove a uniform lower bound on the variance of the relevant fields. (Note that the gap can be taken to zero slowly with $n \to \infty$.)

Moving back to the main line of argument underlying the proof of Theorem 11.1, next we observe:

Lemma 11.11 *Conditionally on* S_k *and* S_{n-k} *and the event* $S_{n+1} = 0$,

(1) the "inner" field ϕ_k,
(2) the random variables $\{S_\ell \colon \ell = k, \dots, n-k\}$, *and*
(3) the "outer" field $\{h_x^{D_N} \colon x \in D_N \smallsetminus \Delta^{n-k}\}$

are independent of one another whenever $n > 2k$.

Proof. Inspecting (11.15), ϕ_k is measurable with respect to

$$\sigma\bigl(\varphi_1(0), \dots, \varphi_k(0), \chi_1, \dots, \chi_k, h_0, \dots, h_k'\bigr), \tag{11.25}$$

while the random variables in (2) are measurable with respect to

$$\sigma\bigl(\{S_k\} \cup \{\varphi_\ell(0) \colon \ell = k+1, \dots, n-k\}\bigr). \tag{11.26}$$

The definition of the concentric decomposition in turn ensures that the random variables in (3), as well as the event $\{S_{n+1} = 0\}$, are measurable with respect to

$$\sigma\left(\{S_{n-k}\} \cup \bigcup_{\ell=n-k}^{n} \{\varphi_\ell(0), \chi_\ell, h_\ell'\}\right). \tag{11.27}$$

The claim follows from the independence of the random objects constituting the concentric decomposition. □

We note that the Lemma 11.11 is the prime reason why we had to replace $h^{\Delta^n} - h_0^{\Delta^n}$ by ϕ_k in the "inner" event. (Indeed, $h^{\Delta^n} - h_0^{\Delta^n}$ still depends on the fields χ_ℓ, $\ell = n-k, \dots, n$.) Our next step is to take expectation conditional on S_k and S_{n-k}. Here we will use:

Lemma 11.12 *For each $t_0 > 0$ there is $c > 0$ such that for all $1 \le k \le n^{1/6}$,*

$$\left| P\left(\bigcap_{\ell=k}^{n-k}\{S_\ell \ge 0\} \,\middle|\, \sigma(S_k, S_{n-k})\right) - \frac{2}{g\log 2}\frac{S_k S_{n-k}}{n}\right| \le c\frac{k^4}{n}\frac{S_k S_{n-k}}{n} \tag{11.28}$$

holds everywhere on $\{S_k, S_{n-k} \in [k^{1/6}, k^2]\}$.

Proof (*idea*). We will only explain the form of the leading term leaving the error to a reference to [27, Lemma 5.6]. Abbreviating $x := S_k$ and $y := S_{n-k}$, the conditional probability in (11.28) is lower bounded by

$$P\big(B_t \ge 0\colon t \in [t_k, t_{n-k}] \,\big|\, B_{t_k} = x,\, B_{t_{n-k}} = y\big), \tag{11.29}$$

where we used the embedding (8.76) of the walk into a path of Brownian motion. In light of Lemma 8.10, we know that

$$t_\ell - t_k = \big(g\log 2 + o(1)\big)(\ell - k), \quad \ell \ge k, \tag{11.30}$$

with $o(1) \to 0$ as $k \to \infty$ uniformly in $\ell \ge k$. Exercise 7.10 then gives

$$\begin{aligned} P\left(\bigcap_{\ell=k}^{n-k}\{S_\ell \ge 0\} \,\middle|\, \sigma(S_k, S_{n-k})\right) &\gtrsim \frac{2S_k S_{n-k}}{t_{n-k} - t_k} \\ &= \frac{2}{g\log 2}\frac{S_k S_{n-k}}{n}\big(1 + o(1)\big) \end{aligned} \tag{11.31}$$

on $\{S_k, S_{n-k} \in [k^{1/6}, k^2]\}$ whenever $k^4 \ll n$.

To get a similar upper bound, one writes the Brownian motion on interval $[t_\ell, t_{\ell+1}]$ as a linear curve connecting S_ℓ to $S_{\ell+1}$ plus a Brownian bridge. Then we observe that the entropic repulsion pushes the walk far away from the positivity constraint so that the Brownian bridges do not affect the resulting probability much. See [27, Lemma 4.15] for details. □

The main consequence of the above observations is an asymptotic formula for the numerator (and thus also the denominator) in (11.13). Indeed, denote

$$\Xi_k^{\text{in}}(f) := E\left(f\left(\phi_k + \tfrac{2}{\sqrt{g}}\mathfrak{a}\right) 1_{\left\{\phi_k + \frac{2}{\sqrt{g}}\mathfrak{a} \ge 0 \text{ in } \Delta^k\right\}} 1_{\{S_k \in [k^{1/6}, k^2]\}} S_k \right) \tag{11.32}$$

and

$$\Xi_{N,k}^{\text{out}}(t) := E\left(1_{\{h^{D_N} \le m_N(t,\cdot) \text{ in } D_N \smallsetminus \Delta^{n-k}\}} 1_{\{S_{n-k} \in [k^{1/6}, k^2]\}} S_{n-k} \,\middle|\, h_0^{D_N} = 0 \right). \tag{11.33}$$

As a result of the above manipulations, we then get:

Lemma 11.13 (Main asymptotic formula) *For $o(1) \to 0$ in the limits as $N \to \infty$ and $k \to \infty$, uniformly on compact sets of t and compact families of $f \in C_c(\overline{\mathbb{R}}^{\mathbb{Z}^2})$ depending on a given finite number of variables,*

$$E\Big(f\big(m_N(t,\cdot) - h^{D_N}\big) 1_{\{h^{D_N} \le m_N(t,\cdot)\}} \,\Big|\, h_0^{D_N} = 0\Big) = \frac{2}{n} \frac{\Xi_k^{\rm in}(f)\Xi_{N,k}^{\rm out}(t)}{g \log 2} + \frac{o(1)}{n}. \tag{11.34}$$

Proof. This follows by plugging Lemma 11.12 in the second expectation in (11.14) and using the independence stated in Lemma 11.11. □

In order to control the $N \to \infty$ and $k \to \infty$ limits of the right-hand side of (11.34), we will also need to observe:

Lemma 11.14 *For each $t_0 > 0$ there are $c_1, c_2 \in (0, \infty)$ such that for all $t \in [-t_0, t_0]$, all $N \ge 1$ and all $k \le n^{1/6}$ (with N and n related as above),*

$$c_1 < \Xi_{N,k}^{\rm out}(t) < c_2 \tag{11.35}$$

and

$$c_1 < \Xi_k^{\rm in}(1) < c_2. \tag{11.36}$$

We postpone the proof of these bounds until the next section. Instead, we note the following consequence thereof:

Corollary 11.15 *Uniformly in f and t as stated in Lemma 11.13,*

$$\lim_{N\to\infty} E\Big(f\big(h_0^{D_N} - h^{D_N}\big) \,\Big|\, h_0^{D_N} = m_N + t,\ h^{D_N} \le h_0^{D_N}\Big) = \lim_{k\to\infty} \frac{\Xi_k^{\rm in}(f)}{\Xi_k^{\rm in}(1)}, \tag{11.37}$$

where, in particular, both limits exist.

Proof. The bounds (11.35) allow us to write the right-hand side of (11.34) as

$$\frac{2\Xi_{N,k}^{\rm out}(t)}{g \log 2} \, \frac{\Xi_k^{\rm in}(f) + o(1)}{n}. \tag{11.38}$$

The ratio in (11.13) thus simplifies into the ratio of $\Xi_k^{\rm in}(f) + o(1)$ and $\Xi_k^{\rm in}(1) + o(1)$. This depends on N only through the $o(1)$ terms which tend to zero as $N \to \infty$ and $k \to \infty$. Using the lower bound in (11.36), the claim follows. □

11.3 Full Process Convergence

Moving towards the proof of of Theorem 11.1, thanks to the representation of the pinned DGFF in Exercise 8.15, the above derivation implies, even in a somewhat simpler form, also the limit of the probabilities in (11.2). The difference is that here the random walk is not constrained to $S_{n+1} = 0$ and also there is no t to worry about. This affects the asymptotics of the relevant probability as follows:

Lemma 11.16 *For $f \in C_c(\overline{\mathbb{R}}^{\mathbb{Z}^2})$ depending on a finite number of coordinates,*

$$E\left(f\left(\phi + \tfrac{2}{\sqrt{g}}\mathfrak{a}\right) 1_{\{\phi + \frac{2}{\sqrt{g}}\mathfrak{a} \geq 0 \text{ in } \Delta^r\}}\right) = \frac{1}{\sqrt{\log 2}} \frac{\Xi_k^{\text{in}}(f)}{\sqrt{r}} + \frac{o(1)}{\sqrt{r}}, \tag{11.39}$$

where $o(1) \to 0$ as $r \to \infty$ followed by $k \to \infty$.

Similarly to $1/n$ in (11.34) arising from the asymptotic of the probability that a random-walk bridge of time length n to stay positive, the asymptotic $1/\sqrt{r}$ stems from the probability that an (unconditioned) random walk stays positive for the first r steps. Indeed, the reader will readily check:

Exercise 11.17 *Let $\{B_t : t \geq 0\}$ be the standard Brownian motion with P^x denoting the law started from $B_0 = x$. Prove that for all $x > 0$ and all $t > 0$,*

$$\sqrt{\frac{2}{\pi}} \frac{x}{\sqrt{t}} \left(1 - \frac{x^2}{2t}\right) \leq P^x\big(B_s \geq 0 \colon 0 \leq s \leq t\big) \leq \sqrt{\frac{2}{\pi}} \frac{x}{\sqrt{t}}. \tag{11.40}$$

We will not give further details concerning the proof of Lemma 11.16 as that would amount to repetitions that the reader may not find illuminating. (The reader may consult [27, Proposition 5.1] for that.) Rather we move on to:

Proof of Lemma 11.14. We begin with (11.36). We start by introducing a variant $\widetilde{K}$ of the control variable K. Let $\widetilde{\theta}_k(\ell) := 1 + [\log(k \vee \ell)]^2$ and define $\widetilde{K}$ to be the least positive natural k such that Definition 8.17(1,2)—with $\theta_{n,k}(\ell)$ replaced by $\widetilde{\theta}_k(\ell)$—as well as

$$\left|\min_{x \in \Delta^\ell \smallsetminus \Delta^{\ell-1}} \big[\chi_\ell(x) + \chi_{\ell+1}(x) + h'_\ell(x)\big] + \frac{2}{\sqrt{g}}\mathfrak{a}(x)\right| \leq \theta_k(\ell) \tag{11.41}$$

and

$$\left|\max_{x \in \Delta^\ell \smallsetminus \Delta^{\ell-1}} \big[\chi_\ell(x) + \chi_{\ell+1}(x) + h'_\ell(x)\big] - \frac{2}{\sqrt{g}}\mathfrak{a}(x)\right| \leq \theta_k(\ell) \tag{11.42}$$

hold true for all $\ell \geq k$. (That $\widetilde{K} < \infty$ a.s. follows by our earlier estimates and the Borel–Cantelli lemma. The condition (11.42) is introduced for later convenience.)

Recall that Exercise 8.15 expresses the pinned DGFF using the objects in the concentric decomposition. Hence we get

$$\bigcap_{k=1}^{r} \{S_k \geq C\widetilde{\theta}_{\widetilde{K}}(k)\} \subseteq \left\{\phi + \frac{2}{\sqrt{g}}\mathfrak{a} \geq 0 \text{ in } \Delta^r\right\} \subseteq \bigcap_{k=1}^{r} \{S_k \geq -C\widetilde{\theta}_{\widetilde{K}}(k)\}, \tag{11.43}$$

for some sufficiently large absolute constant $C > 0$. Ballot problem/entropy repulsion arguments invoked earlier then show that the probability of both sides decays proportionally to that of the event $\bigcap_{k=1}^{r}\{S_\ell \geq -1\}$. Lemma 11.16 and Exercise 11.17 then give (11.36).

Plugging (11.36) to (11.34), the inclusions in Lemma 11.5 along with the bounds in Lemmas 11.7–11.8 then imply (11.35) as well. □

We are finally ready to give:

Proof of Theorem 11.1. From Lemmas 11.16 and 11.14 we have

$$\lim_{r\to\infty} E\left(f\left(\phi+\tfrac{2}{\sqrt{g}}\mathfrak{a}\right)\,\Big|\,\phi_x+\frac{2}{\sqrt{g}}\mathfrak{a}(x)\geq 0\colon |x|\leq r\right) = \lim_{k\to\infty}\frac{\Xi_k^{\text{in}}(f)}{\Xi_k^{\text{in}}(1)}. \tag{11.44}$$

Jointly with Corollary 11.15, this proves equality of the limits in the statement. To see that ν concentrates on $\mathbb{R}^{\mathbb{Z}^2}$ (and, in fact, on $[0,\infty)^{\mathbb{Z}^2}$) we observe that, since $\ell \mapsto \widetilde{\theta}_k(\ell)$ from Lemma 11.14 grows only polylogarithmically while $\mathfrak{a}$ is order k on $\Delta^k \smallsetminus \Delta^{k-1}$, once k is sufficiently large we get

$$\left\{\phi+\frac{2}{\sqrt{g}}\mathfrak{a}\not\leq k^2 \text{ on } \Delta^k\right\} \subseteq \{\widetilde{K} > k\}. \tag{11.45}$$

(It is here where we make use of (11.42).) By an analogue of Lemma 11.7 for random variable $\widetilde{K}$, we have $\nu(\widetilde{K} > k) \leq c_1 \mathrm{e}^{-c_2(\log k)^2}$. The Borel–Cantelli lemma then implies $\nu(\mathbb{R}^{\mathbb{Z}^2}) = 1$, as desired. □

The above techniques also permit us to settle the asymptotic (11.4):

Proof of Theorem 11.3. Observe that after multiplying both sides of (11.39) by $\sqrt{r}$, the left-hand side is independent of k while the right-hand side depends on r only through the $o(1)$-term. In light of (11.36), $r \to \infty$ and $k \to \infty$ can be taken independently (albeit in this order) to get that

$$\Xi_\infty^{\text{in}}(1) := \lim_{k\to\infty} \Xi_k^{\text{in}}(1) \tag{11.46}$$

exists, is positive and finite. Since $r \log 2$ is, to the leading order, the logarithm of the diameter of Δ^r, the claim follows from (11.39) with $c_\star := \Xi_\infty^{\text{in}}(1)$. □

We are now finally ready to move to the proof of Theorem 9.3 dealing with the weak convergence of the full, three coordinate process $\eta_{N,r}^D$ defined in (9.7). In light of Theorem 11.1, all we need to do is to come up with a suitable localization method that turns a large local maximum (in a large domain) to the actual maximum (in a smaller domain). We will follow a different line of reasoning than [27] as the arguments there seem to contain flaws whose removal would take us through lengthy calculations that we prefer to avoid.

Proof of Theorem 9.3. Every $f \in C_c(D \times \mathbb{R} \times \overline{\mathbb{R}}^{\mathbb{Z}^2})$ is uniformly close to a compactly supported function that depends only on a finite number of "cluster" coordinates. So let us assume that $x, h, \phi \mapsto f(x, h, \phi)$ is continuous and depends only on $\{\phi_y : y \in \Lambda_r(0)\}$ for some $r \geq 1$ and vanishes when $|h|, \max_{x\in\Lambda_r} |\phi_y| \geq \lambda$ for some $\lambda > 0$ or if $\operatorname{dist}(x, D^c) < \delta$ for some $\delta > 0$.

The proof hinges on approximation of η_N^D by three auxiliary processes. Given an integer $K \geq 1$, let $\{S^i : i = 1, \dots, m\}$ be the set of all the squares of the form $z/K + (0, 1/K)^2$ with $z \in \mathbb{Z}^2$ that fit entirely into D^δ. For each $i = 1, \dots, m$, let S_N^i be the lattice box of side-length (roughly) N/K contained in NS^i so that every pair of neighboring squares S_N^i and S_N^j keep a "line of sites" in-between. Given $\delta > 0$ small, let $S_{N,\delta}^i := \{x \in S_N^i : \operatorname{dist}_\infty(x, S_N^i) > \delta N\}$. For $x \in D_N$ such that $x \in S_{N,\delta}^i$ for some $i = 1, \dots, m$, set

$$\Theta_{N,K,\delta}(x) := \Bigl\{h_x^{D_N} = \max_{y\in S_N^i} h_y^{D_N}\Bigr\} \tag{11.47}$$

and let $\Theta_{N,K,\delta}(x) := \emptyset$ for $x \in D_N$ where no such i exists. Setting

$$\eta_N^{D,K,\delta} := \sum_{x\in D_N} 1_{\Theta_{N,K,\delta}(x)}\, \delta_{x/N} \otimes \delta_{h_x^{D_N} - m_N} \otimes \delta_{\{h_x^{D_N} - h_{x+z}^{D_N} :\, z\in\mathbb{Z}^2\}}, \tag{11.48}$$

Lemma 10.4 and Theorem 9.2 show that, for any f as above,

$$\lim_{\delta\downarrow 0} \limsup_{K\to\infty} \limsup_{N\to\infty} \Bigl| E(\mathrm{e}^{-\langle \eta_{N,r_N}^D, f\rangle}) - E(\mathrm{e}^{-\langle \eta_N^{D,K,\delta}, f\rangle}) \Bigr| = 0. \tag{11.49}$$

Next we consider the Gibbs-Markov decomposition $h^{D_N} = h^{\widetilde{D}_N} + \varphi^{D_N,\widetilde{D}_N}$, where $\widetilde{D}_N := \bigcup_{i=1}^m S_N^i$. Denoting, for $x \in \widetilde{D}_N$ and $i = 1, \dots, m$ such that $x \in S_{N,\delta}^i$,

$$\widetilde{\Theta}_{N,K,\delta}(x) := \Bigl\{h_x^{\widetilde{D}_N} = \max_{y\in S_N^i} h_y^{\widetilde{D}_N}\Bigr\}, \tag{11.50}$$

we then define $\widetilde{\eta}_N^{D,K,\delta}$ by the same formula as $\eta_N^{D,K,\delta}$ but with $\Theta_{N,K,\delta}(x)$ replaced by $\widetilde{\Theta}_{N,K,\delta}(x)$ and the sum running only over $x \in \widetilde{D}_N$. The next point to observe is that, for K sufficiently large, the maximum of h^{D_N} in S_N^i coincides (with high probability) with that of $h^{\widetilde{D}_N}$. (We used a weaker version of this already in Lemma 10.18.) This underpins the proof of:

Lemma 11.18 *For any f as above,*

$$\lim_{\delta\downarrow 0} \limsup_{K\to\infty} \limsup_{N\to\infty} \Bigl| E(\mathrm{e}^{-\langle \eta_N^{D,K,\delta}, f\rangle}) - E(\mathrm{e}^{-\langle \widetilde{\eta}_N^{D,K,\delta}, f\rangle}) \Bigr| = 0. \tag{11.51}$$

Postponing the proof to Lecture 12, we now define a third auxiliary process $\widehat{\eta}_N^{D,K,\delta}$ by replacing h^{D_N} in the cluster variables of $\widetilde{\eta}_N^{D,K,\delta}$ by $h^{\widetilde{D}_N}$,

$$\widehat{\eta}_N^{D,K,\delta} := \sum_{x\in \widetilde{D}_N} 1_{\widetilde{\Theta}_{N,K,\delta}(x)}\, \delta_{x/N} \otimes \delta_{h_x^{D_N} - m_N} \otimes \delta_{\{h_x^{\widetilde{D}_N} - h_{x+z}^{\widetilde{D}_N} :\, z\in\mathbb{Z}^2\}}. \tag{11.52}$$

By the uniform continuity of f and the fact that $\varphi^{D_N,\widetilde{D}_N}$ converges locally-uniformly to the (smooth) continuum binding field $\Phi^{D,\widetilde{D}}$ (see Lemma 4.4) the reader will readily verify:

Exercise 11.19 *Show that for each $K \geq 1$, each $\delta > 0$ and each f as above,*

$$\lim_{N\to\infty}\left|E(\mathrm{e}^{-\langle\widehat{\eta}_N^{D,K,\delta},f\rangle}) - E(\mathrm{e}^{-\langle\widetilde{\eta}_N^{D,K,\delta},f\rangle})\right| = 0. \tag{11.53}$$

Now comes the key calculation of the proof. Let X_i, for $i = 1,\dots,m$, denote the (a.s.-unique) maximizer of $h^{\widetilde{D}_N}$ in S^i_N. For the given f as above, $x \in S^i$ for some $i = 1,\dots,m$ and any $t \in \mathbb{R}$, abbreviate $x_N := \lfloor xN\rfloor$ and let

$$f_{N,K}(x,t) := -\log E\Bigl(\mathrm{e}^{-f(x,t,h^{\widetilde{D}_N}_{x_N}-h^{\widetilde{D}_N}_{x_N+z}\colon z\in\mathbb{Z}^2)}\Bigm| X_i = x_N,\ h_{x_N} = m_N + t\Bigr). \tag{11.54}$$

Thanks to the independence of the field $h^{\widetilde{D}_N}$ over the boxes S^i_N, $i = 1,\dots,m$, by conditioning on the binding field $\varphi^{D_N,\widetilde{D}_N}$ and using its independence of $h^{\widetilde{D}_N}$ (and thus also of the X_i's), and then also conditioning on the X_i's and the values $h^{\widetilde{D}_N}_{X_i}$, we get

$$\begin{aligned}E(\mathrm{e}^{-\langle\widehat{\eta}_N^{D,K,\delta},f\rangle}) &= E\left(\prod_{i=1}^m \mathrm{e}^{-f(X_i/N,\,h^{\widetilde{D}_N}_{X_i}-m_N+\varphi^{D_N,\widetilde{D}_N}_{X_i},\,\{h^{\widetilde{D}_N}_{X_i}-h^{\widetilde{D}_N}_{X_i+z}\colon z\in\mathbb{Z}^2\})}\right)\\ &= E\left(\prod_{i=1}^m \mathrm{e}^{-f_{N,K}(X_i/N,\,h^{\widetilde{D}_N}_{X_i}-m_N+\varphi^{D_N,\widetilde{D}_N}_{X_i})}\right) = E(\mathrm{e}^{-\langle\widehat{\eta}_N^{D,K,\delta},f_{N,K}\rangle}).\end{aligned} \tag{11.55}$$

Since X_i marks the location of the absolute maximum of $h^{\widetilde{D}_N}$ in S^i_N, recalling the notation ν for the cluster law from (11.2), Theorem 11.1 yields

$$f_{N,K}(x,t) \underset{N\to\infty}{\longrightarrow} f_\nu(x,t) := -\log E_\nu(\mathrm{e}^{-f(x,t,\phi)}) \tag{11.56}$$

uniformly in t and $x \in \bigcup_{i=1}^m \widetilde{S}^i_\delta$, where $\widetilde{S}^i_\delta$ is the shift of $(0,(1-\delta)/K)^2$ centered the same point as S^i. Using this in (11.55), the series of approximations invoked earlier shows

$$E\bigl(\mathrm{e}^{-\langle\eta^D_{N,r_N},f\rangle}\bigr) = E\bigl(\mathrm{e}^{-\langle\eta^D_{N,r_N},f_\nu\rangle}\bigr) + o(1) \tag{11.57}$$

with $o(1) \to 0$ as $N \to \infty$. As $f_\nu \in C_c(D \times \mathbb{R})$, the convergence of the two-coordinate process proved in Sect. 10.2 yields

$$E\bigl(\mathrm{e}^{-\langle\eta^D_{N,r_N},f_\nu\rangle}\bigr) \underset{N\to\infty}{\longrightarrow} E\Bigl(\exp\Bigl\{-\int_{D\times\mathbb{R}} Z^D(\mathrm{d}x)\otimes\mathrm{e}^{-\alpha h}\mathrm{d}h\,\bigl(1-\mathrm{e}^{-f_\nu(x,h)}\bigr)\Bigr\}\Bigr). \tag{11.58}$$

To conclude, it remains to observe that

$$\int_{D\times\mathbb{R}} Z^D(\mathrm{d}x)\otimes \mathrm{e}^{-\alpha h}\mathrm{d}h\,\bigl(1-\mathrm{e}^{-f_\nu(x,h)}\bigr)$$
$$=\int_{D\times\mathbb{R}\times\mathbb{R}^{\mathbb{Z}^2}} Z^D(\mathrm{d}x)\otimes \mathrm{e}^{-\alpha h}\mathrm{d}h\otimes\nu(\mathrm{d}\phi)\bigl(1-\mathrm{e}^{-f(x,h,\phi)}\bigr) \tag{11.59}$$

turns (11.58) into the Laplace transform of $\mathrm{PPP}(Z^D(\mathrm{d}x)\otimes \mathrm{e}^{-\alpha h}\mathrm{d}h\otimes\nu(\mathrm{d}\phi))$. □

11.4 Some Corollaries

Having established the limit of the structured point measure, we proceed to state a number of corollaries of interest. We begin with the limit of the "ordinary" extreme value process (9.4):

Corollary 11.20 (Cluster process) *For Z^D and ν as in Theorem 9.3,*

$$\sum_{x\in D_N}\delta_{x/N}\otimes\delta_{h_x^{D_N}-m_N}\ \underset{N\to\infty}{\overset{\text{law}}{\longrightarrow}}\ \sum_{i\in\mathbb{N}}\sum_{z\in\mathbb{Z}^2}\delta_{(x_i,\,h_i-\phi_z^{(i)})}, \tag{11.60}$$

where the right-hand side is defined using the following independent objects:

(1) $\{(x_i,h_i)\colon i\in\mathbb{N}\}:=$ *sample from* $\mathrm{PPP}(Z^D(\mathrm{d}x)\otimes\mathrm{e}^{-\alpha h}\mathrm{d}h)$,
(2) $\{\phi^{(i)}\colon i\in\mathbb{N}\}:=$ *i.i.d. samples from* ν.

The measure on the right is locally finite on $D\times\mathbb{R}$ a.s.

We relegate the proof to:

Exercise 11.21 *Derive* (11.60) *from the convergence statement in Theorem 9.3 and the tightness bounds in Theorems 9.1–9.2.*

The limit object on the right of (11.60) takes the form of a *cluster process*. This term generally refers to a collection of random points obtained by taking a sample of a Poisson point process and then associating with each point thereof an independent cluster of (possibly heavily correlated) points. See again Fig. 17. We note that a cluster process naturally appears in the limit description of the extreme-order statistics of the Branching Brownian Motion [11–13].

Another observation that is derived from the above limit law concerns the Gibbs measure on D_N associated with the DGFF on D_N as follows:

$$\mu^D_{\beta,N}\bigl(\{x\}\bigr):=\frac{1}{\mathfrak{Z}_N(\beta)}\,\mathrm{e}^{\beta h_x^{D_N}}\quad\text{where}\quad \mathfrak{Z}_N(\beta):=\sum_{x\in D_N}\mathrm{e}^{\beta h_x^{D_N}}. \tag{11.61}$$

In order to study the scaling limit of this object, we associate the value $\mu^D_{\beta,N}(\{x\})$ with a point mass at x/N. From the convergence of the suitably-normalized measure

$\sum_{x\in D_N} e^{\beta h_x^{D_N}} \delta_{x/N}$ to the Liouville Quantum Gravity for $\beta < \beta_c := \alpha$ it is known (see, e.g., Rhodes and Vargas [105, Theorem 5.12]) that

$$\sum_{z\in D_N} \mu_{\beta,N}^D(\{z\})\delta_{z/N}(dx) \underset{N\to\infty}{\overset{\text{law}}{\longrightarrow}} \frac{Z_\lambda^D(dx)}{Z_\lambda^D(D)}, \tag{11.62}$$

where $\lambda := \beta/\beta_c$ and where Z_λ^D is the measure we saw in the discussion of the intermediate level sets (for $\lambda < 1$). The result extends (see [105, Theorem 5.13], although the proof details seem scarce) to the case $\beta = \beta_c$, where thanks to Theorem 10.30 we get $\widehat{Z}^D(dx)$ from (9.9) instead.

The existence of the limit in the supercritical cases $\beta > \beta_c$ has been open for quite a while (and was subject to specific conjectures; e.g., [66, Conjecture 11]). Madaule, Rhodes and Vargas [92] first proved it for star-scale invariant fields as well as certain specific cutoffs of the CGFF. For the DGFF considered here, Arguin and Zindy [10] proved convergence of the *overlaps* of $\mu_{\beta,N}^D$ to those of Poisson–Dirichlet distribution PD(s) with parameter $s := \beta_c/\beta$. This only identified the law of the *sizes* of the atoms in the limit measure; the full convergence including the spatial distribution was settled in [27]:

Corollary 11.22 (Poisson–Dirichlet limit for the Gibbs measure) *For all $\beta > \beta_c := \alpha$ we then have*

$$\sum_{z\in D_N} \mu_{\beta,N}^D(\{z\})\delta_{z/N}(dx) \underset{N\to\infty}{\overset{\text{law}}{\longrightarrow}} \sum_{i\in\mathbb{N}} p_i\, \delta_{X_i}, \tag{11.63}$$

where

(1) $\{X_i\}$ are (conditionally on Z^D) i.i.d. with law $\widehat{Z}^D$, while
(2) $\{p_i\}_{i\in\mathbb{N}}$ is independent of Z^D and $\{X_i\}$ with $\{p_i\}_{i\in\mathbb{N}} \overset{\text{law}}{=} PD(\beta_c/\beta)$.

We remark that PD(s) is a law on non-increasing sequences of non-negative numbers with unit total sum obtained as follows: Take a sample $\{q_i\}_{i\in\mathbb{N}}$ from the Poisson process on $(0,\infty)$ with intensity $t^{-1-s}dt$, order the points decreasingly and normalize them by their total sum (which is finite a.s. when $s < 1$).

Corollary 11.22 follows from the description of the Liouville Quantum Gravity measure for $\beta > \beta_c$ that we will state next. For $s > 0$ and Q a Borel probability measure on $\mathbb{C}$, let $\{q_i\}_{i\in\mathbb{N}}$ be a sample from the Poisson process on $(0,\infty)$ with intensity $t^{-1-s}dt$ and let $\{X_i\}_{i\in\mathbb{N}}$ be independent i.i.d. samples from Q. Use these to define the random measure

$$\Sigma_{s,Q}(dx) := \sum_{i\in\mathbb{N}} q_i\, \delta_{X_i}. \tag{11.64}$$

We then have:

Theorem 11.23 (Liouville measure in the glassy phase) *Let Z^D and ν be as in Theorem 9.3. For each $\beta > \beta_c := \alpha$ there is $c(\beta) \in (0,\infty)$ such that*

$$\sum_{z\in D_N} e^{\beta(h_z-m_N)}\delta_{z/N}(dx) \xrightarrow[N\to\infty]{\text{law}} c(\beta)\, Z^D(D)^{\beta/\beta_c}\, \Sigma_{\beta_c/\beta,\,\widehat{Z}^D}(dx), \tag{11.65}$$

where $\Sigma_{\beta_c/\beta,\,\widehat{Z}^D}$ *is defined conditionally on* Z^D. *Moreover,*

$$c(\beta) = \beta^{-\beta/\beta_c}\big[E_\nu(Y^\beta(\phi)^{\beta_c/\beta})\big]^{\beta/\beta_c} \quad \text{with} \quad Y^\beta(\phi) := \sum_{x\in\mathbb{Z}^2} e^{-\beta\phi_x}. \tag{11.66}$$

In particular, $E_\nu(Y^\beta(\phi)^{\beta_c/\beta}) < \infty$ *for each* $\beta > \beta_c$.

Note that the limit laws in (11.63) and (11.65) are *purely atomic*, in contrast to the limits of the subcritical measures (11.62) which, albeit a.s. singular with respect to the Lebesgue measure, are non-atomic a.s.

Proof of Theorem 11.23 (*main computation*). We start by noting that the Laplace transform of the above measure $\Sigma_{s,Q}$ is explicitly computable:

Exercise 11.24 *Show that for any measurable* $f\colon \mathbb{C}\to[0,\infty)$,

$$E\big(e^{-\langle\Sigma_{s,Q},f\rangle}\big) = \exp\Big\{-\int_{\mathbb{C}\times(0,\infty)} Q(dx)\otimes t^{-1-s}dt\,\,(1-e^{-tf(x)})\Big\}. \tag{11.67}$$

Pick a continuous $f\colon\mathbb{C}\to[0,\infty)$ with support in D and (abusing our earlier notations) write M_N to denote the measure on the left of (11.65). Through suitable truncations, the limit proved in Theorem 9.3 shows that $E(e^{-\langle M_N,f\rangle})$ tends to

$$E\Big(\exp\Big\{-\int Z^D(dx)\otimes e^{-\alpha h}dh\otimes\nu(d\phi)(1-e^{-g(x,h,\phi)})\Big\}\Big), \tag{11.68}$$

where $g(x,h,\phi) := f(x)\,e^{\beta h}\,Y^\beta(\phi)$. The change of variables $t := e^{\beta h}$ then gives

$$\begin{aligned}\int dh\, e^{-\alpha h}(1-e^{-g(x,h,\phi)}) &= \int_0^\infty dt\frac{1}{\beta t}\,t^{-\alpha/\beta}\big(1-e^{-f(x)Y^\beta(\phi)t}\big)\\ &= \frac{1}{\beta}\big(Y^\beta(\phi)\big)^{\alpha/\beta}\int_0^\infty dt\,t^{-1-\alpha/\beta}(1-e^{-tf(x)}).\end{aligned} \tag{11.69}$$

The integral with respect to ν affects only the term $(Y^\beta(\phi))^{\alpha/\beta}$; scaling t by $c(\beta)Z^D(D)^{\beta/\alpha}$ then absorbs all prefactors and identifies the integral on the right of (11.68) with that in (11.67) for $Q := \widehat{Z}^D$ and $f(x)$ replaced by $c(\beta)Z^D(D)^{\beta/\alpha}f(x)$. This now readily gives the claim (11.65). (The finiteness of the expectation of $Y^\beta(\phi)^{\beta_c/\beta}$ requires a separate tightness argument.) □

The reader might wonder at this point how it is possible that the rather complicated (and correlated) structure of the cluster law ν does not at all appear in the limit measure on the right of (11.65)—that is, not beyond the expectation of $Y^\beta(\phi)^{\beta_c/\beta}$ in $c(\beta)$. This may, more or less, be traced to the following property of the Gumbel law:

Exercise 11.25 *(Derandomizing i.i.d. shifts of Gumbel PPP) Consider a sample* $\{h_i : i \in \mathbb{N}\}$ *from* $\text{PPP}(e^{-\alpha h}dh)$ *and let* $\{T_i : i \in \mathbb{N}\}$ *be independent, i.i.d. random variables. Prove that*

$$\sum_{i\in\mathbb{N}} \delta_{h_i+T_i} \overset{\text{law}}{=} \sum_{i\in\mathbb{N}} \delta_{h_i+\alpha^{-1}\log c} \tag{11.70}$$

whenever $c := Ee^{\alpha T_1} < \infty$.

Our final corollary concerns the behavior of the function

$$G_{N,\beta}(t) := E\left(\exp\left\{-e^{-\beta t}\sum_{x\in D_N} e^{\beta h_x^{D_N}}\right\}\right), \tag{11.71}$$

which, we observe, is a re-parametrization of the Laplace transform of the normalizing constant $\mathfrak{Z}_N(\beta)$ from (11.61). In their work on the Branching Brownian Motion, Derrida and Spohn [55] and later Fyodorov and Bouchaud [74] observed that, in a suitable limit, an analogous quantity ceases to depend on β once β crosses a critical threshold. They referred to this as *freezing*. Our control above is sufficient to yield the same phenomenon for the quantity arising from DGFF:

Corollary 11.26 (Freezing) *For all* $\beta > \beta_c := \alpha$ *there is* $\tilde{c}(\beta) \in \mathbb{R}$ *such that*

$$G_{N,\beta}\big(t + m_N + \tilde{c}(\beta)\big) \underset{N\to\infty}{\longrightarrow} E\big(e^{-Z^D(D)e^{-\alpha t}}\big). \tag{11.72}$$

Proof. Noting that $e^{-\beta m_N}\mathfrak{Z}_N(\beta)$ is the total mass of the measure on the left of (11.65), from Theorem 11.23 we get (using the notation of (11.64) and (11.66))

$$e^{-\beta m_N}\mathfrak{Z}_N(\beta) \underset{N\to\infty}{\overset{\text{law}}{\longrightarrow}} c(\beta)Z^D(D)^{\beta/\alpha}\sum_{i\in\mathbb{N}} q_i\,. \tag{11.73}$$

The Poisson nature of the $\{q_i\}_{i\in\mathbb{N}}$ shows, for any $\lambda > 0$,

$$\begin{aligned} E\big(e^{-\lambda\sum_{i\in\mathbb{N}} q_i}\big) &= \exp\left\{-\int_0^\infty dt\, t^{-1-\beta/\alpha}(1-e^{-\lambda t})\right\} \\ &= \exp\left\{-\lambda^{\alpha/\beta}\int_0^\infty dt\, t^{-1-\beta/\alpha}(1-e^{-t})\right\}. \end{aligned} \tag{11.74}$$

Thanks to the independence of $\{q_i\}_{i\in\mathbb{N}}$ and Z^D, the Laplace transform of the random variable on the right of (11.73) at parameter $e^{-\beta t}$ is equal to the right-hand side of (11.72), modulo a shift in t by a constant that depends on $c(\beta)$ and the integral on the second line in (11.74). □

We refer to [27] for further consequences of the above limit theorem and additional details. The (apparently quite deep) connection between freezing and Gumbel laws has recently been further explored by Subag and Zeitouni [120].

Lecture 12: Limit Theory for DGFF Maximum

In this final lecture on the extremal process associated with the DGFF we apply the concentric decomposition developed in Lectures 8 and 11 to give the proofs of various technical results scattered throughout Lectures 9–11. We start by Theorems 9.1–9.2 and Lemma 10.4 that we used to control the spatial tightness of the extremal level sets. Then we establish the convergence of the centered maximum from Theorem 10.3 by way of proving Theorem 10.7 and extracting from this the uniqueness (in law) of the limiting Z^D-measures from Theorem 9.6. At the very end we state (without proof) a local limit result for both the position and the value of the absolute maximum.

12.1 Spatial Tightness of Extremal Level Sets

Here we give the proofs of Theorems 9.1–9.2 and Lemma 10.4 that we relied heavily on in the previous lectures. Recall the notation $\Gamma_N^D(t)$ from (9.1) for the extremal level set "t units below m_N" and the definition $D_N^\delta := \{x \in D_N\colon \operatorname{dist}_\infty(x, D_N^{\mathrm{c}}) > \delta N\}$. We start by:

Lemma 12.1 *For each $D \in \mathfrak{D}$, $\delta > 0$ and $c' > 0$ there is $c > 0$ such that for all $N \geq 1$, all $x \in D_N^\delta$ and all t and r with $|t|, r \in [0, c'(\log N)^{1/5}]$,*

$$P\Big(h^{D_N} \leq m_N + r \,\Big|\, h_x^{D_N} = m_N + t\Big) \leq \frac{c}{\log N}\big(1 + r + |t|\big)^2. \tag{12.1}$$

Proof. Assume without loss of generality that $0 \in D$ and let us first address the case $x = 0$. Lemma 8.5 shifts the conditioning to $h_0^{D_N} = 0$ at the cost of reducing the field by $(m_N + t)\mathfrak{g}^{D_N}$. Invoking the concentric decomposition and Lemma 11.5, the resulting event can be covered as

$$\begin{aligned}\big\{h^{D_N} \leq m_N(1 - \mathfrak{g}^{D_N}) + r - t\mathfrak{g}^{D_N}\big\} &\subseteq \{K = \lfloor n/2 \rfloor + 1\} \\ &\cup \bigcap_{k=1}^{n} \Big(\big\{S_k - S_{n+1} \geq -R_K(k) - r - |t|\big\} \cap \big\{K \leq \lfloor n/2 \rfloor\big\}\Big),\end{aligned} \tag{12.2}$$

where we used $0 \leq \mathfrak{g}^{\Delta^n} \leq 1$. Summing the bound in Lemma 11.7 we then get (12.1). (The restriction to $t \leq n^{1/5}$ can be replaced by $t \leq c'n^{1/5}$ at the cost of changing the constants.) Noting that constant c depends only on k_1 from (8.46), shifting D_N around extends the bound to all $x \in D_N^\delta$. □

With Lemma 12.1 at our disposal, we can settle the tightness of the cardinality of the extremal level set $\Gamma_N^D(t)$:

Proof of Theorem 9.1 (*upper bound*). Building on the idea underlying Exercises 3.3–3.4, we first ask the reader to solve:

Exercise 12.2 *Let $\widetilde{D} \subseteq D$. Then, assuming $\widetilde{D}_N \subseteq D_N$, for any $n \in \mathbb{N}$,*

$$P\Big(\big|\Gamma_N^{\widetilde{D}}(t)\big| \geq 2n\Big) \leq 2P\Big(\big|\Gamma_N^D(t) \cap \widetilde{D}_N\big| \geq n\Big). \tag{12.3}$$

In light of this, it suffices to bound the size of $\Gamma_N^D(t) \cap D_N^\delta$. Here we note that, by the Markov inequality

$$P\Big(|\Gamma_N^D(t) \cap D_N^\delta| \geq e^{Ct}\Big) \leq P\Big(\max_{x \in D_N} h_x^{D_N} > m_N + t\Big) \\ + e^{-Ct} \sum_{x \in D_N^\delta} P\Big(h^{D_N} \leq m_N + t,\, h_x^{D_N} \geq m_N - t\Big). \tag{12.4}$$

The first probability is bounded by $e^{-\tilde{a}t}$ using the exponential tightness of absolute maximum proved in see Lemma 8.3. The probability under the sum is in turn bounded by conditioning on $h_x^{D_N}$ and applying Lemma 12.1 along with the standard Gaussian estimate (for the probability of $h_x^{D_N} \geq m_N - t$). This bounds the probability by a quantity of order $(1+t)^2 t e^{\alpha t}/N^2$. Since $|D_N^\delta| = O(N^2)$, for $C > \alpha$, we get the upper bound in (9.2). □

We leave the lower bound to:

Exercise 12.3 *Use the geometric setting, and the ideas underlying the proof of Lemma 8.3 to prove the lower bound on $|\Gamma_N^D(t)|$ in (9.2) for any $c < \alpha/2$.*

We remark that the bound proposed in this exercise is far from optimal. Indeed, once Theorem 9.3 has been established in full (for which we need only the upper bound in Theorem 9.1), we get $|\Gamma_N(t)| = e^{(\alpha+o(1))t}$. See Conjecture 16.5 for even a more precise claim.

Next we will give:

Proof of Lemma 10.4. First we note that, thanks to Exercise 3.4 we may assume that $A \subseteq D_N^\delta$ for some $\delta > 0$ independent of N. Lemma 12.1, the union bound, the definition of m_N and the exponential tightness of the absolute maximum from Lemma 8.3 then shows, for all $r, |t| \in [0, (\log N)^{1/5}]$,

$$\begin{aligned} P\Big(\max_{x \in A} h_x^{D_N} \geq m_N + t\Big) &\leq P\Big(\max_{x \in D_N} h_x^{D_N} \geq m_N + r\Big) \\ &\quad + P\Big(\max_{x \in A} h_x^{D_N} \geq m_N + t,\, \max_{x \in D_N} h_x^{D_N} \leq m_N + r\Big) \\ &\leq e^{-\tilde{a}r} + c\frac{|A|}{N^2} e^{-\alpha t}\big(1 + r + |t|\big)^2. \end{aligned} \tag{12.5}$$

Setting $r := 5\tilde{a}^{-1}\log(1+N^2/|A|) + 2\alpha\tilde{a}^{-1}|t|$ then proves the claim for all A such that $|A| \ge N^2 e^{-\tilde{a}(\log N)^{1/5}}$ (as that ensures $r \le c'(\log N)^{1/5}$). For the complementary set of A's we apply the standard Gaussian tail estimate along with the uniform bound on the Green function and a union bound to get

$$P\bigl(\max_{x\in A} h_x^{D_N} \ge m_N + t\bigr) \le c e^{-\alpha t}\frac{|A|}{N^2}\log N. \tag{12.6}$$

Then we note that $\log N \le c''[\log(1+N^2/|A|)]^5$ for $c'' > 0$ sufficiently large. □

Our next item of business is the proof of Theorem 9.2 dealing with spatial tightness of the set $\Gamma_N^D(t)$. Here we will also need:

Lemma 12.4 *For each $D \in \mathfrak{D}$, each $\delta > 0$ and each $t \ge 0$ there is $c > 0$ such that for all $t' \in [-t, t]$, all $N \ge 1$, all $x \in D_N^\delta$ and all r with $1/\delta < r < \delta N$,*

$$P\Bigl(\max_{y\in A_{r,N/r}(x)} h_y^{D_N} \ge h_x^{D_N},\, h^{D_N} \le m_N + t \,\Big|\, h_x^{D_N} = m_N + t'\Bigr) \le c\frac{(\log r)^{-\frac{1}{16}}}{\log N}, \tag{12.7}$$

where we abbreviated $A_{a,b}(x) := \{y \in \mathbb{Z}^2 : a \le \operatorname{dist}_\infty(x, y) \le b\}$.

Proof. Let us again assume $0 \in D$ and start with $x = 0$. The set under the maximum is contained in $\Delta^{n-k} \smallsetminus \Delta^k$ for a natural k proportional to $\log r$. Lemma 8.5 shifts the conditioning to $h_0^{D_N} = 0$ at the cost of adding $(m_N + t')\mathfrak{g}^{D_N}$ to all occurrences of h^{D_N}. This embeds the event under consideration into

$$\bigcup_{y\in\Delta^{n-k}\smallsetminus\Delta^k} \bigl\{h_y^{D_N} \ge (m_N + t')(1 - \mathfrak{g}^{D_N}(y))\bigr\} \cap \bigl\{h^{D_N} \le (m_N + t)(1 - \mathfrak{g}^{D_N}) + (t - t')\mathfrak{g}^{D_N}\bigr\}. \tag{12.8}$$

For $|t'| \le t$, the event is the largest when $t' = -t$ so let us assume that from now on. Using that $\mathfrak{g}^{D_N} \le 1$ and invoking the estimates underlying Lemma 11.5, the resulting event is then covered by the union of

$$\{K > k\} \cap \bigcap_{\ell=1}^{n} \{S_\ell \ge -R_K(\ell) - 2t\} \tag{12.9}$$

and

$$\Biggl(\bigcup_{\ell=k+1}^{n-k} \{S_\ell \le R_k(\ell) + t\}\Biggr) \cap \bigcap_{\ell=1}^{n} \{S_\ell \ge -R_k(\ell) - 2t\}. \tag{12.10}$$

By Lemma 11.7, the conditional probability (given $S_{n+1} = 0$) of the event in (12.9) is at most order $e^{-c_2(\log k)^2}/\log N$. Lemma 11.8 in turn bounds the corresponding conditional probability of the event in (12.10) by a constant times $k^{-\frac{1}{16}}/\log N$. The constants in these bounds are uniform on compact sets of t and over the shifts of D_N such that $0 \in D_N^\delta$. Using the translation invariance of the DGFF, the claim thus extends to all $x \in D_N^\delta$. □

With Lemma 12.4 in hand, we are ready to give:

Proof of Theorem 9.2. Given $\delta > 0$ and $s \geq t \geq 0$, the event in the statement is the subset of the union of

$$\bigl\{\Gamma_N^D(t) \smallsetminus D_N^\delta \neq \emptyset\bigr\} \cup \bigl\{\max_{x \in D_N} h_x^{D_N} > m_N + s\bigr\} \tag{12.11}$$

and the event

$$\bigcup_{x \in D_N^\delta} \Bigl\{\max_{y \in A_{r,N/r}(x)} h_y^{D_N} \geq h_x^{D_N} \geq m_N - t,\, h^{D_N} \leq m_N + s\Bigr\} \tag{12.12}$$

provided $1/\delta < r < \delta N$. Lemmas 10.4 and 8.3 show that the probability of the event in (12.11) tends to zero in the limits $N \to \infty$, $\delta \downarrow 0$ and $s \to \infty$. The probability of the event in (12.12) is in turn estimated by the union bound and conditioning on $h_x^{D_N} = m_N + t'$ for $t' \in [-t, s]$. Lemma 12.4 then dominates the probability by an s-dependent constant times $(\log r)^{-\frac{1}{16}}$. The claim follows by taking $N \to \infty$, $r \to \infty$ followed by $\delta \downarrow 0$ and $s \to \infty$. □

The arguments underlying Lemma 12.4 also allow us to prove that large local extrema are preserved (with high probability) upon application of the Gibbs-Markov property:

Proof of Lemma 11.18. Let D_N be a lattice approximation of a continuum domain $D \in \mathfrak{D}$ and let $\widetilde{D}_N$ be the union of squares of side length (roughly) N/K that fit into D_N^δ with distinct squares at least two lattice steps apart. For $x \in \widetilde{D}_N$, let $S_{N,K}(x)$ denote the square (of side-length roughly N/K) in $\widetilde{D}_N$ containing x and let $S_{N,K}^\delta(x)$ be the square of side-length $(1-\delta)(N/K)$ centered at the same point as $S_{N,K}(x)$. Abusing our earlier notation, let $\widetilde{D}_N^\delta := \bigcup_{x \in \widetilde{D}_N} S_{N,K}^\delta(x)$.

Consider the coupling of h^{D_N} and $h^{\widetilde{D}_N}$ in the Gibbs-Markov property. The claim in Lemma 11.18 will follow by standard approximation arguments once we prove that, for every $t \geq 0$,

$$P\Bigl(\exists x \in \Gamma_N^D(t) \cap \widetilde{D}_N^\delta\colon \max_{y \in S_{N,K}(x)} h_y^{D_N} \leq h_x^{D_N},\ \max_{y \in S_{N,K}(x)} h_y^{\widetilde{D}_N} > h_x^{\widetilde{D}_N}\Bigr) \tag{12.13}$$

tends to zero as $N \to \infty$ and $K \to \infty$. For this it suffices to show that for all $t > 0$ there is $c > 0$ such that for all $N \gg K \geq 1$, all $s \in [-t, t]$, all $x \in \widetilde{D}_N^\delta$ and all $y \in S_{N,K}(x)$,

$$P\Bigl(\max_{y\in S_{N,K}(x)} h_y^{D_N}\le h_x^{D_N},\ \max_{y\in S_{N,K}(x)} h_y^{\widetilde{D}_N}>h_x^{\widetilde{D}_N},$$
$$h^{D_N}\le m_N+t\,\Big|\,h_x^{D_N}=m_N+s\Bigr)\le c\frac{(\log K)^{-\frac{1}{16}}}{\log N}. \tag{12.14}$$

Indeed, multiplying (12.14) by the probability density of $h_x^{D_N}-m_N$, which is of order $N^{-2}\log N$, integrating over $s\in[-t,t]$ and summing over $x\in\widetilde{D}_N^\delta$, we get a bound on (12.13) with $\Gamma_N^D(t)\cap\widetilde{D}_N$ replaced by $\Gamma_N^D(t)\cap\widetilde{D}_N^\delta$ and the event restricted to $h^{D_N}\le m_N+t$. These defects are handled via Lemma 10.4, which shows that $P(\Gamma_N^D(t)\smallsetminus\widetilde{D}_N^\delta=\emptyset)\to1$ as $N\to\infty$, $K\to\infty$ and $\delta\downarrow0$, and Lemma 8.3, which gives $P(h^{D_N}\not\le m_N+t)\le\mathrm{e}^{-\tilde{a}t}$ uniformly in $N\ge1$.

In order to prove (12.14), we will shift the domains so that x is at the origin. We will again rely on the concentric decomposition with the following caveat: $S_{N,K}(0)$ is among the Δ^i's. More precisely, we take

$$\Delta^j:=\begin{cases}\{x\in\mathbb{Z}^2\colon|x|_\infty\le2^j\}, & \text{if } j=0,\dots,r-1,\\ S_{N,K}^\delta(0), & \text{if } j=r,\\ \{x\in\mathbb{Z}^2\colon|x|_\infty\le2^{j+\widetilde{r}-r}\}, & \text{if } j=r+1,\dots,n-1,\\ D_N, & \text{if } j=n,\end{cases} \tag{12.15}$$

where

$$\begin{aligned}r&:=\max\bigl\{j\ge0\colon\{x\in\mathbb{Z}^2\colon|x|_\infty\le2^j\}\subseteq S_{N,K}^\delta(0)\bigr\},\\ \widetilde{r}&:=\min\bigl\{j\ge0\colon S_{N,K}(0)\subseteq\{x\in\mathbb{Z}^2\colon|x|_\infty\le2^j\}\bigr\},\\ n&:=\max\bigl\{j\ge0\colon\{x\in\mathbb{Z}^2\colon|x|_\infty\le2^{j+\widetilde{m}-m+1}\}\subseteq D_N\bigr\}.\end{aligned} \tag{12.16}$$

We leave it to the reader to check that all the estimates pertaining to the concentric decomposition remain valid, albeit with constants that depend on δ.

Denoting $k:=n-r$, we have $k=\log_2(K)+O(1)$ and $n=\log_2(N)+O(1)$. Assuming that the fields $\{h^{\Delta^j}\colon j=0,\dots,n\}$ are coupled via the concentric decomposition, the probability in (12.14) becomes

$$P\Bigl(h^{\Delta^{n-k}}\not\le h_0^{\Delta^{n-k}},\ h^{\Delta^n}\le h_0^{\Delta^n}\text{ in }\Delta^{n-k},\ h^{\Delta^n}\le m_N+t\,\Big|\,h_0^{\Delta^n}=m_N+s\Bigr). \tag{12.17}$$

Observe that, for all $x\in\Delta^{n-k}$,

$$h_x^{\Delta^n}-h_0^{\Delta^n}=h_x^{\Delta^{n-k}}-h_0^{\Delta^{n-k}}+\sum_{j=n-k+1}^{n}\bigl[\mathfrak{b}_j(x)\varphi_j(0)+\chi_j(x)\bigr]. \tag{12.18}$$

Assuming $K\le k$, the estimates in the concentric decomposition bound the jth term in the sum by $(\log(k\vee j))^2\mathrm{e}^{-c(n-k-j)}$ on Δ^j for all $j=0,\dots,n-k-1$. This will suffice for $j \leq k$; for the complementary j we instead get

$$\sum_{j=n-k+1}^{n} \bigl[\mathfrak{b}_j(x)\varphi_j(0) + \chi_j(x)\bigr] + m_N(s,x) \geq -R_k(j) - t \tag{12.19}$$

for $x \in \Delta^j \smallsetminus \Delta^{j-1}$, where we used that $|s| \leq t$ and invoked the shorthand (11.12). The event (12.17) is thus contained in the intersection of $\{h^{\Delta^n} \leq m_N + t\}$ with the union of three events: $\{K > k\}$,

$$\bigl\{h_0^{\Delta^n} \geq h^{\Delta^n} \text{ on } \Delta^k\bigr\} \cap \bigcup_{x\in\Delta^k} \bigl\{h_x^{\Delta^n} > h_0^{\Delta^n} - (\log k)^2 \mathrm{e}^{-c(2n-k)}\bigr\} \tag{12.20}$$

and

$$\{K \leq k\} \cap \bigcup_{x\in\Delta^{n-k}\smallsetminus\Delta^k} \Bigl\{h_x^{\Delta^n} - h_0^{\Delta^n} + m_N(s,x) \geq -R_k(j) - t\Bigr\}. \tag{12.21}$$

Lemma 11.7 bounds the contribution of the first event by $c_1 \mathrm{e}^{-c_2(\log k)^2}/\log N$. For (12.20) we drop the event $h^{D^n} \leq m_N + t$ and then invoke the "gap estimate" in Exercise 11.10 to bound the requisite probability by a quantity of order $|\Delta^k|(\log k)^2 \mathrm{e}^{-c(2n-k)}$. For (12.21) we shift the conditioning to $h_0^{\Delta^n} = 0$, which turns the event under the union to $\{h_x^{\Delta^n} - h_0^{\Delta^n} \geq -R_k(j) - t\}$. The resulting event is then contained in (12.10) which has probability less than order $k^{-\frac{1}{16}}/\log N$. As $k \approx \log_2(K)$ and $n \gg k$ we get (12.14). $\square$

12.2 Limit of Atypically Large Maximum

Our next item of business here is the proof of Theorem 10.7 dealing with the limit law of the maximum (and the corresponding maximizer) conditioned on atypically large values. This will again rely on the concentric decomposition of the DGFF; specifically, the formula

$$P\Bigl(h^{D_N} \leq h_0^{D_N} \,\Big|\, h_0^{D_N} = m_N + t\Bigr) = \frac{\Xi_k^{\mathrm{in}}(1)\,\Xi_{N,k}^{\mathrm{out}}(t)}{g \log N}\bigl(2 + o(1)\bigr) \tag{12.22}$$

from Lemma 11.13, where the quantities $\Xi_k^{\mathrm{in}}(1)$ and $\Xi_{N,k}^{\mathrm{out}}(t)$ are as in (11.32)–(11.33) and where we incorporated the $o(1)$-term to the expression thanks to the bounds in Lemma 11.14. The novel additional input is:

Proposition 12.5 (Asymptotic contribution of outer layers) *$\Xi_{N,k}^{\mathrm{out}}(t) \sim t$ in the limit as $N \to \infty$, $k \to \infty$ and $t \to \infty$. More precisely,*

$$\lim_{t\to\infty}\,\limsup_{k\to\infty}\,\limsup_{N\to\infty}\left|\frac{\Xi^{\rm out}_{N,k}(t)}{t}-1\right|=0. \tag{12.23}$$

For each $\delta>0$, the limit is uniform in the shifts of D_N such that $0\in D_N^\delta$.

Deferring the proof to later in this lecture, we will now show how this implies Theorem 10.7. Recall that $r_D(x)$ denotes the conformal radius of D from x and that $\Xi^{\rm in}_\infty(1)$ is the limit of $\Xi^{\rm in}_k(1)$ whose existence was established in the proof of Theorem 11.3; see (11.46). We start by:

Lemma 12.6 *For any $\delta>0$ and any $q>1$,*

$$\lim_{t\to\infty}\,\limsup_{N\to\infty}\,\max_{x\in D_N^\delta}\left|N^2\,\frac{{\rm e}^{\alpha t}}{t}\,P\Bigl(h^{D_N}\le h^{D_N}_x,\,h^{D_N}_x-m_N\in[t,qt]\Bigr)\right.$$
$$\left.-{\rm e}^{2c_0/g}\sqrt{\frac{\text{ß}}{8}}\,\Xi^{\rm in}_\infty(1)\,{\rm r}_{\rm D}({\rm x/N})^2\right|=0\,, \tag{12.24}$$

where c_0 as in (1.36).

Proof. Denote by $f_{N,x}$ the density with respect to the Lebesgue measure of the distribution of $h^{D_N}_x-m_N$ and write

$$P\Bigl(h^{D_N}\le h^{D_N}_x,\,h^{D_N}_x-m_N\in[t,qt]\Bigr)$$
$$=\int_t^{qt}{\rm d}s\,f_{N,x}(s)\,P\Bigl(h^{D_N}\le h^{D_N}_x\,\Big|\,h^{D_N}_x=m_N+s\Bigr). \tag{12.25}$$

A calculation based on the Green function asymptotic in Theorem 1.17 now shows that, with $o(1)\to0$ as $N\to\infty$ uniformly in $x\in D^\delta$ and $s\in[t,qt]$,

$$\frac{N^2}{\log N}\,f_{N,\lfloor Nx\rfloor}(s)=\bigl(1+o(1)\bigr)\frac{1}{\sqrt{2\pi g}}\,{\rm e}^{-{\rm s}}\,{\rm e}^{2c_0/g}\,{\rm r}_{\rm D}({\rm x})^2. \tag{12.26}$$

Combining this with (12.22), (12.23) and (11.46) gives

$$N^2\,\frac{{\rm e}^{\alpha t}}{t}\,P\Bigl(h^{D_N}\le h^{D_N}_x,\,h^{D_N}_x-m_N\in[t,qt]\Bigr)$$
$$=\bigl(1+o(1)\bigr)\,\frac{2}{g}\,\Xi^{\rm in}_\infty(1)\,\frac{1}{\sqrt{2\pi g}}\,{\rm e}^{2c_0/g}\,{\rm r}_{\rm D}({\rm x/N})^2\left[\frac{{\rm e}^{\rm t}}{\rm t}\int_{\rm t}^{\rm qt}{\rm d}s\,{\rm e}^{-{\rm s}}{\rm s}\right], \tag{12.27}$$

where $o(1)\to0$ as $N\to\infty$ followed by $t\to\infty$ uniformly in $x\in\mathbb{Z}^2$ such that $x/N\in D^\delta$. The expression inside the square brackets tends to α^{-1} as $t\to\infty$. Since $(2/g)(2\pi g)^{-1/2}\alpha^{-1}=(\sqrt{2\pi}\,g)^{-1}=\sqrt{\pi/8}$, we get (12.24). □

In order to rule out the occurrence of multiple close points with excessive values of the field, and thus control certain boundary issues in the proof of Theorem 10.7, we will also need:

Lemma 12.7 *For each $\delta > 0$ small enough there is $c > 0$ such that for all $N \ge 1$, all $x \in D_N$ and all $0 \le s, t \le \log N$,*

$$P\Bigl(\max_{\substack{y\in D_N\\ \text{dist}_\infty(x,y)>\delta N}} h_y^{D_N} \ge m_N + s \,\Big|\, h^{D_N} \le h_x^{D_N} = m_N + t\Bigr) \le c\,\mathrm{e}^{-\tilde a s}, \tag{12.28}$$

where $\tilde a > 0$ is as in Lemma 8.3.

Proof. The FKG inequality for $h_x^{D_N}$ conditioned on $h_x^{D_N} = m_N + t$ bounds the desired probability by

$$P\Bigl(\max_{\substack{y\in D_N\\ \text{dist}_\infty(x,y)>\delta N}} h_y^{D_N} \ge m_N + s \,\Big|\, h_x^{D_N} = m_N + t\Bigr). \tag{12.29}$$

Lemma 8.5 along with Exercise 8.6 then permit us to rewrite this as the probability that there exists $y \in D_N$ with $\text{dist}_\infty(x, y) > \delta N$ such that

$$h_y^{D_N\smallsetminus\{0\}} \ge m_N + s - (m_N + t)\mathfrak{g}^{-x+D_N}(y - x). \tag{12.30}$$

The Maximum Principle shows that $A \mapsto \mathfrak{g}^A$ is non-decreasing with respect to the set inclusion and so $\mathfrak{g}^{-x+D_N}(\cdot) \le \mathfrak{g}^{\widetilde D_N}(\cdot)$, where $\widetilde D_N$ is a box of side $4\text{diam}_\infty(D_N)$ centered at the origin. Under our assumptions on t, the term $(m_N + t)\mathfrak{g}^{\widetilde D_N}(y - x)$ is at most a constant times $\log(1/\delta)$. The claim then follows from the upper-tail tightness of the maximum in Lemma 8.3. □

We are ready to give:

Proof of Theorem 10.7. We may as well give the proof for a general sequence of approximating domains D_N. Set

$$\bar c := \sqrt{\frac{\pi}{8}}\,\mathrm{e}^{2c_0/g}\, r^{\text{in}}_{\infty}(1). \tag{12.31}$$

Let $A \subseteq D$ be open and denote by $X_N^\star$ the (a.s.-unique) maximizer of h^{D_N} over D_N. Abbreviate $M_N := \max_{x\in D_N} h_x^{D_N}$. Summing (12.24) then yields

$$\lim_{\delta\downarrow 0}\limsup_{t\to\infty}\limsup_{N\to\infty}\left|\frac{\mathrm{e}^{\alpha t}}{t}P\Bigl(\frac{1}{N}X_N^\star \in D^\delta\cap A,\ M_N - m_N \in [t, qt]\Bigr) - \bar c\int_A \mathrm{d}x\, r_D(x)^2\right| = 0. \tag{12.32}$$

Assuming $q > 1$ obeys $q\tilde a > 2\alpha$, for $\tilde a$ as in Lemma 8.3, the probability of $M_N \ge m_N + qt$ decays as $\mathrm{e}^{-2\alpha t}$ and so (12.32) holds with $M_N - m_N \in [t, qt]$ replaced by $M_N \ge m_N + t$. This would already give us a lower bound in (10.11) but, in order to prove the complementary upper bound as well as (10.12), we need to address the possibility of the maximizer falling outside D_N^δ.

Let $\widetilde{D}_N$ be the square of side $4\mathrm{diam}_\infty(D_N)$ centered at an arbitrary point of D_N. Exercise 3.4 then gives

$$P\Big(\max_{x\in D_N\smallsetminus D_N^\delta} h_x^{D_N} > m_N + t\Big) \le 2P\Big(\max_{x\in D_N\smallsetminus D_N^\delta} h_x^{\widetilde{D}_N} > m_N + t\Big). \tag{12.33}$$

Denote $B_N^\delta := \{x \in \widetilde{D}_N\colon \mathrm{dist}_\infty(x, D_N \smallsetminus D_N^\delta) < \delta N\}$ and write $(\widetilde{X}_N^\star, \widetilde{M}_N)$ for the maximizer and the maximum of $h^{\widetilde{D}_N}$ in $\widetilde{D}_N$. The probability on the right-hand side of (12.33) is then bounded by the sum of

$$P\big(\widetilde{X}_N^\star \in B_N^\delta,\, \widetilde{M}_N > m_N + t\big) \tag{12.34}$$

and

$$\sum_{x\in\widetilde{D}_N} P\Big(h^{\widetilde{D}_N} \le h_x^{\widetilde{D}_N},\, h_x^{\widetilde{D}_N} > m_N + t,\, \max_{\substack{y\in\widetilde{D}_N\\ \mathrm{dist}_\infty(x,y)>\delta N}} h_y^{D_N} > m_N + t\Big). \tag{12.35}$$

Once δ is sufficiently small, we have $B_N^\delta \subseteq \widetilde{D}_N^{\delta'}$ for some $\delta' > 0$ and so we may bound (12.34) using (12.32) by a quantity of order $t\mathrm{e}^{-\alpha t}|B_N^\delta|/N^2$. Since $|B_N^\delta|/N^2$ is bounded by a constant multiple of $\mathrm{Leb}(D \smallsetminus D^\delta)$, this term is $o(t\mathrm{e}^{-\alpha t})$ in the limits $N \to \infty$, $t \to \infty$ and $\delta \downarrow 0$.

The term in (12.35) is bounded using Lemma 12.7 by $c\mathrm{e}^{-\tilde{a}t}$ times the probability $P(\widetilde{M}_N > m_N + t)$. By Lemma 10.4, this probability is (for $t \ge 1$) at most a constant times $t^2\mathrm{e}^{-\alpha t}$ and so (12.35) is $o(t\mathrm{e}^{-\alpha t})$ as well. Hence, for $o(1) \to 0$ as $N \to \infty$ and $t \to \infty$,

$$P\left(\frac{1}{N}X_N^\star \in A,\, M_N \ge m_N + t\right) = t\mathrm{e}^{-\alpha t}\left(o(1) + \bar{c}\int_A \mathrm{d}x\, r_D(x)^2\right). \tag{12.36}$$

This yields the claim with ψ equal to the normalized conformal radius squared (and with $c_\star$ defined by $c_\star := \bar{c}\int_S \mathrm{d}x\, r_S(x)^2$ for $S := (0,1)^2$). □

As noted earlier, our proof identifies ψ with (a constant multiple of) $r_D(x)^2$ directly; unlike the proof of part (5) of Theorem 10.15 which takes the existence of some ψ as an input and infers equality with the conformal radius squared using the Gibbs-Markov property of the Z^D-measures (whose uniqueness we have yet to prove). For future use, we pose a minor extension of Theorem 10.7:

Exercise 12.8 *For any bounded continuous* $f\colon \overline{D} \to [0,\infty)$,

$$P\Big(h_x^{D_N} \le m_N + t - \alpha^{-1}\log f(x/N)\colon x \in D_N\Big) = \exp\Big\{-te^{-\alpha t}\Big(o(1) + \bar{c}\int_D \mathrm{d}x\, r_D(x)^2 f(x)\Big)\Big\}, \tag{12.37}$$

where $o(1) \to 0$ in the limits as $N \to \infty$ and $t \to \infty$ uniformly on compact families of non-negative $f \in C(\overline{D})$.

12.3 Precise Upper Tail Asymptotic

We now move to the proof of Proposition 12.5. We begin by removing the conditioning on $h_0^{D_N} = 0$ from the quantity of interest. To this end we note that

$$h^{D_N} - h_0^{D_N}\mathfrak{g}^{D_N} \overset{\text{law}}{=} h^{D_N \smallsetminus \{0\}} \tag{12.38}$$

which (by Exercise 8.6) also has the law of h^{D_N} conditioned on $h_0^{D_N} = S_{n+1} = 0$. Since $\mathfrak{g}^{D_N}(x) \le c/\log N$ on $D_N \smallsetminus \Delta^{n-k}$ with some $c = c(k)$ while (by a first moment bound) $P(h_0^{D_N} \ge \epsilon\log N) \to 0$ as $N \to \infty$, the term $h_0^{D_N}\mathfrak{g}^{D_N}$ can be absorbed into a negligible change of t. Since $\Xi_{N,k}^{\text{out}}(t)$ is non-decreasing in t, this gives the main idea how to solve:

Exercise 12.9 *Abbreviate $\widetilde{m}_N(x,t) := m_N(1 - \mathfrak{g}^{D_N}(x)) + t$ and let*

$$\widetilde{\Xi}_{N,k}^{\text{out}}(t) := E\Big(1_{\{h^{D_N} \le \widetilde{m}_N(t,\cdot) \text{ in } D_N \smallsetminus \Delta^{n-k}\}} 1_{\{\widetilde{S}_k \in [k^{1/6},\, k^2]\}} \widetilde{S}_k\Big), \tag{12.39}$$

where we abbreviated

$$\widetilde{S}_\ell := S_{n-\ell} - S_{n+1}, \quad \ell = 0, \dots, k. \tag{12.40}$$

Prove that if, for some $\delta > 0$,

$$\lim_{t\to\infty}\limsup_{k\to\infty}\limsup_{N\to\infty}\left|\frac{\widetilde{\Xi}_{N,k}^{\text{out}}(t)}{t} - 1\right| = 0 \tag{12.41}$$

holds uniformly in the shifts of D_N such that $0 \in D_N^\delta$, then the same applies to $\Xi_{N,k}^{\text{out}}(t)$—meaning that (12.24) *holds.*

We thus have to prove the claim for $\widetilde{\Xi}_{N,k}^{\text{out}}(t)$ instead of $\Xi_{N,k}^{\text{out}}(t)$. A key point is that the removal of the conditioning on $S_{n+1} = 0$ makes the increments of $\widetilde{S}$ independent (albeit still slightly time-inhomogeneous) and so we can analyze $\widetilde{\Xi}_{N,k}^{\text{out}}(t)$ using the ballot problem/entropic-repulsion methods used earlier.

Recall the basic objects φ_j, χ_j and h'_j from the concentric decomposition and define a (new) control random variable L as the minimal index in $\{1,\dots,k-1\}$ such that for all $j=0,\dots,k$,

$$\bigl|\varphi_{n-j}(0)\bigr| \le [\log(L\vee j)]^2, \tag{12.42}$$

$$\max_{j+1\le r\le k}\ \max_{x\in\Delta^{n-r}} \bigl|\chi_{n-j}(x)\bigr| 2^{(j-r)/2} \le [\log(L\vee j)]^2 \tag{12.43}$$

and

$$\Bigl|\max_{x\in\Delta^{n-j}\smallsetminus\Delta^{n-j-1}} \bigl[\chi_{n-j}(x)+\chi_{n-j+1}(x)+h'_{n-j}(x)\bigr]-m_{2^{n-j}}\Bigr| \le [\log(L\vee j)]^2. \tag{12.44}$$

If no such index exists, we set $L:=k$. We then have:

Lemma 12.10 *There is an absolute constant $C\in(0,\infty)$ such that for all $t\in\mathbb{R}$, all $n\ge 1$, all $k=1,\dots,n$ obeying*

$$0\le tk/n\le 1 \tag{12.45}$$

and all $\ell=0,\dots,k$, on the event $\{L<k\}$ we have

$$\Bigl|\max_{x\in\Delta^{n-\ell}\smallsetminus\Delta^{n-\ell-1}} \bigl[h_x^{D_N}-\widetilde{m}_N(t,x)\bigr]+(\widetilde{S}_\ell+t)\Bigr| \le \zeta(\ell\vee L), \tag{12.46}$$

where

$$\zeta(s):=C\bigl[1+\log(s)\bigr]^2. \tag{12.47}$$

Proof. From (8.41), (8.48) and the fact that h'_j is supported on $\Delta^j\smallsetminus\Delta^{j-1}$ while χ_j is supported on Δ^j we get, for all $\ell=0,\dots,k$ and all $x\in\Delta^{n-\ell}\smallsetminus\Delta^{n-\ell-1}$,

$$h_x^{D_N}=S_{n+1}-S_{n-\ell}+h'_{n-\ell}(x)+\sum_{j=n-\ell-1}^{n}\bigl(\mathfrak{b}_j(x)\varphi_j(0)+\chi_j(x)\bigr). \tag{12.48}$$

The definition of L along with the decay of $j\mapsto\mathfrak{b}_j(x)$ in (8.49) bound the sum by a quantity proportional to $1+\log(L\vee\ell)^2$. The claim follows from the definition of $\widetilde{m}_N(t,\cdot)$, the tightness of the absolute maximum and Exercise 11.6. □

As in our earlier use of the concentric decomposition, Lemma 12.10 permits us to squeeze the event of interest between two random-walk events:

Exercise 12.11 *Prove that, under* (12.45),

$$
\begin{aligned}
\{L<k\}\cap\bigcap_{\ell=0}^{k}\{\widetilde{S}_\ell\ge\zeta(\ell\vee L)-t\}&\\
\subseteq\{L<k\}\cap\bigl\{h^{D_N}\le\widetilde{m}_N(t,\cdot)\text{ in }D_N\smallsetminus\Delta^{n-k}\bigr\}&\\
\subseteq\bigcap_{\ell=0}^{k}\bigl\{\widetilde{S}_\ell\ge-\zeta(\ell\vee L)-t\bigr\}.&
\end{aligned}
\tag{12.49}
$$

Notice that, unlike in Lemma 8.19, the t-dependent term has the same form and sign on both ends of this inclusion. The next item to address are suitable estimates on the relevant probabilities for the control variable L and the random walk $\widetilde{S}$. First, a variation on Exercise 8.18 shows

$$
P\bigl(L=\ell\bigr)\le c_1\mathrm{e}^{-c_2(\log\ell)^2},\quad \ell=1,\dots,k,
\tag{12.50}
$$

for some $c_1,c_2>0$. In addition, we will need:

Lemma 12.12 *There is a constant $c>0$ such that for all $\ell,k\in\mathbb{N}$ with $\ell\le k^{\frac{1}{12}}$ and $k\le n/2$ and all $0\le t\le k^{1/6}$,*

$$
P\left(\bigcap_{j=\ell}^{k}\{\widetilde{S}_j-\widetilde{S}_\ell\ge-\zeta(j)-t-\ell^2\}\right)\le c\frac{\ell^2+t}{\sqrt{k}}.
\tag{12.51}
$$

This is showed by invoking again estimates on the inhomogeneous ballot problem encountered earlier in these lectures. We refer to [27] for details. We use Lemma 12.12 to prove:

Lemma 12.13 (Controlling the control variable) *There are $c_1,c_2>0$ such that for all $\ell,k,n\in\mathbb{N}$ satisfying $k\le n/2$ and $1\le\ell\le k^{\frac{1}{12}}$, and for all $1\le t\le k^{1/6}$,*

$$
E\left((\widetilde{S}_k\vee k^{1/6})1_{\{L=\ell\}}\prod_{j=\ell}^{k}1_{\{\widetilde{S}_j\ge-\zeta(j)-t\}}\right)\le c_1(\ell^2+t)\mathrm{e}^{-c_2(\log\ell)^2}
\tag{12.52}
$$

and also

$$
E\left(\widetilde{S}_k1_{\{0\le\widetilde{S}_k\le k^{1/6}\}}\,1_{\{L\le\ell\}}\prod_{j=\ell}^{k}1_{\{\widetilde{S}_j\ge-\zeta(j)-t\}}\right)\le c_1(\ell^2+t)k^{-1/3}.
\tag{12.53}
$$

Proof. Since $|\widetilde{S}_\ell|\le\ell^2$ when $L=\ell$, we have

$$
\widetilde{S}_k\vee k^{1/6}\le\ell^2+(\widetilde{S}_k-\widetilde{S}_\ell)\vee k^{1/6},\quad\text{on }\{L=\ell\}
\tag{12.54}
$$

and

$$1_{\{L=\ell\}}1_{\{\widetilde{S}_j \geq -\zeta(j)-t\}} \leq 1_{\{L=\ell\}}1_{\{\widetilde{S}_j-\widetilde{S}_\ell \geq -\zeta(j)-t-\ell^2\}} \tag{12.55}$$

Substituting these in (12.52), the indicator of $\{L = \ell\}$ is independent of the rest of the expression. Invoking (12.50), to get (12.52) we thus have to show

$$E\left(((\widetilde{S}_k - \widetilde{S}_\ell) \vee k^{1/6}) \prod_{j=\ell}^{k} 1_{\{\widetilde{S}_j-\widetilde{S}_\ell \geq -\zeta(j)-t-\ell^2\}}\right) \leq c(\ell^2 + t) \tag{12.56}$$

for some constant $c > 0$. Abbreviate

$$\widehat{S}_j := \widetilde{S}_{\ell+j} - \widetilde{S}_\ell. \tag{12.57}$$

We will need:

Exercise 12.14 *Prove that the law of* $\{\widehat{S}_j : j = 1, \dots, k - \ell\}$ *on* $\mathbb{R}^{k-\ell}$ *is strong FKG. Prove the same for the law of the standard Brownian motion on* $\mathbb{R}^{[0,\infty)}$.

Exercise 12.15 *Prove that if the law of random variables* $X_1, \dots, X_n$ *is strong FKG, then for any increasing function* Y,

$$a_1, \dots, a_n \mapsto E\big(Y \,|\, X_1 \geq a_1, \dots, X_n \geq a_n\big) \tag{12.58}$$

is increasing in each variable.

Embedding the random walk into the standard Brownian motion $\{B_s : s \geq 0\}$ via $\widehat{S}_j := B_{s_j}$, where $s_j := \mathrm{Var}(\widehat{S}_j)$, and writing P^0 is the law of the Brownian motion started at 0 and E^0 is the associated expectation, Exercise 12.15 dominates the expectation in (12.56) by

$$E^0\Big((B_{s_{k-\ell}} \vee k^{1/6})\, 1_{\{B_s \geq -1:\, s\in[0,s_{k-\ell}]\}}\Big)\, \frac{P\big(\bigcap_{j=0}^{k-\ell+1}\{\widehat{S}_j \geq -\zeta(j+\ell) - t - \ell^2\}\big)}{P^0\big(B_s \geq -1 : s \in [0, s_{k-\ell}]\big)}. \tag{12.59}$$

The expectation in (12.59) is bounded by conditioning on B_t and invoking Exercise 7.10 along with the bound $1 - \mathrm{e}^{-a} \leq a$ to get

$$E^0\big((B_{s_{k-\ell}} \vee k^{1/6})\, 1_{\{B_s \geq -1:\, s\in[0,s_{k-\ell}]\}}\big) \leq E^0\left((B_{s_{k-\ell}} \vee k^{1/6}) \frac{2(B_{s_{k-\ell}} + 1)}{s_{k-\ell}}\right). \tag{12.60}$$

As $s_{k-\ell}$ is of the order of $k - \ell$, the expectation on the right is bounded uniformly in $k > \ell \geq 0$. Thanks to Lemma 12.12 and Exercise 11.17 (and the fact that $\zeta(\ell) \leq \ell^2$ and that $j \mapsto \zeta(j + \ell) - \zeta(\ell)$ is slowly varying) the ratio of probabilities in (12.59) is at most a constant times $\ell^2 + t$.

The proof of (12.53) easier; we bound $\widetilde{S}_k$ by $k^{1/6}$ and thus estimate the probability by passing to $\widehat{S}_j$ and invoking Lemma 12.12. □

The above proof shows that the dominating contribution to $\widetilde{\Xi}^{\text{out}}_{N,k}(t)$ comes from the event when $\widetilde{S}_k$ is order $\sqrt{k}$. This is just enough to cancel the factor $1/\sqrt{k}$ arising from the probability that $\widetilde{S}$ stays above a slowly varying curve for k steps. This is what yields the desired asymptotic as $t \to \infty$ as well:

Proof of Proposition 12.5. Note that Lemma 12.13 and the fact that $|S_k| \le k^2$ on $\{L < k\}$ permits us to further reduce the computation to the asymptotic of

$$\widehat{\Xi}^{\text{out}}_{N,k}(t) := E\Bigl(1_{\{h^{D_N} \le \widetilde{m}_N(t,\cdot) \text{ in } D_N \smallsetminus \Delta^{n-k}\}}(\widetilde{S}_k \vee 0)\Bigr). \tag{12.61}$$

Indeed, writing $\epsilon_\ell(t)$ for the right-hand side of (12.52), this bound along with (12.49) yield

$$\begin{aligned} E\Biggl((\widetilde{S}_k \vee 0)\prod_{j=0}^{k} 1_{\{\widetilde{S}_j \ge \zeta(j\vee\ell)-t\}}\Biggr) &- \epsilon_\ell(t) \\ &\le \widehat{\Xi}^{\text{out}}_{N,k}(t) \le E\Biggl((\widetilde{S}_k \vee 0)\prod_{j=0}^{k} 1_{\{\widetilde{S}_j \ge -\zeta(j\vee\ell)-t\}}\Biggr) + \epsilon_\ell(t). \end{aligned} \tag{12.62}$$

Similarly as in (12.59), and denoting $\tilde{s}_k := \mathrm{Var}(\widehat{S}_k)$, Exercise 12.15 bounds the expectation on the right of (12.62) by

$$E^0\bigl((B_{\tilde{s}_k} \vee 0)1_{\{B_s \ge -t:\ s\in[0,\tilde{s}_k]\}}\bigr) \tag{12.63}$$

times the ratio

$$\frac{P\bigl(\bigcap_{j=1}^{k+1}\{\widetilde{S}_j \ge -\zeta(j\vee\ell)-t\bigr)}{P^0\bigl(B_s \ge -t:\ s \in [0,\tilde{s}_k]\bigr)} \tag{12.64}$$

and that on the left by (12.63) times

$$\frac{P^0\bigl(B_s \ge \tilde{\zeta}(s\vee\tilde{s}_\ell)-t:\ s\in[0,\tilde{s}_k]\bigr)}{P^0\bigl(B_s \ge -t:\ s\in[0,\tilde{s}_k]\bigr)}, \tag{12.65}$$

where $\tilde{\zeta}$ is a piecewise-linear (non-decreasing) function such that $\tilde{\zeta}(\tilde{s}_j) = \zeta(j)$ for each $j \ge 1$. The precise control of inhomogeneous Ballot Theorem (cf [27, Proposition 4.7 and 4.9]) then shows that the ratios (12.64)–(12.65) converge to one in the limits $k \to \infty$ followed by $t \to \infty$. Exercise 7.10 shows that the expectation in (12.63) is asymptotic to t and so the claim follows. □

12.4 Convergence of the DGFF Maximum

We are now ready to address the convergence of the centered DGFF maximum by an argument that will also immediately yield the convergence of the extremal value process. The key point is the proof of:

Theorem 12.16 (Uniqueness of Z^D-measure) *Let $D \in \mathfrak{D}$ and let Z^D be the measure related to a subsequential weak limit of $\{\eta^D_{N,r_N} : N \geq 1\}$ as in Theorem 9.3. Then Z^D is the weak limit of slight variants of the measures in* (10.53). *In particular, its law is independent of the subsequence used to define it and the processes $\{\eta^D_{N,r_N} : N \geq 1\}$ thus converge in law.*

The proof will follow closely the proof of Theorem 10.21 that characterizes the Z^D-measures by a list of natural conditions. These were supplied in Theorem 10.15 for all domains of interest through the existence of the limit of processes $\{\eta^D_{N,r_N} : N \geq 1\}$. Here we allow ourselves only subsequential convergence and so we face the following additional technical difficulties:

(1) We can work only with a countable set of domains at a time, and so the Gibbs-Markov property needs to be restricted accordingly.
(2) We lack the transformation rule for the behavior of the Z^D measures under dilations of D in property (3) of Theorem 10.15.
(3) We also lack the Laplace transform asymptotic in property (5) of Theorem 10.15.

The most pressing are the latter two issues so we start with them. It is the following lemma where the majority of the technical work is done:

Lemma 12.17 *Fix $R > 1$ and $D \in \mathfrak{D}$ and for any positive $n \in \mathbb{N}$ abbreviate $D^n := n^{-1}D$. Suppose that (for a given sequence of approximating domains for each $n \in \mathbb{N}$), there is $N_k \to \infty$ such that*

$$\eta^{D^n}_{N_k,r_{N_k}} \overset{\text{law}}{\longrightarrow} \eta^{D^n}, \quad n \geq 1. \tag{12.66}$$

Let Z^{D^n} be the measure related to η^{D_n} as in Theorem 9.6. Then for any choice of $\lambda_n > 0$ satisfying

$$\lambda_n n^{-4} \underset{n\to\infty}{\longrightarrow} 0 \tag{12.67}$$

and any choice of Rn-Lipschitz functions $f_n : D^n \to [0, R]$, we have

$$\log E\bigl(\mathrm{e}^{-\lambda_n \langle Z^{D^n}, f_n\rangle}\bigr) = -\lambda_n \log(n^4/\lambda_n)\left[o(n^{-4}) + \bar{c}\int_{D^n} \mathrm{d}x\, r_{D^n}(x)^2 f_n(x)\right], \tag{12.68}$$

where $n^4 o(n^{-4}) \to 0$ as $n \to \infty$ uniformly in the choice of the f_n's. The constant $\bar{c}$ is as in (12.31).

Proof. We start by some observations for a general domain D. Let $\varrho\colon \overline{D} \to [0, \infty)$ be measurable and denote $A_\varrho := \{(x, h)\colon x \in D,\, h > -\alpha^{-1} \log \varrho(x)\}$. Then for any $a > 0$, any subsequential limit process η^D obeys

$$(1 - \mathrm{e}^{-a}) P\bigl(\eta^D(A_\varrho) > 0\bigr) \le 1 - E\bigl(\mathrm{e}^{-a\eta^D(A_\varrho)}\bigr) \le P\bigl(\eta^D(A_\varrho) > 0\bigr). \tag{12.69}$$

Assuming the probability on the right-hand side is less than one, elementary estimates show

$$\Bigl|\log E\bigl(\mathrm{e}^{-a\eta^D(A_\varrho)}\bigr) - \log P\bigl(\eta^D(A_\varrho) = 0\bigr)\Bigr| \le \frac{\mathrm{e}^{-a}}{P(\eta^D(A_\varrho) = 0)}. \tag{12.70}$$

We also note that, setting

$$f(x) := (1 - \mathrm{e}^{-a})\alpha^{-1}\varrho(x), \tag{12.71}$$

implies

$$E(\mathrm{e}^{-a\eta^D(A_\varrho)}) = E(\mathrm{e}^{-\langle Z^D, f\rangle}) \tag{12.72}$$

thanks to the Poisson structure of η^D proved in Theorem 9.6.

Given $D \in \mathfrak{D}$, for each $n \in \mathbb{N}$ let $D^n := n^{-1}D$ and let $f_n\colon D^n \to [0, R]$ be an Rn-Lipschitz function. Fix an auxiliary sequence $\{a_n\}$ of numbers $a_n \ge 1$ such that $a_n \to \infty$ subject to a minimal growth condition to be stated later. Define $\varrho_n\colon D^n \to [0, \infty)$ so that

$$f_n(x) = (1 - \mathrm{e}^{-a_n})\alpha^{-1}\varrho_n(x) \tag{12.73}$$

holds for each $n \in \mathbb{N}$ and note that $\varrho_n(x)$ is $R'n$-Lipschitz for some R' depending only on R. Let $\{\lambda_n\colon n \ge 1\}$ be positive numbers such that (12.67) holds and note that, by (12.70) and (12.72), in order to control the asymptotic of $E(\mathrm{e}^{-\lambda_n\langle Z^{D^n}, f_n\rangle})$ in the limit as $n \to \infty$ we need to control the asymptotic of $P(\eta^{D^n}(A_{\lambda_n\varrho_n}) = 0)$.

Writing $\{D^n_N\colon N \ge 1\}$ for the approximating domains for D^n, the assumed convergence yields

$$P\bigl(\eta^{D^n}(A_{\lambda_n\varrho_n}) = 0\bigr) = \lim_{k\to\infty} P\Bigl(h_x^{D^n_N} \le m_N - \alpha^{-1}\log\bigl(\lambda_n\varrho_n(x/N)\bigr)\colon x \in D^n_N\Bigr)\Big|_{N=N_k}. \tag{12.74}$$

A key point is that, by Theorem 10.7 (or, more precisely, Exercise 12.8) the limit of the probabilities on the right of (12.74) exists for all N; not just those in the subsequence $\{N_k\}$. We may thus regard D^n_N as an approximating domain of D at scale $\lfloor N/n\rfloor$ and note that, in light of $2\sqrt{g} = 4\alpha^{-1}$,

$$m_N - \alpha^{-1}\log\lambda_n = m_{\lfloor N/n\rfloor} + \alpha^{-1}\log(n^4/\lambda_n) + o(1), \tag{12.75}$$

where $o(1) \to 0$ as $N \to \infty$. Note also that $\tilde{\varrho}_n \colon D \to \mathbb{R}$ defined by $\tilde{\varrho}_n(x) := \varrho_n(x/n)$ is R'-Lipschitz for some R' depending only on R. Using (12.67), Exercise 12.8 with $t = \alpha^{-1}\log(n^4/\lambda_n) + o(1)$ shows

$$\log P\Big(h_x^{D_N^n} \le m_N - \alpha^{-1}\log\lambda_n - \alpha^{-1}\log\varrho_n(x/N) \colon x \in D_N^n\Big)$$
$$= -\alpha^{-1}\lambda_n\, n^{-4}\log(n^4/\lambda_n)\Big[\,o(1) + \tilde{c}\int_D \mathrm{d}x\, r_D(x)^2\tilde{\varrho}_n(x)\Big], \tag{12.76}$$

where $o(1) \to 0$ as $N \to \infty$ and $n \to \infty$. Changing the variables back to $x \in D^n$ absorbs the term n^{-4} into the integral and replaces $\tilde{\varrho}_n$ by ϱ_n, which can be then related back to f_n via (12.73). Since $a_n \to \infty$, we get

$$\log P\big(\eta^{D^n}(A_{\lambda_n\varrho_n}) = 0\big) = -\frac{\lambda_n\log(n^4/\lambda_n)}{1-\mathrm{e}^{-a_n}}\Big[\,o(n^{-4}) + \tilde{c}\int_{D^n}\mathrm{d}x\; r_{D^n}(x)^2 f_n(x)\Big], \tag{12.77}$$

where $o(n^{-4})n^4 \to 0$ as $n \to \infty$ uniformly in the choice of $\{f_n\}$ with the above properties.

In remains to derive (12.68) from (12.77). Note that the boundedness of $f_n(x)$ ensures that the right-hand side of (12.77) is of order $\lambda_n n^{-4}\log(n^4/\lambda_n)$ which tends to zero by (12.67). In particular, the probability on the left of (12.77) tends to one. If a_n grows so fast that

$$\mathrm{e}^{a_n}\lambda_n n^{-4}\log(n^4/\lambda_n) \underset{n\to\infty}{\longrightarrow} \infty, \tag{12.78}$$

then (12.70) and (12.72) equate the leading order asymptotic of the quantities on the left-hand side of (12.68) and (12.77). The claim follows. □

We are now ready to give:

Proof of Theorem 12.16. Let $\mathfrak{D}_0 \subseteq \mathfrak{D}$ be a countable set containing the domain of interest along with the collection of all open equilateral triangles of side length 2^{-n}, $n \in \mathbb{Z}$, with (at least) two vertices at points with rational coordinates, and all finite disjoint unions thereof. Let $\{N_k \colon k \ge 1\}$ be a subsequence along which η^D_{N,r_N} converges in law to some η^D for all $D \in \mathfrak{D}_0$. We will never need the third coordinate of η^D_{N,r_N} so we will henceforth think of these measures as two-coordinate processes only.

Let Z^D, for each $D \in \mathfrak{D}_0$, denote the random measure associated with the limit process η^D. Pick $f \in C(\overline{D})$ and assume that f is positive and Lipschitz. As we follow closely the proof of Theorem 10.21, let us review it geometric setting: Given a $K \in \mathbb{N}$, which we will assume to be a power of two, and $\delta \in (0, 1/100)$ and let $\{T^i \colon i = 1,\dots,n_K\}$ be those triangles in the tiling of $\mathbb{R}^2$ by triangles of side length K^{-1} that fit into D^δ. Denote by x_i the center of T^i and let T^i_δ be the triangle of side length $(1-\delta)K^{-1}$ centered at x_i and oriented the same way as T^i. Write $\widetilde{D}_\delta := \bigcup_{i=1}^{n_K} T^i_\delta$ and let $\chi_\delta \colon D \to [0,1]$ be a function that is one on $\widetilde{D}_{2\delta}$, zero on $D \smallsetminus \widetilde{D}_\delta$ and is $(2K\delta^{-1})$-Lipschitz elsewhere.

Set $f_\delta := f\chi_\delta$. In light of Lemma 10.4 (proved above) and Exercise 10.10, we then have (10.58) and so it suffices to work with $\langle Z^D, f_\delta\rangle$. First we ask the reader to solve:

Exercise 12.18 *Prove that the measures* $\{Z^{\widetilde{D}}\colon \widetilde{D}\in\mathfrak{D}_0\}$ *satisfy the Gibbs-Markov property in the form* (10.59) *(for* M^D *replaced by* Z^D*).*

Moving along the proof of Theorem 10.21, next we observe that (10.60) and Proposition 10.24 apply because they do not use any particulars of the measures M^D there. Letting $c>0$ is the constant from (10.60) and writing

$$\widetilde{A}^i_{K,R} := \bigl\{\operatorname{osc}_{T^i_\delta}\Phi^{D,\widetilde{D}}\le R\bigr\}\cap\bigl\{\Phi^{D,\widetilde{D}}(x_i)\le 2\sqrt{g}\log K - c\log\log K\bigr\}. \tag{12.79}$$

for a substitute of event $A^i_{K,R}$ from (10.52), we thus have

$$\langle Z^D, f_\delta\rangle = o(1) + \sum_{i=1}^{n_K} 1_{\widetilde{A}^i_{K,R}}\,\mathrm{e}^{\alpha\Phi^{D,\widetilde{D}}(x_i)}\int_{T^i} Z^{T^i}(\mathrm{d}x)\,\mathrm{e}^{\alpha[\Phi^{D,\widetilde{D}}(x)-\Phi^{D,\widetilde{D}}(x_i)]}\,f_\delta(x), \tag{12.80}$$

where $o(1)\to 0$ in probability as $K\to\infty$ and $R\to\infty$ and where the family of measures $\{Z^{T^i}: i=1,\dots,n_K\}$ are independent on one another (and equidistributed, modulo shifts) and independent of $\Phi^{D,\widetilde{D}}$.

Proceeding along the argument from the proof of Theorem 10.21, we now wish to compute the negative exponential moment of the sum on the right of (12.80), conditional on $\Phi^{D,\widetilde{D}}$. The indicator of $\widetilde{A}^i_{K,R}$ enforces that

$$\lambda_K := \mathrm{e}^{\alpha\Phi^{D,\widetilde{D}}(x_i)}\quad\text{obeys}\quad K^4\lambda_K\le(\log K)^{-\alpha c} \tag{12.81}$$

while, in light of the regularity assumptions on f and χ_δ and the harmonicity and boundedness of oscillation of $\Phi^{D,\widetilde{D}}$,

$$f_K(x) := f(x)\chi_\delta(x)\,\mathrm{e}^{\alpha[\Phi^{D,\widetilde{D}}(x)-\Phi^{D,\widetilde{D}}(x_i)]} \tag{12.82}$$

is $R'K$-Lipschitz for some R' depending only on R. Lemma 12.17 (which is our way to by-pass Proposition 10.26) then yields the analogue of (10.70): For any $i=1,\dots,n_K$ where $\widetilde{A}^i_{K,R}$ occurs,

$$\begin{aligned}
E\biggl(\exp\Bigl\{-\mathrm{e}^{\alpha\Phi^{D,\widetilde{D}}(x_i)}Z^{T^i}(f_K)\Bigr\}\,\Bigg|\,\Phi^{D,\widetilde{D}}\biggr)
&= \exp\Bigl\{\log\bigl(K^{-4}\mathrm{e}^{\alpha\Phi^{D,\widetilde{D}}(x_i)}\bigr)\,\bar{c}\int_{T^i}\mathrm{d}x\, f_\delta(x)\,\mathrm{e}^{\alpha\Phi^{D,\widetilde{D}}(x)}\,r_{T^i}(x)^2\Bigr\}\\
&\quad\times\exp\Bigl\{\widetilde{\epsilon}_K K^{-4}\mathrm{e}^{\alpha\Phi^{D,\widetilde{D}}(x_i)}\log\bigl(K^{-4}\mathrm{e}^{\alpha\Phi^{D,\widetilde{D}}(x_i)}\bigr)\Bigr\},
\end{aligned} \tag{12.83}$$

where $\widetilde{\epsilon}_K$ is a random ($\Phi^{D,\widetilde{D}}$-measurable) quantity taking values in some $[\epsilon_K,\epsilon_K]$ for some deterministic ϵ_K with $\epsilon_K\to 0$ as $K\to\infty$.

Assuming (at the cost of a routine limit argument at the end of the proof) that $f\ge\delta$ on D, the bound on the oscillation of $\Phi^{D,\widetilde{D}}$ permits us to absorb the error term into

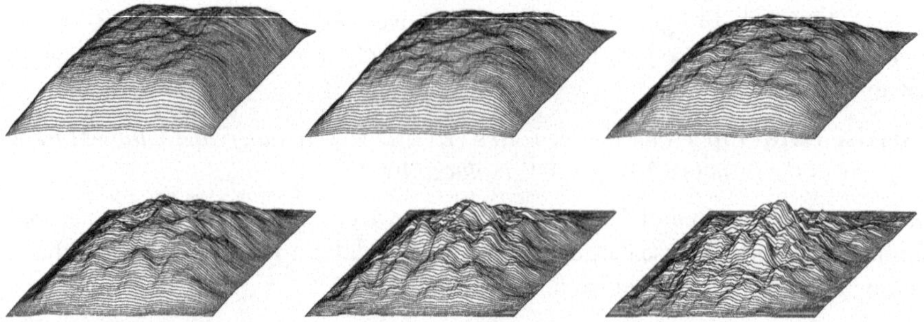

Fig. 22 Empirical plots of $x \mapsto \rho^D(x,t)$ obtained from a set of about 1,00,000 samples of the maximum of the DGFF on a 100×100 square. The plots (labeled left to right starting with the top row) correspond to t increasing by uniform amounts over an interval of length 3 with t in the fourth figure set to the empirical mean

an additive modification of f by a term of order $\epsilon_K e^{\alpha R}$. Invoking (10.71)–(10.72), we conclude that the measure

$$\alpha\bar{c}\, r_D(x)^2 \sum_{i=1}^{n_K} 1_{\widetilde{A}^i_{K,R}}\bigl(\alpha\text{Var}(\Phi^{D,\widetilde{D}}(x)) - \Phi^{D,\widetilde{D}}(x)\bigr) e^{\alpha\Phi^{D,\widetilde{D}}(x) - \frac{1}{2}\alpha^2 \text{Var}(\Phi^{D,\widetilde{D}}(x))} 1_{T^i_\delta}(x)\,\mathrm{d}x \tag{12.84}$$

tends to Z^D in the limit $K \to \infty$, $R \to \infty$ and $\delta \downarrow 0$. In particular, the law of Z^D is independent of the subsequence $\{N_k : k \geq 1\}$ used to define it and the processes $\{\eta^D_{N,r_N} : N \geq 1\}$ thus converge in law. □

Theorem 12.16 implies, via Lemma 10.1, the convergence of the centered maximum from Theorem 10.3 and, by Lemma 10.9, also the joint convergence of maxima in disjoint open subsets as stated in Lemma 10.8. The convergence statement of the full extremal process in Theorem 9.3 is now finally justified.

12.5 The Local Limit Theorem

Our way of control of the absolute maximum by conditioning on its position can be used to give also a local limit theorem for both the position and value of the maximum:

Theorem 12.19 (Local limit theorem for absolute maximum) *Using the setting of Theorem 9.3, for any $x \in D$ and any $a < b$,*

$$\lim_{N\to\infty} N^2 P\Bigl(\operatorname*{argmax}_{D_N} h^{D_N} = \lfloor xN \rfloor,\ \max_{x\in D_N} h^{D_N}_x - m_N \in (a,b)\Bigr) = \int_a^b \rho^D(x,t)\mathrm{d}t\,, \tag{12.85}$$

where $x \mapsto \rho^D(x,t)$ is the Radon–Nikodym derivative of the measure

$$A \mapsto e^{-\alpha t} E\bigl(Z^D(A) e^{-\alpha^{-1} e^{-\alpha t} Z^D(D)}\bigr) \tag{12.86}$$

with respect to the Lebesgue measure on D.

We only state the result here and refer the reader to [27] for details of the proof. Nothing explicit is known about ρ^D except its $t \to \infty$ asymptotics; see (16.3). Empirical plots of ρ^D for different values of t are shown in Fig. 22.

Lecture 13: Random Walk in DGFF Landscape

In this lecture we turn our attention to a rather different problem than discussed so far: a random walk in a random environment. The connection with the main theme of these lectures is through the specific choice of the random walk dynamics which we interpret as the motion of a random particle in a DGFF landscape. We first state the results, obtained in a recent joint work with Jian Ding and Subhajit Goswami [29], on the behavior of this random walk. Then we proceed to develop the key method of the proof, which is based on recasting the problem as a random walk among random conductances and applying methods of electrostatic theory.

13.1 A Charged Particle in an Electric Field

Let us start with some physics motivation. Suppose we are given the task to describe the motion of a charged particle in a rapidly varying electric field. A natural choice is to fit this into the framework of the theory of random walks in random environment (RWRE) as follows. The particle is confined to the hypercubic lattice $\mathbb{Z}^d$ and the electric field is represented by a configuration $h = \{h_x : x \in \mathbb{Z}^d\}$ with h_x denoting the *electrostatic potential* at x. Given a realization of h, the charged particle then performs a "random walk" which, technically, is a discrete-time Markov chain on $\mathbb{Z}^d$ with the transition probabilities

$$\mathsf{P}_h(x, y) := \frac{\mathrm{e}^{\beta(h_y - h_x)}}{\sum_{z\colon (x,z)\in E(\mathbb{Z}^d)} \mathrm{e}^{\beta(h_z - h_x)}} 1_{(x,y)\in E(\mathbb{Z}^d)}, \tag{13.1}$$

where β is a parameter playing the role of the *inverse temperature*. We will assume $\beta > 0$ which means that the walk is more likely to move in the direction where the electrostatic potential increases. Alternatively, we may think of β as the value of the charge of the moving particle.

Let $X = \{X_n : n \geq 0\}$ denote a sample path of the Markov chain. We will write P_h^x for the law of X with $P_h^x(X_0 = x) = 1$, use E_h^x to denote expectation with respect to P_h^x and write $\mathbb{P}$ to denote the law of the DGFF on $\mathbb{Z}^2 \smallsetminus \{0\}$. As usual in RWRE theory, we will require that

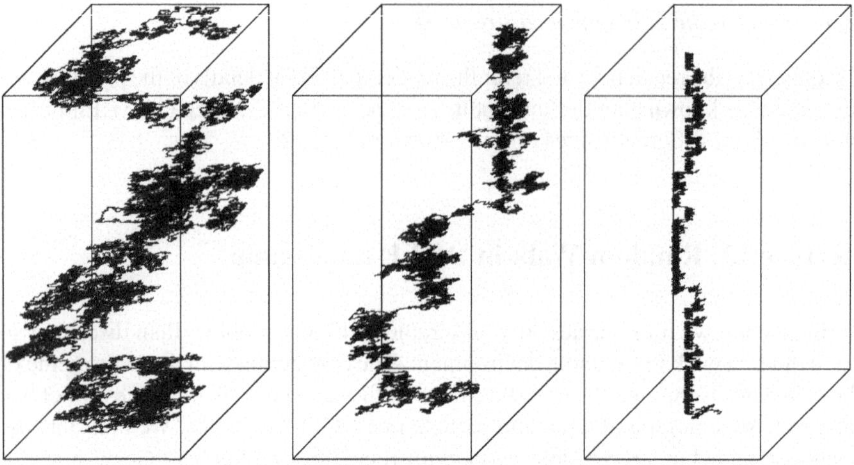

Fig. 23 Runs of 100,000 steps of the random walk with transition probabilities (13.1) and β equal to 0.2, 0.6 and 1.2 multiples of $\tilde{\beta}_{\mathrm{c}}$. Time runs upwards along the vertical axis. Trapping effects are quite apparent

$$\{\mathsf{P}_h(x,\cdot)\colon x\in\mathbb{Z}^d\}\text{ is stationary, ergodic under the shifts of }\mathbb{Z}^d \tag{13.2}$$

as that is typically the minimal condition needed to extract a limit description of the path properties. However, since $\mathsf{P}_h(x,\cdot)$ depends only on the differences of the field, in our case (13.2) boils down to the requirement:

$$\bigl\{h_x-h_y\colon (x,y)\in E(\mathbb{Z}^d)\bigr\}\text{ is stationary, ergodic under the shifts of }\mathbb{Z}^d\,. \tag{13.3}$$

A number of natural examples may be considered, with any i.i.d. random field or, in fact, any stationary and ergodic random field obviously satisfying (13.3). However, our desire in the lectures is to work with the fields that exhibit logarithmic correlations. A prime example of such a field is the two-dimensional DGFF.

The motivation for our focus on log-correlated fields comes from the 2004 paper of Carpentier and Le Doussal [39], who discovered, on the basis of physics arguments, that such environments exhibit the following phenomena:

(1) trapping effects make the walk behave subdiffusively with the diffusive exponent ν, defined via $|X_n|=n^{\nu+o(1)}$, depending non-trivially on β, and
(2) $\beta\mapsto\nu(\beta)$ undergoes a phase transition (i.e., a change in analytic dependence) as β varies through a critical point $\tilde{\beta}_{\mathrm{c}}$.

Log-correlated fields are in fact deemed critical for the above phase transition to occur (with weaker correlations generically corresponding to $\beta\downarrow 0$ regime and stronger correlations to $\beta\to\infty$) although we will not try to make this precise. The purpose of these lectures it to demonstrate that (1–2) indeed happen in at least one example; namely, the two-dimensional DGFF.

We will thus henceforth focus on $d = 2$. To get around the fact that the DGFF on $\mathbb{Z}^2$ does not exist, we will work with

$$h := \text{ DGFF in } \mathbb{Z}^2 \smallsetminus \{0\} \tag{13.4}$$

as the driving "electric" field for the rest of these lectures. This does fall into the class of systems introduced above; indeed, we have:

Exercise 13.1 *(Pinned DGFF has stationary and ergodic gradients) Show that h in* (13.4) *obeys* (13.3).

13.2 Statement of Main Results

Having elucidated the overall context of the problem at hand, we are now ready to describe the results for the particular choice (13.4). These have all been proved in a joint 2016 paper with Ding and Goswami [29]. Our first result concerns the decay of return probabilities (a.k.a. the *heat kernel*):

Theorem 13.2 (Heat-kernel decay) *For each $\beta > 0$ and each $\delta > 0$,*

$$\mathbb{P}\biggl(\frac{1}{T}\text{e}^{-(\log T)^{1/2+\delta}} \le P_h^0(X_{2T} = 0) \le \frac{1}{T}\text{e}^{(\log T)^{1/2+\delta}}\biggr) \underset{T\to\infty}{\longrightarrow} 1. \tag{13.5}$$

A noteworthy point is that the statement features no explicit dependence on β. (In fact, it applies even to $\beta = 0$ when X is just the simple random walk on $\mathbb{Z}^2$.) Hence, as far as the leading order of the return probabilities is concerned, the walk behaves just as the simple random walk. Notwithstanding, the $\text{e}^{\pm(\log T)^{1/2+\delta}}$ terms are too large to let us decide whether X is recurrent or transient, a question that we will eventually resolve as well albeit by different means.

Although the propensity of the walk to move towards larger values of the field does not seem to affect the (leading order) heat kernel decay, the effect on the path properties is more detectable. For each set $A \subseteq \mathbb{Z}^2$, define

$$\tau_A := \inf\bigl\{n \ge 0\colon X_n \in A\bigr\}. \tag{13.6}$$

Denote also

$$B(N) := [-N, N]^2 \cap \mathbb{Z}^2. \tag{13.7}$$

Then we have:

Theorem 13.3 (Subdiffusive expected exit time) *For each $\beta > 0$ and each $\delta > 0$,*

$$\mathbb{P}\biggl(N^{\theta(\beta)}\text{e}^{-(\log N)^{1/2+\delta}} \le E_h^0(\tau_{B(N)^c}) \le N^{\theta(\beta)}\text{e}^{(\log N)^{1/2+\delta}}\biggr) \underset{N\to\infty}{\longrightarrow} 1, \tag{13.8}$$

where, for $\tilde{\beta}_c := \sqrt{\pi/2}$,

$$\theta(\beta) := \begin{cases} 2 + 2(\beta/\tilde{\beta}_c)^2, & \text{if } \beta \le \tilde{\beta}_c, \\ 4\beta/\tilde{\beta}_c, & \text{if } \beta \ge \tilde{\beta}_c. \end{cases} \tag{13.9}$$

The functional dependence of $\theta(\beta)$ on β actually takes a rather familiar form: Denoting $\lambda := \beta/\tilde{\beta}_c$, for $\lambda \in (0, 1)$ we have $\theta(\beta) = 2 + 2\lambda^2$, which is the scaling exponent associated with the intermediate level set at height λ-multiple of the absolute maximum. The dependence on β changes at $\tilde{\beta}_c$, just as predicted by Carpentier and Le Doussal [39]. Moreover, for $\beta > 0$, we have $\theta(\beta) > 0$. The walk thus takes considerably longer (in expectation) to exit a box than the simple random walk. This can be interpreted as a version of *subdiffusive behavior*. See Fig. 23.

The standard definition of subdiffusive behavior is via the typical spatial scale of the walk at large times. Here we can report only a one-way bound:

Corollary 13.4 (Subdiffusive lower bound) *For all* $\beta > 0$ *and all* $\delta > 0$,

$$P_h^0\Big(|X_T| \ge T^{1/\theta(\beta)} e^{-(\log N)^{1/2+\delta}}\Big) \underset{N\to\infty}{\longrightarrow} 1, \quad \text{in } \mathbb{P}\text{-probability}. \tag{13.10}$$

Unfortunately, the more relevant upper bound is elusive at this point (although we believe that our methods can be boosted to include a matching leading-order upper bound as well).

Our method of proof of the above results relies on the following elementary observation: The transition probabilities from (13.1) can be recast as

$$\mathsf{P}_h(x, y) := \frac{e^{\beta(h_y + h_x)}}{\pi_h(x)} 1_{(x,y)\in E(\mathbb{Z}^d)}, \tag{13.11}$$

where

$$\pi_h(x) := \sum_{z\colon (x,z)\in E(\mathbb{Z}^d)} e^{\beta(h_z + h_x)}. \tag{13.12}$$

This, as we will explain in the next section, phrases the problem as a *random walk among random conductances*, with the conductance of edge $(x, y) \in E(\mathbb{Z}^d)$ given by

$$c(x, y) := e^{\beta(h_y + h_x)}. \tag{13.13}$$

As is readily checked, π_h is a reversible, and thus stationary, measure for X.

Over the past couple of decades, much effort went into the understanding of random walks among random conductances (see, e.g., a review by the present author [22] or Kumagai [83]). A standing assumption of these is that the conductance configuration,

$$\big\{c(x, y)\colon (x, y) \in E(\mathbb{Z}^d)\big\}, \tag{13.14}$$

is stationary and ergodic with respect to the shifts of $\mathbb{Z}^d$. For $\mathbb{Z}^d$ in $d \geq 3$, the full-lattice DGFF (with zero boundary conditions) is stationary and so the problem falls under this umbrella. The results of Andres, Deuschel and Slowik [9] then imply scaling of the random walk to a non-degenerate Brownian motion.

When $d = 2$ and $h =$ DGFF on $\mathbb{Z}^2 \smallsetminus \{0\}$ the conductances (13.13) are no longer stationary. Thus we have a choice to make: either work with stationary transition probabilities with no help from reversibility techniques or give up on stationarity and earn reversibility. (A similar situation occurs for Sinai's RWRE on $\mathbb{Z}$ [117].) It is the latter option that has (so far) led to results.

The prime benefit of reversibility is that it makes the Markov chain amenable to analysis via the methods of *electrostatic theory*. This is a standard technique; cf Doyle and Snell [61] or Lyons and Peres [90]. We thus interpret the underlying graph as an electric network with resistance $r(x, y) := 1/c(x, y)$ assigned to edge (x, y). The key notion to consider is the *effective resistance* $R_{\text{eff}}(0, B(N)^{\text{c}})$ from 0 to $B(N)^{\text{c}}$. We will define this quantity precisely in the next section; for the knowledgeable reader we just recall that $R_{\text{eff}}(0, B(N)^{\text{c}})$ is the voltage difference one needs to put between 0 and $B(N)^{\text{c}}$ to induce unit (net) current through the network. Using sophisticated techniques, for the effective resistance we then get:

Theorem 13.5 (Effective resistance growth) *For each* $\beta > 0$,

$$\limsup_{N\to\infty} \frac{\log R_{\text{eff}}(0, B(N)^{\text{c}})}{(\log N)^{1/2}(\log\log N)^{1/2}} < \infty, \qquad \mathbb{P}\text{-a.s.} \tag{13.15}$$

and

$$\liminf_{N\to\infty} \frac{\log R_{\text{eff}}(0, B(N)^{\text{c}})}{(\log N)^{1/2}/(\log\log\log N)^{1/2}} > 0, \qquad \mathbb{P}\text{-a.s.} \tag{13.16}$$

Both conclusions of Theorem 13.5 may be condensed into one (albeit weaker) statement as

$$R_{\text{eff}}(0, B(N)^{\text{c}}) = \mathrm{e}^{(\log N)^{1/2+o(1)}}, \quad N \to \infty. \tag{13.17}$$

In particular, $R_{\text{eff}}(0, B(N)^{\text{c}}) \to \infty$ as $N \to \infty$, a.s. The standard criteria of recurrence and transience of Markov chains (to be discussed later) then yield:

Corollary 13.6 *For* $\mathbb{P}$*-a.e. realization of* h*, the Markov chain* X *is recurrent.*

With the help of Exercise 13.1 we get:

Exercise 13.7 *Show that the limits in (13.15)–(13.16) depend only on the gradients of the pinned DGFF and are thus constant almost surely.*

The remainder of these lectures will be spent on describing the techniques underpinning the above results. Several difficult proofs will only be sketched or outright omitted; the aim is to communicate the ideas rather then give a full fledged account that the reader may just as well get by reading the original paper.

13.3 A Crash Course on Electrostatic Theory

We begin by a review of the connection between Markov chains and electrostatic theory. Consider a finite, unoriented, connected graph $\mathfrak{G} = (\mathsf{V}, \mathsf{E})$ where both orientations of edge $e \in \mathsf{E}$ are identified as one. An assignment of *resistance* $r_e \in (0, \infty)$ to each edge $e \in \mathsf{E}$ then makes $\mathfrak{G}$ an electric network. An alternative description uses *conductances* $\{c_e : e \in \mathsf{E}\}$ where

$$c_e := \frac{1}{r_e}. \tag{13.18}$$

We will exchangeably write $r(x, y)$ for r_e when $e = (x, y)$, and similarly for $c(x, y)$. Note that these are symmetric quantities, $r(x, y) = r(y, x)$ and $c(x, y) = c(y, x)$ whenever $(x, y) = (y, x) \in \mathsf{E}$.

Next we define some key notions of the theory. For any two distinct $u, v \in \mathsf{V}$, let

$$\mathcal{F}(u, v) := \bigl\{f \text{ function } \mathsf{V} \to \mathbb{R} \colon\ f(u) = 1,\ f(v) = 0\bigr\}. \tag{13.19}$$

We interpret $f(x)$ as an assignment of a *potential* to vertex $x \in \mathsf{V}$; each $f \in \mathcal{F}(u, v)$ then has unit potential difference (a.k.a. *voltage*) between u and v. For any potential $f : \mathsf{V} \to \mathbb{R}$, define its Dirichlet energy by

$$\mathcal{E}(f) := \sum_{(x,y)\in\mathsf{E}} c(x, y)\bigl[f(y) - f(x)\bigr]^2, \tag{13.20}$$

where, by our convention, each edge contributes only once.

Definition 13.8 *(Effective conductance) The infimum*

$$C_{\mathrm{eff}}(u, v) := \inf\bigl\{\mathcal{E}(f) \colon\ f \in \mathcal{F}(u, v)\bigr\} \tag{13.21}$$

is the effective conductance from u to v.

Note that $C_{\mathrm{eff}}(u, v) > 0$ since $\mathfrak{G}$ is assumed connected and finite and the conductances are assumed to be strictly positive.

Next we define the notion of (electric) current as follows:

Definition 13.9 *(Current) Let* E *denote the set of oriented edges in* $\mathfrak{G}$*, with both orientations included. A current from u to v is an assignment* $i(e)$ *of a real number to each* $e \in \mathsf{E}$ *such that, writing* $i(x, y)$ *for* $i(e)$ *with* $e = (x, y)$*,*

$$i(x, y) = -i(y, x), \quad (x, y) \in \mathsf{E} \tag{13.22}$$

and

$$\sum_{y \colon (x,y)\in\mathsf{E}} i(x, y) = 0, \quad x \in \mathsf{V} \smallsetminus \{u, v\}. \tag{13.23}$$

The first condition reflects on the fact that the current flowing along (x, y) is the opposite of the current flowing along (y, x). The second condition then forces that the current be conserved at all vertices except u and v. Next we observe:

Lemma 13.10 (Value of current) *For each current i from u to v,*

$$\sum_{x\colon (u,x)\in \mathsf{E}} i(u, x) = \sum_{x\colon (x,v)\in \mathsf{E}} i(x, v)\,. \tag{13.24}$$

Proof. Conditions (13.22)–(13.23) imply

$$\begin{aligned} 0 = \sum_{(x,y)\in \mathsf{E}} i(x, y) &= \sum_{x\in \mathsf{V}} \sum_{y\colon (x,y)\in \mathsf{E}} i(x, y) \\ &= \sum_{y\colon (u,y)\in \mathsf{E}} i(u, y) + \sum_{y\colon (v,y)\in \mathsf{E}} i(v, y). \end{aligned} \tag{13.25}$$

Employing (13.22) one more time, we then get (13.24). □

A natural interpretation of (13.24) is that the current incoming to the network at u equals the outgoing current at v. (Note that this may be false in infinite networks.) We call the common value in (13.24) the *value of current* i, with the notation $\text{val}(i)$. It is natural to single out the currents with unit value into

$$\mathcal{I}(u, v) := \big\{i\colon \text{ current from } u \text{ to } v \text{ with } \text{val}(i) = 1\big\}\,. \tag{13.26}$$

For each current i, its Dirichlet energy is then given by

$$\widetilde{\mathcal{E}}(i) := \sum_{e\in \mathsf{E}} r_e i(e)^2\,, \tag{13.27}$$

where we again note that each edge contributes only one term to the sum.

Definition 13.11 *(Effective resistance) The infimum*

$$R_{\text{eff}}(u, v) := \inf\big\{\widetilde{\mathcal{E}}(i)\colon i \in \mathcal{I}(u, v)\big\} \tag{13.28}$$

is the effective resistance from u to v.

Note that $\mathcal{I}(u, v) \neq \emptyset$ and thus also $R_{\text{eff}}(u, v) < \infty$ thanks to the assumed finiteness and connectivity of the underlying network $\mathfrak{G}$.

It is quite clear that the effective resistance and effective conductance must somehow be closely related. For instance, by (13.18) they are the reciprocals of each other in the network with two vertices and one edge. To address this connection in general networks, we first observe:

Lemma 13.12 *For any two distinct $u, v \in \mathsf{V}$,*

$$\mathcal{E}(f)\widetilde{\mathcal{E}}(i) \geq 1, \quad f \in \mathcal{F}(u,v),\ i \in \mathcal{I}(u,v). \tag{13.29}$$

In particular, $R_{\text{eff}}(u,v)C_{\text{eff}}(u,v) \geq 1$.

Proof. Let $f \in \mathcal{F}(u,v)$ and $i \in \mathcal{I}(u,v)$. By a symmetrization argument and the definition of unit current,

$$\begin{aligned}\sum_{(x,y)\in\mathsf{E}} i(x,y)\bigl[f(x)-f(y)\bigr] &= \frac{1}{2}\sum_{x\in\mathsf{V}}\ \sum_{y\colon (x,y)\in\mathsf{E}} i(x,y)\bigl[f(x)-f(y)\bigr] \\ &= \sum_{x\in\mathsf{V}} f(x) \sum_{y\colon (x,y)\in\mathsf{E}} i(x,y) = f(u)-f(v) = 1.\end{aligned} \tag{13.30}$$

On the other hand, (13.18) and the Cauchy–Schwarz inequality yield

$$\begin{aligned}&\sum_{(x,y)\in\mathsf{E}} i(x,y)\bigl[f(x)-f(y)\bigr] \\ &\quad= \sum_{(x,y)\in\mathsf{E}} \sqrt{r(x,y)}\, i(x,y)\sqrt{c(x,y)}\,\bigl[f(x)-f(y)\bigr] \leq \widetilde{\mathcal{E}}(i)^{1/2}\mathcal{E}(f)^{1/2}.\end{aligned} \tag{13.31}$$

This gives (13.29). The second part follows by optimizing over f and i. □

We now claim:

Theorem 13.13 (Electrostatic duality) *For any distinct* $u, v \in \mathsf{V}$,

$$C_{\text{eff}}(u,v) = \frac{1}{R_{\text{eff}}(u,v)}. \tag{13.32}$$

Proof. Since $\mathcal{I}(u,v)$ can be identified with a closed convex subset of $\mathbb{R}^{\mathsf{E}}$ and $i \mapsto \widetilde{\mathcal{E}}(i)$ with a strictly convex function on $\mathbb{R}^{\mathsf{E}}$ that has compact level sets, there is a unique minimizer $i_\star$ of (13.28). We claim that $i_\star$ obeys the *Kirchhoff cycle law*: For each $n \geq 1$ and each $x_0, x_1, \dots, x_n = x_0 \in \mathsf{V}$ with $(x_i, x_{i+1}) \in \mathsf{E}$ for each $i = 0, \dots, n-1$,

$$\sum_{k=1}^{n} r(x_k, x_{k+1}) i_\star(x_k, x_{k+1}) = 0. \tag{13.33}$$

To show this, let j be defined by $j(x_k, x_{k+1}) = -j(x_{k+1}, x_k) := 1$ for $k = 1, \dots, n$ and $j(x,y) := 0$ on all edges not belonging to the cycle $(x_0, \dots, x_n)$. Then $i_\star + aj \in \mathcal{I}(u,v)$ for any $a \in \mathbb{R}$ and so, since $i_\star$ is the minimizer,

$$\widetilde{\mathcal{E}}(i_\star + aj) = \widetilde{\mathcal{E}}(i_\star) + a\sum_{k=1}^{n} r(x_k, x_{k+1}) i_\star(x_k, x_{k+1}) + a^2\widetilde{\mathcal{E}}(j) \geq \widetilde{\mathcal{E}}(i_\star). \tag{13.34}$$

Taking $a \downarrow 0$ then shows "$\geq$" in (13.33) and taking $a \uparrow 0$ then proves equality.

The fact that $e \mapsto r_e i_\star(e)$ obeys (13.33) implies that it is a gradient of a function. Specifically, we claim that there is $f \colon \mathsf{V} \to \mathbb{R}$ such that $f(v) = 0$ and

$$f(y) - f(x) = r(x, y) i_\star(x, y), \quad (x, y) \in \mathsf{E}. \tag{13.35}$$

To see this, consider any path $x_0 = v, x_1, \dots, x_n = x$ with $(x_k, x_{k+1}) \in \mathsf{E}$ for each $k = 0, \dots, n-1$ and let $f(x)$ be the sum of $-r_e i_\star(e)$ for edges along this path. The condition (13.33) then ensures that the value of $f(x)$ is independent of the path chosen. Hence we get also (13.35).

Our next task is to compute $f(u)$. Here we note that (13.35) identifies $\widetilde{\mathcal{E}}(i_\star)$ with the quantity on the left of (13.30) and so

$$R_{\text{eff}}(u, v) = \widetilde{\mathcal{E}}(i_\star) = f(u) - f(v) = f(u). \tag{13.36}$$

The function $\tilde f(x) := f(x)/R_{\text{eff}}(u, v)$ thus belongs to $\mathcal{F}(u, v)$ and since, as is directly checked, $\mathcal{E}(f) = \widetilde{\mathcal{E}}(i_\star) = R_{\text{eff}}(u, v)$, we get

$$C_{\text{eff}}(u, v) \le \mathcal{E}(\tilde f) = \frac{1}{R_{\text{eff}}(u, v)^2}\mathcal{E}(f) = \frac{1}{R_{\text{eff}}(u, v)}. \tag{13.37}$$

This gives $C_{\text{eff}}(u, v) R_{\text{eff}}(u, v) \le 1$, complementing the inequality from Lemma 13.12. The claim follows. □

The above proof is based on optimizing over currents although one can also start from the minimizing potential. That this is more or less equivalent is attested by:

Exercise 13.14 *(Ohm's law) Prove that if $f^\star$ is a minimizer of $f \mapsto \mathcal{E}(f)$ over $f \in \mathcal{F}(u, v)$, then*

$$i(x, y) := c(x, y)\big[f^\star(y) - f^\star(x)\big], \quad (x, y) \in \mathsf{E}, \tag{13.38}$$

defines a current from u to v with $val(i) = C_{\text{eff}}(u, v)$. (Compare (13.38) *with* (13.35).*)*

There is a natural extension of the effective resistance/conductance from pairs of vertices to pairs of sets. Indeed, for any pair of disjoint sets $A, B \subseteq \mathsf{V}$, we define $R_{\text{eff}}(A, B)$ to be the effective resistance $R_{\text{eff}}(\langle A\rangle, \langle B\rangle)$ in the network where all edges between the vertices in A as well as those between the vertices in B have been dropped and the vertices in A then merged into a single vertex $\langle A\rangle$ and those in B merged into a vertex $\langle B\rangle$. (The outgoing edges from A then emanate from $\langle A\rangle$.) Similarly, we may define $C_{\text{eff}}(A, B)$ as $C_{\text{eff}}(\langle A\rangle, \langle B\rangle)$ or directly by

$$C_{\text{eff}}(A, B) := \inf\Big\{\mathcal{E}(f) \colon f|_A = 1,\ f|_B = 0\Big\}. \tag{13.39}$$

Note that we simply set the potential to constants on A and B. In the engineering vernacular, this amounts to *shorting* the vertices in A and in B; see Fig. 24 for an illustration. The electrostatic duality still applies and so we have

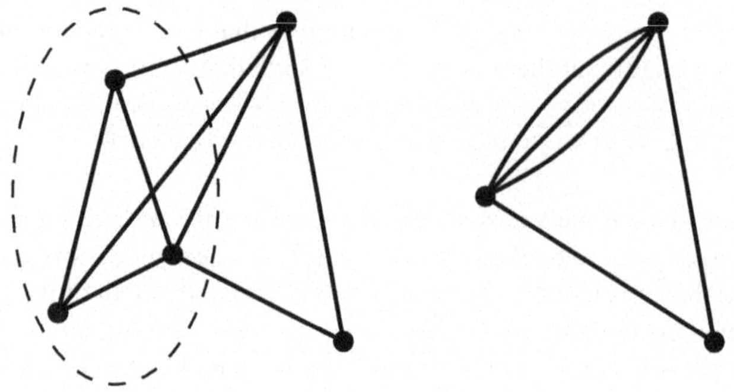

Fig. 24 An example of shorting three vertices in a network (the one on the left) and collapsing them to one thus producing a new, effective network (the one on the right). All the outgoing edges are kept along with their original resistances

$$C_{\text{eff}}(A, B) = \frac{1}{R_{\text{eff}}(A, B)} \tag{13.40}$$

whenever $A, B \subseteq \mathsf{V}$ with $A \cap B = \emptyset$.

13.4 Markov Chain Connections and Network Reduction

With each electric network one can naturally associate a Markov chain on V with transition probabilities

$$\mathsf{P}(x, y) := \frac{c(x, y)}{\pi(x)} 1_{(x,y)\in\mathsf{E}} \quad \text{where} \quad \pi(x) := \sum_{y\colon (x,y)\in\mathsf{E}} c(x, y). \tag{13.41}$$

The symmetry condition $c(x, y) = c(y, x)$ then translates into

$$\pi(x)\mathsf{P}(x, y) = \pi(y)\mathsf{P}(y, x) \tag{13.42}$$

thus making π a reversible measure. Since G is connected, and the conductances are strictly positive, the Markov chain is also irreducible. Writing P^x for the law of the Markov chain started at x, we then have:

Proposition 13.15 (Connection to Markov chain) *The variational problem* (13.21) *has a unique minimizer f which is given by*

$$f(x) = P^x(\tau_u < \tau_v) \tag{13.43}$$

where $\tau_z := \inf\{n \geq 0\colon X_n = z\}$.

Proof. Define an operator $\mathcal{L}$ on $\ell^2(\mathsf{V})$ by

$$\mathcal{L}f(x) := \sum_{y\colon (x,y)\in \mathsf{E}} c(x,y)\big[f(y) - f(x)\big]. \tag{13.44}$$

(This is an analogue of the discrete Laplacian we encountered earlier in these lectures.) As is easy to check by differentiation, the minimizer of (13.21) obeys $\mathcal{L}f(x) = 0$ for all $x \neq u, v$, with boundary values $f(y) = 1$ and $f(v) = 0$. The function $x \mapsto P^x(\tau_u < \tau_v)$ obeys exactly the same set of conditions; the claim thus follows from the next exercise. □

Exercise 13.16 *(Dirichlet problem and Maximum Principle) Let $\mathsf{U} \subsetneq \mathsf{V}$ and suppose that $f\colon \mathsf{V} \to \mathbb{R}$ obeys $\mathcal{L}f(x) = 0$ for all $x \in \mathsf{U}$, where $\mathcal{L}$ is as in* (13.44). *Prove that*

$$f(x) = E^x\big(f(X_{\tau_{\mathsf{V}\smallsetminus\mathsf{U}}})\big), \quad x \in \mathsf{U}, \tag{13.45}$$

where $\tau_{\mathsf{V}\smallsetminus\mathsf{U}} := \inf\{n \geq 0\colon X_n \notin \mathsf{U}\}$. In particular, we have

$$\max_{x\in\mathsf{U}} \big|f(x)\big| \leq \max_{x\in\mathsf{V}\smallsetminus\mathsf{U}} \big|f(y)\big| \tag{13.46}$$

and f is thus uniquely determined by its values on $\mathsf{V} \smallsetminus \mathsf{U}$.

We note that the above conclusion generally fails when U is allowed to be infinite. Returning to our main line of thought, from Proposition 13.15 we immediately get:

Corollary 13.17 *Denoting $\hat\tau_x := \inf\{n \geq 1\colon X_n = x\}$, for each $u \neq v$ we have*

$$\frac{1}{R_{\text{eff}}(u,v)} = \pi(u) P^u(\hat\tau_u > \tau_v). \tag{13.47}$$

Proof. Let f be the minimizer of (13.21). In light of $\mathcal{L}f(x) = 0$ for all $x \neq u, v$, symmetrization arguments and $f \in \mathcal{F}(u,v)$ show

$$\mathcal{E}(f) = -\sum_{x\in\mathsf{V}} f(x)\mathcal{L}f(x) = -f(u)\mathcal{L}f(u) - f(v)\mathcal{L}f(v) = -\mathcal{L}f(u). \tag{13.48}$$

The representation (13.43) yields

$$\begin{aligned}-\mathcal{L}f(u) &= \pi(u) - \sum_{x\colon (u,x)\in\mathsf{E}} c(u,x) P^x(\tau_x < \tau_v) \\ &= \pi(u)\left[1 - \sum_{x\colon (u,x)\in\mathsf{E}} \mathsf{P}(u,x) P^x(\tau_x < \tau_v)\right] = \pi(u) P^u(\hat\tau_u > \tau_v),\end{aligned} \tag{13.49}$$

where we used the Markov property and the fact that $u \neq v$ implies $\hat\tau_u \neq \tau_v$. The claim follows from the Electrostatic Duality. □

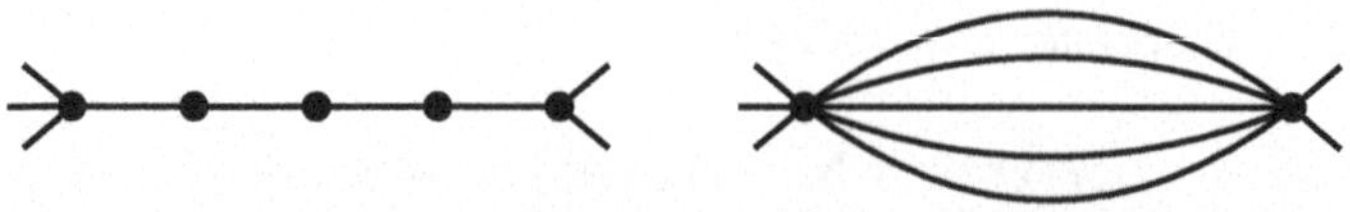

Fig. 25 Examples of the situations handled by (left) the series law from Exercise 13.21 and (right) the parallel law from Exercise 13.22

The representation (13.47) leads to a criterion for recurrence/transience of a Markov chain X on an infinite, locally finite, connected electric network with positive resistances. Let $B(x,r)$ denote the ball in the graph-theoretical metric of radius r centered at x. (The local finiteness ensures that $B(x,r)$, as well as the set of all edges emanating from it, are finite.) First we note:

Exercise 13.18 *Write $C_{\text{eff}}(x, B(x,r)^{\text{c}})$, resp., $R_{\text{eff}}(x, B(x,r)^{\text{c}})$ for the effective conductance, resp., effective resistance in the network where $B(x,r)^{\text{c}}$ has been collapsed to a single vertex. Prove, by employing a shorting argument, that $r \mapsto C_{\text{eff}}(x, B(x,r)^{\text{c}})$ is non-increasing.*

This and the Electrostatic Duality ensure that

$$R_{\text{eff}}(x,\infty) := \lim_{r\to\infty} R_{\text{eff}}\big(x, B(x,r)^{\text{c}}\big) \tag{13.50}$$

is well defined, albeit possibly infinite. The quantity $R_{\text{eff}}(x,\infty)$, which we call effective resistance from x to infinity, depends on x, although (by irreducibility) if $R_{\text{eff}}(x,\infty)$ diverges for one x, then it diverges for all x. We now have:

Corollary 13.19 (Characterization of recurrence/transience)

$$X \text{ is recurrent} \quad \Leftrightarrow \quad R_{\text{eff}}(\cdot,\infty) = \infty. \tag{13.51}$$

Proof. By Corollary 13.17, $P^x(\hat\tau_x > \tau_{B(x,r)^{\text{c}}})$ is proportional to $C_{\text{eff}}(u, B(x,r)^{\text{c}})$. Since $\tau_{B(x,r)^{\text{c}}} \ge r$, $P^x(\hat\tau_x = \infty)$ is proportional to $R_{\text{eff}}(x,\infty)^{-1}$. □

The advantage of casting properties of Markov chains in electrostatic language is that we can manipulate networks using operations that do not always have a natural counterpart, or type of underlying monotonicity, in the context of Markov chains. We will refer to these operations using the (somewhat vague) term *network reduction*. Shorting part of the network serves as an example. Another example is (Fig. 25):

Exercise 13.20 *(Restriction to a subnetwork) Let $\mathsf{V}' \subseteq \mathsf{V}$ and, for any function $f\colon \mathsf{V}' \to \mathbb{R}$, set*

$$\mathcal{E}'(f) := \inf\big\{\mathcal{E}(g)\colon g(x) = f(x)\ \forall x \in \mathsf{V}'\big\}. \tag{13.52}$$

Prove that $\mathcal{E}'(f)$ is still a Dirichlet energy of the form

$$\mathcal{E}'(f) = \frac{1}{2}\sum_{x,y\in\mathsf{V}'} c'(x,y)\big[f(y)-f(x)\big]^2, \tag{13.53}$$

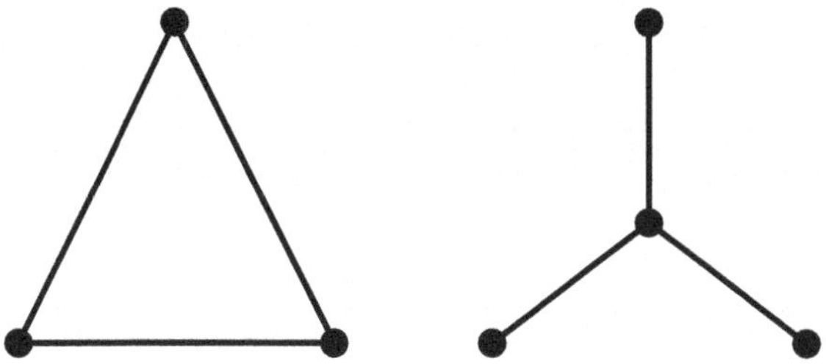

Fig. 26 The "triangle" and "star" networks from Exercise 13.23

where

$$c'(x, y) := \pi(x) P^x \big(X_{\hat{\tau}_{\mathsf{V}'}} = y\big) \tag{13.54}$$

with $\hat{\tau}_A := \inf\{n \geq 1 \colon X_n \in A\}$.

We remark that the restriction to a subnetwork does have a counterpart for the underlying Markov chain; indeed, it corresponds to observing the chain only at times when it is at V'. (This gives rise to the formula (13.54).) The simplest instance is when V' has only two vertices. Then, for any $u \neq v$, we have

$$\mathsf{V}' = \{u, v\} \quad \Rightarrow \quad c'(u, v) = C_{\text{eff}}(u, v). \tag{13.55}$$

A another relevant example is the content of:

Exercise 13.21 *(Series law) Suppose* $\mathfrak{G}$ *contains vertices* $x_0, \dots, x_n$ *such that* $(x_{i-1}, x_i) \in \mathsf{E}$ *for each* $i = 1, \dots, n$ *and such that, for* $i = 1, \dots, n-1$, *the vertex* x_i *has no other neighbors than* x_{i-1} *and* x_{i+1}. *Prove that in the reduced network with* $\mathsf{V}' := \mathsf{V} \smallsetminus \{x_1, \dots, x_{n-1}\}$ *the string is replaced by an edge* (x_0, x_n) *with resistance*

$$r'(x_0, x_n) := \sum_{i=1}^n r(x_{i-1}, x_i)\,. \tag{13.56}$$

There are other operations that produce equivalent networks which are not of the type discussed in Exercise 13.20:

Exercise 13.22 *(Parallel law) Suppose* $\mathfrak{G}$ *contains* n *edges* $e_1, \dots, e_n$ *between vertices* x *and* y *with* e_i *having conductance* $c(e_i)$. *Prove that we can replace these by a single edge* e *with conductance*

$$c'(e) := \sum_{i=1}^n c(e_i). \tag{13.57}$$

Here is another operation that does fall under the scheme of Exercise 13.20:

Exercise 13.23 *(Star-triangle transformation) Consider an electric network that contains (among others) three nodes* $\{1, 2, 3\}$ *and an edge between every pair of these nodes. Write* c_{ij} *for the conductance of edge* (i, j)*. Prove that an equivalent network is produced by replacing the "triangle" on these vertices by a "star" which consists of four nodes* $\{0, 1, 2, 3\}$ *and three edges* $\{(0, i): i = 1, 2, 3\}$*, and no edges between the remaining vertices. See Fig. 26. [We call this replacement the "star-triangle transformation."]*

The start-triangle transformation is a very powerful tool in network theory. Same tools as in the previous exercise can be used to conclude:

Exercise 13.24 *For the network with vertices* $\mathsf{V} := \{1, 2, 3\}$ *and edges* $\mathsf{E} := \{(i, j): 1 \le i < j \le 3\}$*, let* c_{ij} *denote the conductance of edge* (i, j)*. Denoting* $R_{ij} := R_{\text{eff}}(i, j)$*, prove that*

$$\frac{c_{12}}{c_{12} + c_{13}} = \frac{R_{13} + R_{23} - R_{12}}{2R_{23}}. \tag{13.58}$$

The network reduction ideas will continue to be used heavily through the remaining lectures.

Lecture 14: Effective Resistance Control

The aim of this lecture is to develop further methods to control effective resistance/conductance in networks related to the DGFF as discussed above. We start by proving geometric representations which allow us to frame estimates of these quantities in the language of percolation theory. We then return to the model at hand and employ duality considerations to control the effective resistance across squares in $\mathbb{Z}^2$. A Russo–Seymour–Welsh (type of) theory permits similar control for the effective resistance across rectangles. Finally, we use these to control the upper tail of the effective resistance between far-away points.

14.1 Path-Cut Representations

The network reduction ideas mentioned at the end of the previous lecture naturally lead to representations on the effective resistance/conductance by quantities involving geometric objects such as paths and cuts. As these representations are the cornerstone of our approach, we start by explaining them in detail.

A path P from u to v is a sequence of edges $e_1, \dots, e_n$, which we think of as oriented for this purpose, with e_1 having the initial point at u and e_n the terminal endpoint at v and with the initial point of e_{i+1} equal to terminal point of e_i for each $i = 1, \dots, n-1$. We will often identify P with the set of these edges; the notation $e \in P$

thus means that the path P crosses the unoriented edge e. The following lemma is classical:

Lemma 14.1 *Suppose $\mathcal{P}$ is a finite set of edge disjoint paths—i.e., those for which $P \neq P'$ implies $P \cap P' = \emptyset$—from u to v. Then*

$$R_{\text{eff}}(u,v) \le \left[\sum_{P\in\mathcal{P}} \frac{1}{\sum_{e\in P} r_e}\right]^{-1}. \tag{14.1}$$

Proof Dropping loops from paths decreases the quantity on the right of (14.1), so we may and will assume that each path in $\mathcal{P}$ visits each edge at most once. The idea is to route current along the paths in $\mathcal{P}$ to arrange for a unit net current to flow from u to v. Let $\overline{R}$ denote the quantity on the right of (14.1) and, for each $P \in \mathcal{P}$, set $i_P := \overline{R}/\sum_{e\in P} r_e$. Define $i(e) := i_P$ if e lies in, and is oriented in the same direction as P (recall that the paths are edge disjoint and each edge is visited by each path at most once) and $i(e) := 0$ if e belongs to none of the paths in $\mathcal{P}$. From $\sum_{P\in\mathcal{P}} i_P = 1$ we then infer $i \in \mathcal{I}(u,v)$. A calculation shows $\widetilde{\mathcal{E}}(i) = \overline{R}$ and so $R_{\text{eff}}(u,v) \le \overline{R}$. □

A natural question to ask is whether the upper bound (14.1) can possibly be sharp. As it turns out, all that stands in the way of this is the edge-disjointness requirement. This is overcome in:

Proposition 14.2 (Path representation of effective resistance) *Let $\mathfrak{P}_{u,v}$ denote the set of finite collections of paths from u to v. Then*

$$R_{\text{eff}}(u,v) = \inf_{\mathcal{P}\in\mathfrak{P}_{u,v}} \; \inf_{\{r_{e,P}\colon P\in\mathcal{P},\, e\in P\}\in\mathfrak{R}_{\mathcal{P}}} \left[\sum_{P\in\mathcal{P}} \frac{1}{\sum_{e\in P} r_{e,P}}\right]^{-1}, \tag{14.2}$$

where $\mathfrak{R}_{\mathcal{P}}$ is the set of all collections of positive numbers $\{r_{e,P}\colon P\in\mathcal{P},\ e\in P\}$ such that

$$\sum_{P\in\mathcal{P}} \frac{1}{r_{e,P}} \le \frac{1}{r_e}, \quad e \in \mathsf{E}. \tag{14.3}$$

The infima in (14.2) *are (jointly) achieved.*

Proof. Pick a collection of paths $\mathcal{P} \in \mathfrak{P}_{u,v}$ and positive numbers $\{r_{e,P}\colon P \in \mathcal{P}\}$ satisfying (14.3). Then split each edge e into a collection of edges $\{e_P\colon P \in \mathcal{P}\}$ and assign resistance $r_{e,P}$ to e_P. If the inequality in (14.3) is strict, introduce also a dummy copy $\tilde{e}$ of e and assign conductance $c_{\tilde{e}} := 1/r_e - \sum_{P\in\mathcal{P}} 1/r_{e,P}$ to $\tilde{e}$. The Parallel Law shows that this operation produces an equivalent network in which, by way of interpretation, the paths in $\mathcal{P}$ are mutually edge disjoint. Lemma 14.1 then gives "$\le$" in (14.2).

To get equality in (14.2), let $i_\star \in \mathcal{I}(u,v)$ be such that $\widetilde{\mathcal{E}}(i_\star) = R_{\text{eff}}(u,v)$. We will now recursively define a sequence of currents i_k (not necessarily of unit value) and paths P_k from u and v. First solve:

Exercise 14.3 *Suppose* $e \mapsto i(e)$ *is a current from* u *to* v *with* $val(i) > 0$. *Show that there is a simple path* P *from* u *to* v *such that* $i(e) > 0$ *for each* $e \in P$ *(which we think of as oriented in the direction of* P*).*

We cast the recursive definition as an algorithm: INITIATE by setting $i_0 := i_\star$. Assuming that i_{k-1} has been defined for some $k \geq 1$, if $\text{val}(i_{k-1}) = 0$ then STOP, else use Exercise 14.3 to find a path P_k from u to v where $i_{k-1}(e) > 0$ for each $e \in P_k$ oriented along the path. Then set $\alpha_k := \min_{e \in P_k} |i_{k-1}(e)|$, let

$$i_k(e) := i_{k-1}(e) - \alpha_k \,\text{sgn}(i_{k-1}(e)) 1_{\{e \in P_k\}} \tag{14.4}$$

and, noting that i_k is a current from u to v with $\text{val}(i) \geq 0$, REPEAT.

The construction ensures that $\{e \in E : i_k(e) \neq 0\}$ is strictly decreasing in k and so the algorithm will terminate after a finite number of steps. Also $k \mapsto |i_k(e)|$ and $k \mapsto \text{val}(i_k)$ are non-increasing and, as is checked from (14.4),

$$\forall e \in \mathsf{E}: \quad \sum_{k\colon e \in P_k} \alpha_k \leq |i_\star(e)| \;\text{ and }\; \sum_k \alpha_k = \text{val}(i_\star) = 1. \tag{14.5}$$

We now set $r_{e,P_k} := |i_\star(e)| r_e / \alpha_k$ and note that (14.5) shows $\sum_k 1/r_{e,P_k} \leq 1/r_e$ for each $e \in \mathsf{E}$. Moreover, from (14.5) we also get

$$\begin{aligned} R_{\text{eff}}(u,v) &= \sum_{e \in \mathsf{E}} r_e i_\star(e)^2 \geq \sum_{e \in \mathsf{E}} r_e |i_\star(e)| \sum_{k\colon e \in P_k} \alpha_k \\ &= \sum_{e \in \mathsf{E}} \sum_{k\colon e \in P_k} r_{e,P_k} \alpha_k^2 = \sum_k \alpha_k^2 \Bigg(\sum_{e \in P_k} r_{e,P_k} \Bigg). \end{aligned} \tag{14.6}$$

Denoting the quantity in the large parentheses by R_k, among all positive α_k's satisfying (14.5), the right-hand side is minimized by $\alpha_k := \frac{1/R_k}{\sum_j 1/R_j}$. Plugging this in, we get "$\geq$" in (14.2) and thus the whole claim. □

As it turns out, the effective conductance admits an analogous geometric variational characterization as well. Here one needs the notion of a *cut*, or a *cutset* form u to v which is a set of edges in E such that every path from u to v must contain at least one edge in this set. We again start with a classical bound:

Lemma 14.4 (Nash-Williams estimate) *For any collection* Π *of edge-disjoint cutsets from* u *to* v,

$$C_{\text{eff}}(u,v) \leq \Bigg[\sum_{\pi \in \Pi} \frac{1}{\sum_{e \in \pi} c_e} \Bigg]^{-1}. \tag{14.7}$$

Proof. Let $i \in \mathcal{I}(u,v)$. The proof is based on:

Exercise 14.5 *For any cutset* π *from* u *to* v, $\sum_{e \in \pi} |i(e)| \geq 1$.

Indeed, the Cauchy–Schwarz inequality and (13.18) tell us

$$1 \le \left[\sum_{e\in\pi} |i(e)|\right]^2 \le \left[\sum_{e\in\pi} r_e i(e)^2\right]\left[\sum_{e\in\pi} c_e\right]. \tag{14.8}$$

The assumed edge-disjointness of the cutsets in Π then yields

$$\widetilde{\mathcal{E}}(i) \ge \sum_{\pi\in\Pi}\sum_{e\in\pi} r_e i(e)^2 \ge \sum_{\pi\in\Pi} \frac{1}{\sum_{e\in\pi} c_e}. \tag{14.9}$$

The claim follows by the Electrostatic Duality. □

Remark 14.6 Lemma 14.4 is easy to prove by network reduction arguments when the cutsets are *nested* meaning that they can be ordered in a sequence $\pi_1, \dots, \pi_n$ such that π_i separates π_{i-1} (as well as u) from π_{i+1} (as well as v). However, as the above proof shows, this geometric restriction is not needed (and, in fact, would be inconvenient to carry around).

The bound (14.7) was first proved by Nash-Williams [102] as a tool for proving recurrence of an infinite network. Indeed, to get $C_{\text{eff}}(0,\infty) := 1/R_{\text{eff}}(0,\infty) = 0$, it suffices to present a disjoint family of cutsets whose reciprocal total conductances add up to infinity. However, as far as the actual computation of $C_{\text{eff}}(u, v)$ is concerned, (14.7) is generally not sharp; again, mostly due to the requirement of edge-disjointness. The following proposition provides the needed fix:

Proposition 14.7 (Cutset representation of effective conductance) *Let $\mathfrak{S}_{u,v}$ be the set of all finite collections of cutsets between u and v. Then*

$$C_{\text{eff}}(u, v) = \inf_{\Pi\in\mathfrak{S}_{u,v}} \ \inf_{\{c_{e,\pi}\colon \pi\in\Pi,\, e\in\pi\}\in\mathfrak{C}_\Pi} \left[\sum_{\pi\in\Pi} \frac{1}{\sum_{e\in\pi} c_{e,\pi}}\right]^{-1}, \tag{14.10}$$

where $\mathfrak{C}_\Pi$ is the set of all families of positive numbers $\{c_{e,\pi}\colon \pi \in \Pi,\ e \in \pi\}$ such that

$$\sum_{\pi\in\Pi} \frac{1}{c_{e,\pi}} \le \frac{1}{c_e}, \quad e \in \mathsf{E}. \tag{14.11}$$

The infima in (14.10) *are (jointly) achieved.*

Proof. Let Π be a family of cutsets from u to v and let $\{c_{e,\pi}\} \in \mathfrak{C}_\Pi$. We again produce an equivalent network as follows: Replace each edge e involved in these cutsets by a series of edges e_π, $\pi \in \Pi$. Assign resistance $1/c_{e,\pi}$ to e_π and, should the inequality in (14.11) be strict for e, add a dummy edge $\tilde{e}$ with resistance $r_{\tilde{e}} := 1/c_e - \sum_{\pi\in\Pi} 1/c_{e,\pi}$. The cutsets can then be deemed edge-disjoint; and so we get "≤" in (14.10) by Lemma 14.4.

To prove equality in (14.10), we consider the minimizer $f^\star$ of the variational problem defining $C_{\text{eff}}(u, v)$. Using the operator in (13.44), note that $\mathcal{L}f^\star(x) = 0$ for $x \ne u, v$. This is an important property in light of:

Exercise 14.8 *Let $f: \mathsf{V} \to \mathbb{R}$ be such that $f(u) > f(v)$ and abbreviate*

$$D := \{x \in \mathsf{V} \colon f(x) = f(u)\}. \tag{14.12}$$

Assume $\mathcal{L} f(x) = 0$ for $x \notin D \cup \{v\}$. Prove that $\pi := \partial D$ defines a cutset π from u to v such that $f(x) > f(y)$ holds for each edge $(x, y) \in \pi$ oriented so that $x \in D$ and $y \notin D$.

We will now define a sequence of functions $f_k \colon \mathsf{V} \to \mathbb{R}$ and cuts π_k by the following algorithm: INITIATE by $f_0 := f^\star$. If f_{k-1} is constant then STOP, else use Exercise 14.8 with D_{k-1} related to f_{k-1} as D is to f in (14.12) to define $\pi_k := \partial D_{k-1}$. Noting that $f_{k-1}(x) > f_{k-1}(y)$ for each edge $(x, y) \in \pi$ (oriented to point from u to v), set $\alpha_k := \min_{(x,y)\in\pi} |f_{k-1}(x) - f_{k-1}(y)|$ and let

$$f_k(z) := f_{k-1}(z) - \alpha_k 1_{D_{k-1}}(z). \tag{14.13}$$

Then REPEAT.

As is checked by induction, $k \mapsto D_k$ is strictly increasing and $k \mapsto f_k$ is non-increasing with $f_k = f^\star$ on $\mathsf{V} \smallsetminus D_k$. In particular, we have $\mathcal{L} f_k(x) = 0$ for all $x \notin D_k \cup \{v\}$ and so Exercise 14.8 can repeatedly be used. The premise of the strict inequality $f_k(u) > f_k(v)$ for all but the final step is the consequence of the Maximum Principle.

Now we perform some elementary calculations. Let Π denote the set of the cutsets π_k identified above. For each edge (x, y) and each $\pi_k \in \Pi$, define

$$c_{e,\pi_k} := \frac{|f^\star(y) - f^\star(x)|}{\alpha_k} c_e. \tag{14.14}$$

The construction and the fact that $f^\star(u) = 1$ and $f^\star(v) = 0$ imply

$$\sum_k \alpha_k = 1\,. \tag{14.15}$$

For any $e = (x, y)$, we also get

$$\sum_{k\colon e\in\pi_k} \alpha_k = \bigl|f^\star(y) - f^\star(x)\bigr|. \tag{14.16}$$

In particular, the collection $\{c_{e,\pi} \colon \pi \in \Pi,\ e \in \pi\}$ obeys (14.11). Moreover, (14.14) also shows, for any $e = (x, y)$,

$$\sum_{k\colon e\in\pi_k} c_{e,\pi_k}\alpha_k^2 = c_e\bigl|f^\star(y) - f^\star(x)\bigr| \sum_{k\colon e\in\pi_k} \alpha_k = c_e\bigl|f^\star(y) - f^\star(x)\bigr|^2. \tag{14.17}$$

Summing over $e \in \mathsf{E}$ and rearranging the sums yields

$$C_{\text{eff}}(u,v) = \mathcal{E}(f^\star) = \sum_{e\in \mathsf{E}} \sum_{k:\, e\in\pi_k} c_{e,\pi_k}\alpha_k^2 = \sum_k \alpha_k^2 \Biggl(\sum_{e:\, e\in\pi_k} c_{e,\pi_k}\Biggr). \tag{14.18}$$

Denoting the quantity in the large parentheses by C_k, among all $\alpha_k \ge 0$ satisfying (14.15), the right-hand side is minimal for $\alpha_k := \frac{1/C_k}{\sum_j 1/C_j}$. This shows "$\ge$" in (14.10) and finishes the proof. □

We note that the above derivations are closely related to characterizations of the effective resistance/conductance based on optimizing over *random* paths and cuts. These are rooted in T. Lyons' random-path method [89] for proving finiteness of the effective resistance. Refined versions of these characterizations can be found in Berman and Konsowa [21].

14.2 Duality and Effective Resistance Across Squares

Although the above derivations are of independent interest, our main reason for giving their full account is to demonstrate the *duality* ideas that underlie the derivations in these final lectures. This is based on the close similarity of the variational problems (14.2)–(14.3) and (14.10)–(14.11). Indeed, these are identical provided we somehow manage to

(1) swap paths for cuts, and
(2) exchange resistances and conductances.

Here (2) is achievable readily assuming symmetry of the underlying random field: For conductances (and resistances) related to a random field h via (13.13),

$$h \overset{\text{law}}{=} -h \quad\Rightarrow\quad \{c_e\colon e\in\mathsf{E}\} \overset{\text{law}}{=} \{r_e\colon e\in\mathsf{E}\}. \tag{14.19}$$

For (1) we can perhaps hope to rely on the fact that paths on *planar* graphs between the left and right sides of a rectangle are in one-to-one correspondence with cuts in the *dual* graph separating the top and bottom sides of that rectangle. Unfortunately, swapping primal and dual edges seems to void the dependence (13.13) on the underlying field which we need for (14.19). We will thus stay on the primal lattice and implement the path-cut duality underlying (1) only approximately.

Consider a rectangular box S of the form

$$B(M,N) := ([-M,M]\times[N,N])\cap\mathbb{Z}^2 \tag{14.20}$$

and let $\partial^{\text{left}}S$, $\partial^{\text{down}}S$, $\partial^{\text{right}}S$, $\partial^{\text{up}}S$ denote the sets of vertices in S that have a neighbor in S^{c} to the left, downward, the right and upward thereof, respectively. Regarding S, along with the edges with both endpoints in S, as an electric network with conductances depending on a field h as in (13.13), let

$$R_S^{\mathrm{LR}}(h) := R_{\mathrm{eff}}\bigl(\partial^{\mathrm{left}} S, \partial^{\mathrm{right}} S\bigr) \tag{14.21}$$

and

$$R_S^{\mathrm{UD}}(h) := R_{\mathrm{eff}}\bigl(\partial^{\mathrm{up}} S, \partial^{\mathrm{down}} S\bigr) \tag{14.22}$$

denote the left-to-right and up-down effective resistances across the box S, respectively. To keep our formulas short, we will also abbreviate

$$B(N) := B(N, N). \tag{14.23}$$

A key starting point of all derivations is the following estimate:

Proposition 14.9 (Effective resistance across squares) *There is $\hat{c} > 0$ and, for each $\epsilon > 0$, there is $N_0 \geq 1$ such that for all $N \geq N_0$ and all $M \geq 2N$,*

$$\mathbb{P}\Bigl(R_{B(N)}^{\mathrm{LR}}(h^{B(M)}) \leq \mathrm{e}^{\hat{c}\log\log M}\Bigr) \geq \frac{1}{2} - \epsilon, \tag{14.24}$$

where $h^{B(M)} :=$ DGFF in $B(M)$.

The proof will involve a sequence of lemmas. We begin by a general statement that combines usefully the *electrostatic* duality along with approximate *path-cut* duality. Consider a finite network $\mathsf{G} = (\mathsf{V}, \mathsf{E})$ with resistances $\{r_e : e \in \mathsf{E}\}$. A dual, or reciprocal, network is that on the same graph but with resistances $\{r_e^\star : e \in \mathsf{E}\}$ where

$$r_e^\star := \frac{1}{r_e}, \quad e \in \mathsf{E}. \tag{14.25}$$

We will denote to the dual network as $\mathsf{G}^\star$. We will write $R_{\mathrm{eff}}^\star$, resp., $C_{\mathrm{eff}}^\star$ to denote the effective resistances in $\mathsf{G}^\star$. We also say that edges e and e' are adjacent, with the notation $e \sim e'$, if they share exactly one endpoint. We then have:

Lemma 14.10 *Let (A, B) and (C, D) be pairs of non-empty disjoint subsets of V such that every path from A to B has a vertex in common with every path from C to D. Then*

$$R_{\mathrm{eff}}(A, B) R_{\mathrm{eff}}^\star(C, D) \geq \frac{1}{4\mathfrak{d}^2 \rho_{\max}}, \tag{14.26}$$

where $\mathfrak{d}$ is the maximum vertex degree in G and

$$\rho_{\max} := \max\bigl\{r_e/r_{e'} : e \sim e'\bigr\}. \tag{14.27}$$

Proof. The proof relies on the fact that every path P between C and D defines a cutset π_P between A and B by taking π_P to be the set of all edges adjacent to the edges in P, but excluding the edges in $(A \times A) \cup (B \times B)$. In light of the Electrostatic Duality, we need to show

$$C_{\text{eff}}(A,B) \le 4\mathfrak{d}^2\rho_{\max} R^\star_{\text{eff}}(C,D)\,. \tag{14.28}$$

Aiming to use the variational characterizations (14.2)–(14.3) and (14.10)–(14.11), fix a collection of paths $\mathcal{P} \in \mathfrak{P}_{C,D}$ and positive numbers $\{r'_{e,P}\colon P \in \mathcal{P}, e \in P\}$ such that

$$\sum_{P\in\mathcal{P}} \frac{1}{r'_{e,P}} \le \frac{1}{r^\star_e} = \frac{1}{c_e}, \quad e \in \mathsf{E}, \tag{14.29}$$

where the equality is a rewrite of (14.25). For each $e \in \mathsf{E}$ and each $P \in \mathcal{P}$, consider the cut π_P defined above and let

$$c_{e,\pi_P} := 2\mathfrak{d}\rho_{\max}\Biggl(\sum_{e'\in P\colon e'\sim e} \frac{1}{r'_{e',P}}\Biggr)^{-1}. \tag{14.30}$$

Then, for each $e \in \mathsf{E}$, (14.29) along with the definitions of $\rho_{\max}$ and $\mathfrak{d}$ yield

$$\begin{aligned}\sum_{P\in\mathcal{P}} \frac{1}{c_{e,\pi_P}} &= \frac{1}{2\mathfrak{d}\rho_{\max}} \sum_{P\in\mathcal{P}} \sum_{e'\in P\colon e'\sim e} \frac{1}{r'_{e',P}} \\ &\le \frac{1}{2\mathfrak{d}\rho_{\max}} \sum_{e'\sim e} \frac{1}{c_{e'}} \le \frac{1}{2\mathfrak{d}} \sum_{e'\sim e} \frac{1}{c_e} \le \frac{1}{c_e}\,.\end{aligned} \tag{14.31}$$

Using that $(\sum_i x_i)^{-1} \le \sum_i 1/x_i$ for any positive x_i's in turn shows

$$\sum_{e\in\pi_P} c_{e,\pi_P} \le 2\mathfrak{d}\rho_{\max} \sum_{e\in\pi_P} \sum_{e'\in P\colon e'\sim e} r'_{e',P} \le 4\mathfrak{d}^2\rho_{\max} \sum_{e'\in P} r'_{e',P}\,. \tag{14.32}$$

In light of (14.31), from (14.10) we thus have

$$C_{\text{eff}}(A,B) \le \Biggl(\sum_{P\in\mathcal{P}} \frac{1}{\sum_{e\in\pi_P} c_{e,\pi_P}}\Biggr)^{-1} \le 4\mathfrak{d}^2\rho_{\max}\Biggl(\sum_{P\in\mathcal{P}} \frac{1}{\sum_{e'\in P} r'_{e',P}}\Biggr)^{-1}. \tag{14.33}$$

This holds for all $\mathcal{P}$ and all positive $\{r'_{e,P}\colon P \in \mathcal{P}, e \in P\}$ subject to (14.29) and so (14.28) follows from (14.2). □

We also have the opposite inequality, albeit under somewhat different conditions on (A,B) and (C,D):

Lemma 14.11 *For $\mathfrak{d}$ and $\rho_{\max}$ as in Lemma 14.10, let (A,B) and (C,D) be pairs of non-empty disjoint subsets of V such that for every cutset π between C and D, the set of edges with one or both vertices in common with some edge in π contains a path from A to B. Then*

$$R_{\text{eff}}(A,B)R^\star_{\text{eff}}(C,D) \le 4\mathfrak{d}^2\rho_{\max}\,. \tag{14.34}$$

Proof (*sketch*). For any cutset π between C and D, the assumptions ensure the existence of a path P_π from u to v that consists of edges that have one or both endpoints in common with some edge in π. Pick a family of cutsets Π between C and D and positive numbers $\{c'_{e,\pi}: \pi \in \Pi,\ e \in P_\pi\}$ such that

$$\sum_{\pi\in\Pi} \frac{1}{c'_{e,\pi}} \le \frac{1}{c^\star_e} = \frac{1}{r_e}. \tag{14.35}$$

Following the exact same sequence of steps as in the proof of Lemma 14.10, the reader will readily construct $\{r_{e,P_\pi}: \pi \in \Pi, e \in P_\pi\}$ satisfying (14.3) such that

$$R_{\text{eff}}(A, B) \le \left(\sum_{\pi\in\Pi} \frac{1}{\sum_{e\in P_\pi} r_{e,P_\pi}}\right)^{-1} \le 4\mathfrak{d}^2 \rho_{\max} \left(\sum_{\pi\in\Pi} \frac{1}{\sum_{e\in\pi} c'_{e,\pi}}\right)^{-1}, \tag{14.36}$$

where the first inequality is by Proposition 14.2. As this holds for all Π and all positive $\{c'_{e,\pi}: \pi \in \Pi, e \in \pi\}$ satisfying (14.35), we get (14.34). □

The previous lemma shows that, in bounded degree planar graphs, primal and dual effective resistances can be compared as long as we can bound the ratio of the resistances on adjacent edges. Unfortunately, this would not work for resistances derived from the DGFF in $B(M)$ via (13.13); indeed, there $\rho_{\max}$ can be as large as a power of M due to the large *maximal* local roughness of the field. We will resolve this by decomposing the DGFF into a smooth part, where the associated $\rho_{\max}$ is sub-polynomial in M, and a rough part whose influence can be controlled directly. This is the content of:

Lemma 14.12 (Splitting DGFF into smooth and rough fields) *Recall that $h^{B(N)}$ denotes the DGFF in $B(N)$. There is $c > 0$ and, for each $N \ge 1$, there are Gaussian fields φ and χ on $B(N)$ such that*

(1) φ and χ are independent with $\varphi + \chi \overset{\text{law}}{=} h^{B(N)}$,
(2) $\text{Var}(\chi_x) \le c \log\log N$ for each $x \in B(N)$, and
(3) $\text{Var}(\varphi_x - \varphi_y) \le c/\log N$ for every adjacent pair $x, y \in B(N/2)$.

Moreover, the law of φ is invariant under the rotations of $B(N)$ by multiples of $\pi/2$.

Proof. Let $\{Y_n : n \ge 0\}$ denote the discrete time simple symmetric random walk on $\mathbb{Z}^2$ with holding probability 1/2 at each vertex and, given any finite $\Lambda \subseteq \mathbb{Z}^2$, let τ_{Λ^c} be the first exit time from Λ. Writing P^x for the law of the walk started at x and denoting

$$\mathsf{Q}(x, y) := P^x(Y_1 = y,\ \tau_{\Lambda^c} \ge 1), \tag{14.37}$$

we first ask reader to solve:

Exercise 14.13 *Prove the following:*

(1) the nth matrix power of Q obeys $\mathsf{Q}^n(x, y) = P^x(Y_n = y,\ \tau_{\Lambda^c} \ge n)$,

(2) *the matrices* $\{\mathsf{Q}^n(x,y)\colon x,y\in\Lambda\}$ *are symmetric and positive semi-definite,*
(3) *the Green function in* Λ *obeys*

$$G^{\Lambda}(x,y)=\sum_{n\ge 0}\frac{1}{2}\mathsf{Q}^n(x,y). \tag{14.38}$$

We now apply the above for $\Lambda:=B(N)$. Writing $C_1(x,y)$ for the part of the sum in (14.38) corresponding to $n\le\lfloor\log N\rfloor^2$ and $C_2(x,y)$ for the remaining part of the sum, Exercise 14.13(2) ensures that the kernels C_1 and C_2 are symmetric and positive semidefinite with $G^D=C_1+C_2$. The fields

$$\chi:=\mathcal{N}(0,C_1)\quad\text{and}\quad\varphi:=\mathcal{N}(0,C_2)\quad\text{with}\quad\chi\perp\!\!\!\perp\varphi \tag{14.39}$$

then realize (1) in the statement. To get (2), we just sum the standard heat-kernel bound $\mathsf{Q}^n(x,x)\le c/n$ (valid uniformly in $x\in B(N)$). For (3) we pick neighbors $x,y\in B(N/2)$ and use the Strong Markov Property to estimate

$$\begin{aligned}
&\Bigl|E\bigl[\varphi_x(\varphi_x-\varphi_y)\bigr]\Bigr|\\
&\quad\le\sum_{n>\lfloor\log N\rfloor^2}\Bigl|P^x(Y_n=y,\,\tau_{B(N)^c}>n)-P^x(Y_n=x,\,\tau_{B(N)^c}>n)\Bigr|\\
&\quad\le\sum_{n>\lfloor\log N\rfloor^2}\Bigl|P^x(Y_n=y)-P^x(Y_n=x)\Bigr|\\
&\qquad\quad+\sum_{n\ge1}E^x\Bigl|P^{X_{\tau_{B(N)^c}}}(Y_n=y)-P^{X_{\tau_{B(N)^c}}}(Y_n=x)\Bigr|.
\end{aligned} \tag{14.40}$$

By [85, Theorem 2.3.6] there is $c>0$ such that for all $(x',y')\in E(\mathbb{Z}^2)$ and all $z\in\mathbb{Z}^2$,

$$\bigl|P^z(Y_n=y')-P^z(Y_n=x')\bigr|\le cn^{-3/2}. \tag{14.41}$$

This shows that the first sum is $O(1/\log N)$ and that, in light of $x,y\in B(N/2)$, the second sum is $O(1/\sqrt{N})$. This readily yields the claim. □

Let $R_{\mathrm{eff},h}(u,v)$ mark the explicit h-dependence of the effective resistance from u to v in a network with conductances related to a field h as in (13.13). In order to control the effect of the rough part of the decomposition of the DGFF from Lemma 14.12, we will also need:

Lemma 14.14 *For any fields* φ *and* χ*,*

$$R_{\mathrm{eff},\varphi+\chi}(u,v)\le R_{\mathrm{eff},\varphi}(u,v)\max_{(x,y)\in\mathsf{E}}\mathrm{e}^{-\beta(\chi_x+\chi_y)}. \tag{14.42}$$

Moreover, if $\varphi\perp\!\!\!\perp\chi$ *then also*

$$\mathbb{E}\big(R_{\text{eff},\varphi+\chi}(u,v)\,\big|\,\varphi\big) \le R_{\text{eff},\varphi}(u,v)\max_{(x,y)\in\mathsf{E}}\mathbb{E}\big(\mathrm{e}^{-\beta(\chi_x+\chi_y)}\big). \tag{14.43}$$

Proof. Let $i \in \mathcal{I}(u,v)$. Then (13.28) and (13.13) yield

$$R_{\text{eff},\varphi+\chi}(u,v) \le \sum_{e=(x,y)\in\mathsf{E}} \mathrm{e}^{-\beta(\varphi_x+\varphi_y)}\mathrm{e}^{-\beta(\chi_x+\chi_y)}i(e)^2. \tag{14.44}$$

Bounding the second exponential by its maximum and optimizing over i yields (14.42). For (14.43) we first take the conditional expectation given φ and then proceed as before. □

Lemma 14.14 will invariably be used through the following bound:

Exercise 14.15 *(Removal of independent field) Suppose φ and χ are independent Gaussian fields on V. Denote $\sigma^2 := \max_{x\in\mathsf{V}}\operatorname{Var}(\chi_x)$. Show that then for each $a > 0$ and each $r > 0$,*

$$\Big|\mathbb{P}\big(R_{\text{eff},\varphi+\chi}(u,v) \le ar\big) - \mathbb{P}\big(R_{\text{eff},\varphi}(u,v) \le r\big)\Big| \le a^{-1}\mathrm{e}^{2\beta^2\sigma^2}. \tag{14.45}$$

We are ready to give:

Proof of Proposition 14.9. Lemma 14.12 permits us to realize $h^{B(M)}$ as the sum of independent fields φ and χ where, by the union bound and the standard Gaussian tail estimate (see Exercise 2.2), for each $\epsilon > 0$ there is $c_1 > 0$ such that

$$\sup_{M\ge1}\mathbb{P}\Big(\max_{\substack{x,y\in B(M/2)\\ |x-y|_1\le2}}|\varphi_x-\varphi_y| > c_1\Big) < \epsilon. \tag{14.46}$$

Hence, $\rho_{\max}$ associated with field φ in the box $B(N) \subseteq B(M/2)$ via (13.13) obeys $\mathbb{P}(\rho_{\max} \le \mathrm{e}^{c_1\beta}) \ge 1-\epsilon$. Next we observe that, abbreviating $S := B(N)$, the pairs

$$(A,B) := (\partial^{\text{left}}S, \partial^{\text{right}}S) \quad\text{and}\quad (C,D) := (\partial^{\text{up}}S, \partial^{\text{down}}S) \tag{14.47}$$

obey the conditions in Lemma 14.11. Since $\mathfrak{d} = 4$, (14.25) and (14.34) give

$$\mathbb{P}\Big(R^{\text{LR}}_{B(N)}(\varphi)R^{\text{UD}}_{B(N)}(-\varphi) \le 64\mathrm{e}^{c_1\beta}\Big) \ge 1-\epsilon. \tag{14.48}$$

The rotational symmetry of $B(N)$, $B(M)$ and φ along with the distributional symmetry $\varphi \overset{\text{law}}{=} -\varphi$ imply

$$R^{\text{UD}}_{B(N)}(-\varphi) \overset{\text{law}}{=} R^{\text{LR}}_{B(N)}(\varphi). \tag{14.49}$$

A union bound then gives

$$\mathbb{P}\big(R^{\text{LR}}_{B(N)}(\varphi) \le 8\mathrm{e}^{c_1\beta/2}\big) \ge \frac{1-\epsilon}{2}. \tag{14.50}$$

Exercise 14.15 with $r := 8e^{c_1\beta/2}$ and $a := e^{\hat{c}\log\log M}(8e^{c_1\beta/2})^{-1}$ for any choice of $\hat{c} > 2\beta^2 c$ in conjunction with Lemma 14.12 then imply the claim for N sufficiently large. □

14.3 RSW Theory for Effective Resistance

The next (and the most challenging) task is to elevate the statement of Proposition 14.9 from squares to rectangles. The desired claim is the content of:

Proposition 14.16 (Effective resistance across rectangles) *There are $c, C \in (0,\infty)$ and $N_1 \geq 1$ such that for all $N \geq N_1$, all $M \geq 16N$ and any translate S of $B(4N, N)$ satisfying $S \subseteq B(M/2)$,*

$$\mathbb{P}\Big(R_S^{\mathrm{LR}}(h^{B(M)}) \leq C e^{\hat{c}\log\log M}\Big) \geq c. \tag{14.51}$$

The same holds for $R_S^{\mathrm{UD}}(h^{B(M)})$ and translates S of $B(N, 4N)$ with $S \subseteq B(M/2)$. (The constant $\hat{c}$ is as in Proposition 14.9.)

The proof (which we will only sketch) is rooted in the Russo–Seymour–Welsh (RSW) argument for critical percolation, which is a way to to bootstrap uniform lower bounds on the probability of an occupied crossing for squares to those for an occupied crossing for rectangles (in the "longer" direction) of a given aspect ratio. The technique was initiated in Russo [107], Seymour and Welsh [112] and Russo [108] for percolation and later adapted to dependent models as well (e.g., Duminil-Copin, Hongler and Nolin [64], Beffara and Duminil-Copin [15]). We will follow the version that Tassion [123] developed for Voronoi percolation.

Proposition 14.2 links the effective resistance to crossings by collections of paths. In order to mimic arguments from percolation theory, we will need a substitute for the trivial geometric fact that, if a path from u to v crosses a path from u' to v', then the union of these paths contains a path from u to v'. For this, given a set $\mathcal{A}$ of paths, let $R_{\mathrm{eff}}(\mathcal{A})$ denote quantity in (14.2) with the first infimum restricted to $\mathcal{P} \subseteq \mathcal{A}$ (not necessarily a subset of $\mathfrak{P}_{u,v}$). We then have:

Lemma 14.17 (Subadditivity of effective resistance) *Let $\mathcal{A}_1, \dots, \mathcal{A}_n$ be sets of paths such that for each selection $P_i \in \mathcal{A}_i$, $i = 1, \dots, n$, the graph union $P_1 \cup \dots \cup P_n$ contains a path from u to v. Then*

$$R_{\mathrm{eff}}(u, v) \leq \sum_{i=1}^{n} R_{\mathrm{eff}}(\mathcal{A}_i). \tag{14.52}$$

Another property we will need can be thought of as a generalization of the Parallel Law: If we route current from u to v along separate families of paths, then $C_{\mathrm{eff}}(u, v)$ is at most the sum of the conductances of the individual families. This yields:

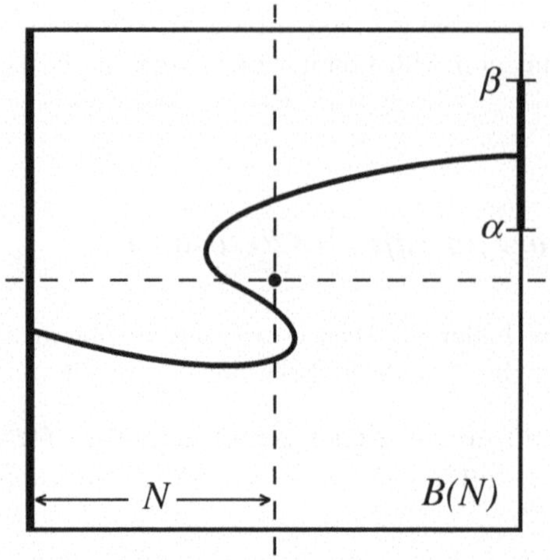

Fig. 27 A path in the set $\mathcal{A}_{N,[\alpha,\beta]}$ underlying the definition of the effective resistivity $R^{\mathrm{LR}}_{N,[\alpha,\beta]}(h)$ between the left side of the square and the portion of the right-side marked by the interval $[\alpha,\beta]$

Lemma 14.18 (Subadditivity of effective conductance) *Let $\mathcal{A}_1,\dots,\mathcal{A}_n$ be sets of paths such that every path from u to v lies in $\mathcal{A}_1\cup\dots\cup\mathcal{A}_n$. Then*

$$C_{\mathrm{eff}}(u,v)\le\sum_{i=1}^n R_{\mathrm{eff}}(\mathcal{A}_i)^{-1}. \tag{14.53}$$

We will not supply proofs of these lemmas as that amounts to further variations on the calculations underlying Propositions 14.2 and 14.7; instead, we refer the reader to [29]. The use of these lemmas will be facilitated by the following observation:

Exercise 14.19 *Show that, for any collection of paths $\mathcal{A}$ and for resistances related to field h as in* (13.13), *$h\mapsto R_{\mathrm{eff},h}(\mathcal{A})$ is decreasing (in each coordinate). Prove that, for h given as DGFF and each $a>0$, under the setting of Lemma 14.17 we have*

$$\mathbb{P}\big(R_{\mathrm{eff}}(u,v)\le a\big)\ge\prod_{i=1}^n\mathbb{P}\big(R_{\mathrm{eff}}(\mathcal{A}_i)\le a/n\big) \tag{14.54}$$

while under the setting of Lemma 14.18 we have

$$\max_{i=1,\dots,n}\mathbb{P}\big(R_{\mathrm{eff}}(\mathcal{A}_i)\le a\big)\ge 1-\Big[1-\mathbb{P}\big(R_{\mathrm{eff}}(u,v)\le a/n\big)\Big]^{1/n}. \tag{14.55}$$

Hint: Use the FKG inequality.

Tassion's version of the RSW argument proceeds by considering crossings of $B(N)$ from the left-side of $B(N)$ to (only) a *portion* of the right side—namely that corresponding to the interval $[\alpha, \beta]$; see Fig. 27 for the geometric setting. Writing $\mathcal{A}_{N,[\alpha,\beta]}$ for the set of all such crossing paths, we abbreviate

$$R^{\text{LR}}_{N,[\alpha,\beta]}(h) := R_{\text{eff},h}(\mathcal{A}_{N,[\alpha,\beta]}). \tag{14.56}$$

We then have:

Lemma 14.20 *For each $\epsilon > 0$ and all $M \geq 2N$ with N sufficiently large,*

$$\mathbb{P}\big(R^{\text{LR}}_{N,[0,N]}(h^{B(M)}) \leq 2e^{\hat{c}\log\log M}\big) > 1 - \frac{1}{\sqrt{2}} - \epsilon. \tag{14.57}$$

Proof. Lemma 14.18 and the Electrostatic Duality show

$$\frac{1}{R^{\text{LR}}_{B(N)}(h)} \leq \frac{1}{R^{\text{LR}}_{N,[0,N]}(h)} + \frac{1}{R^{\text{LR}}_{N,[-N,0]}(h)} \tag{14.58}$$

and the rotation (or reflection) symmetry yields

$$R^{\text{LR}}_{N,[0,N]}(h^{B(M)}) \overset{\text{law}}{=} R^{\text{LR}}_{N,[-N,0]}(h^{B(M)}). \tag{14.59}$$

The bound (14.55) along with Proposition 14.9 then show (14.57). □

Equipped with these techniques, we are ready to give:

Proof of Proposition 14.16 *(rough sketch).* Since $\alpha \mapsto R^{\text{LR}}_{N,[\alpha,N]}(h)$ is increasing on $[0, N]$ with the value at $\alpha = N$ presumably quite large, one can identify (with some degree of arbitrariness) a value α_N where this function first exceeds a large multiple of $e^{\hat{c}\log\log(2N)}$ with high-enough probability. More precisely, set

$$\phi_N(\alpha) := \mathbb{P}\Big(R^{\text{LR}}_{N,[\alpha,N]}(h^{B(2N)}) > Ce^{\hat{c}\log\log(2N)}\Big) \tag{14.60}$$

for a suitably chosen $C > 2$ and note that, by (14.57), $\phi_N(0) < 0.99$. Then set

$$\alpha_N := \lfloor N/2 \rfloor \wedge \min\Big\{\alpha \in \{0, \dots, \lfloor N/2 \rfloor\} \colon \phi_N(\alpha) > 0.99\Big\}. \tag{14.61}$$

We now treat separately the cases $\alpha_N = \lfloor N/2 \rfloor$ and $\alpha_N < \lfloor N/2 \rfloor$ using the following sketch of the actual proof, for which we refer to the original paper:
CASE $\alpha_N = \lfloor N/2 \rfloor$: Here the fact that $\phi_N(\alpha_N - 1) \leq 0.99$ implies (via perturbation arguments that we suppress) that $\phi_N(\alpha_N) < 0.999$ and so

$$\mathbb{P}\big(R^{\text{LR}}_{N,[\alpha_N,N]} \leq Ce^{\hat{c}\log\log(2N)}\big) \geq 0.001. \tag{14.62}$$

Invoking suitable shifts of the box $B(N)$, Lemma 14.17 permits us to bound the left-to-right effective resistance $R^{\text{LR}}_{B(4N,N)}$ by the sum of seven copies of effective

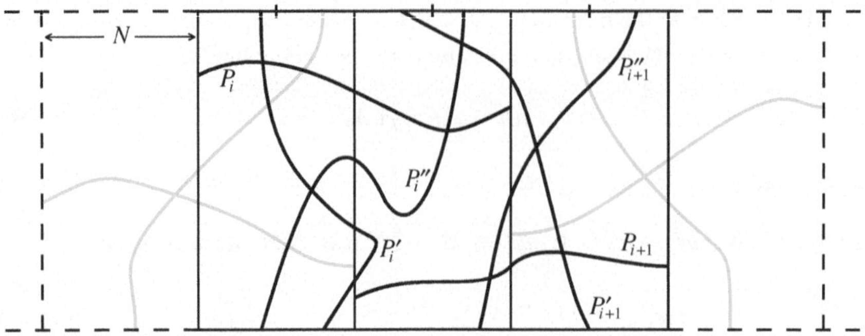

Fig. 28 An illustrations of paths P_i, P_i' and P_i'', all of the form in Fig. 27 with $\alpha := \lfloor N/2 \rfloor$ and $\beta := N$ (or just left-to-right crossing for P_i) that ensure the existence a left-right crossing of the $4N \times N$ rectangle when $\alpha_N = \lfloor N/2 \rfloor$. The effective resistance in the rectangle is then bounded by the sum of (suitably shifted) copies of $R^{\text{LR}}_{N,[0,N]}(h)$

resistances of the form $R^{\text{LR}}_{N,[\alpha_N,N]}$ (and rotations thereof) and four copies of $R^{\text{LR}}_{B(N)}$ (and rotations thereof); see Fig. 28. The inequalities (14.54), (14.62) and (14.24) then yield (14.51). A caveat is that these squares/rectangles are (generally) not centered at the same point as $B(M)$, which we need in order to apply Proposition 14.9 and the definition of α_N. This is remedied by invoking the Gibbs-Markov decomposition and removing the binding field via Exercise 14.15.

CASE $\alpha_N < \lfloor N/2 \rfloor$: In this case $\phi_N(\alpha_N) > 0.99$. Since, by Lemma 14.18,

$$\frac{1}{R^{\text{LR}}_{N,[0,N]}} \le \frac{1}{R^{\text{LR}}_{N,[\alpha_N,N]}} + \frac{1}{R^{\text{LR}}_{N,[0,\alpha_N]}} \tag{14.63}$$

Lemma 14.20 (and $C > 2$) show that $R^{\text{LR}}_{N,[0,\alpha_N]} \le C' \mathrm{e}^{\hat{c} \log\log(2N)}$ with a uniformly positive probability. Assuming in addition that

$$\alpha_N \le 2\alpha_L \quad \text{for} \quad L := \lfloor 4N/7 \rfloor \tag{14.64}$$

another path crossing argument (see Fig. 29) is used to ensure a left-to-right crossing of the rectangle $B(2N - L, N)$. This is then readily bootstrapped to the crossing of $B(4N, N)$, and thus a bound on the effective resistance, by way of Lemma 14.17 and the inequality (14.54).

It remains to validate the assumption (14.64). Here one proceeds by induction assuming that the statement (14.51) already holds for N and proving that, if $\alpha_L \le N$, then the statement holds (with slightly worse constants) for L as well. This is based on a path-crossing argument whose geometric setting is depicted in Fig. 30. This permits the construction of a sequence $\{N_k : k \ge 1\}$ such that $\alpha_{N_{k+1}} \le N_k$ and such that N_{k+1}/N_k is bounded as $k \to \infty$. (Another path crossing argument is required here for which we refer the reader to the original paper.) The claim is thus proved for $N \in \{N_k : k \ge 1\}$; general N are handled by monotonicity considerations. □

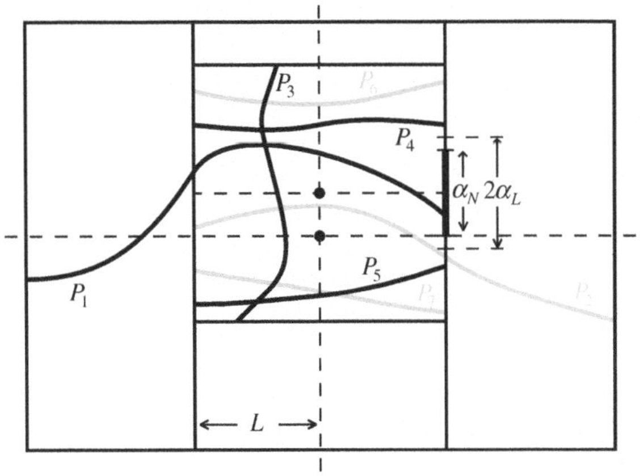

Fig. 29 Arranging a left-to-right crossing of $B(2N - L, N)$, with $L := \lfloor 4N/7 \rfloor$, under the assumption that $\alpha_N \leq 2\alpha_L$. Some of the paths have been drawn in gray to make parsing of the connections easier

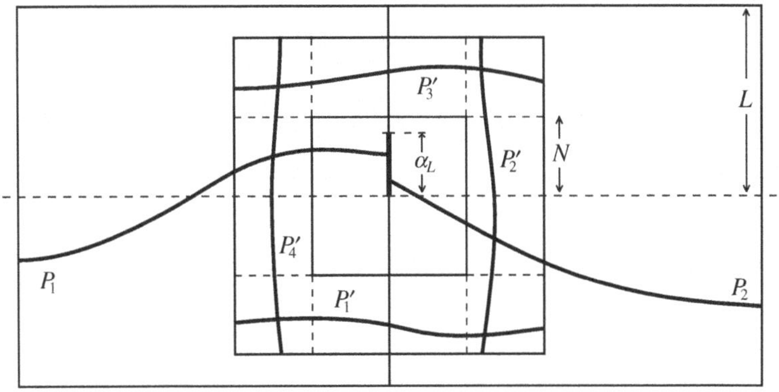

Fig. 30 A collection of paths that enforce a left-to-right crossing of $B(4L, L)$. The paths $P_1', \dots, P_4'$ arise from our assumption that (14.51) holds for N and the corresponding quantity $R^{\mathrm{LR}}_{B(4N,N)}$ is moderate. The paths P_1 and P_2 arise from the assumption $\alpha_L \leq N$. Note that since $\lfloor L/2 \rfloor > N \geq \alpha_L$, we know that also $R^{\mathrm{LR}}_{L,[0,\alpha_L]}$ is moderate

14.4 Upper Tail of Effective Resistance

With Proposition 14.16 in hand, we are now ready to give the proof of:

Theorem 14.21 (Effective resistance, upper tail) *Given integers $M \geq N \geq 1$, let $R_{\mathrm{eff}}(u, v)$ denote the effective resistance from u to v in the network on $B(N)$ with conductances related to $h = h^{B(M)}$ via (13.13). There are $c_1, c_2 \in (0, \infty)$ such that*

$$\max_{u,v \in B(N)} \mathbb{P}\Big(R_{\mathrm{eff}}(u, v) \geq c_1(\log M)\mathrm{e}^{t\sqrt{\log M}}\Big) \leq c_1(\log M)\mathrm{e}^{-c_2 t^2} \qquad (14.65)$$

holds for all $N \geq 1$, all $M \geq 32N$ and all $t \geq 1$.

As we shall see, the proof is based on the following concentration estimate for the effective resistance:

Lemma 14.22 *For any collection of paths $\mathcal{A}$, let $f(h) := \log R_{\text{eff},h}(\mathcal{A})$ (which entails that the resistances depend on h via* (13.13)*). Then*

$$\sup_h \sum_x \left|\frac{\partial f}{\partial h_x}(h)\right| \leq 2\beta. \tag{14.66}$$

Proof. We will prove this under the simplifying assumption that $\mathcal{A}$ is the set of all paths from u to v and, therefore, $R_{\text{eff},h}(\mathcal{A}) = R_{\text{eff},h}(u,v)$. Let $i_\star$ be the current realizing the minimum in (13.28). Then (13.13) yields

$$\begin{aligned}\frac{\partial}{\partial h_x} \log R_{\text{eff},h}(u,v) &= \frac{1}{R_{\text{eff},h}(u,v)} \sum_{e\in \mathsf{E}} i_\star(e)^2 \frac{\partial}{\partial h_x} r_e \\ &= -\frac{\beta}{R_{\text{eff},h}(u,v)} \sum_{y\colon (x,y)\in \mathsf{E}} i_\star(x,y)^2\, r_e\,.\end{aligned} \tag{14.67}$$

Summing the last sum over all x yields $2R_{\text{eff},h}(u,v)$. The claim thus follows. □

Exercise 14.23 *Use the variational representation of $R_{\text{eff}}(\mathcal{A})$ as in* (14.2) *to prove Lemma 14.22 in full generality.*

As a consequence we get:

Corollary 14.24 *There is a constant $c > 0$ such that, for any $M \geq 1$, any collection $\mathcal{A}$ of paths in $B(M)$ and any $t \geq 0$,*

$$\mathbb{P}\Big(\big|\log R_{\text{eff},h^{B(M)}}(\mathcal{A}) - \mathbb{E}\log R_{\text{eff},h^{B(M)}}(\mathcal{A})\big| > t\sqrt{\log M}\Big) \leq 2\mathrm{e}^{-ct^2}. \tag{14.68}$$

Proof. Lemma 14.22 implies the premise (6.16) (with $M := 2\beta$) of the general Gaussian concentration estimate in Corollary 6.6. The claim follows by the union bound and the uniform control of the variance of the DGFF (cf, e.g., (2.3)). □

In order to localize the value of the expectation in (14.68), we also need:

Lemma 14.25 (Effective resistance across rectangles, lower bound) *There are constants $c', C' > 0$ such that for all $N \geq 1$, all $M \geq 16N$ and any translate S of $B(4N,N)$ with $S \subseteq B(M/2)$,*

$$\mathbb{P}\Big(R_S^{\text{LR}}(h^{B(M)}) \geq C'\mathrm{e}^{-\hat{c}\log\log M}\Big) \geq c'. \tag{14.69}$$

The same holds for $R_S^{\text{UD}}(h^{B(M)})$ and translates S of $B(N,4N)$ with $S \subseteq B(M/2)$. (The constant $\hat{c}$ is as in Proposition 14.9.)

Proof (sketch). By swapping resistances for conductances (and relying on Lemma 14.10 instead of Lemma 14.11), we find out that (14.69) holds for $S := B(N)$. Invoking the Gibbs-Markov property, Exercise 14.15 yields a similar bound for all translates of $B(N)$ contained in $B(M/4)$. Since every translate S of $B(4N, N)$ with $S \subseteq B(M/2)$ contains a translate S' of $B(N)$ contained in $B(M/4)$, the claim follows from the fact that $R_{S'}^{\mathrm{LR}} \le R_S^{\mathrm{LR}}$. □

With Lemma 14.25 in hand, we can finally prove the upper bound on the point-to-point effective resistances:

Proof of Theorem 14.21. Using Corollary 14.24 along with Proposition 14.16 and Lemma 14.25 (and the fact that $\log\log M \le \sqrt{\log M}$ for M large) we get that

$$E\big|\log R_S^{\mathrm{LR}}(h^{B(M)})\big| \le \tilde{c}\sqrt{\log M} \tag{14.70}$$

and

$$\mathbb{P}\Big(\log R_S^{\mathrm{LR}}(h^{B(M)}) \ge \tilde{c}(1+t)\sqrt{\log M}\Big) \le \mathrm{e}^{-\tilde{c}' t^2}, \quad t \ge 0, \tag{14.71}$$

hold with the same constants $\tilde{c}, \tilde{c}' \in (0, \infty)$ for all translates S of $B(4N, N)$ or $B(N, 4N)$ contained in $B(M/2)$, uniformly in $N \ge 1$ and $M \ge 32N$.

Next we invoke the following geometric fact: Given $u, v \in B(N)$, there are at most order-$\log N$ translates of $B(4K, K)$ or $B(K, 4K)$ with $K \in \{1, \dots, N\}$ such that the existence of a crossing (in the longer direction) in each of these rectangles implies the existence of a path from u to v; see Fig. 31. Assuming that the corresponding resistance (left-to-right in horizontal rectangles and top-to-bottom in vertical ones) is at most $\mathrm{e}^{\tilde{c}(1+t)\sqrt{\log M}}$, Lemma 14.17 bounds $R_{\mathrm{eff}}(u, v)$ by a constant times $(\log M)\mathrm{e}^{\tilde{c}(1+t)\sqrt{\log M}}$. The claim then follows from (14.71) and the union bound. □

Theorem 14.21 now in turn implies the upper bound on the growth rate of resistances from 0 to $B(N)^{\mathrm{c}}$ in an infinite network on $\mathbb{Z}^2$:

Proof of (13.15) *in Theorem* 13.5 *(sketch).* First note that $R_{\mathrm{eff}}(0, B(N)^{\mathrm{c}}) \le R_{\mathrm{eff}}(0, v)$ for any vertex $v \in \partial B(N)$. The Gibbs-Markov property along with Exercise 14.15 allow us to estimate the tail of $R_{\mathrm{eff}}(0, B(N)^{\mathrm{c}})$ for $h :=$ DGFF in $\mathbb{Z}^2 \smallsetminus \{0\}$ by essentially that for $h :=$ DGFF in $B(2N)$. Theorem 14.21 gives

$$\mathbb{P}\Big(\log R_{\mathrm{eff}}(0, B(N)^{\mathrm{c}}) \ge a(\log N)^{1/2}(\log\log N)^{1/2}\Big) \le c(\log N)^{1-c_2 a^2} \tag{14.72}$$

for all $a > 0$ as soon as N is sufficiently large. Once a is sufficiently large, this is summable for $N \in \{2^n : n \in \mathbb{N}\}$. The claim thus follows from the Borel–Cantelli lemma and the monotonicity of $N \mapsto R_{\mathrm{eff}}(0, B(N)^{\mathrm{c}})$. □

The lower corresponding lower bound (13.16) requires the use of the concentric decomposition and is therefore deferred to the next lecture.

Fig. 31 An illustration of a sequence of $4K \times K$ rectangles (for K taking values in powers of 2) such that the occurrence of a crossing of each of these rectangles implies the existence of a path from u to v. For vertices within distance N, order-log N rectangles suffice

Lecture 15: From Resistance to Random Walk

We are left with the task to apply the technology for controlling the resistance to establish the desired conclusions about the random walk in DGFF landscape. It is here where we will tie the random walk problem to the overall theme of these lectures: the large values of the DGFF. We commence by some elementary identities from Markov chain theory which readily yield upper bounds on the expected exit time and the heat kernel. The control of the corresponding lower bounds is considerably more involved. We again focus on conveying the main ideas while referring the reader to the original paper for specific technical details.

15.1 Hitting and Commute-Time Identities

We begin by considerations that apply to a general discrete-time Markov chain X on a countable state space V with transition probability P and a reversible measure π. This problem always admits the electric network formulation by setting the conductances to $c(x, y) := \pi(x)\mathsf{P}(x, y)$—whose symmetry is equivalent to reversibility of π—and defining the resistances accordingly. The set of edges E is then the set of pairs (x, y) where $c(x, y) > 0$. Our main tool will be:

Lemma 15.1 (Hitting time identity) *Let $A \subseteq \mathsf{V}$ be finite and let us denote, as before, $\tau_{A^c} := \inf\{n \geq 0 \colon X_n \notin A\}$. Then for each $x \in A$,*

$$E^x(\tau_{A^c}) = R_{\text{eff}}(x, A^c) \sum_{y \in A} \pi(y)\phi(y), \tag{15.1}$$

where $\phi(y) := P^x(\tau_x < \tau_{A^c})$ *has the interpretation of the potential (a.k.a. voltage) that minimizes the variational problem defining* $C_{\text{eff}}(x, A^c)$.

Proof. The Markov property ensures that the Green function associated with the Markov chain, defined by the formula (1.3) or via $G^A := (1 - \mathsf{P})^{-1}$, is P-harmonic in the first coordinate. The uniqueness of the solution to the Dirichlet problem in finite sets (see Exercise 13.16) implies

$$\phi(y) = \frac{G^A(y,x)}{G^A(x,x)}. \tag{15.2}$$

Reversibility then yields

$$\begin{aligned} \sum_{y\in A} \pi(y)\phi(y) &= G^A(x,x)^{-1} \sum_{y\in A} \pi(y) G^A(y,x) \\ &= G^A(x,x)^{-1}\pi(x) \sum_{y\in A} G^A(x,y) \\ &= G^A(x,x)^{-1}\pi(x) E^x(\tau_{A^c}). \end{aligned} \tag{15.3}$$

The claim follows from

$$G^A(x,x) = \pi(x) R_{\text{eff}}(x, A^c) \tag{15.4}$$

as implied by Corollary 13.17 and (1.17). □

The above lemma is well known particularly in the form of the following corollaries. The first one of these is folklore:

Corollary 15.2 (Hitting time estimate) *For all finite* $A \subseteq \mathsf{V}$ *and all* $x \in A$,

$$E^x(\tau_{A^c}) \le R_{\text{eff}}(x, A^c)\pi(A). \tag{15.5}$$

Proof. Apply $\phi(y) \le 1$ for all $y \in A$ in (15.1). □

The second corollary, which we include mostly for completeness of exposition, appeared first in Chandra et al. [40]:

Corollary 15.3 (Commute-time identity) *For all* V *finite and all distinct vertices* $u, v \in \mathsf{V}$,

$$E^u(\tau_v) + E^v(\tau_u) = R_{\text{eff}}(u, v)\pi(\mathsf{V}). \tag{15.6}$$

Proof. Let ϕ be the minimizer of $C_{\text{eff}}(u, v)$ and ϕ' the minimizer of $C_{\text{eff}}(v, u)$. Noting that $\phi(x) + \phi'(x) = 1$ for all $x = u, v$ the uniqueness of the solution to the Dirichlet problem implies that $\phi(x) + \phi'(x) = 1$ all $x \in \mathsf{V}$. The claim follows from (15.1). □

15.2 Upper Bounds on Expected Exit Time and Heat Kernel

The upshot of the above corollaries is quite clear: In order to bound the expected hitting time from above, we only need tight upper bounds on the effective resistance and the total volume of the reversible measure. Drawing heavily on our earlier work in these notes, this now permits to give:

Proof of Theorem 13.3, *upper bound in* (13.8). Let $\delta > 0$ and let h be the DGFF in $\mathbb{Z}^2 \smallsetminus \{0\}$. The upper bound in Theorem 13.5 ensures

$$R_{\text{eff},h}\bigl(0, B(N)^{\text{c}}\bigr) \le \text{e}^{(\log N)^{1/2+\delta}} \tag{15.7}$$

with probability tending to one as $N \to \infty$. The analogue of (2.16) for the DGFF in $\mathbb{Z}^2 \smallsetminus \{0\}$ along with Markov inequality and Borel–Cantelli lemma yields

$$\pi_h\bigl(B(N)\bigr) \le N^{\theta(\beta)}\text{e}^{(\log N)^{\delta}} \tag{15.8}$$

with probability tending to one as $N \to \infty$ when $\beta \le \tilde{\beta}_{\text{c}}$; for $\beta > \tilde{\beta}_{\text{c}}$ we instead use the Gibbs-Markov property and the exponential tails of the maximum proved in Lemma 8.3. Corollary 15.2 then gives

$$E_h^0(\tau_{B(N)^{\text{c}}}) \le N^{\theta(\beta)}\text{e}^{2(\log N)^{1/2+2\delta}} \tag{15.9}$$

with probability tending to one as $N \to \infty$. □

Next we will address the decay of the heat kernel:

Proof of Theorem 13.2, *upper bound in* (13.5). For $T > 0$ define the set

$$\Xi_T := \{0\} \cup \Bigl\{x \in B(2T) \smallsetminus B(T) \colon \widetilde{R}_{\text{eff}}(0, x) \le \text{e}^{(\log T)^{1/2+\delta}}\Bigr\}, \tag{15.10}$$

where $\widetilde{R}_{\text{eff}}$ stands for the effective resistance in the network on $B(2T)$. From Theorem 14.21 and the Markov inequality we conclude that $|\Xi_T|$ will contain an overwhelming fraction of all vertices in the annulus $B(2T) \smallsetminus B(T)$. Let $\widetilde{X}$ be the Markov chain on $B(4T)$ defined by the same conductances as X except those corresponding to the jumps out of $B(4T)$ which are set to zero (and these jumps are thus suppressed) and write $\widetilde{\pi}_h$ for the correspondingly modified measure π_h. Let Y_k be the position of the kth visit of $\widetilde{X}$ to Ξ_T, set $\tau_0 := 0$ and let τ_1, τ_2, etc be times of the successive visits of $\widetilde{X}$ to 0. Let

$$\hat{\sigma} := \inf\bigl\{k \ge 1 \colon \tau_k \ge T\bigr\}. \tag{15.11}$$

We ask the reader to verify:

Exercise 15.4 *Prove that* $Y := \{Y_k \colon k \ge 0\}$ *is a Markov chain on* Ξ_T *with stationary distribution*

$$\nu(x) := \frac{\widetilde{\pi}_h(x)}{\widetilde{\pi}_h(\Xi_T)}. \tag{15.12}$$

Prove also that $\hat{\sigma}$ is a stopping time for the natural filtration associated with Y and that $E_h^x(\hat{\sigma}) < \infty$ (and thus $\hat{\sigma} < \infty$ P_h^x-a.s) hold for each $x \in \Xi_T$.

We also recall an exercise from a general theory of reversible Markov chains:

Exercise 15.5 *Prove that $T \mapsto P^0(\widetilde{X}_{2T} = 0)$ is non-increasing.*

This permits us to write, for T even (which is all what matters),

$$\frac{1}{2} T P_h^0(\widetilde{X}_T = 0) \leq E_h^0\left(\sum_{n=0}^{T-1} 1_{\{\widetilde{X}_n=0\}}\right) \leq E_h^0\left(\sum_{k=0}^{T-1} 1_{\{Y_k=0\}}\right) \leq E_h^0\left(\sum_{k=0}^{\hat{\sigma}-1} 1_{\{Y_k=0\}}\right), \tag{15.13}$$

where the second inequality relies on the fact that $0 \in \Xi_T$. Next we observe:

Exercise 15.6 *Given a Markov chain Y with stationary distribution ν and a state x, suppose σ is a stopping time for Y such that $Y_\sigma = x$ a.s. Then for each y*

$$E^x\left(\sum_{k=0}^{\sigma-1} 1_{\{Y_k=y\}}\right) = E^x(\sigma)\nu(y). \tag{15.14}$$

Hint: The left-hand side is, as a function of y, a stationary measure for Y.

By conditioning on Y_T we now have

$$E_h^0(\hat{\sigma}) \leq T + E_h^0\big(E^{\widetilde{X}_T}(\tau_1)\big) \leq T + \max_{u\in\Xi_T} E_h^u(\tau_1) \tag{15.15}$$

and the commute-time identity (Corollary 15.3) and the definition of Ξ_T give

$$E_h^u(\tau_1) \leq \widetilde{\pi}_h(\Xi_T)\widetilde{R}_{\mathrm{eff}}(u,0) \leq \pi_h(\Xi_T)\mathrm{e}^{(\log T)^{1/2+\delta}}. \tag{15.16}$$

The nearest-neighbor nature of the walk permits us to couple X and $\widetilde{X}$ so that X coincides with $\widetilde{X}$ at least up to time $4T$. The above then bounds $P_h^0(X_T = 0)$ by $2\pi_h(0)T^{-1}\mathrm{e}^{(\log T)^{1/2+\delta}}$. □

We can now also settle:

Proof of Corollary 13.4. Using the Markov property, reversibility and the Cauchy–Schwarz inequality we get

$$\begin{aligned} P_h^0(X_{2T} = 0) &\geq \sum_{x\in B(N)} P_h^0(X_T = x)P_h^x(X_T = 0) \\ &= \pi_h(0) \sum_{x\in B(N)} \frac{P_h^0(X_T = x)^2}{\pi_h(x)} \geq \pi_h(0)\frac{P_h^0\big(X_T \in B(N)\big)^2}{\pi_h\big(B(N)\big)}. \end{aligned} \tag{15.17}$$

Invoking the (already proved) upper bound on the heat-kernel and the bound (15.8) we get that, with high probability,

$$P_h^0\big(X_T \in B(N)\big) \le \left[\frac{1}{T}\mathrm{e}^{(\log T)^{1/2+\delta}} N^{\theta(\beta)}\mathrm{e}^{(\log N)^\delta}\right]^{1/2}. \tag{15.18}$$

Setting $T := N^{\theta(\beta)}\mathrm{e}^{(\log N)^{1/2+2\delta}}$ gives the desired claim. □

15.3 Bounding Voltage from Below

We now turn to the lower bounds in Theorems 13.2 and 13.3. As an inspection of (15.1) shows, this could be achieved by proving a lower bound on the potential difference (a.k.a. voltage) minimizing $C_{\mathrm{eff}}(0, B(N)^{\mathrm{c}})$. This will be based on:

Lemma 15.7 *In the notation of Lemma 15.1, for all $x \in A$ and $y \in A \smallsetminus \{x\}$,*

$$2R_{\mathrm{eff}}(x, A^{\mathrm{c}})\phi(y) = R_{\mathrm{eff}}(x, A^{\mathrm{c}}) + R_{\mathrm{eff}}(y, A^{\mathrm{c}}) - R_{A,\mathrm{eff}}(x, y)\,, \tag{15.19}$$

where $R_{A,\mathrm{eff}}$ denotes the effective resistance in the network with A^{c} collapsed to a point.

Proof. We will apply the network reduction principle from Exercise 13.20 and reduce the problem to the network with three nodes 1, 2 and 3 corresponding to x, y and A^{c}, respectively. Since ϕ is harmonic, so is its restriction to the reduced network. But $\phi(y)$ also has the interpretation of the probability that the reduced Markov chain at 2 jumps to 1 before 3. Writing the conductance between the ith and the jth node as c_{ij}, this probability is given as $\frac{c_{12}}{c_{12}+c_{23}}$. Exercise 13.24 equates this to the ratio of the right-hand side of (15.19) and $2R_{\mathrm{eff}}(0, A^{\mathrm{c}})$. □

For the setting at hand, the quantity on the right-hand side of (15.19) becomes:

$$D_N(x) := R_{\mathrm{eff}}\big(0, B(N)^{\mathrm{c}}\big) + R_{\mathrm{eff}}\big(x, B(N)^{\mathrm{c}}\big) - R_{B(N),\mathrm{eff}}(0, x)\,. \tag{15.20}$$

We claim:

Proposition 15.8 *For any $\delta \in (0, 1)$*

$$\lim_{N\to\infty} \mathbb{P}\Big(\min_{x\in B(N\mathrm{e}^{-(\log N)^\delta})} D_N(x) \ge \log N\Big) = 1. \tag{15.21}$$

We will only give a sketch of the main idea which is also key for the proof of the lower bound in Theorem 13.5 that we will prove in some more detail. The proof relies on the concentric decomposition of the pinned DGFF from Exercise 8.15 which couples the field to the Gaussian random walk $\{S_n : n \ge 0\}$ defined in (8.54) and some other Gaussian fields. We claim:

Proposition 15.9 *Let $h :=$ DGFF in $\mathbb{Z}^2 \smallsetminus \{0\}$ and let $\{S_n : n \ge 1\}$ be the Gaussian random walk related to h via Exercise 8.15. There is a constant $c > 0$ such that*

$$R_{\text{eff},h}\big(0, B(N)^{\text{c}}\big) \ge \max_{k=1,\dots,\lfloor \log_4 N\rfloor} \mathrm{e}^{2\beta S_k - c(\log\log N)^2} \tag{15.22}$$

fails for at most finitely many N's, $\mathbb{P}$-a.s.

For the proof, recall the definition of Δ^k from (8.36) and write

$$A_k := \Delta^k \smallsetminus \overline{\Delta^{k-1}}, \qquad k \ge 1, \tag{15.23}$$

for the annuli used to define the concentric decomposition. Invoking, in turn, the notation $B(N)$ from (13.7), define also the thinned annuli

$$A_k' := B(3\cdot 2^{k-2}) \smallsetminus B(2^{k-2}) \tag{15.24}$$

and note that $A_k' \subseteq A_k$ with $\text{dist}_\infty(A_k', A_k^{\text{c}}) \ge 2^{k-2}$. Network reduction arguments (underlying the Nash-Williams estimate) imply

$$R_{\text{eff},h}\big(0, B(N)^{\text{c}}\big) \ge R_{\text{eff},h}\big(\partial^{\text{in}} A_k', \partial^{\text{out}} A_k'\big), \tag{15.25}$$

where $\partial^{\text{in}} A$, resp., $\partial^{\text{out}} A$ denote the inner, resp., outer external boundary of the annulus A. We first observe:

Lemma 15.10 *There are $c, C > 0$ such that for each $k \ge 1$,*

$$\mathbb{P}\Big(R_{\text{eff},\,h^{A_k}}\big(\partial^{\text{in}} A_k', \partial^{\text{out}} A_k'\big) \ge C\mathrm{e}^{-\hat{c}\log\log(2^k)}\Big) \ge c\,, \tag{15.26}$$

where h^{A_k} := DGFF in A_k.

Proof. Let $N := 2^{k-1}$ and let $U_1, \dots, U_4$ denote the four (either $4N \times N$ or $N \times 4N$) rectangles that fit into A_k'. We label these in the clockwise direction starting from the one on the right. Since every path from the inner boundary of A_k' to the outer boundary of A_k' crosses (and connects the longer sides of) one of these rectangles, Lemma 14.18 and the variational representation (14.2)–(14.3) imply

$$R_{\text{eff},\,h}\big(\partial^{\text{in}} A_k', \partial^{\text{out}} A_k'\big) \ge \frac{1}{4}\min\Big\{R_{U_1}^{\text{LR}}(h),\, R_{U_2}^{\text{UD}}(h),\, R_{U_3}^{\text{LR}}(h),\, R_{U_4}^{\text{UD}}(h)\Big\}. \tag{15.27}$$

(Note that the resistances are between the *longer* sides of the rectangles.) Invoking also the Gibbs-Markov property along with Lemma 14.14, it suffices to prove that, for some $C, c > 0$ and all $N \ge 1$,

$$\mathbb{P}\Big(R_U^{\text{LR}}(h^{B(16N)}) \ge C\mathrm{e}^{-\hat{c}\log\log N}\Big) \ge c \tag{15.28}$$

for any translate U of $B(N, 4N)$ contained in $B(8N)$.

We will proceed by the duality argument underlying the proof of Proposition 14.9. Abbreviate $M := 16N$ and consider the decomposition of $h^{B(M)} = \varphi + \chi$ from

Lemma 14.12. Then for each $r, a > 0$, Exercise 14.15 yields

$$\mathbb{P}\bigl(R_U^{\mathrm{LR}}(h^{B(M)}) \ge r\bigr) \ge \mathbb{P}\bigl(R_U^{\mathrm{LR}}(\varphi) \ge r/a\bigr) - a^{-1}\mathrm{e}^{\hat{c}\log\log M}. \tag{15.29}$$

The path-cut approximate duality shows

$$\mathbb{P}\Bigl(R_U^{\mathrm{LR}}(\varphi)R_U^{\mathrm{UD}}(-\varphi) \ge \mathrm{e}^{-2\beta c_1}/64\Bigr) \ge 1-\epsilon. \tag{15.30}$$

For any $r' > 0$, Exercise 14.15 also gives

$$\mathbb{P}\bigl(R_U^{\mathrm{UD}}(\varphi) \le r'\bigr) \ge \mathbb{P}\bigl(R_U^{\mathrm{UD}}(h^{B(M)}) \le r'/a\bigr) - a^{-1}\mathrm{e}^{\hat{c}\log\log M}. \tag{15.31}$$

Setting $r'/a := \mathrm{e}^{\hat{c}\log\log M}$ with $a := C'\mathrm{e}^{\hat{c}\log\log M}$ for some $C' > 0$ large enough, Lemma 14.25 bounds the probability on the right of (15.31) by a positive constant. Via (15.30) for ϵ small, this yields a uniform lower bound on $\mathbb{P}(R_U^{\mathrm{LR}}(\varphi) \ge r/a)$ for $r/a := (\mathrm{e}^{-2\beta c_1}/64)/r'$. The bound (15.29) then gives (15.28). □

Proof of Proposition 15.9. Using the representation of $h :=$ DGFF in $\mathbb{Z}^2 \smallsetminus \{0\}$ from Exercise 8.15, the estimates from Lemmas 8.10, 8.11, 8.12 and 11.4 on the various "bits and pieces" constituting the concentric decomposition yield

$$h \le -S_k + (\log n)^2 + h'_k \quad \text{on } A'_k \tag{15.32}$$

for all $k = 1, \dots, n$ as soon as $n := \lfloor \log_4 N \rfloor$ is sufficiently large. The inequality (15.25) then gives

$$R_{\mathrm{eff},h}\bigl(0, B(N)^{\mathrm{c}}\bigr) \ge \mathrm{e}^{2\beta[S_k-(\log n)^2]} R_{\mathrm{eff},h'_k}(\partial^{\mathrm{in}} A'_k, \partial^{\mathrm{out}} A'_k). \tag{15.33}$$

The fields $\{h'_k : k \ge 1\}$ are independent and hence so are the events

$$E_k := \Bigl\{R_{\mathrm{eff},h'_k}(\partial^{\mathrm{in}} A'_k, \partial^{\mathrm{out}} A'_k) \ge C\mathrm{e}^{-\hat{c}\log\log(2^k)}\Bigr\}, \quad k \ge 1. \tag{15.34}$$

Moreover, $h'_k \overset{\mathrm{law}}{=} h^{A_k}$ and so Lemma 15.10 shows $\mathbb{P}(E_k) \ge c$ for all $k \ge 1$.

A standard use of the Borel–Cantelli lemma implies that, almost surely for n sufficiently large, the longest interval of $k \in \{1, \dots, n\}$ where E_k fails is of length at most order $\log n$. The Gaussian nature of the increments of S_k permits us to assume that $\max_{1\le k\le n} |S_{k+1} - S_k| \le \log n$ for n large and so the value of S_k changes by at most another factor of order $(\log n)^2$ over any interval of k's where E_k fails. Since $\log\log(2^n) \ll (\log n)^2$ for n large, this and the fact that $\log n = \log\log N + O(1)$ readily implies the claim. □

We are now ready to complete our proof of Theorem 13.5:

Proof of (13.16) *in Theorem* 13.5. Since $\{S_k : k \ge 1\}$ is a random walk with Gaussian increments of bounded and positive variance, Chung's Law of the Iterated Logarithm (see [46, Theorem 3]) implies that, for some constant $c > 0$,

$$\liminf_{n\to\infty} \frac{\max_{k\le n} S_k}{\sqrt{n/\log\log n}} \ge c, \quad \mathbb{P}\text{-a.s.} \tag{15.35}$$

As $\log n = \log\log N + O(1)$, the claim follows from Proposition 15.9. □

It remains to give:

Proof of Proposition 15.8, *main idea.* Consider the annuli A_k as above and notice the following consequences of the network reduction arguments discussed earlier. Fix n large and denote $N := 2^n$. A shorting argument then implies that for any $k^\star \in \{1,\dots,n\}$ such that $x \in \Delta^{k^\star-1}$,

$$R_{\text{eff}}\big(x, B(N)^{\text{c}}\big) \ge R_{\text{eff}}\big(x, \partial^{\,\text{in}} A'_{k^\star}\big) + R_{\text{eff}}(\partial^{\,\text{in}} A'_{k^\star}, \partial^{\,\text{out}} A'_{k^\star}). \tag{15.36}$$

Next consider the four rectangles constituting the annulus A'_k and let $R^{(i)}_{A'_k}$ denote the effective resistance between the *shorter* sides of the ith rectangle. Lemma 14.17 and the path-crossing argument from Fig. 32 show that, for each $k_\star \in \{1,\dots,n\}$ such that $x \in \Delta^{k_\star-1}$,

$$R_{\text{eff},B(N)}(0,x) \le R_{\text{eff}}\big(0, \partial^{\,\text{out}} A'_{k_\star}\big) + R_{\text{eff}}\big(x, \partial^{\,\text{out}} A'_{k_\star}\big) + \sum_{i=1}^{4} R^{(i)}_{A'_{k_\star}}. \tag{15.37}$$

Assuming that $k_\star < k_\star$, also have

$$R_{\text{eff}}\big(x, \partial^{\,\text{in}} A'_{k^\star}\big) \ge R_{\text{eff}}\big(x, \partial^{\,\text{out}} A'_{k_\star}\big) \tag{15.38}$$

and so

$$R_{\text{eff}}\big(x, B(N)^{\text{c}}\big) - R_{\text{eff},B(N)}(0,x) \ge R_{\text{eff}}(\partial^{\,\text{in}} A'_{k^\star}, \partial^{\,\text{out}} A'_{k^\star}) - \sum_{i=1}^{4} R^{(i)}_{A'_{k_\star}}. \tag{15.39}$$

The point is to show that, with probability tending to one as $n \to \infty$, one can find $k_\star, k^\star \in \mathbb{N}$ with

$$(\log n)^\delta < k_\star < k_\star \le n \tag{15.40}$$

such that the first resistance on the right of (15.39) significantly exceeds the sum of the four resistances therein. For this we note that, thanks to Lemma 15.10 and Proposition 14.16, the overall scale of these resistances is determined by the value of the random walk S_k at $k := k^\star$ (for the first resistance) and $k := k_\star$ (for the remaining resistances). In light of the observations made in the proof of Proposition 15.9, it will suffice to find $k_\star, k^\star$ obeying (15.40) such that

$$S_{k^\star} \ge 2(\log n)^2 \quad \text{and} \quad S_{k_\star} \le -2(\log n)^2 \tag{15.41}$$

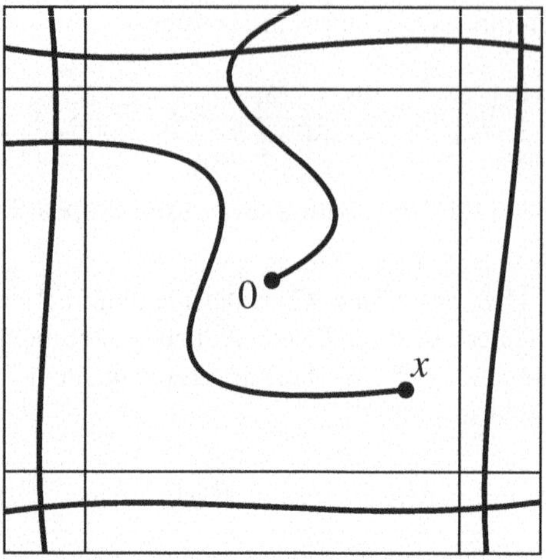

Fig. 32 A path-crossing event underlying (15.37). Notice that the existence of a path from 0 to the outer square, another path from x to the outer square and the crossings of the four rectangles as shown imply the existence of a path from 0 to x within the outer square

occur with overwhelming probability. This requires a somewhat more quantitative version the Law of the Iterated Logarithm for which we refer the reader to the original paper. □

15.4 Wrapping Up

We will now finish by presenting the proofs of the desired lower bounds on the expected exit time and the heat kernel.

Proof of Theorem 13.3, *lower bound in* (13.8). Let $\delta \in (0, 1)$. The hitting-time identity from Lemma 15.1 along with Proposition 15.8 imply

$$E_h^0(\tau_{B(N)^c}) \geq \pi_h\big(B(N\mathrm{e}^{-(\log N)^\delta})\big) \log N \tag{15.42}$$

with probability tending to one as $N \to \infty$. Theorem 2.7 on the size of the intermediate level sets and Theorem 7.3 on the tightness of the absolute maximum yield

$$\pi_h\big(B(N\mathrm{e}^{-(\log N)^\delta})\big) \geq N^{\theta(\beta)}\mathrm{e}^{-(\log N)^{2\delta}} \tag{15.43}$$

with probability tending to one as $N \to \infty$. The claim follows. □

It remains to show:

Proof of Theorem 13.2, *lower bound in* (13.5). Let $\Xi_T^\star$ be the union of $\{0\} \cup B(N)^c$ with the set of all $x \in B(N\mathrm{e}^{-(\log N)^\delta})$ such that

$$R_{\text{eff},B(N)}(0,x) \vee R_{\text{eff}}\bigl(x, B(N)^{\text{c}}\bigr) \le \mathrm{e}^{(\log N)^{1/2+\delta}} \tag{15.44}$$

Abusing our earlier notation slightly, let Y_k be the kth visit of X to $\Xi_T^\star$ (counting the starting point as $k = 0$ case). Denote $\hat{\tau} := \inf\{k > 0\colon Y_k \in B(N)^{\text{c}}\}$. Then

$$\begin{aligned} E_h^0(\hat{\tau}) &\le T P_h^0(\hat{\tau} \le T) + P_h^0(\hat{\tau} > T)\Bigl(T + \max_{x\in\Xi_N} E_h^x(\hat{\tau})\Bigr) \\ &= T + P_h^0(\hat{\tau} > T) \max_{x\in\Xi_N} E_h^x(\hat{\tau}). \end{aligned} \tag{15.45}$$

The hitting time estimate (Corollary 15.2) ensures

$$E_h^x(\hat{\tau}) \le \pi_h\bigl(\Xi_T^\star \cap B(N)\bigr)\mathrm{e}^{(\log N)^{1/2+\delta}}, \quad x \in \Xi_T^\star, \tag{15.46}$$

implying

$$P_h^0(\hat{\tau} > T) \ge \pi_h\bigl(\Xi_T^\star \cap B(N)\bigr)^{-1}\mathrm{e}^{-(\log N)^{1/2+\delta}}\bigl[E_h^0(\hat{\tau}) - T\bigr]. \tag{15.47}$$

By our choice of $\Xi_T^\star$, the lower bound on the voltage from Proposition 15.8 applies to all $x \in \Xi_T^\star \cap B(N)$ and so

$$E_h^0(\hat{\tau}) \ge \pi_h\bigl(\Xi_T^\star \cap B(N)\bigr)\log N. \tag{15.48}$$

We now need to solve:

Exercise 15.11 *Prove that*

$$\mathbb{P}\Bigl(\pi_h\bigl(\Xi_N^\star \cap B(N)\bigr) \le N^{\theta(\beta)}\mathrm{e}^{-(\log N)^{2\delta}}\Bigr) \underset{N\to\infty}{\longrightarrow} 0. \tag{15.49}$$

Hint: Estimate $\mathbb{E}\pi_h(B(N\mathrm{e}^{(\log N)^\delta}) \smallsetminus \Xi_N^\star)$ *using the FKG inequality and the fact that* $h \mapsto R_{\text{eff},h}(u,v)$ *is decreasing.*

For $N := T^{1/\theta(\beta)}\mathrm{e}^{(\log T)^\delta}$ this implies $E_h^0(\hat{\tau}) \ge 2T$ and via (15.47)–(15.48) also

$$P_h^0(\hat{\tau} > T) \ge \mathrm{e}^{-(\log N)^{1/2+\delta}}. \tag{15.50}$$

But $\hat{\tau} \le \tau_{B(N)^{\text{c}}} := \inf\{k \ge 0\colon X_k \in B(N)^{\text{c}}\}$ and so

$$P^0\bigl(X_T \in B(N)\bigr) \ge P^0\bigl(\tau_{B(N)^{\text{c}}} > T\bigr) \ge \mathrm{e}^{-(\log N)^{1/2+\delta}}. \tag{15.51}$$

Plugging this in (15.17) and invoking (15.8) then gives the claim. □

Lecture 16: Questions, Conjectures and Open Problems

In this final lecture we discuss conjectures and open problems. These are all considerably harder than the exercises scattered throughout these notes. In fact, most of the problems below constitute non-trivial projects for future research. The presentation is at times (and sometimes deliberately) vague on detail and may even require concepts that lie outside the scope of these lecture notes.

16.1 DGFF Level Sets

We open our discussion by questions that are related to the description of the DGFF level sets. Let h^{D_N} be the DGFF in D_N, for a sequence $\{D_N : N \geq 1\}$ of admissible lattice approximation of domain $D \in \mathfrak{D}$. We commence by:

Problem 16.1 *(Joint limit of intermediate level sets) Find a way to extract a joint distributional limit of the level sets*

$$\big\{x \in D_N : h_x \geq 2\sqrt{g}\lambda \log N\big\} \tag{16.1}$$

or their associated point measures (2.29)*, simultaneously for all* $\lambda \in (0, 1)$*. Use this to design a coupling of the corresponding* Z_λ^D*'s and show that they all arise (via the construction of the LQG measure) from the same underlying CGFF.*

We remark that we see a way to control any finite number of the level sets by building on the moment calculations underpinning Theorem 2.7. The problem is that, the more level sets we wish to include, the higher moments we seem to need. The key is to find a formulation that works simultaneously for all λ's and to show that that the limit measures can be linked to the same copy of the CGFF.

The insistence on connecting the limit process to the CGFF may in fact suggest a solution: Use the same sample of CGFF to define the DGFF on D_N for all $N \geq 1$ and thus all level sets simultaneously. Then show that this CGFF is what one gets (via the construction of Gaussian Multiplicative Chaos) as a limit measure. The same strategy will perhaps also make the next question accessible:

Problem 16.2 *(Fluctuations and deviations) Describe the rate of convergence in, e.g., Corollary 2.9. Identify the limit law of the fluctuations of the level set away from its limit value. Study (large) deviations estimates for the intermediate level sets.*

By "large deviations" we mean, intentionally vaguely, any kind of deviation whose probability tends to zero as $N \to \infty$. We note that this may be relevant for instance already in attempts to connect the limit law of the intermediate level sets to the construction of the LQG measure. Indeed, one is effectively asking to integrate the process η_N^D against the function

$$f(x,h) := \mathrm{e}^{\beta h} \quad \text{where} \quad \beta := \lambda\alpha. \tag{16.2}$$

Plugging this formally into the limit (which is of course not mathematically justified) yields a diverging integral. This indicates that the problem requires some control of the deviations and/or the support of the LQG measure.

16.2 At and Near the Absolute Maximum

Moving to the extremal values, we start by a conjecture which, if true, might greatly simplify the existing proof of the uniqueness of the subsequential limits η^D of $\{\eta^D_{N,r_N} : N \geq 1\}$. Indeed, in the version of the proof presented in Lecture 12, a great deal of effort is spent on the verification of the Laplace transform tail (10.33) for the measures Z^D extracted from η^D via Theorem 9.6. This would not be necessary if we could resolve affirmatively:

Conjecture 16.3 (By-passing the Laplace transform tail) *Prove that any family of non-trivial random measures $\{Z^D : D \in \mathfrak{D}\}$ obeying (1–4) in Theorem 10.15 satisfies also property (5) for some $\hat{c} > 0$.*

The reason why we believe this to be possible stems (somewhat vaguely) from the analogy of the Gibbs-Markov property along, say, partitions of squares into smaller squares, with fixed-point equations for the so-called *smoothing transformations.* These have been classified in Durrett and Liggett [68].

Our next point of interest is Theorem 12.19 which, we recall, states the local limit theorem for the absolute maximum. The limiting position (scaled by N) for the maximum restricted above $m_N + t$ is then distributed according to the density $x \mapsto \rho^D(x,t)$ in (12.86). From the estimate (10.33) we are able to conclude

$$\rho^D(x,t) \sim c_\star t \mathrm{e}^{-\alpha t} r_D(x)^2, \quad t \to \infty, \tag{16.3}$$

but we have not been able to characterize ρ^D explicitly for any finite t. Still, ρ^D is determined by Z^D, which should be universal (up to a constant multiple) for a whole class of logarithmically correlated fields in $d = 2$. By analogy with the role of the KPP equation plays in the analysis of the Branching Brownian Motion (cf Bramson [35]), we hope the following is perhaps achievable:

Problem 16.4 *(Nailing density explicitly) Determine ρ^D explicitly or at least characterize it via, e.g., a PDE that has a unique solution.*

Another question related to the extremal values is the crossover between the regime of intermediate level sets and those within order-unity of the absolute maximum. We state this as:

Conjecture 16.5 (Size of deep extremal level set) *There is $c \in (0,\infty)$ such that, for each open $A \subseteq D$ (with* $\mathrm{Leb}(\partial A) = 0$*)*

$$\frac{1}{t}\mathrm{e}^{-\alpha t}\,\#\bigl\{x\in\Gamma_N^D(t)\colon x/N\in A\bigr\}\underset{\substack{N\to\infty\\ t\to\infty}}{\overset{\text{law}}{\longrightarrow}} cZ^D(A), \tag{16.4}$$

with convergence in law valid even jointly for any finite number of disjoint open sets $A_1,\dots,A_k\subseteq D$ *(with* $\mathrm{Leb}(\partial A_j)=0$ *for all* j*).*

Note that this constitutes a stronger version of Theorem 9.1 because (16.4) implies

$$\lim_{r\downarrow 0}\,\liminf_{t\to\infty}\,\liminf_{N\to\infty}\,P\Bigl(rt\mathrm{e}^{\alpha t}\le|\Gamma_N^D(t)|\le r^{-1}t\mathrm{e}^{\alpha t}\Bigr)=1. \tag{16.5}$$

Writing $\Gamma_N(A,t)$ for the set on the left-hand side of (16.4), Theorem 9.2 and truncation of the absolute maximum to lie below $m_N+\sqrt{t}$ implies, for each $\lambda>0$,

$$E\bigl(\mathrm{e}^{-\lambda|\Gamma_N^D(A,t)|}\bigr)=E\Biggl(\exp\Bigl\{-Z^D(A)\mathrm{e}^{\alpha t}\int_0^{t+\sqrt{t}}\mathrm{d}h\,\mathrm{e}^{-\alpha h}E_\nu(1-\mathrm{e}^{-\lambda f_h(\phi)})\Bigr\}\Biggr)+o(1)\,, \tag{16.6}$$

where $o(1)\to 0$ as $N\to\infty$ followed by $t\to\infty$ and

$$f_h(\phi):=\sum_{z\in\mathbb{Z}^2}1_{[0,h]}(\phi_z). \tag{16.7}$$

Using the above for $\lambda:=t^{-1}\mathrm{e}^{-\alpha t}$, Conjecture 16.5 seems closely linked to:

Conjecture 16.6 *For* f_h *as above,*

$$\lim_{h\to\infty}\mathrm{e}^{-\alpha h}E_\nu(f_h)\ \textit{exists in}\ (0,\infty)\,. \tag{16.8}$$

Note that (16.4) is similar to the limit law for the size of Daviaud's level sets proved in Corollary 2.9.

Another conjecture concerning deep extremal level sets is motivated by the question on at what levels is the critical LQG measure mainly supported and how is the DGFF distributed thereabout. The Seneta–Heyde normalization indicates that this happens at levels order $\sqrt{\log N}$ below the absolute maximum. Discussions of this problem with O. Louidor during the summer school led to:

Conjecture 16.7 (Profile of near-extremal level sets) *There is a constant* $c>0$ *such that (writing* h *for the DGFF in* D_N*),*

$$\eta_N^D:=\frac{1}{\log N}\sum_{x\in D_N}\mathrm{e}^{\alpha(h_x-m_N)}\,\delta_{x/N}\otimes\delta_{\frac{m_N-h_x}{\sqrt{\log N}}}\,, \tag{16.9}$$

where m_N *is as in (7.9), obeys*

$$\eta_N^D\underset{N\to\infty}{\overset{\text{law}}{\longrightarrow}} c\,Z^D(\mathrm{d}x)\otimes 1_{[0,\infty)}(h)\,h\,\mathrm{e}^{-\frac{h^2}{2g}}\,\mathrm{d}h\,. \tag{16.10}$$

Here Z^D is the measure in Theorem 9.3.

In particular, the density

$$h \mapsto 1_{[0,\infty)}(h)\, h\, \mathrm{e}^{-\frac{h^2}{2g}} \tag{16.11}$$

gives the asymptotic "profile" of the values of the DGFF that contribute to the critical LQG measure at scales order-$\sqrt{\log N}$ below m_N.

16.3 Universality

A very natural question to ask is to what extent are the various results reported here universal with respect to various changes of the underlying setting. As a starter, we pose:

Problem 16.8 *(Log-correlated Gaussian fields in $d = 2$) For each $N \geq 1$, consider a Gaussian field in D_N whose covariance matrix C_N scales, as described in Theorem 1.17, to the continuum Green function in the underlying (continuum) domain and such that for all $z \in \mathbb{Z}^2$, the limit*

$$\lim_{N\to\infty} \bigl[C_N(x, x+z) - C_N(x,x)\bigr] \tag{16.12}$$

exists, is independent of x and is uniform in x with $\operatorname{dist}_\infty(x, D_N^{\mathrm{c}}) > \epsilon N$, *for every $\epsilon >$* 0. *Prove the analogues of Theorems 2.7 and 9.3.*

Some instances of this can be resolved based on the results already presented here. For instance, if C_N is given as

$$C_N(x, y) = G^N(x, y) + F(x - y), \tag{16.13}$$

where $F\colon \mathbb{Z}^2 \to \mathbb{R}$ a bounded, positive semi-definite function that decays fast to zero (as the argument tends to infinity), one can realize $\mathcal{N}(0, C_N)$ as the sum of the DGFF plus a Gaussian field with fast-decaying correlations. The effect of the additional Gaussian field can be handled within the structure of Theorems 2.7 and 9.3. The point is thus to go beyond such elementary variations.

Another question is that of generalizations to log-correlated Gaussian processes in dimensions $d \neq 2$. Naturally, here the choice of the covariance is considerably less canonical. Notwithstanding, convergence of the centered maximum to a non-degenerate limit law has already been established in a class of such fields (in any $d \geq 1$) by Ding, Roy and Zeitouni [60]. (This offers some headway on Problem 16.8 as well.) The next goal is an extension to the full extremal process as stated for the DGFF in Theorem 9.3.

The question of universality becomes perhaps even more profound once we realize that there are other non-Gaussian models where one expects the results reported in these notes to be relevant. The first one of these is the class of *Gradient Models*. These are random fields with law in finite sets $V \subseteq \mathbb{Z}^2$ given by

$$P(h^V \in A) := \frac{1}{\text{norm.}} \int_A \mathrm{e}^{-\sum_{(x,y)\in \mathrm{E}(\mathbb{Z}^d)} \mathcal{V}(h_x - h_y)} \prod_{x\in V} \mathrm{d}h_x \prod_{x\notin V} \delta_0(\mathrm{d}h_x), \qquad (16.14)$$

where $\mathcal{V}\colon \mathbb{R} \to \mathbb{R}$ is the potential which is assumed even, continuous, bounded from below and with superlinear growth at infinity. An inspection of (1.1) shows that the DGFF is included in this class via

$$\mathcal{V}(h) := \frac{1}{4d} h^2. \qquad (16.15)$$

The formula (16.14) defines the field with zero boundary conditions, although other boundary conditions can be considered as well (although they cannot be easily reduced to zero boundary conditions outside the Gaussian case).

Much is known about these models when $\mathcal{V}$ is uniformly strictly convex (i.e., for $\mathcal{V}''$ positive and bounded away from zero and infinity). Here through the work of Funaki and Spohn [73], Naddaf and Spencer [101], Giacomin, Olla and Spohn [76] we know that the field tends to a linear transform of CGFF in the thermodynamic limit (interpreted in the sense of gradients in $d = 1, 2$), and by Miller [94] also in the scaling limit in $d = 2$. Recently, Belius and Wu [17] have proved the counterpart of Theorem 2.1 by identifying the leading-order growth of the absolute maximum. Wu and Zeitouni [125] have then extended the Dekking–Host subsequential tightness argument from Lemma 7.1 to this case as well.

The situation seems ripe to tackle more complicated questions such as those discussed in these notes. We believe that the easiest point of entry is via:

Problem 16.9 *(Intermediate level sets for gradient fields) Prove the scaling limit of intermediate level sets for gradient models with uniformly strictly convex potentials.*

This should be manageable as the main technical input in these are moment calculations that lie at the heart of [17] which deals with the case where they should, in fact, be hardest to carry out.

Perhaps even closer to the subject of these notes is the problem of DGFF generated by a *Random Conductance Model* (see [22]): Assign to each edge e in $\mathbb{Z}^2$ a non-negative conductance $c(e)$. We may in fact assume that the conductances are *uniformly elliptic*, meaning that

$$\exists \lambda \in (0,1)\ \forall e \in E(\mathbb{Z}^2)\colon \qquad c(e) \in [\lambda, \lambda^{-1}] \ \text{ a.s.} \qquad (16.16)$$

Given a realization of the conductances, we can define an associated DGFF in domain $D \subsetneq \mathbb{Z}^2$ as $\mathcal{N}(0, G^D)$ where $G^D := (1 - \mathsf{P})^{-1}$ is, for P related to the conductances as discussed at the beginning of Sect. 15.1, the Green function of the random walk among random conductances. We then pose:

Problem 16.10 *(DGFF over Random Conductance Model) Assume that the conductances are uniformly elliptic and their law is invariant and ergodic with respect to the shifts of $\mathbb{Z}^2$. Prove that for almost every realization of the conductances, the analogues of Theorems 2.7 and 9.3 hold.*

This could in principle be easier than Problem 16.9 due to the underlying Gaussian nature of the field. Notwithstanding, the use of homogenization theory, a key tool driving many studies of gradient models, can hardly be avoided. Incidentally, Problem 16.10 still falls under the umbrella of gradient models, albeit with non-convex interactions. The connection to the DGFF over random conductances yields much information about the corresponding gradient models as well; see the papers of the author with Kotecký [23] and Spohn [28].

Another problem of interest that should be close to the DGFF is that of *local time* of the simple random walk. The connection to the DGFF arises via the *Dynkin isomorphism* (usually attributed to Dynkin [69] though going back to Symanzik [121]) or, more accurately, the *second Ray-Knight theorem* (see, e.g., Eisenbaum et al. [71]). This tool has been useful in determining the asymptotic behavior of the *cover time* (namely, in the studies of the leading-order behavior by Dembo, Peres, Rosen and Zeitouni [53], Ding [56], Ding, Lee and Peres [59], in attempts to nail the subleading order in Belius and Kistler [16] and Abe [2] and, very recently, in a proof of tightness under proper centering in Belius, Rosen and Zeitouni [18]).

We wish to apply the connection to the study of the local time ℓ_t for a continuous-time (constant-speed) random walk on $N \times N$ torus $\mathbb{T}_N$ in $\mathbb{Z}^2$ started at 0. We choose the parametrization by the actual time the walk spends at 0; i.e., $\ell_t(0) = t$ for all $t \geq 0$. For $\theta > 0$, define the time scale

$$t_\theta := \theta[g \log N]^2. \tag{16.17}$$

Then the total time of the walk, $\sum_x \ell_{t_\theta}(x)$, is very close to a θ-multiple of the cover time. It is known that $(\ell_t - t)/\sqrt{t}$ scales to a DGFF on $\mathbb{T}_N \smallsetminus \{0\}$ in the limit $t \to \infty$. This motivated Abe [1] to show that, as $N \to \infty$, the maximum of ℓ_{t_θ} scales as

$$\frac{\max_{x\in\mathbb{T}_N} \ell_{t_\theta}(x) - t_\theta}{\sqrt{t_\theta}} = \left(1 + \frac{1}{2\sqrt{\theta}} + o(1)\right) 2\sqrt{g} \log N\,. \tag{16.18}$$

He also studied the cardinality of the level set

$$\left\{x \in \mathbb{T}_N \colon \frac{\ell_{t_\theta}(x) - t_\theta}{\sqrt{t_\theta}} \geq 2\eta\sqrt{g} \log N\right\} \tag{16.19}$$

and proved a result analogous, albeit with *different* exponents, to Daviaud's [51] (see Theorem 2.3). Some correspondence with the DGFF at the level of intermediate level sets and maxima thus exists but is far from straightforward.

The proofs of [1] are based on moment calculations under a suitable truncation, which is also the method used for proving (2.7). It thus appears that a good starting point would be to solve:

Problem 16.11 *Show that a suitably normalized level set in* (16.19), *or a point measure of the kind* (2.29), *admits a scaling limit for all* $0 < \eta < 1 + \frac{1}{2\sqrt{\theta}}$.

We remark that Abe [2] studied the corresponding problem also for the random walk on a homogeneous tree of depth n. By embedding the tree ultrametrically into a unit interval, he managed to describe the full scaling limit of the process of extremal local maxima, much in the spirit of Theorem 9.6. Remarkably, the limit process coincides (up to a constant multiple of the intensity measure) with that for the DGFF on the tree. A description of the cluster process is in the works [3].

Update in revision: As shown in a recent posting by Abe and the author [4], the intermediate level sets for the local time have the same scaling limit as the DGFF, albeit in a different (and considerably more convenient) parametrization than suggested in (16.19). Concerning the distributional limit of the cover time; this has recently been achieved for the random walk on a homogeneous tree by Cortines, Louidor and Saglietti [48].

16.4 Random Walk in DGFF Landscape

The next set of problems we wish to discuss deals with the random walk driven by the DGFF. Here the first (and obvious) follow-up question is the complementary inequality to that stated in Corollary 13.4:

Problem 16.12 *(Subdiffusive upper bound) Using the notation from* (13.9), *show that, for each* $\beta > 0$ *and each* $\delta > 0$,

$$P_h^0\Big(|X_T| \le T^{1/\psi(\beta)}\mathrm{e}^{(\log T)^{1/2+\delta}}\Big) \underset{T\to\infty}{\longrightarrow} 1\,, \quad \textit{in } \mathbb{P}\textit{-probability}. \tag{16.20}$$

A key issue in here is to show that the exit time $\tau_{B(N)^c}$ is (reasonably) concentrated around its expectation. We believe that this can perhaps be resolved by having better control of the concentration of the effective resistance.

Another question concerns a possible scaling limit of the random walk. Heuristic considerations make it fairly clear that the limit process cannot be the *Liouville Brownian Motion* (LBM), introduced in Berestycki [19] and Garban, Rhodes and Vargas [75]. Indeed, this process exists only for $\beta \le \beta_c$ (with the critical version constructed by Rhodes and Vargas [106]) and simulations of the associated Liouville Random Walk, which is a continuous-time simple symmetric random walk with exponential holding time with parameter $\mathrm{e}^{\beta h_x}$ at x, show a dramatic change in the behavior as β increases through β_c; see Fig. 33. (The supercritical walk is trapped for a majority of its time with transitions described by a K-process; cf Cortines, Gold

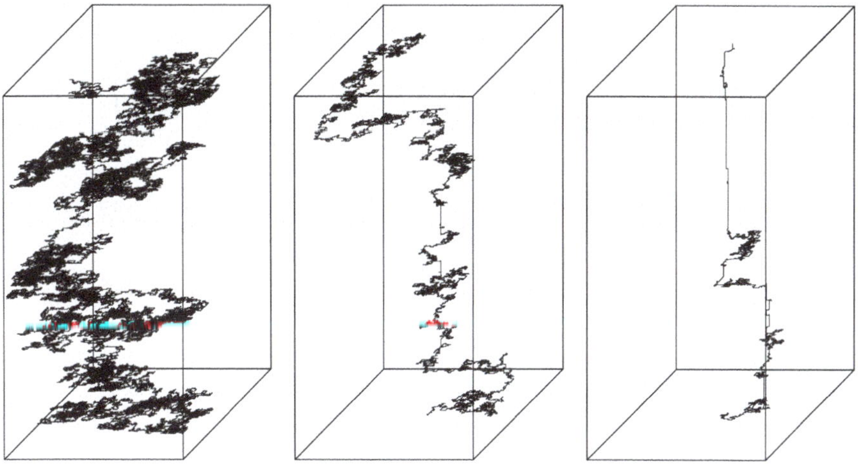

Fig. 33 Runs of 1,00,000 steps of the Liouville random walk with β equal to 0.2, 0.6 and 1.2 multiples of β_c. Time runs upwards along the vertical axis. For $\beta > \beta_c$ the walk becomes trapped for a majority of its time

and Louidor [47].) No such dramatic change seems to occur for the random walk with the transition probabilities in (13.1), at any $\beta > 0$.

The reason why (we believe) the LBM is not the limit process is because its paths, once stripped of their specific parametrization, are those of the two-dimensional Brownian motion. In particular, their law is completely decoupled from the underlying CGFF. On the other hand, the limit of our random walk must feel the drift towards larger "values" of the CGFF which will in turn couple the law of the paths (still regarded as geometric objects) to the (limiting) CGFF.

We thus have to define the limit process through the important features of the discrete problem. Here we observe that the random walk is reversible with respect to the measure π_h defined in (13.12). As $\pi_h(x) \approx e^{2\beta h_x}$, whose limit is (for $\beta < \tilde{\beta}_c = \beta_c/2$) a LQG measure, one might expect that the resulting process should be reversible with respect to the LQG measure. Although this remains problematic because of the restriction on β, we can perhaps make sense of this by imposing reversibility only for a regularized version of the CGFF and taking suitable limits.

Consider a sample of a smooth, centered Gaussian process $\{h_\epsilon(x) : x \in D\}$ such that, for $N := \lfloor \epsilon^{-1} \rfloor$, we have

$$\mathrm{Cov}\big(h_\epsilon(x), h_\epsilon(y)\big) = G^{D_N}\big(\lfloor Nx \rfloor, \lfloor Ny \rfloor\big) + o(1), \quad \epsilon \downarrow 0. \tag{16.21}$$

Then h_ϵ tends in law to the CGFF as $\epsilon \downarrow 0$. Define a diffusion X^ϵ via the Langevin equation

$$\mathrm{d}X_t^\epsilon = \beta \nabla h_\epsilon(X_t^\epsilon)\mathrm{d}t + \sqrt{2}\,\mathrm{d}B_t\,, \tag{16.22}$$

where $\{B_t : t \geq 0\}$ is a standard Brownian motion. The (unnormalized) Gibbs measure $e^{\beta h_\epsilon(x)}\mathrm{d}x$ is then stationary and reversible for X^ϵ. Moreover, under a suitable

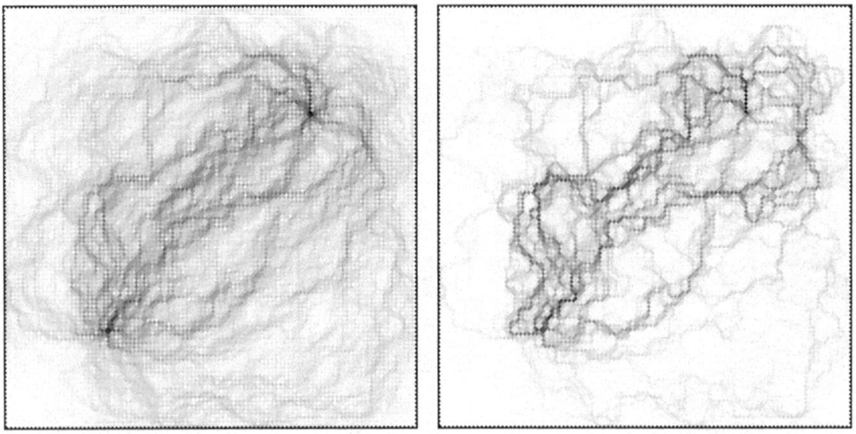

Fig. 34 Samples of the current minimizing the effective resistance between two points on the diagonal of the square (the dark spots in the left figure), for the network on the square with conductances related to the DGFF on the square as in (13.13) at inverse temperatures $\beta = 0.4\,\beta_c$ (left) and $\beta = \beta_c$ (right). The shade of each pixel is proportional to intensity of the current. The current seems to be carried along well defined paths, particularly, in the figure on the right

time change, the process X^ϵ mimics closely the dynamics of the above random walk on boxes of side-length ϵ^{-1} (and β replaced by 2β). We then pose:

Problem 16.13 *(Liouville Langevin Motion) Under a suitable time change for X^ϵ, prove that the pair (h_ϵ, X^ϵ) converges as $\epsilon \downarrow 0$ jointly in law to (h, X), where h is a CGFF and X is a process with continuous (non-constant) paths (correlated with h).*

We propose to call the limit process X the *Liouville Langevin Motion*, due to the connection with the Langevin equation (16.22) and LQG measure. One strategy for tackling the above problem is to prove characterizing all possible limit processes through their behavior under conformal maps and/or restrictions to subdomain via the Gibbs-Markov property.

16.5 DGFF Electric Network

The final set of problems to be discussed here concern the objects that we used to control the random walk in DGFF landscape. Consider first the effective resistivity $R_{\text{eff}}(x, B(N)^c)$. Proposition 15.9 suggests that, for $x := 0$, the logarithm thereof is to the leading order the maximum of the random walk S_n associated with the concentric decomposition. Upon scaling by $\sqrt{n}$, this maximum tends in law to absolute value of a normal random variable. The question is:

Problem 16.14 *For any sequence $\{D_N : N \geq 1\}$ of admissible approximations of domain $D \in \mathfrak{D}$, characterize the limit distribution of*

$$x \mapsto (\log N)^{-1/2} \log R_{\text{eff}}(\lfloor xN \rfloor, D_N^c) \tag{16.23}$$

as $N \to \infty$.

In light of the concentric decomposition, a natural first guess for the limit process is the CGFF but proving this explicitly seems far from clear.

Next consider the problem of computing $R_{\text{eff}}(u, v)$ for the network in a square $[-N, N]^2 \cap \mathbb{Z}^2$, the resistances associated to the DGFF via (13.13) and u and v two generic points in the square. The question we ask: Is there a way to describe the scaling limit of the minimizing current $i_\star$? Tackling this directly may end up to be somewhat hard, because (as seen in Fig. 34) the current is supported on a sparse (fractal) set.

A possible strategy to address this problem is based on the path-representation of the effective resistance from Proposition 14.2. Indeed, the proof contains an algorithm that identifies a set of simple paths $\mathcal{P}^\star$ from u to v and resistances $\{r^\star_{e,P} : e \in \mathsf{E}, P \in \mathcal{P}^\star\}$ that achieve the infima in (14.2). Using these objects, $i_\star$ can be decomposed into a family of currents $\{i_P : P \in \mathcal{P}^\star\}$ from u to v, such that $i_P(e) = 0$ for $e \notin P$ and with $i_P(e)$ equal to

$$i_P := \frac{R_{\text{eff}}(u, v)}{\sum_{e \in P} r^\star_{e,P}} \tag{16.24}$$

on each edge $e \in P$ oriented in the direction of P (recall that P is simple). Noting that

$$\sum_{P \in \mathcal{P}^\star} i_P = \text{val}(i_\star) = 1\,, \tag{16.25}$$

this recasts the computation of, say, the net current flowing through a linear segment $[a, b]$ between two points $a, b \in D$ as the difference of the probability that a *random path* crosses $[a, b]$ in the given direction and the probability that it crosses $[a, b]$ in the opposite direction, in the distribution where path P (from u to v) is chosen with probability i_P. We pose:

Problem 16.15 *(Scaling limit of minimizing current) Prove that the joint law of the DGFF and the random path P admits a non-degenerate scaling limit as* $N \to \infty$.

We envision tackling this using the strategy that we have used already a few times throughout these notes: First prove tightness, then extract subsequential limits and then characterize the limit law uniquely by, e.g., use of the Gibbs-Markov property and/or conformal maps. We suspect that the limit path law (in Problem 16.15) will be closely related to the scaling limit of the actual random walk process (in Problem 16.13); e.g., through a suitable loop erasure.

Once the continuum version of the above picture has been elucidated, the following additional questions (suggested by J. Ding) come to mind:

Problem 16.16 *What is the Hausdorff dimension of the (continuum) random path? And, say, for the electric current between opposite boundaries of a square box, what is the dimension of the support for the energy?*

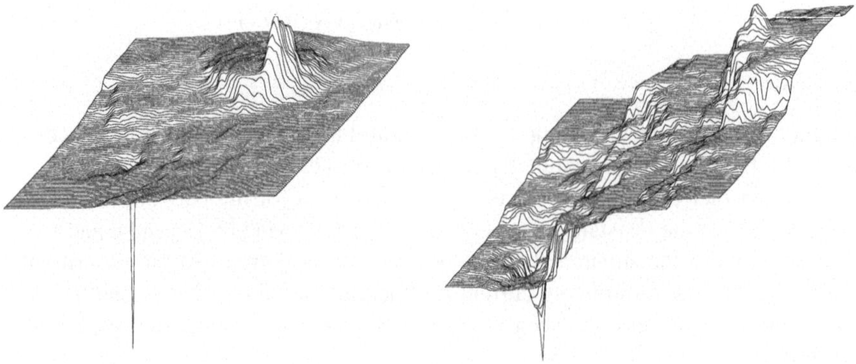

Fig. 35 Two samples of the voltage realizing the effective conductances $C_{\text{eff}}(u, v)$ in a box of side 100 and u and v two points on the diagonal. The conductances are related to the DGFF in the box via (13.13)

Similar geometric ideas can perhaps be used to address the computation of the voltage minimizing the effective conductance $C_{\text{eff}}(u, v)$; see Fig. 35 for some illustrations. There collections of paths will be replaced by collections of nested cutsets (as output by the algorithm in the proof of Proposition 14.7).

An alternative approach to the latter problem might thus be based on the observation that the cutsets are nested, and thus can be thought of as time snapshots of a *growth process*. The use of Loewner evolution for the description of the kinematics of the process naturally springs to mind.

References

1. Y. Abe (2015). Maximum and minimum of local times for two-dimensional random walk. *Electron. Commun. Probab.* **20**, paper no. 22, 14 pp.
2. Y. Abe (2018). Extremes of local times for simple random walks on symmetric trees. *Electron. J. Probab.* **23**, paper no. 40, 41 pp.
3. Y. Abe and M. Biskup (2017). In preparation.
4. Y. Abe and M. Biskup (2019). Exceptional points of two-dimensional random walks at multiples of the cover time. arXiv:1903.04045.
5. R.J. Adler (1990). *An introduction to continuity, extrema, and related topics for general Gaussian processes*. Institute of Mathematical Statistics Lecture Notes–Monograph Series, vol. 12. Institute of Mathematical Statistics, Hayward, CA, x+160 pp.
6. R.J. Adler and J.E. Taylor (2007). *Random fields and geometry*. Springer Monographs in Mathematics. Springer, New York.
7. E. Aïdékon (2013). Convergence in law of the minimum of a branching random walk. *Ann. Probab.* **41**, no. 3A, 1362–1426.
8. E. Aïdékon, J. Berestycki, E. Brunet and Z. Shi (2013). The branching Brownian motion seen from its tip. *Probab. Theory Rel. Fields* **157**, no. 1–2, 405–451.
9. S. Andres, J.-D. Deuschel, M. Slowik (2015). Invariance principle for the random conductance model in a degenerate ergodic environment, *Ann. Probab.*, **43**, no. 4, 1866–1891.

10. L.-P. Arguin and O. Zindy (2015). Poisson-Dirichlet Statistics for the extremes of the two-dimensional discrete Gaussian Free Field. *Electron. J. Probab.* **20**, paper no. 59, 19 pp.
11. L.-P. Arguin, A. Bovier, and N. Kistler (2011). Genealogy of extremal particles of branching brownian motion. *Commun. Pure Appl. Math.* **64**, 1647–1676.
12. L.-P. Arguin, A. Bovier, and N. Kistler (2011). The extremal process of branching Brownian motion. *Probab. Theory Rel. Fields* **157**, no. 3–4, 535–574.
13. L.-P. Arguin, A. Bovier, and N. Kistler (2012). Poissonian statistics in the extremal process of branching brownian motion. *Ann. Appl. Probab.* **22**, no. 4, 1693–1711.
14. S. Armstrong, T. Kuusi, J.-C. Mourrat (2017). Quantitative stochastic homogenization and large-scale regularity. arXiv:1705.05300.
15. V. Beffara and H. Duminil-Copin (2012). The self-dual point of the two-dimensional random-cluster model is critical for $q \geq 1$. *Probab. Theory Rel. Fields* **153**, no. 3–4, 511–542.
16. D. Belius and N. Kistler (2017). The subleading order of two dimensional cover times. *Probab. Theory Rel. Fields* **167**, no. 1–2, 461–552.
17. D. Belius and W. Wu (2016). Maximum of the Ginzburg-Landau fields, arXiv:1610.04195.
18. D. Belius, J. Rosen and O. Zeitouni (2017). Tightness for the cover time of S^2. arXiv:1711.02845.
19. N. Berestycki (2015). Diffusion in planar Liouville quantum gravity. *Ann. Inst. Henri Poincaré Probab. Stat.* **51**, no. 3, 947–964.
20. N. Berestycki (2017). An elementary approach to Gaussian multiplicative chaos. *Electron. J. Probab.* **22**, paper no. 27, 12 pp.
21. K. A. Berman and M. H. Konsowa (1990). Random paths and cuts, electrical networks, and reversible Markov chains. *SIAM J. Discrete Math.* **3**, no. 3, 311–319.
22. M. Biskup (2011). Recent progress on the Random Conductance Model. *Prob. Surveys* **8** 294–373.
23. M. Biskup and R. Kotecký (2007). *Phase coexistence of gradient Gibbs states*. Probab. Theory Rel. Fields **139**, no. 1–2, 1–39.
24. M. Biskup and O. Louidor (2014). Conformal symmetries in the extremal process of two-dimensional discrete Gaussian Free Field. arXiv:1410.4676.
25. M. Biskup and O. Louidor (2016). Extreme local extrema of two-dimensional discrete Gaussian free field. *Commun. Math. Phys.* **345**, no. 1, 271–304.
26. M. Biskup and O. Louidor (2019). On intermediate level sets of two-dimensional discrete Gaussian free field. *Ann. Inst. Henri Poincaré* **55**, no. 4, 1948–1987.
27. M. Biskup and O. Louidor (2018). Full extremal process, cluster law and freezing for two-dimensional discrete Gaussian free field. *Adv. Math.* **330** 589–687.
28. M. Biskup and H. Spohn (2011). Scaling limit for a class of gradient fields with non-convex potentials. *Ann. Probab.* **39**, no. 1, 224–251.
29. M. Biskup, J. Ding and S. Goswami (2016). Return probability and recurrence for the random walk driven by two-dimensional Gaussian free field. *Commun. Math. Phys.* (to appear) arXiv:1611.03901.
30. E. Bolthausen, J.-D. Deuschel and G. Giacomin (2001). Entropic repulsion and the maximum of the two-dimensional harmonic crystal. *Ann. Probab.* **29**, no. 4, 1670–1692.
31. E. Bolthausen, J.-D. Deuschel, and O. Zeitouni (2011). Recursions and tightness for the maximum of the discrete, two dimensional gaussian free field. *Elect. Commun. Probab.* **16**, 114–119.
32. C. Borell (1975). The Brunn-Minkowski inequality in Gauss space. *Invent. Math.* **30**, no. 2, 207–216.
33. A. Bovier (2015). From spin glasses to Branching Brownian Motion – and back?. In: M. Biskup, J. Černý, R. Kotecký (eds). *Random Walks, Random Fields, and Disordered Systems*. Lecture Notes in Mathematics, vol 2144. Springer, Cham.
34. A. Bovier and L. Hartung (2017). Extended convergence of the extremal process of branching Brownian motion. *Ann. Appl. Probab.* **27**, no. 3, 1756–1777.
35. M. Bramson (1978). Maximal displacement of branching Brownian motion. *Commun. Pure Appl. Math.* **31**, no. 5, 531–581.

36. M. Bramson (1983). Convergence of solutions of the Kolmogorov equation to traveling waves. *Mem. Amer. Math. Soc.* **44**, no. 285, iv+190.
37. M. Bramson and O. Zeitouni (2012). Tightness of the recentered maximum of the two-dimensional discrete Gaussian free field. *Comm. Pure Appl. Math.* **65**, 1–20.
38. M. Bramson, J. Ding and O. Zeitouni (2016). Convergence in law of the maximum of the two-dimensional discrete Gaussian free field. *Commun. Pure Appl. Math* **69**, no. 1, 62–123.
39. D. Carpentier and P. Le Doussal (2001). Glass transition of a particle in a random potential, front selection in nonlinear renormalization group, and entropic phenomena in Liouville and sinh-Gordon models, *Phys. Rev. E* **63**, 026110.
40. A. K. Chandra, P. Raghavan, W. L. Ruzzo, R. Smolensky, and P. Tiwari (1996). The electrical resistance of a graph captures its commute and cover times. *Comput. Complexity*, **6**, no. 4, 312–340.
41. S. Chatterjee, A. Dembo and J. Ding (2013). On level sets of Gaussian fields. arXiv:1310.5175.
42. A. Chiarini, A. Cipriani, R.S. Hazra (2015). A note on the extremal process of the supercritical Gaussian Free Field. *Electron. Commun. Probab.* **20**, paper no. 74, 10 pp.
43. A. Chiarini, A. Cipriani, R.S. Hazra (2016). Extremes of some Gaussian random interfaces. *J. Statist. Phys.* **165**, no. 3, 521–544.
44. A. Chiarini, A. Cipriani, R.S. Hazra (2016). Extremes of the supercritical Gaussian Free Field. *ALEA, Lat. Am. J. Probab. Math. Stat.* **13** 711–724.
45. G. Choquet and J. Deny (1960). Sur l'équation de convolution $\mu = \mu * \sigma$. *C.R. Acad. Sci. Paris* **250** 799–801.
46. K.L. Chung (1948). On the maximum partial sums of sequences of independent random variables. *Trans. Amer. Math. Soc.* **64** 205–233.
47. A. Cortines, J. Gold and O. Louidor (2018). Dynamical freezing in a spin glass system with logarithmic correlations. *Electron. J. Probab.* **23**, paper no. 59, 31 pp.
48. A. Cortines, O. Louidor and S. Saglietti (2018). A scaling limit for the cover time of the binary tree. arXiv:1812.10101.
49. A. Cortines, L. Hartung and O. Louidor (2019). Decorated random walk restricted to stay below a curve (supplement material). arXiv:1902.10079.
50. T.J. Cox (1977). Entrance laws for Markov chains. *Ann. Probab.* **5**, no. 4, 533–549.
51. O. Daviaud (2006). Extremes of the discrete two-dimensional Gaussian free field. *Ann. Probab.* **34**, 962–986.
52. F. M. Dekking and B. Host (1991). Limit distributions for minimal displacement of branching random walks, *Probab. Theory Rel. Fields* **90**, 403–426.
53. A. Dembo, Y. Peres, J. Rosen, and O. Zeitouni (2004). Cover times for Brownian motion and random walks in two dimensions. *Ann. of Math. (2)* **160**, 433–464.
54. J. Deny. (1960). Sur l'équation de convolution $\mu = \mu * \sigma$. *Séminaire Brelot-Choquet-Deny. Théorie du potentiel* **4**, 1–11.
55. B. Derrida and H. Spohn (1988). Polymers on disordered trees, spin glasses, and traveling waves. *J. Statist. Phys.* **51**, no. 5–6, 817–840.
56. J. Ding (2012). On cover times for 2D lattices. *Electron. J. Probab.* **17**, Paper no. 45.
57. J. Ding (2013). Exponential and double exponential tails for maximum of two-dimensional discrete Gaussian free field. *Probab. Theory Rel. Fields* **157**, no. 1–2, 285–299.
58. J. Ding and O. Zeitouni (2014). Extreme values for two-dimensional discrete Gaussian free field. *Ann. Probab.* **42**, no. 4, 1480–1515.
59. J. Ding, J. R. Lee, and Y. Peres (2012). Cover times, blanket times, and majorizing measures. *Ann. of Math. (2)* **175**, no. 3, 1409–1471.
60. J. Ding, R. Roy and O. Zeitouni (2017). Convergence of the centered maximum of log-correlated Gaussian fields. *Ann. Probab.* **45**, no. 6A, 3886–3928.
61. P.G. Doyle and J.L. Snell (1984). *Random walks and electric networks*. Carus Mathematical Monographs, 22. Mathematical Association of America, Washington, DC.
62. R.M. Dudley (1967). The sizes of compact subsets of Hilbert space and continuity of Gaussian processes. *J. Funct. Anal.* **1**, no. 3, 290–330.

63. H. Duminil-Copin (2017). Lectures on the Ising and Potts models on the hypercubic lattice. arXiv:1707.00520.
64. H. Duminil-Copin, C. Hongler, and P. Nolin (2011). Connection probabilities and RSW-type bounds for the two-dimensional FK Ising model. *Comm. Pure Appl. Math.* **64**, no. 9, 1165–1198.
65. B. Duplantier and S. Sheffield (2011). Liouville quantum gravity and KPZ. *Invent. Math.* **185**, no. 2, 333–393.
66. B. Duplantier, R. Rhodes, S. Sheffield and V. Vargas (2014). Critical Gaussian multiplicative chaos: Convergence of the derivative martingale. *Ann. Probab.* **42**, no. 5, 1769–1808.
67. B. Duplantier, R. Rhodes, S. Sheffield and V. Vargas (2014). Renormalization of critical Gaussian multiplicative chaos and KPZ formula. *Commun. Math. Phys.* **330**, no. 1, 283–330.
68. R. Durrett and T.M. Liggett (1983). Fixed points of the smoothing transformation. *Probab. Theory Rel. Fields* **64**, no. 3, 275–301.
69. E.B. Dynkin (1984). Gaussian and non-Gaussian random fields associated with Markov processes. *J. Funct. Anal.* **55**, no. 3, 344–376.
70. F. J. Dyson (1962). A Brownian-motion model for the eigenvalues of a random matrix. *J. Math. Phys.* **3** 1191–1198.
71. N. Eisenbaum, H. Kaspi, M.B. Marcus, J. Rosen and Z. Shi (2000). A Ray-Knight theorem for symmetric Markov processes. *Ann. Probab.* **28**, no. 4, 1781–1796.
72. C.W. Fortuin, P.W. Kasteleyn, J. Ginibre (1971). Correlation inequalities on some partially ordered sets. *Commun. Math. Phys.* **22**, no. 2, 89–103.
73. T. Funaki and H. Spohn (1997). Motion by mean curvature from the Ginzburg-Landau $\nabla\phi$ interface model. *Commun. Math. Phys.* **185**, no. 1, 1–36.
74. Y. V. Fyodorov and J.-P. Bouchaud (2008). Freezing and extreme-value statistics in a random energy model with logarithmically correlated potential. *J. Phys. A* **41**, no. 37, 372001.
75. C. Garban, R. Rhodes, and V. Vargas (2016). Liouville Brownian motion. *Ann. Probab.* **44**, no 4, 3076–3110.
76. G. Giacomin, S. Olla and H. Spohn (2001). Equilibrium fluctuations for $\nabla\varphi$ interface model. *Ann. Probab.* **29**, no. 3, 1138–1172.
77. G. Grimmett (2006). *The Random-Cluster Model.* Grundlehren der Mathematischen Wissenschaften, vol. 333. Springer-Verlag, Berlin.
78. X. Hu, J. Miller and Y. Peres (2010). Thick point of the Gaussian free field. *Ann. Probab.* **38**, no. 2, 896–926.
79. S. Janson (1997). *Gaussian Hilbert spaces.* Cambridge University Press.
80. J. Junnila and E. Saksman (2017). Uniqueness of critical Gaussian chaos. *Elect. J. Probab* **22**, 1–31.
81. J.-P. Kahane (1985). Sur le chaos multiplicatif. *Ann. Sci. Math. Québec* **9**, no.2, 105–150.
82. G. Kozma and E. Schreiber (2004). An asymptotic expansion for the discrete harmonic potential. *Electron. J. Probab.* **9**, Paper no. 1, pages 10–17.
83. T. Kumagai (2014). *Random walks on disordered media and their scaling limits.* Lecture Notes in Mathematics vol. 2101. École d'Été de Probabilités de Saint-Flour. Springer, Cham, 2014. x+147 pp.
84. S. P. Lalley and T. Sellke (1987). A conditional limit theorem for the frontier of a branching Brownian motion. *Ann. Probab.* **15**, no. 3, 1052–1061.
85. G.F. Lawler and V. Limić (2010). *Random walk: a modern introduction.* Cambridge Studies in Advanced Mathematics, vol. 123. Cambridge University Press, Cambridge, xii+364.
86. M. Ledoux (2001). *The Concentration of Measure Phenomenon.* Mathematical Surveys and Monographs, American Mathematical Society.
87. T.M. Liggett (1978). Random invariant measures for Markov chains, and independent particle systems. *Z. Wahrsch. Verw. Gebiete* **45** 297–313.
88. T.M. Liggett (1985). *Interacting Particle Systems.* Springer Verlag. New York.
89. T. Lyons (1983). A simple criterion for transience of a reversible Markov chain. *Ann. Probab.* **11**, no. 2, 393–402.

90. R. Lyons and Y. Peres (2016). *Probability on Trees and Networks*. Cambridge Series in Statistical and Probabilistic Mathematics, vol. 42, Cambridge University Press, New York.
91. T. Madaule (2015). Maximum of a log-correlated Gaussian field. *Ann. Inst. H. Poincaré Probab. Statist.* **51**, no. 4, 1369–1431.
92. T. Madaule, R. Rhodes and V. Vargas (2016). Glassy phase and freezing of log-correlated Gaussian potentials. *Ann. Appl. Probab.* **26**, no. 2, 643–690.
93. H.P. McKean (1975). Application of Brownian motion to the equation of Kolmogorov-Petrovskii-Piskunov. *Comm. Pure Appl. Math.* **28** 323–331.
94. J. Miller (2010). Universality of SLE(4). arXiv:1010.1356.
95. J. Miller and S. Sheffield (2013). Imaginary geometry IV: interior rays, whole-plane reversibility, and space-filling trees. arXiv:1302.4738.
96. J. Miller and S. Sheffield (2015). Liouville quantum gravity and the Brownian map I: The QLE(8/3,0) metric. arXiv:1507.00719.
97. J. Miller and S. Sheffield (2016). Imaginary Geometry I: Interacting SLEs. *Probab. Theory Rel. Fields* **164**, no. 3, 553–705.
98. J. Miller and S. Sheffield (2016). Imaginary geometry II: reversibility of $\mathrm{SLE}_\kappa(\rho_1;\rho_2)$ for $\kappa\in(0,4)$. *Ann. Probab* **44**, no. 3, 1647–1722.
99. J. Miller and S. Sheffield (2016). Imaginary geometry III: reversibility of SLE_κ for $\kappa\in(4,8)$. *Ann. Math.* **184**, no. 2, 455–486.
100. J. Miller and S. Sheffield (2016). Liouville quantum gravity and the Brownian map II: geodesics and continuity of the embedding. arXiv:1605.03563.
101. A. Naddaf and T. Spencer (1997). On homogenization and scaling limit of some gradient perturbations of a massless free field. *Commun. Math. Phys.* **183**, no. 1, 55–84.
102. C.S.J.A. Nash-Williams (1959). Random walk and electric currents in networks. *Proc. Cambridge Philos. Soc.* **55** 181–194.
103. S. Orey and S.J. Taylor (1974). How often on a Brownian path does the law of the iterated logarithm fail? *Proc. London Math. Soc. (3)* **28** 174–192.
104. E. Powel (2018). Critical Gaussian chaos: convergence and uniqueness in the derivative normalisation. *Electron. J. Probab.* **23**, paper no. 31, 26 pp.
105. R. Rhodes and V. Vargas (2014). Gaussian multiplicative chaos and applications: A review. *Probab. Surveys* **11**, 315–392.
106. R. Rhodes and V. Vargas (2015). Liouville Brownian Motion at criticality. *Poten. Anal.* **43**, no. 2, 149–197.
107. L. Russo (1978). A note on percolation. *Z. Wahrsch. Verw. Gebiete* **43**, no. 1, 39–48.
108. L. Russo (1981). On the critical percolation probabilities. *Z. Wahrsch. Verw. Gebiete* **56**, no. 2, 229–237.
109. A. Ruzmaikina and M. Aizenman (2005). Characterization of invariant measures at the leading edge for competing particle systems. *Ann. Probab.* **33**, no. 1, 82–113.
110. O. Schramm and S. Sheffield (2009). Contour lines of the two-dimensional discrete Gaussian free field. *Acta Math.* **202**, no. 1, 21–137.
111. M. Secci (2014). *Random Measures and Independent Particle Systems*. Diploma thesis, Università degli Studi di Padova.
112. P.D. Seymour and D.J.A. Welsh (1978). Percolation probabilities on the square lattice. *Ann. Discr. Math.* **3** 227–245.
113. A. Shamov (2014). On Gaussian multiplicative chaos. arXiv:1407.4418.
114. S. Sheffield (2007). Gaussian free fields for mathematicians. *Probab. Theory Rel. Fields* **139**, no. 3–4, 521–541.
115. S. Sheffield (2009). Exploration trees and conformal loop ensembles. *Duke Math J.* **147**, no. 1, 79–129.
116. S. Sheffield and W. Werner (2010). Conformal Loop Ensembles: The Markovian characterization and the loop-soup construction. *Ann. Math.* **176** (2012), no. 3, 1827–1917.
117. Y.G. Sinai (1982). The limit behavior of a one-dimensional random walk in a random environment. *Teor. Veroyatnost. i Primenen.* **27** 247–258.

118. D. Slepian (1962). The one-sided barrier problem for Gaussian noise. *Bell System Tech. J.* **41** 463–501.
119. A. Stöhr (1950). Uber einige lineare partielle Differenzengleichungen mit konstanten Koeffizienten III. *Math. Nachr.* **3**, 330–357.
120. E. Subag and O. Zeitouni (2015). Freezing and decorated Poisson point processes. *Commun. Math. Phys.* **337**, no. 1, 55–92.
121. K. Symanzik (1969). *Euclidean quantum field theory*. In: Scuola internazionale di Fisica Enrico Fermi, XLV Corso, pp. 152–223, Academic Press.
122. M. Talagrand (1987). Regularity of Gaussian processes. *Acta Math.* **159** 99–149.
123. V. Tassion (2016). Crossing probabilities for Voronoi percolation. *Ann. Probab.* **44**, no. 5, 3385–3398.
124. B.S. Tsirelson, I.A. Ibragimov and V.N. Sudakov (1976). Norms of Gaussian sample functions. In: *Proceedings of the 3rd Japan, USSR Symposium on Probability Theory (Tashkent, 1975)*. Lecture Notes in Mathematics, vol. 550, Springer-Verlag, Berlin, pp. 20–41.
125. W. Wu and O. Zeitouni (2018). Subsequential tightness of the maximum of two dimensional Ginzburg-Landau fields. arXiv:1802.09601.
126. O. Zeitouni (2012). *Branching Random Walk and Gaussian Fields*. Lecture notes (see the author's website).

The manufacturer's authorised representative in the EU is Springer Nature Customer Service Centre GmbH, Europaplatz 3, 69115 Heidelberg, Germany. If you have any concerns regarding our products, please contact ProductSafety@springernature.com

Printed and bound by CPI Group (UK) Ltd, Croydon, CR0 4YY
05/10/2026
02228130-0017